Lecture Notes in Computer Science 4547

Commenced Publication in 1973
Founding and Former Series Editors:
Gerhard Goos, Juris Hartmanis, and Jan van Leeuwen

T0241056

Claude Carlet Berk Sunar (Eds.)

Arithmetic of Finite Fields

First International Workshop, WAIFI 2007
Madrid, Spain, June 2007
Proceedings

 Springer

Volume Editors

Claude Carlet
Université Paris 8, Département de mathématiques
2, rue de la Liberté; 93526 - SAINT-DENIS Cedex 02, France
E-mail: claude.carlet@inria.fr

Berk Sunar
Worcester Polytechnic Institute
100 Institute Road, Worcester, MA 01609, USA
E-mail: sunar@wpi.edu

Library of Congress Control Number: 2007928526

CR Subject Classification (1998): E.4, I.1, E.3, G.2, F.2

LNCS Sublibrary: SL 1 – Theoretical Computer Science and General Issues

ISSN   0302-9743
ISBN-10   3-540-73073-7 Springer Berlin Heidelberg New York
ISBN-13   978-3-540-73073-6 Springer Berlin Heidelberg New York

Springer is a part of Springer Science+Business Media

springer.com

© Springer-Verlag Berlin Heidelberg 2007
Printed in Germany

Typesetting: Camera-ready by author, data conversion by Scientific Publishing Services, Chennai, India
Printed on acid-free paper   SPIN: 12077106   06/3180   5 4 3 2 1 0

Preface

These are the proceedings of WAIFI 2007. The conference was held in Madrid, Spain, during June 21–22, 2007. We are very grateful to the Program Committee members and to the external reviewers for their hard work! The conference received 94 submissions out of which 27 were finally selected for presentation. Each paper was refereed by at least two reviewers, and at least by three in the case of papers (co)-authored by Program Committee members. All final decisions were taken only after a clear position was clarified through additional reviews and comments. The Committee also invited Harald Niederreiter and Richard E. Blahut to speak on topics of their choice and we thank them for having accepted.

Special compliments go out to José L. Imaña, the general Co-chair and local organizer of WAIFI 2007, who brought the workshop to beautiful Madrid, Spain. WAIFI 2007 was organized by the department Computer Architecture of Facultad de Informática of the Universidad Complutense, in Madrid. We also would like to thank the General Co-chair Çetin K. Koç for his guidance. Finally, we would like to thank the Steering Committee for providing us with this wonderful opportunity.

The submission and selection of papers were done using the iChair software, developed at EPFL by Thomas Baignères and Matthieu Finiasz. Many thanks for their kind assistance! We also thank Gunnar Gaubatz for his precious help in this matter.

June 2007

Claude Carlet
Berk Sunar

Organization

Steering Committee

Jean-Pierre Deschamps   University Rovira i Virgili, Spain
José L. Imaña   Complutense University of Madrid, Spain
Çetin K. Koç   Oregon State University, USA
Christof Paar   Ruhr University of Bochum, Germany
Jean-Jacques Quisquater   Université Katholique de Louvain, Belgium
Berk Sunar   Worcester Polytechnic Institute, USA
Gustavo Sutter   Autonomous University of Madrid, Spain

Executive Committee

General Co-chairs
José L. Imaña   Complutense University of Madrid, Spain
Çetin K. Koç   Oregon State University, USA

Program Co-chairs
Claude Carlet   University of Paris 8, France
Berk Sunar   Worcester Polytechnic Institute, USA

Financial, Local Arrangements Chairs
Luis Piñuel   Complutense University of Madrid, Spain
Manuel Prieto   Complutense University of Madrid, Spain

Publicity Chair
Gustavo Sutter   Autonomous University of Madrid, Spain

Program Committee

Jean-Claude Bajard   CNRS-LIRMM in Montpellier, France
Ian F. Blake   University of Toronto, Canada
Marc Daumas   CNRS-LIRMM in Perpignan, France
Jean-Pierre Deschamps   University Rovira i Virgili, Spain
Josep Domingo-Ferrer   University Rovira i Virgili, Spain
Philippe Gaborit   University of Limoges, France
Joachim von zur Gathen   B-IT, University of Bonn, Germany
Pierrick Gaudry   LORIA-INRIA, France
Guang Gong   University of Waterloo, Canada
Jorge Guajardo   Philips Research, Netherlands
Anwar Hasan   University of Waterloo, Canada

Çetin K. Koç	Oregon State University, USA
Tanja Lange	Technische Universiteit Eindhoven, Netherlands
Julio López	UNICAMP, Brazil
Gary Mullen	Pennsylvania State University, USA
Harald Niederreiter	National University of Singapore, Singapore
Ferruh Ozbudak	Middle East Technical University, Turkey
Erkay Savaş	Sabanci University, Turkey
Igor Shparlinski	Macquarie University, Australia
Horacio Tapia-Recillas	UAM-Iztapalapa, D.F., Mexico
Apostol Vourdas	University of Bradford, UK

Referees

O. Ahmadi	M. Finiasz	A. Martínez-Ballesté
J. Aragonés	D. Freeman	N. Méloni
R.M. Avanzi	T. Güdü	Y. Nawaz
O. Barenys	C. Güneri	C. Negre
I. Barenys	K. Gupta	T.B. Pedersen
L. Batina	G. Hanrot	M.N. Plasencia
D.J. Bernstein	F. Hess	D. Pointcheval
P. Birkner	K. Horadam	T. Plantard
M. Cenk	L. Imbert	C. Ritzenthaler
J. Chung	S. Jiang	G. Saldamli
V. Daza	T. Kerins	Z. Saygı
C. Ding	D. Kohel	F. Sebe
A. Doğanaksoy	G. Kömürcü	B. Schoenmakers
N. Ebeid	G. Kyureghyan	A. Tisserand
N. El Mrabet	G. Leander	F. Vercauteren
H. Fan	J. Lutz	W. Willems

Sponsoring Institutions

Real Sociedad Matemática Española, Spain.
Ministerio de Educación y Ciencia, Spain.
Facultad de Informática de la Universidad Complutense de Madrid, Spain.
ArTeCs: Architecture and Technology of Computing Systems Group,
    Universidad Complutense de Madrid, Spain.
Universidad Complutense de Madrid, Spain.

Table of Contents

Structures in Finite Fields

Efficient Implementation and Architectures

Efficient Finite Field Arithmetic

Classification and Construction of Mappings over Finite Fields

Curve Algebra

Cryptography

Codes

Discrete Structures

Explicit Factorizations of Cyclotomic and Dickson Polynomials over Finite Fields

Robert W. Fitzgerald and Joseph L. Yucas

Southern Illinois University Carbondale

Abstract. We give, over a finite field F_q, explicit factorizations into a product of irreducible polynomials, of the cyclotomic polynomials of order $3 \cdot 2^n$, the Dickson polynomials of the first kind of order $3 \cdot 2^n$ and the Dickson polynomials of the second kind of order $3 \cdot 2^n - 1$.

Keywords: finite field, cyclotomic polynomial, Dickson polynomial.

1 Introduction

Explicit factorizations, into a product of irreducible polynomials, over F_q of the cyclotomic polynomials $Q_{2^n}(x)$ are given in [4] when $q \equiv 1 \pmod 4$. The case $q \equiv 3 \pmod 4$ is done in [5]. Here we give factorizations of $Q_{2^n r}(x)$ where r is prime and $q \equiv \pm 1 \pmod r$. In particular, this covers $Q_{2^n 3}(x)$ for all F_q of characteristic not 2, 3. We apply this to get explicit factorizations of the first and second kind Dickson polynomials of order $2^n 3$ and $2^n 3 - 1$ respectively.

Explicit factorizations of certain Dickson polynomials have been used to compute Brewer sums [1]. But our basic motivation is curiosity, to see what factors arise. Of interest then is how the generalized Dickson polynomials $D_n(x, b)$ arise in the factors of the cyclotomic polynomials and how the Dickson polynomials of the first kind appear in the factors of both kinds of Dickson polynomials.

Let q be a power of an odd prime and let $v_2(k)$ denote the highest power of 2 dividing k. We will only consider the case where r is prime and $q \equiv \pm 1 \pmod r$. We recall the general form of the factors of cyclotomic polynomials in this case (see [4] 3.35 and 2.47).

Proposition 1. Let $L = v_2(q^2 - 1)$, and work over F_q.

1. Suppose $q \equiv 1 \pmod r$. Then:
   (a) For $0 \leq n \leq v_2(q - 1)$, $Q_{2^n r}(x)$ is a product of linear factors.
   (b) For $v_2(q - 1) < n \leq L$, $Q_{2^n r}(x)$ is a product of irreducible quadratic polynomials.
   (c) For $n > L$, $Q_{2^n r}(x) = \prod f_i(x^{2^{n-L}})$, where $Q_{2^L r}(x) = \prod f_i(x)$.
2. Suppose $q \equiv -1 \pmod r$. Then:
   (a) For $0 \leq n \leq L$, $Q_{2^n r}(x)$ is a product of irreducible quadratic factors.
   (b) For $n > L$, $Q_{2^n r}(x) = \prod f_i(x^{2^{n-L}})$, where $Q_{2^L r}(x) = \prod f_i(x)$.

C. Carlet and B. Sunar (Eds.): WAIFI 2007, LNCS 4547, pp. 1–10, 2007.

2   Factors of Cyclotomic Polynomials

As before, $L = v_2(q^2 - 1)$ and $r = 2s + 1$ be a prime. Let $\Omega(k)$ denote the primitive kth roots of unity in F_{q^2}.

We will often use the following, which is equation 7.10 in [4]. For $m \geq 0$

$$D_{2m}(x, c) = D_m(x^2 - 2c, c^2).$$

Lemma 1. *Suppose $q \equiv -1 \pmod{r}$. Let N denote the norm $F_{q^2} \to F_q$.*

1. *If $q \equiv 3 \pmod 4$ and $\rho \in \Omega(2^n)$, for $n \leq L$, then*

$$N(\rho) = \begin{cases} 1, & \text{if } 2 \leq n < L \\ -1, & \text{if } n = L. \end{cases}$$

2. *If $\omega \in \Omega(r)$ then $N(\omega) = 1$.*
3. *If $\alpha \in F_{q^2}$ and $N(\alpha) = a$ then $\alpha + a/\alpha \in F_q$.*

Proof. (1) Since $q - 1 \equiv 2 \pmod 4$, $L - 1$ is the highest power of 2 dividing $q + 1$. Let $\rho \in \Omega(2^L)$. Now $N(\rho) = \rho^{q+1}$ so that $N(\rho)^2 = \rho^{2(q+1)} = 1$ and $N(\rho) = \pm 1$. If $N(\rho) = 1$ then $\rho^{q+1} = 1$ and $2^L = o(\rho)$ divides $q + 1$, a contradiction. Hence $N(\rho) = -1$. If $\omega \in \Omega(2^n)$ for $n < L$, then ω is an even power of ρ and so $N(\omega) = 1$.

(2) $N(\omega)^r = N(\omega^r) = 1$ and, as r is prime, the only rth root of unity in F_q is 1. So $N(\omega) = 1$.

(3) We have $\alpha \alpha^q = a$ so that $\alpha + a/\alpha = \text{tr}(\alpha) \in F_q$. $\square$

Theorem 1.   *1. Suppose $q \equiv -1 \pmod r$ and $q \equiv 3 \pmod 4$.*
    (a) *$Q_r(x) = \prod_{a \in S_1}(x^2 - ax + 1)$ and $Q_{2r}(x) = \prod_{a \in S_1}(x^2 + ax + 1)$, where S_1 is the set of roots of $1 + \sum_{i=1}^s D_i(x, 1)$*
    (b) *For $2 \leq n < L$, $Q_{2^n r}(x) = \prod_{a \in S_n}(x^2 + ax + 1)$, where S_n is the set of roots of $1 + \sum_{i=1}^s (-1)^s D_{2^{n-1}i}(x, 1)$.*
    (c) *For $n \geq L$, $Q_{2^n r}(x) = \prod_{b \in T_L}(x^{2^{n-L+1}} + bx^{2^{n-L}} - 1)$, where T_L is the set of roots of $1 + \sum_{i=1}^s (-1)^s D_{2^{L-1}i}(x, -1)$.*
2. *Suppose $q \equiv -1 \pmod r$ and $q \equiv 1 \pmod 4$.*
    (a) *$Q_r(x) = \prod_{a \in S_1}(x^2 - ax + 1)$ and $Q_{2r}(x) = \prod_{a \in S_1}(x^2 + ax + 1)$.*
    (b) *For $2 \leq n \leq L$,*

$$Q_{2^n r}(x) = \prod_{\rho \in \Omega(2^{n-1})} \prod_{b \in T(\rho)} (x^2 + bx + \rho),$$

    where $T(\rho)$ is the set of roots in F_q of $1 + \sum_{i=1}^s (-1)^i D_{2^{n-1}i}(x, \rho)$.
    (c) *For $n > L$, $Q_{2^n r}(x) = \prod_{\rho \in \Omega(2^{L-1})} \prod_{b \in T(\rho)} (x^{2^{n-L+1}} + bx^{2^{n-L}} + \rho)$.*
3. *Suppose $q \equiv 1 \pmod r$ and $q \equiv 3 \pmod 4$.*
    (a) *$Q_r(x) = \prod(x - \omega)$, $Q_{2r}(x) = \prod(x + \omega)$ and $Q_{4r}(x) = \prod(x^2 + \omega)$, with each product over $\Omega(r)$.*

(b) For $3 \leq n < L$,

$$Q_{2^n r}(x) = \prod_{w \in \Omega(r)} \prod_{c \in U_n} (x^2 + cwx + w^2)$$

where U_n is the set of roots in F_q of $D_{2^{n-2}}(x, 1)$.

(c) For $n \geq L$,

$$Q_{2^n r}(x) = \prod_{w \in \Omega(r)} \prod_{d \in V_L} (x^{2^{n-L+1}} + dwx - w^2)$$

where V_L is the set of roots in F_q of $D_{2^{L-2}}(x, -1)$.

4. *Suppose $q \equiv 1 \pmod{r}$ and $q \equiv 1 \pmod{4}$.*

(a) $Q_r(x) = \prod_{w \in \Omega(r)} (x - w)$.

(b) For $1 \leq n < L$,

$$Q_{2^n r}(x) = \prod_{w \in \Omega(r)} \prod_{\rho \in \Omega(2^n)} (x + w\rho)$$

(c) For $n \geq L$,

$$Q_{2^n r}(x) = \prod_{w \in \Omega(r)} \prod_{\rho \in \Omega(2^{L-1})} (x^{2^{n-L+1}} + w\rho).$$

Proof. (1) If $w \in \Omega(r)$ then $N(w) = 1$ and $w + 1/w \in F_q$ by Lemma 1. So

$$Q_r(x) = \prod_{w \in \Omega(r)} (x - w) = \prod (x^2 - ax + 1),$$

is a factorization over F_q, where a runs over all distinct $w + w^{-1}$. The quadratic factors are irreducible by Corollary 1. Also,

$$1 + \sum_{i=1}^{s} D_i(a, 1) = 1 + \sum_{i=1}^{s} (w^i + w^{-i})$$

$$= w^{-s} \left(\sum_{j=0}^{2s} w^j \right) = 0.$$

As $\deg(1 + \sum D_i(x, 1)) = s$, the a are all of the roots. Further,

$$Q_{2r}(x) = Q_r(-x) = \prod (x^2 + ax + 1),$$

which completes the proof of (1)(a).

For (1)(b), the case $n = 2$ can be checked directly. So suppose $3 \leq n < L$. Note that $a_2 = \rho w + (\rho w)^{-1}$ as $\rho^{-1} = -\rho$. Let $\rho_n \in \Omega(2^n)$ and set $a_n = \rho \rho_n w + (\rho \rho_n w)^{-1}$. We claim that $a_n \in F_q$ and that $a_n^2 = 2 - a_{n-1}$ (with a_{n-1}

defined via a different choice of ω). Namely, $N(\rho \rho_n \omega) = 1$ as $n < L$ and so $a_n \in F_q$. And

$$
\begin{aligned}
a_n^2 &= \rho^2 \rho_n^2 \omega^2 + (\rho^2 \rho_n^2 \omega^2)^{-1} + 2 \\
&= -\rho_{n-1} \omega^2 - (\rho_{n-1} \omega^2)^{-1} + 2 = 2 - a_{n-1}.
\end{aligned}
$$

Then inductively,

$$
Q_{2^n r}(x) = \prod (x^4 + a_{n-1} x^2 + 1) = \prod (x^2 + a_n x + 1)(x^2 - a_n x + 1),
$$

where again the quadratic factors are irreducible over F_q. Lastly, again by induction, the a_n are roots of

$$
1 + \sum_{i=1}^{s} (-1)^i D_{2^{n-2}i}(-(x^2 - 2), 1) = 1 + \sum_{i=1}^{s} (-1)^i D_{2^{n-1}i}(x, 1).
$$

This has degree $2^{n-1}s$ and there are $2^{n-1}s = \frac{1}{2} \deg Q_{2^n r}(x)$ many a_n's. So the a_n's are all of the roots of the above polynomial.

We finish the proof of (1) by checking the case $n = L$ (the cases $n > L$ then follow from Corollary 1). Now $N(\rho \rho_L \omega) = -1$ by Lemma 1, so that $b = \rho \rho_L \omega - (\rho \rho_L \omega)^{-1} \in F_q$. And $b^2 = -a_{L-1} - 2$. Hence

$$
x^4 + a_{L-1} x^2 + 1 = (x^2 + bx - 1)(x^2 - bx - 1)
$$

is an irreducible factorization over F_q. Lastly, b is a root of

$$
1 + \sum_{i=1}^{s} (-1)^i D_{2^{L-2}i}(-(x^2 + 2), 1) = 1 + \sum_{i=1}^{s} (-1)^i D_{2^{L-1}i}(x, -1).
$$

As before, the b's are all of the roots.

(2) First note that $L = v_2(q - 1) + 1$ so that $\Omega(2^n) \subset F_q$ for $n < L$. The factorization of $Q_r(x)$ and $Q_{2r}(x)$ is the same as in (1). For (2)(b), again the case $n = 2$ can be checked directly. For $2 < n < L$ we work by induction. Set $b_n = \rho_n(\omega + \omega^{-1})$, for $\rho_n \in \Omega(2^n)$. Then $b_n \in F_q$ and $b_n^2 = b_{n-1} + 2\rho_{n-1}$. Note that the set of b_{n-1}'s is closed under multiplication by -1. Hence we need only check that

$$
x^4 - a_{n-1} x^2 + \rho_{n-1} = (x^2 + b_n x + \rho_n)(x^2 - b_n x + \rho_n).
$$

Further, b_n is a root of

$$
1 + \sum_{i=1}^{s} (-1)^i D_{2^{n-2}i}(x^2 - 2\rho_{n-1}, \rho_{n-2}) = 1 + \sum_{i=1}^{s} (-1)^i D_{2^{n-1}i}(x, \rho_{n-1}).
$$

Set $\delta_{\rho_{n-1}}(x) = 1 + \sum_{i=1}^{s} (-1)^i D_{2^{n-1}i}(x, \rho_{n-1})$. Fix a ρ_{n-1} and pick a ρ_n with $\rho_n^2 = \rho_{n-1}$. To complete the proof of (2)(b) we need to check that the b_n's are all of the roots of $\delta_{\rho_{n-1}}(x)$ in F_q.

For $n = 2$, deg $\delta_{\rho_1} = 2s$ which is the number of b_2's so δ_{ρ_1} has no other roots. Inductively assume that

$$\delta_{\rho_{n-1}}(x) = \prod (x - b_n) \cdot h(x),$$

where $h(x)$ is a product of non-linear factors. Then

$$\delta_{\rho_n}(x) = \delta_{rho_{n-1}}(x^2 - 2\rho_n) = \prod (x^2 - 2\rho_n - b_n) \cdot h(x^2 - 2\rho_n).$$

Now $x^2 - 2\rho_n - b_n$ splits in F_q iff $2\rho_n + b_n$ is a square in F_q. The b_n's in $T(\rho_n)$ are $\pm\rho_n(\omega + \omega^{-1})$. And $2\rho_n + \rho_n(\omega + \omega^{-1}) = \rho_n(\omega^r + \omega^{-r})^2$ is a square (in fact, the square of a b_{n+1}) while $2\rho_n - \rho_n(\omega + \omega^{-1}) = -\rho_n(\omega^r - \omega^{-r})^2$ is not a square (as $\omega^r - \omega^{-r} \notin F_q^2$). Hence the roots of δ_{ρ_n} in F_q are precisely the b_{n+1}'s.

(2)(c) The case $n = L$ must be done separately as $\rho_L \notin F_q$. Set $b_L = \rho\rho_L(\omega - \omega^{-1})$. As in the proof of (a), $(\omega - \omega^{-1})^2 \in F_q \setminus F_q^2$. And $\rho_{L-1} \in F_q \setminus F_q^2$. Hence $\rho_{L-1}(\omega - \omega^{-1})^2 \in F_q^2$ and its square root, b_L, is in F_q. Also $b_L^2 = -b_{L-1} + 2\rho_{L-1}$. Then

$$x^4 + b_{L-1}x^2 + \rho_{L-2} = (x^2 + b_L x + \rho_{L-1})(x^2 - b_L x + \rho_{L-1}),$$

giving the desired factorization. Further, $b_{L-1} = -b_L^2 + 2\rho_{L-1}$ so that b_L is a root of

$$1 + \sum_{i=1}^{s} (-1)^i D_{2^{L}-2i}(-(x^2 - 2\rho_{L-1}), \rho_{L-2}) = 1 + \sum_{i=1}^{s} (-1)^i D_{2^{L}-1i}(x, \rho_{L-1}).$$

As before, these are all of the roots in F_q. Finally, the cases $n > L$ follow from Corollary 1.

(3) As $q \equiv 1 \pmod{r}$, we have $\Omega(r) \subset F_q$. The factorizations for Q_r and Q_{2r} are clear and that of Q_{4r} follows from Corollary 1. We do the case $n = 3 < L$ (the case $n = 3 = L$ will follow from the case $n = L$ to be done later). Let $\rho_3 \in \Omega(2^3)$. Then $\rho_3 \in F_{q^2} \setminus F_q$, $N(\rho_3) = 1$ as $n < L$, and $c_3 = \rho_3 + \rho_3^{-1} \in F_q$. Also $c_3^2 = \rho + \rho^{-1} + 2 = 2$. A typical factor of Q_{4r} can be written as $x^2 + \omega^4$ and we have

$$x^4 + \omega^4 = (x^2 + c_2\omega x + \omega^2)(x^2 - c_2\omega x + \omega^2),$$

giving the desired factorization of $Q_{2^3 r}(x)$. Note that $c_3 = \pm\sqrt{2}$, the roots of $D_2(x, 1) = x^2 - 2$.

Now suppose $3 < n < L$ and work inductively. We have $N(\rho_n) = 1$ so that $c_n = \rho_n + \rho_n^{-1} \in F_q$. And $c_n^2 = c_{n-1} + 2$. A typical factor of $Q_{2^{n-1}r}(x)$ can be written as $x^2 - c_{n-1}\omega^2 x + \omega^4$ and we have

$$x^4 - c_{n-1}\omega^2 x^2 + \omega^4 = (x^2 + c_n\omega x + \omega^2)(x^2 - c_n\omega x + \omega^2),$$

giving the desired factorization. Further, c_n is a root of $D_{2^{n-3}}(x^2 - 2, 1) = D_{2^{n-2}}(x, 1)$. A counting argument shows the c_n's are all of the roots.

Next suppose $n = L$. We have $N(\rho_L) = -1$ so that $c_L = \rho_L - \rho_L^{-1} \in F_q$. And $c_L^2 = c_{L-1} - 2$. Then

$$x^4 - c_{L-1}\omega^2 x^2 + \omega^4 = (x^2 + c_L\omega x - \omega^2)(x^2 - c_L\omega x - \omega^2),$$

giving the desired factorization. Further, The c_L's are all of the roots of $D_{2^{L-3}}(x^2 + 2, 1) = D_{2^{L-2}}(x, -1)$. The cases $n > L$ follow from Corollary 1.

(4) Note that $L = v_2(q - 1) + 1$. Hence here $\Omega(r), \Omega(2^n) \subset F_q$, for $n < L$. The factorizations of $Q_{2^n r}(x)$ for $n < L$ are clear and the rest follows from Corollary 1. $\square$

3   Cyclotomic Polynomials in the Case $r = 3$

We work out the case $r = 3$ (so that all F_q not of characteristic 2, 3 are covered). By way of comparison, we first recall the result for $r = 1$. L continues to denote $v_2(q^2 - 1)$.

Proposition 2. *The following are factorizations.*

1. *If $q \equiv 1 \pmod 4$ then*
   (a) *For $1 \le n < L$, $Q_{2^n}(x) = \prod(x + a)$, where a runs over all primitive 2^n roots of unity.*
   (b) *For $n \ge L$, $Q_{2^n}(x) = \prod(x^{2^{n-L+1}} + a)$, where a runs over all primitive 2^{L-1} roots of unity.*
2. *If $q \equiv 3 \pmod 4$ then*
   (a) *For $2 \le n < L$, $Q_{2^n}(x) = \prod(x^2 + ux + 1)$ where u runs over all roots of $D_{2^{n-2}}(x, 1)$.*
   (b) *For $n \ge L$, $Q_{2^n}(x) = \prod(x^{2^{n-L+1}} + vx^{2^{n-L}} - 1)$, where v runs over all roots of $D_{2^{L-2}}(x, -1)$.*

Proof. Statement (1) is from [4]. Statement (2) is by Meyn [5]. $\square$

Proposition 3. *The following are factorizations.*

1. *If $q \equiv 1 \pmod{12}$ then let $u, v \in F_q$ be the primitive cube roots of unity.*
   (a) *$Q_3(x) = (x - u)(x - v)$.*
   (b) *For $1 \le n < L$, $Q_{2^n 3}(x) = \prod(x + u\rho)(x + v\rho)$, where $\rho \in \Omega(2^n)$.*
   (c) *For $n \ge L$, $Q_{2^n 3}(x) = \prod(x^{2^{n-L+1}} + u\rho)(x^{2^{n-L+1}} + v\rho)$, where $\rho \in \Omega(2^L)$.*
2. *If $q \equiv 5 \pmod{12}$ then*
   (a) *$Q_3(x) = x^2 + x + 1$ and $Q_6(x) = x^2 - x + 1$ are irreducible.*
   (b) *For $2 \le n \le L$, $Q_{2^n 3}(x) = \prod(x^2 + cx + \rho_{n-1})$, where $\rho_{n-1} \in \Omega(2^{n-1})$ and for each ρ_{n-1}, the c's run over all the solutions to $D_{2^{n-1}}(x, \rho_{n-1}) = 1$.*
   (c) *For $n > L$, $Q_{2^n 3}(x) = \prod(x^{2^{n-L+1}} + cx^{2^{n-L}} + \rho_{L-1})$, with ρ_{L-1} and c as before.*
3. *If $q \equiv 7 \pmod{12}$ then again let $u, v \in F_q$ be the primitive cube roots of unity.*
   (a) *$Q_3(x) = (x - u)(x - v)$, $Q_6(x) = (x + u)(x + v)$ and $Q_{12}(x) + (x^2 + u)(x^2 + v)$.*
   (b) *For $3 \le n < L$, $Q_{2^n 3}(x) = \prod(x^2 + cux + v)(x^2 + cvx + u)$, where c runs over the roots of $D_{2^{n-2}}(x, 1)$.*

(c) For $n \geq L$, $Q_{2^n 3}(x) = \prod(x^{2^{n-L+1}} + dux^{2^{n-L}} - v)(x^{2^{n-L+1}} + dvx^{2^{n-L}} - u)$, where d runs over the roots of $D_{2^{L-2}}(x, -1)$.

4. If $q \equiv 11 \pmod{12}$ then

(a) $Q_3(x) = x^2 + x + 1$ and $Q_6(x) = x^2 - x + 1$ are irreducible.

(b) For $2 \leq n < L$, $Q_{2^n 3}(x) = \prod(x^2 + ax + 1)$, where the a's run over all solutions to $D_{2^{n-1}}(x, 1) = 1$.

(c) For $n \geq L$, $Q_{2^n 3}(x) = \prod(x^{2^{n-L+1}} + bx^{2^{n-L}} - 1)$, where the b's run over all solutions to $D_{2^{L-1}}(x, -1) = 1$.

4 Factors of Dickson Polynomials

The results here for the Dickson polynomials of the first kind are a re-formulation of results in [2]. The results for the Dickson polynomials of the second kind are new. We have included the first kind results to illustrate how the approach taken here covers the two kinds simultaneously.

Recall that the factorization of $x^t + 1$ is

$$x^t + 1 = \prod_{\substack{d \mid t \\ t/d \text{ odd}}} Q_{2d}(x).$$

The following generalization is standard.

Proposition 4.

$$\sum_{i=0}^{w-1} x^{it} = \prod_{\substack{d \mid t, 1 \neq s \mid w \\ (s, t/d) = 1}} Q_{ds}(x).$$

We review the transformations of [2]. Let P_n be the collection of all polynomials over a field F of degree n and let S_n denote the family of all self-reciprocal polynomials over F of degree n. Define

$$\Phi : P_n \to S_{2n} \quad \text{by}$$
$$f(x) \mapsto x^n f(x + x^{-1}),$$

where $n = \deg f$.

A self-reciprocal polynomial $b(x)$ of degree $2n$ can be written as

$$b(x) = \sum_{i=0}^{n-1} b_i(x^{2n-i} + x^i) + b_n x^n.$$

Define

$$\Psi : S_{2n} \to P_n \quad \text{by}$$
$$b(x) \mapsto \sum_{i=0}^{n-1} b_i D_{n-i}(x) + b_n.$$

Φ and Ψ are multiplicative inverses (this was proved only for finite fields in [2], Theorem 3, and for arbitrary fields in [3], Theorem 6.1). We write $D_n(x)$ for $D_n(x, 1)$ and $E_n(x)$ for the nth order Dickson polynomial of the second kind.

Proposition 5. *Write $n = 2^k m$ with m odd. Then:*

$$\Phi(D_n(x)) = \prod_{e|m} Q_{2^{k+2}e}(x)$$

$$\Phi(E_{n-1}(x)) = \prod_{e|m} \prod_{i=0}^{k+1} Q_{2^i e}(x),$$

where we exclude $e = 1, i = 0, 1$ from the second equation.

Proof. Note that by Waring's identity

$$\Phi(D_n(x)) = x^n D_n(x + x^{-1}) = x^{2n} + 1.$$

Take $t = 2n$ and $w = 2$ (and so $s = 2$) in Lemma 4 to get the result. Similarly,

$$\Phi(E_{n-1}(x)) = x^{n-1} E_{n-1}(x + x^{-1}) = (x^{2n} - 1)/(x^2 - 1).$$

Take $t = 2$ and $w = n$ in Lemma 4 to get the result. $\qquad\square$

Corollary 1. *Write $n = 2^k m$, with m odd. The factorizations over $\mathbb{Q}$ are:*

$$D_n(x) = \prod_{e|m} \Psi(Q_{2^{k+2}e}(x))$$

$$E_{n-1} = \prod_{e|m} \prod_{i=0}^{k+1} \Psi(Q_{2^i e}(x)),$$

where we again exclude $e = 1, i = 0, 1$ from the second equation.

Proof. This follows from Proposition 5 and the properties of Φ, Ψ since each $Q_r(x)$, $r > 1$ is irreducible over $\mathbb{Q}$ and self-reciprocal. $\qquad\square$

5   Dickson Polynomials in the Case $r = 3$

We return to the case of finite fields F_q. We use the explicit factorizations of cyclotomic polynomials to get explicit factorizations of the Dickson polynomials of order $2^n r$, via Proposition 1. We begin with the case $r = 1$, where the factorizations of $Q_{2^n}(x)$ were known but the results for Dickson polynomials are new.

Proposition 6. *Set $L = v_2(q^2 - 1)$.*

1. *For $1 \leq n \leq L - 3$, $D_{2^n}(x)$ splits in F_q. For $n \geq L - 2$, we have the factorization*

$$D_{2^n}(x) = \prod (D_{2^{n-L+3}}(x) + a),$$

where a runs over all roots of $D_{2^{L-3}}(x)$.

2. For $1 \leq n \leq L - 2$, $E_{2^n - 1}(x)$ splits in F_q. For $n \geq L - 1$, we have the factorization

$$E_{2^n - 1}(x) = \prod_{i=0}^{L-3}(x + a_i) \cdot \prod_{i=1}^{n-L+2}(D_{2^i}(x) + a_{L-3}),$$

where a_i runs over all the roots of $D_{2^i}(x)$.

We note that, when $L = 3$, the statement (1) means that $D_{2^n}(x)$ is irreducible over F_q for $n \geq 1$.

Theorem 2. Set $L = v_2(q^2 - 1)$.

1. Suppose $q \equiv \pm 1 \pmod{12}$. For $0 \leq n \leq L - 3$, $D_{2^n 3}(x)$ splits in F_q. For $n \geq L - 2$, we have the factorization

$$D_{2^n 3}(x) = \prod (D_{2^{n-L+3}}(x) + a),$$

where a runs over all the roots of $D_{2^{L-3}3}(x)$.
2. Suppose $q \equiv \pm 5 \pmod{12}$. The following are factorizations.
   (a) For $0 \leq n \leq L - 3$,

$$D_{2^n 3}(x) = \prod (D_1(x) + a)(D_2(x) + aD_1(x) + (a^2 - 1)),$$

   where a runs over all the roots of $D_{2^n}(x)$.
   (b) For $n \geq L - 2$,

$$D_{2^n 3}(x) = \prod (D_{2^{n-L+3}}(x) + b)(D_{2^{n-L+3}}(x) + uD_{2^{n-L+2}}(x) + (b + 3)),$$

   where b runs over roots of $D_{2^{L-3}}(x)$ and $u^2 = 3b + 6$.

Proof. The proof is a tedious computation. Take each factor of the appropriate cyclotomic polynomial, pair it with its reciprocal and then apply Ψ. We note that in Case 2, $3 \notin F_q^{*2}$. And $b + 2 \notin F_q^{*2}$ since otherwise $\sqrt{b+2}$ is a root in F_q of $D_{2^{L-3}}(x^2 - 2) = D_{2^{L-2}}(x)$, contradicting Proposition 6. Thus $3b + 6$ has square roots u in F_q. □

The factorizations of $E_{2^n 3 - 1}(x)$ follow from the previous result and the following identity:

Corollary 2. For $n \geq 1$

$$E_{2^n 3 - 1}(x) = (x^2 - 1) \prod_{i=0}^{n-1} D_{2^i 3}(x).$$

Proof. We use induction. For $n = 1$, Proposition 5 gives,

$$E_{2^n 3-1}(x) = E_5(x) = \Psi(Q_4)\Psi(Q_3)\Psi(Q_6)\Psi(Q_{12}).$$

Then $Q_3 = x^2+x+1$ so $\Psi(Q_3) = x+1$, $Q_6 = x^2-x+1$ so $\Psi(Q_6) = x-1$ and,using Proposition 5 again, $\Psi(Q_4)\Psi(Q_{12}) = D_3$. So $E_{2^n 3-1}(x) = (x^2 - 1)D_{2^n-1 3}(x)$.
Proposition 5 gives:

$$\begin{aligned}
E_{2^{n+1}3-1}(x) &= E_{2^n 3-1}\Psi(Q_{2^{n+2}})\Psi(Q_{2^{n+2}3}) \\
&= E_{2^n 3-1}D_{2^n 3},
\end{aligned}$$

which gives the result by induction.   □

References

1. Alaca, Ş.: Congruences for Brewer sums. Finite Fields Appl. 13, 1–19 (2007)
2. Fitzgerald, R.W., Yucas, J.L.: Factors of Dickson polynomials over finite fields. Finite Fields Appl. 11, 724–737 (2005)
3. Fitzgerald, R.W., Yucas, J.L.: A generalization of Dickson polynomials via linear fractional transformations. Int. J. Math. Comput. Sci. 1, 391–416 (2006)
4. Lidl, R., Niederreiter, H.: Finite Fields. In: Encyclopedia of Mathematics and Its Applications, 2nd edn., vol. 20, Cambridge University Press, Cambridge (1997)
5. Meyn, H.: Factorization of the cyclotomic polynomial $x^{2^n} +1$ over finite fields. Finite Fields Appl. 2, 439–442 (1996)

Some Notes on d-Form Functions with Difference-Balanced Property[*]

Tongjiang Yan[1,2], Xiaoni Du[2,4], Enjian Bai[3], and Guozhen Xiao[2]

[1] Math. and Comp. Sci., China Univ. of Petro., Dongying 257061, China
[2] P.O.Box 119, Key Lab.on ISN, Xidian Univ., Xi'an 710071, China
[3] Inform. Sci. and Tech., Donghua Univ., Shanghai 201620, China
[4] Math. and Inform. Sci, Northwest Normal Univ., Lanzhou 730070, China
yantoji@163.com

Abstract. The relation between a cyclic relative difference set and a cyclic difference set is considered. Both the sets are with Singer parameters and can be constructed from a difference-balanced d-form function. Although neither of the inversions of Klapper A.'s and No J. S.'s main theorems is true, we prove that a difference-balanced d-form function can be obtained by the cyclic relative difference set and the cyclic difference set introduced by these two main theorems respectively.

Keywords: Cyclic difference sets, cyclic relative difference sets, d-form functions, difference-balanced.

1 Introduction and Preliminaries

In this paper, we use the following notation: q: power of prime p; m, n: positive integers such that $n > 2, m \mid n$; d: positive integer relatively prime to q; F_q, F_{q^m}, F_{q^n} : finite fields with q, q^m, and q^n elements, respectively; α: a primitive element of F_{q^n}; $\beta = \alpha^T$: a primitive element of F_{q^m}, where $T = \frac{q^n-1}{q^m-1}$. $F_{q^n}^* = F_{q^n} \setminus \{0, 1\}$.

A function $f(x)$ on F_{q^n} over F_q is said to be balanced if the element 0 appears one less time than each nonzero element in F_q in the list $f(\alpha^0), f(\alpha^1), f(\alpha^2), \ldots, f(\alpha^{q^n-2})$. A function $f(x)$ is said to be difference-balanced if $f(xz) - f(x)$ is balanced for any $z \in F_{q^n} \setminus \{0\}$. By replacing x by α^t, a function $f(x)$ can be considered as a q-ary sequence $f(\alpha^t)$ of period $q^n - 1$. Hence, for convenience, we will use the expression "a sequence $f(\alpha^t)$ of period $q^n - 1$" interchangeably with "a function $f(\alpha^t)$" (or $f(x)$) from F_{q^n} to F_q ([13]).

Let $f(x)$ be a function on F_{q^n} over F_q. Define

$$D_f^{(m)} = \{x | f(x) = m, x \in F_{q^n}^*\}.$$

Then $F_{q^n}^* = \cup_{m \in F_q} D_f^{(m)}$.

[*] Project supported by the National Natural Science Foundations of China (No.60473028) and (No.60503009).

C. Carlet and B. Sunar (Eds.): WAIFI 2007, LNCS 4547, pp. 11–17, 2007.

Let G be a multiplicative group of order uv and let N be a normal subgroup of order u. A subset D of k elements of the group G is called a (v, u, k, λ) relative difference set(RDS) in G relative to N if the set of $k(k-1)$ elements given by

$$\{d_1 d_2^{-1} \mid d_1 \neq d_2, d_1, d_2 \in D\}.$$

contains every nonidentity element of $G \setminus N$ exactly λ times and no element in N ([1,3]). Thus, the parameters of relative difference sets satisfy the following equation: $k(k-1) = u(v-1)\lambda$. If $u = 1$, D becomes a (v, k, λ) difference set(DS). If G is a cyclic group, a relative difference set (a difference set) in it is called a cyclic relative difference set(CRDS)(a cyclic difference set(CDS)).

If A and B are finite groups and $g(x)$ is a function from A to B, for arbitrary $z \in A$ and $a, b \in B$, define $n_z(a, b) = |\{x \mid f(xz) = a, f(x) = b\}|$. By the definitions of DS and RDS, it follows that

Lemma 1. *For each $z \in A \setminus \{1\}$, $n_z(a, a)$ is fixed if and only if the set $\{x \mid f(x) = a\}$ is a CDS of the finite group A.*

The concept of relative difference set was defined by Elliott J. E. H., Butson, A. T. ([2,7]). Two cyclic relative difference sets D_1 and D_2 are equivalent if there exists an integer $e, \gcd(e; uv) = 1$, such that $D_1^e = D_2 g$ for some $g \in G$, where $D_1^e = \{d^e \mid d \in D_1\}$ and $D_2 g = \{dg \mid d \in D_2\}$ ([14]).

Every cyclic difference set with Singer parameters is a projection of its corresponding cyclic relative difference set ([3,17]), which are equivalent to the sequences with two-level autocorrelation property ([4,5,10]). Cyclic relative difference sets and cyclic differences sets with Singer parameters are useful for constructions of optical orthogonal codes ([6]), difference families, and Hadamard matrices ([15,16]).

In 1995, Klapper A. introduced the d-form function on F_{q^n} over F_q which is defined as $f(xy) = y^d f(x)$ for any $x \in F_{q^n}$ and $y \in F_q$([11]). Any homogeneous polynomial function on F_{q^n} over F_q is a d-form function. If $f_1(x)$ and $f_2(x)$ both are d-form functions, so is $af_1(x) + bf_2(x)$, where $a, b \in F_q$.

In 2004, No, J. S. ([13]) introduced a method of constructing cyclic difference sets with Singer parameters from a d-form function on F_q with difference-balanced property as in the following theorem.

Theorem 1. *([13]) Let α be a primitive element in F_{q^n}. If $f(x)$ is a d-form function on $F_{q^n}^*$ over F_q with difference-balanced property, where d is relatively prime to $q^n - 1$. then the set integers defined by*

$$Z_f^{(00)} = \{t \mid f(\alpha^t) = 0, 0 \leq t < \frac{q^n - 1}{q - 1}\}$$

forms a cyclic difference set with Singer parameters $(\frac{q^n-1}{q-1}, \frac{q^{n-1}-1}{q-1}, \frac{q^{n-2}-1}{q-1})$ in the additive group of the residue ring $Z_{\frac{q^n-1}{q-1}}$.

Using this method, some new cyclic difference sets with Singer parameters were constructed from Helleseth Kumar Martinsen (HKM) sequences for $p = 3$ ([9]) and HG sequences ([8]).

Also this method was modified by Chandler D., Xiang Q.([3]) to construct new cyclic relative difference sets with parameters $(\frac{q^{3k}-1}{q-1}, q-1; q^{3k}-1; q^{3k}-2)$ for $q = 3e$ from HKM sequences.

Kim S. H., No J. S., Chung H. and Helleseth T.([14]) generalized Chandler and Xiang's construction of cyclic relative difference sets from HKM sequences to a common $d-$ form function on $\mathbf{F}_{q^n}$ with difference-balanced property. Their main result is in the following theorem.

Theorem 2. ([14]) *Let q be a prime power and n a positive integer. Let α be a primitive element in $\mathbf{F}_{q^n}$. If $f(x)$ is a d-form function on $\mathbf{F}_{q^n}^*$ over $\mathbf{F}_q$ with difference-balanced property, where d is relatively prime to $q^n - 1$. Then the set*

$$D_f^{(1)} = \{x \mid f(x) = 1, x \in \mathbf{F}_{q^n}\}$$

is a cyclic relative difference set with parameters $(\frac{q^n-1}{q-1}, q-1; q^n - 1; q^n - 2)$ in the multiplicative group $\mathbf{F}_{q^n}$ relative to its normal subgroup $\mathbf{F}_q$.

It is well-known that there exists a natural function from the quotient group $\mathbf{F}_{q^n}^*/\mathbf{F}_q^*$ to $\mathbf{F}_q$, which is induced by the function $f(x)$. We denote this function to be $\tilde{f}(x)$, namely $\tilde{f}(\overline{x}) = f(x)$, where $\overline{x} = x\mathbf{F}_q^*$.

Define $D_f^{(00)} = \{\alpha^t | \tilde{f}(\overline{\alpha^t}) = 0, 0 \le t < \frac{q^n-1}{q-1}\}$ and $\overline{D_f^{(00)}} = \{\overline{\alpha^t} | \alpha^t \in D_f^{(00)}\}$.

Then $D_f^{(00)} = \alpha^{Z_f^{(00)}} \subset D_f^{(0)}$ and $\overline{D_f^{(00)}} = \overline{\alpha}^{Z_f^{(00)}}$. Since the multiplicative quotient group $\mathbf{F}_{q^n}^*/\mathbf{F}_q^*$ is isomorphic to the additive group of the residue ring $Z_{\frac{q^n-1}{q-1}}$, and $\overline{D_f^{(00)}}$ corresponds to $Z_f^{(00)}$ by this isomorphism, then Theorem 1 can also be expressed by the following theorem.

Theorem 1'. Under the same condition with Theorem 1, the set

$$\overline{D_f^{(00)}} = \{\overline{\alpha^t} | \tilde{f}(\overline{\alpha^t}) = 0, 0 \le t < \frac{q^n - 1}{q - 1}\}$$

is a cyclic difference set with the Singer parameters $(\frac{q^n - 1}{q - 1}, \frac{q^{n-1} - 1}{q - 1}, \frac{q^{n-2} - 1}{q - 1})$ in the quotient group $\mathbf{F}_{q^n}^*/\mathbf{F}_q^*$.

This paper contributes to the relation between the sets $Z_f^{(00)}$ and $D_f^{(1)}$.

2   Properties of d-Form Functions

Lemma 2. *Assume $f(x)$ be a d-form function on $\mathbf{F}_{q^n}$ over $\mathbf{F}_q$. Then $f(x)$ satisfies the following:*

(1) $f(0) = 0$.

(2) If there exists $m \in \mathbf{F}_q^*$ such $f(m) = 0$, then we have $f(y) = 0$ for each $y \in \mathbf{F}_q$, namely $\mathbf{F}_q \subset D_f^{(0)}$.

(3) Assume $\gcd(d, q-1) = 1$ *and there exists* $m \in \mathbf{F}_q^*$ *such that* $f(m) \neq 0$, *then, for* $y_1, y_2 \in \mathbf{F}_q$, $f(y_1) = f(y_2) \Rightarrow y_1 = y_2$, *namely* $y_1 \neq y_2 \Leftrightarrow y_1, y_2$ *are not in a same* $D_f^{(m)}$.

Proof. (1) Since $f(x)$ is a d-form function on $\mathbf{F}_{q^n}$ over $\mathbf{F}_q$, then

$$f(0) = f(00) = 0^d f(0) = 0.$$

(2) If there exists $m \in \mathbf{F}_q^*$ such that $f(m) = 0$, then, for each $y \in \mathbf{F}_q$,

$$f(y) = f(ym^{-1}m) = (ym^{-1})^d f(m) = 0.$$

(3) From (2) of this lemma, if there exists $m \in \mathbf{F}_q^*$ such that $f(m) \neq 0$, then, for each $y \in \mathbf{F}_q^*$, $f(y) \neq 0$. It follows that $f(1) \neq 0$. Assume $f(y_1) = f(y_2) = 0$, from (1) of this lemma, if $y_1 y_2 = 0$, then $y_1 = y_2 = 0$; if $y_1 y_2 \neq 0$, then

$$f(y_1) = f(y_2) \Leftrightarrow y_1^d f(1) = y_2^d f(1) \Leftrightarrow (y_1^d - y_2^d) f(1) = 0.$$

Since $f(1) \neq 0$, then $y_1^d - y_2^d = 0$. And from $\gcd(d, q-1) = 1$, we have $y_1 = y_2$.

And we have

Lemma 3. *For each* $m \in \mathbf{F}_q^*$, $D_f^{(m^d)} = m D_f^{(1)}$, *and* $\mid D_f^{(m^d)} \mid = \mid D_f^{(1)} \mid$.

Proof. Since $f(x)$ is a d-form function, then

$$
\begin{aligned}
D_f^{(m^d)} &= \{x | f(x) = m^d, x \in \mathbf{F}_{q^n}^*\} = \{x | (m^{-d}) f(x) = 1, x \in \mathbf{F}_{q^n}^*\} \\
&= \{x | f(m^{-1}x) = 1, x \in \mathbf{F}_{q^n}^*\} = \{my | f(y) = 1, y \in \mathbf{F}_{q^n}^*\} \\
&= m\{y | f(y) = 1, y \in \mathbf{F}_{q^n}^*\} = m D_f^{(1)}.
\end{aligned}
$$

It follows that $\mid D_f^{(m^d)} \mid = \mid D_f^{(1)} \mid$.

Lemma 4. *Assume* $m_1, m_2 \in \mathbf{F}_q$.
  (1) $m_1 D_f^{(1)} = m_2 D_f^{(1)} \Longleftrightarrow m_1^d = m_2^d$.
  (2) If $\gcd(d, q-1) = 1$, *then*

$$m_1 D_f^{(1)} = m_2 D_f^{(1)} \Longleftrightarrow m_1 = m_2; m_1 \neq m_2 \Longleftrightarrow m_1 D_f^{(1)} \cap m_2 D_f^{(1)} = \emptyset.$$

Proof. (1)From Lemma 3, $m_1 D_f^{(1)} = m_2 D_f^{(1)} \Longleftrightarrow D_f^{(m_1^d)} = D_f^{(m_2^d)} \Longleftrightarrow m_1^d = m_2^d$.

(2)If $\gcd(d, q-1) = 1$, since

$$m_1^d = m_2^d \Longleftrightarrow (m_1 m_2^{-1})^d = 1 \Longleftrightarrow m_1 m_2^{-1} = 1 \Longleftrightarrow m_1 = m_2,$$

then $m_1 D_f^{(1)} = m_2 D_f^{(1)} \Longleftrightarrow m_1 = m_2$. For arbitrary $m_1, m_2 \in \mathbf{F}_q^*$, $m_1 D_f^{(1)} \neq m_2 D_f^{(1)}$. And by Lemma 3, $m_i D_f^{(1)} = D_f^{(m_i^d)}, i = 0, 1$. It follows that $D_f^{(m_1^d)} \neq D_f^{(m_2^d)}$. And by the definitions of $D_f^{(m_1^d)}$ and $D_f^{(m_2^d)}$, $D_f^{(m_1^d)} \cap D_f^{(m_2^d)} = \emptyset$. Then $m_1 D_f^{(1)} \cap m_2 D_f^{(1)} = \emptyset$. The inversion is obvious.

Lemma 5. *If* $\gcd(d, q-1) = 1$, *then the class of sets*

$$\{mD_f^{(1)} \mid m \in F_q^*\} \cup \{D_f^{(0)}\}$$

form a partition of $\mathrm{GF}(q^n)^*$.

Proof. By Lemma 3, $D_f^{(m^d)} = mD_f^{(1)}$. Since $\gcd(d, q-1) = 1$, then m^d runs over F_q^* as m run over F_q^*. It follows that

$$\bigcup_{m \in F_q^*} mD_f^{(1)} \cup D_f^{(0)} = \mathrm{GF}(q^n)^*.$$

And from Lemma 4, the above union is a disjoint union. Thus this lemma is proved.

Remark 1. Since the condition $\gcd(d, q-1) = 1$ can be obtained by $\gcd(d, q^n - 1) = 1$, then, by Lemmas 3 and 5, for each $m \in F_q^*$, there exists an element $m_1 \in F_q^*$ such that $D_f^{(m)} = m_1 D_f^{(1)}$. So $D_f^{(m)}$ is a CRDS equivalent to $D_f^{(1)}$ under the conditions of Theorem 2.

For the sets $D_f^{(00)}$ and $D_f^{(0)}$, we have

Lemma 6. *Let* $f(x)$ *be a* d-*form function on* F_{q^n} *over* F_q. *Then*

$$D_f^{(0)} = D_f^{(00)} \cdot F_q^*.$$

Proof. For each $m\gamma \in D_f^{(00)} \cdot F_q^*$, where $m \in F_q^*$, $\gamma \in D_f^{(00)}$, the fact $f(\gamma) = 0$ is obvious. Since $f(x)$ is a d-form function on F_{q^n} over F_q, then $f(m\gamma) = m^d f(\gamma) = 0$, namely $m\gamma \in D_f^{(0)}$. Thus $D_f^{(00)} \cdot F_q^* \subset D_f^{(0)}$. Let $T = \dfrac{q^n - 1}{q - 1}$, then, for each $t : 0 \le t \le q^n - 2$, there exists only a representation $t = t_1 T + t_2$, $0 \le t_1 \le q - 2$, $0 \le t_2 \le T - 1$. Then $\alpha^t = \alpha^{t_1 T} \alpha^{t_2}$, where $\alpha^{t_1 T} \in F_q^*$. Thus the sequence

$$f(\alpha^t) = f(\alpha^{t_1 T + t_2}) = \alpha^{dt_1 T} f(\alpha^{t_2}).$$

So $f(\alpha^t) = 0$ if and only if $f(\alpha^{t_2}) = 0$, namely $\alpha^{t_2} \in D_f^{(00)}$. It follows that $D_f^{(0)} \subset D_f^{(00)} \cdot F_q^*$.

3   Main Results

Theorem 3. *Let* $f(x)$ *be a* d-*form function on* F_{q^n} *over* F_q, *where* $\gcd(d, q-1) = 1$. *Then* $D_f^{(1)} \cup D_f^{(00)}$ *is a quotient group of* $F_{q^n}^*$ *with respect to subgroup* F_q^*.

Proof. At first, by Lemmas 5 and 6,

$$(D_f^{(1)} \cup D_f^{(00)}) \cdot F_q^* = F_{q^n}^*.$$

Further more, for arbitrary different elements a, b in $D_f^{(1)} \cup D_f^{(00)}$, if there exists an element m in F_q^* such that $a = mb$, by the definition of $D_f^{(00)}$, a, b can not in $D_f^{(00)}$ simultaneously. Assume $b \in D_f^{(1)}$ might as well, then $f(a) = f(mb) = m^d f(b) = m^d$. Since $\gcd(d, q - 1) = 1$, then $m^d \neq 1$, and $m^d \neq 0$ is obvious. It follows that $a \notin D_f^{(1)} \cup D_f^{(00)}$. This contradicts to the assumption. Thus a, b can not in a same equivalent class of the quotient group $F_{q^n}^* / F_q^*$.

The following Lemma 7 is needed to prove Theorem 4.

Lemma 7. If $f(x)$ is a d-form function on F_{q^n} over F_q, where $\gcd(d, q^n - 1) = 1$, then, for arbitrary $a, b \in GF(q^n)^*$, $n_z(a, b) = n_z(am, bm)$, where $m \in GF(q)^*$.

Proof. If x_0 satisfies $f(xz) = a, f(x) = b$, by $\gcd(d, q^n - 1) = 1$, then there exist one and only element m_0 in $GF(q)^*$ such that $m_0^d = m^{-1}$. Since $f(x)$ is d-form, then x_0 satisfies $f(xz) = a, f(x) = b$ if and only if $m_0 x_0$ satisfies the equation system $f(xz) = am, f(x) = bm$. Then the roots of the above two equation systems are corresponding one by one. It follows that $n_z(a, b) = n_z(am, bm)$.

Both inversions of Theorems 1 and 2 are not true obviously. However, consider them simultaneously , we have the following theorem.

Theorem 4. Let $f(x)$ be a d-form function on F_{q^n} over F_q and $\gcd(d, q^n - 1) = 1$. Then $f(x)$ is difference-balanced if and only if

$$Z_f^{(00)} = \{t | f(\alpha^t) = 0, 0 \leq t < \frac{q^n - 1}{q - 1}\}$$

is a CDS with Singer parameters $(\frac{q^n - 1}{q - 1}, \frac{q^{n-1} - 1}{q - 1}, \frac{q^{n-2} - 1}{q - 1})$ in the additive group of $Z_{\frac{q^n-1}{q-1}}$ and $D_f^{(1)}$ is a CRDS with Singer parameters $(\frac{q^n - 1}{q - 1}, q - 1, q^{n-1}, q^{n-2})$.

Proof. Sufficiency of this theorem can be proved easily by Lemmas 1 and 2. Necessity can be proved by the following: To prove the function $f(x)$ is difference-balanced, we have to prove that $f(xz) - f(x)$ is balanced for each $z \in F_{q^n} \setminus \{0, 1\}$. Since $f(x)$ is a d-form function, then $f(xz) - f(x)$ is a d-form function too. And from the fact that $\gcd(d, q^n - 1) = 1$, by Lemma 3, $f(xz) - f(x)$ is balanced on F_q^* for each $z \in F_{q^n} \setminus \{0, 1\}$. Next we prove that $f(xz) - f(x)$ takes the element 0 $q^{n-1} - 1$ times as x runs over $F_{q^n}^*$ once, namely $\sum_{a \in F_q} n_z(a, a) = q^{n-1} - 1$. For each $h, 0 \leq h \leq q - 2$, we have $\alpha^{\frac{q^n-1}{q-1}h} \in F_q$. Thus

$$f(\alpha^{t+\frac{q^n-1}{q-1}h}) = \alpha^{\frac{q^n-1}{q-1}hd} f(\alpha^t) = 0. \tag{1}$$

Let $Z_f^{(0h)} = \{t | f(\alpha^t) = 0, \frac{q^n - 1}{q - 1}h \leq t < \frac{q^n - 1}{q - 1}(h+1)\}$. Then $\bigcup_{h=0}^{q-2} D_f^{(0h)} = D_0$. And by the equation (1), we have $| D_f^{(0h)} | = | D_f^{(00)} |$. Since $D_f^{(00)}$ is a CDS with the Singer parameters $(\dfrac{q^n - 1}{q - 1}, \dfrac{q^{n-1} - 1}{q - 1}, \dfrac{q^{n-2} - 1}{q - 1})$ in the additive group of $Z_{\frac{q^n-1}{q-1}}$, then $n_z(0,0) = q^{n-2} - 1$.

Since $D_f^{(1)}$ is a CRDS with the Singer parameters $(\dfrac{q^n - 1}{q - 1}, q - 1, q^{n-1}, q^{n-2})$, then $n_z(1,1) = q^{n-2}$. By Lemma 7, for each $m \in F_q^*$, it follows that $n_z(1,1) = n_z(m,m)$. This implies that $\sum_{a \in F_q} n_z(a,a) = q^{n-2} - 1 + q^{n-2}(q-1) = q^{n-1} - 1$. Then we proved this theorem.

References

1. Baumert, L.D.: Cyclic Difference Sets. Lecture Notes in Mathematics, vol. 182. Springer-Verlag, Berlin/Heidelberg/New York (1971)
2. Butson, A.T.: Relations among generalized Hadamard matrices, relative difference sets and maximal length linear recurring sequences. Canad. J. Math. 15, 42–48 (1963)
3. Chandler, D., Xiang, Q.: Cyclic relative difference sets and their p-ranks, Des., Codes, Cryptogr. 30, 325–343 (2003)
4. Dillon, J.F., Dobbertin, H.: Cyclic difference sets with singer parameters. Finite Fields Their Appl. 10, 342–389 (2004)
5. Jungnickel, D., Pott, A.: Difference sets: An introduction, in Difference Sets, Sequences and their Correlation Properties. In: Pott, A., Kumar, P., Helleseth, T., Jungnickel, D. (eds.), pp. 259–295. Kulwer, North-Holland, Amsterdam (1999)
6. Chung, F.R.K., Salehi, J.A., Wei, V.K.: Optical orthogonal codes: Design, analysis, and applications. IEEE Trans. Inf. Theory 35(3), 595–604 (1989)
7. Elliott, J.E.H., Butson, A.T.: Relative difference sets. Illinois J. Math. 10, 517–531 (1966)
8. Helleseth, T., Gong, G.: New nonbinary sequences with ideal two-level autocorrelation function. IEEE Trans. Inf. Theory 48(11), 2868–2872 (2002)
9. Helleseth, T., Kumar, P.V., Martinsen, H.M.: A new family of ternary sequences with ideal two-level autocorrelation. Des., Codes, Cryptogr. 23, 157–166 (2001)
10. Jungnickel, D.: Difference sets, in Contemporary Design Theory: A Collection of Surveys. In: Dinitz, J., Stinson, D. (eds.), pp. 241–324. Wiley, New York (1992)
11. Klapper, A.: d-form sequence: Families of sequences with low correlation values and large linear spans. IEEE Trans. Inf. Theory 41(2), 423–431 (1995)
12. No, J.S.: p-ary unified sequences: p-ary extended d-form sequences with ideal autocorrelation property. IEEE Trans. Inf. Theory 48(9), 2540–2546 (2002)
13. No, J.S.: New cyclic difference sets with Singer parameters constructed from $d-$homogeneous function. Des., Codes, Cryptogr. 33, 199–213 (2004)
14. Kim, S.H., No, J.S., Chung, H.: New cyclic relative difference sets constructed from $d-$homogeneous functions with difference-balanced properties. IEEE Transactions on Information Theory 51(3), 1155–1163 (2005)
15. Spence, E.: Hadamard matrices from relative difference sets. J. Combin. Theory 19, 287–300 (1975)
16. Yamada, M.: On a relation between a cyclic relative difference sets associated with the quadratic extensions of a finite field and the szekeres difference sets. Combinatorica 8, 207–216 (1988)
17. Singer, J.: A theorem in finite projective geometry and some applications to number theory. Trans. Amer. Math. Soc. 43, 377–385 (1938)

A Note on Modular Forms on Finite Upper Half Planes

Yoshinori Hamahata*

Department of Mathematics
Tokyo University of Science, Noda, Chiba, 278-8510, Japan
`hamahata_yoshinori@ma.noda.tus.ac.jp`

Abstract. Finite upper half planes are finite field analogues of the Poincaré upper half plane. We introduce modular forms of new type on finite upper half planes, and consider related topics.

Keywords: Finite upper half planes, modular forms.

1 Introduction

Classical modular forms play an important role in every area in number theory. Modular forms on the Poincaré upper half plane $\mathbb{H} = \{z \in \mathbb{C} \mid \operatorname{Im} z > 0\}$ are prototypes, and have been generalized in some directions. These are objects over the field of complex numbers $\mathbb{C}$.

In the mid 1980s, A. Terras defined finite upper half planes, which are defined over finite fields as analogue of $\mathbb{H}$. She and coworkers studied special functions on these planes ([1], [5]). The purpose of the present article is to introduce modular forms of new type and consider related topics on finite upper half planes. In contrast to the classical case, it seems difficult to find good examples to these modular forms in our situation. In [1] the finite field analogues of Eisenstein series are considered. The authors call them Eisenstein sums. In this paper, we proceed observing examples related to them.

2 Modular Forms

In this section we briefly recall modular forms on finite upper half planes.

Let p be an odd prime. We denote $\mathbb{F}_{p^r}$ $(r \geq 1)$ for the finite field with p^r elements. Take a nonsquare element $\delta_r \in \mathbb{F}_{p^r}$ and fix it. Set

$$H_{p^r} = \{z = x + y\sqrt{\delta_r} \mid x, y \in \mathbb{F}_{p^r}, y \neq 0\}.$$

We call it a *finite upper half plane*. This plane is a finite field version of the Poincaré upper half plane.

---

* Partially supported by Grant-in-Aid for Scientific Research, Ministry of Education, Japanese Government, No. 18540050.

C. Carlet and B. Sunar (Eds.): WAIFI 2007, LNCS 4547, pp. 18–24, 2007.

Let $G_{p^r} = GL(2, \mathbb{F}_{p^r})$ be the general linear group over $\mathbb{F}_{p^r}$. The group G_{p^r} acts on H_{p^r} by the linear fractional transformation: for $z \in H_{p^r}$ and $g = \begin{pmatrix} a & b \\ c & d \end{pmatrix} \in G_{p^r}$, define

$$gz = \frac{az + b}{cz + d}.$$

The fixed subgroup of $\sqrt{\delta_r}$ in G_{p^r} is

$$K_{p^r} = \left\{ \begin{pmatrix} a & b\delta_r \\ b & a \end{pmatrix} \mid a, b \in \mathbb{F}_{p^r}, a^2 - \delta_r b^2 \neq 0 \right\}.$$

It is known that the action of G_{p^r} on H_{p^r} is transitive. So that H_{p^r} is expressed as $H_{p^r} = G_{p^r}/K_{p^r}$.

Let Γ be a subgroup of G_{p^r}. The map $m : \Gamma \times H_{p^r} \to \mathbb{C}^\times$ is called a *multiplier system* for Γ if the condition

$$m(\gamma\eta, z) = m(\gamma, \eta z)m(\eta, z)$$

is satisfied for any $\gamma, \eta \in \Gamma$ and $z \in H_{p^r}$. For these Γ and m, the map $f : H_{p^r} \to \mathbb{C}$ is called a *modular form* for Γ with the multiplier system m if for any $\gamma \in \Gamma$,

$$f(\gamma z) = m(\gamma, z)f(z)$$

holds. Denote by $M(\Gamma, m)$ the space of modular forms of this type.

We give an example due to [1] and [3]. Let $\pi : \mathbb{F}_{p^r}(\sqrt{\delta_r})^\times \to \mathbb{C}^\times$ be a multiplicative character. For $\gamma = \begin{pmatrix} a & b \\ c & d \end{pmatrix} \in G_{p^r}$, we put $J_\pi(\gamma, z) := \pi(cz + d)$. Then for any $\gamma, \eta \in G_{p^r}$, we have

$$J_\pi(\gamma\eta, z) = J_\pi(\gamma, \eta z)J_\pi(\eta, z).$$

Hence $J_\pi : G_{p^r} \times H_{p^r} \to \mathbb{C}^\times$ is a multiplier system for G_{p^r}. Define $m : K_{p^r} \times H_{p^r} \to \mathbb{C}^\times$ by

$$m(k, z) = \frac{J_\pi(k, z)}{J_\pi(k, \sqrt{\delta_r})}.$$

We see that m is a multiplier system for K_{p^r}. Put

$$E_{\pi,r}(z) = \frac{1}{|K_{p^r}|} \sum_{k \in K_{p^r}} \frac{J_\pi(k, \sqrt{\delta_r})}{J_\pi(k, z)},$$

where $|K_{p^r}|$ is the number of elements of K_{p^r}. We call $E_{\pi,r}$ the *Eisenstein sum* associated to K_{p^r}, π. This is a finite field analogue of the Eisenstein series on the Poincaré upper half plane. It is possible to prove that $E_{\pi,r} \in M(K_{p^r}, m)$.

3   Mixed Modular Forms

Classical automorphic forms satisfy a transformation formula for a discrete subgroup Γ of $SL(2, \mathbb{R})$. To be more specific, if f is an automorphic form of weight k for Γ, then for $z \in H$ and $\gamma = \begin{pmatrix} a & b \\ c & d \end{pmatrix} \in \Gamma$, we have

$$f(\gamma z) = j(\gamma, z)^k f(z).$$

Here we have used the notation $j(\gamma, z) = cz + d$. Let $\omega : H \to H$ be a holomorphic map that is equivariant with respect to a homomorphism $\chi : \Gamma \to SL(2, \mathbb{R})$. Then a mixed automorphic form is a holomorphic function on H satisfying a transformation formula

$$f(\gamma z) = j(\gamma, z)^k j(\chi(\gamma), \omega(z))^l f(z)$$

for some nonnegative integers k and l as well as the holomorphic condition at the cusps.

Now we define a mixed version for modular forms on H_{p^r}. Let $\omega : H_{p^r} \to H_{p^r}$ be a map, and $\chi : \Gamma \to G_{p^r}$ a homomorphism such that $\chi(\gamma)\omega(z) = \omega(\gamma z)$ ($\gamma \in \Gamma$). We call the pair (ω, χ) an *equivariant pair*.

Definition 1. Let Γ be a subgroup of G_{p^r}, and m a multiplier system for Γ. Then the map $f : H_{p^r} \to \mathbb{C}$ is called a *mixed modular form* for Γ with the multiplier system m and the equivariant pair (ω, χ) if for any $\gamma \in \Gamma$,

$$f(\gamma z) = m(\gamma, z)m(\chi(\gamma), \omega(z))f(z)$$

is satisfied.

We denote by $M(\Gamma, m, \omega, \chi)$ the space of mixed modular forms for $\Gamma, m, (\omega, \chi)$. It is easy to see the following:

(1) If ω is the identity and χ is the inclusion map, then $M(\Gamma, m, \chi, \omega) = M(\Gamma, m^2)$;

(2) If $f, g \in M(\Gamma, m)$, then $f(z)g(\omega(z)) \in M(\Gamma, m, \omega, \chi)$.

Mixed modular form is a generalization of modular form in a sense.

We here give an example for mixed modular forms. Recall that K_{p^r} is the fixed subgroup of $\sqrt{\delta_r}$ in G_{p^r}. Let π denote a multiplicative character of $\mathbb{F}_{p^r}(\sqrt{\delta_r})^\times$, and J_π the multiplier system for G_{p^r} defined in the last section. We define

$$E_{\pi,r}^{\omega,\chi}(z) = \frac{1}{|K_{p^r}|} \sum_{k \in K_{p^r}} \frac{J_\pi(k, \sqrt{\delta_r})J_\pi(\chi(k), \omega(\sqrt{\delta_r}))}{J_\pi(k, z)J_\pi(\chi(k), \omega(z))}.$$

We call $E_{\pi,r}^{\omega,\chi}$ the *mixed Eisenstein sum* associated for K_{p^r}, π, (ω, χ).

Theorem 1. *Let $m : K_{p^r} \times H_{p^r} \to \mathbb{C}^\times$ be the multiplier system for K_{p^r} defined by $m(k, z) = J_\pi(k, z)/J_\pi(k, \sqrt{\delta_r})$. Then $E_{\pi,r}^{\omega,\chi} \in M(K_{p^r}, m, \omega, \chi)$.*

Proof. Since $J_\pi(k, z)$ is a multiplier for K_{p^r}, we have

$$J_\pi(kk', z) = J_\pi(k, k'z)J_\pi(k', z)$$

for $k, k' \in K_{p^r}$. Similarly, we have

$$J_\pi(\chi(k)\chi(k'), \omega(z)) = J_\pi(\chi(k), \chi(k')\omega(z))J_\pi(\chi(k'), \omega(z))$$

for $k, k' \in K_{p^r}$. By this equality, we obtain

$$J_\pi(\chi(kk'), \omega(z)) = J_\pi(\chi(k), \omega(k'z))J_\pi(\chi(k'), \omega(z))$$

for $k, k' \in K_{p^r}$. Using these equalities, we have

$$|K_{p^r}|E_{\pi,r}^{\omega,\chi}(k'z) = \sum_{k \in K} \frac{J_\pi(k, \sqrt{\delta_r})J_\pi(\chi(k), \omega(\sqrt{\delta_r}))}{J_\pi(k, k'z)J_\pi(\chi(k), \omega(k'z))}$$

$$= \sum_{k \in K} \frac{J_\pi(k, \sqrt{\delta_r})J_\pi(k', z)}{J_\pi(kk', z)} \cdot \frac{J_\pi(\chi(k), \omega(\sqrt{\delta_r}))J_\pi(\chi(k'), \omega(z))}{J_\pi(\chi(kk'), \omega(z))}$$

$$= \sum_{k \in K} \frac{J_\pi(kk', \sqrt{\delta_r})J_\pi(k', z)}{J_\pi(kk', z)J_\pi(k', \sqrt{\delta_r})} \cdot \frac{J_\pi(\chi(kk'), \omega(\sqrt{\delta_r}))J_\pi(\chi(k'), \omega(z))}{J_\pi(\chi(kk'), \omega(z))J_\pi(\chi(k'), \omega(\sqrt{\delta_r}))}$$

$$= \frac{J_\pi(k', z)J_\pi(\chi(k'), \omega(z))}{J_\pi(k', \sqrt{\delta_r})J_\pi(\chi(k'), \omega(\sqrt{\delta_r}))} \sum_{k \in K} \frac{J_\pi(kk', z)J_\pi(\chi(kk'), \omega(\sqrt{\delta_r}))}{J_\pi(kk', z)J_\pi(\chi(kk'), \omega(z))}$$

$$= m(k, z)m(\chi(k), \omega(z))|K_{p^r}|E_{\pi,r}^{\omega,\chi}(z).$$

$\square$

4 Modular Embeddings

So far we have worked on a fixed upper half plane H_{p^r}. In what follows let us treat some upper half planes arising from various finite fields.

Take integers s, r with $s \geq r \geq 1$. Let (Φ, ϕ) be a pair of maps $\Phi : G_{p^r} \to G_{p^s}$ and $\phi : H_{p^r} \to H_{p^s}$ satisfying

(1) Φ is an injective group homomorphism;
(2) ϕ is an injection;
(3) $\phi(\gamma z) = \Phi(\gamma)\phi(z)$ $(\gamma \in G_{p^r}, z \in H_{p^r})$.

Then we call the pair (Φ, ϕ) a *modular embedding* from H_{p^r} into H_{p^s}. Modular embedding defines a pullback of modular forms. To be more precise, let Γ, Γ' be the subgroups of G_{p^r}, G_{p^s}, respectively with the condition $\Phi(\Gamma) \subset \Gamma'$. We take a multiplier system $m' : \Gamma' \times H_{p^s} \to \mathbb{C}^\times$. Using it, one gets the multiplier system $m : \Gamma \times H_{p^r} \to \mathbb{C}^\times$ defined by $(\gamma, z) \mapsto m'(\Phi(\gamma), \phi(z))$. For each modular form $f \in M(\Gamma', m')$, we find that $f \circ \phi \in M(\Gamma, m)$. Hence we have a pullback map

$$\phi^* : M(\Gamma', m') \longrightarrow M(\Gamma, m), \quad f \mapsto f \circ \phi.$$

We are going to give an example to a modular embedding. Suppose that p is an odd prime number. We choose a generator δ_r of $\mathbb{F}_{p^r}^\times$. Define some kinds of finite upper half planes $H_{p^r}^{(i)}$ $(i = 1, \ldots, r)$ by

$$H_{p^r}^{(i)} = \{x + y\sqrt{\delta_r^{p^i}} \mid x, y \in \mathbb{F}_{p^r}, y \neq 0\}.$$

Notice that $H_{p^r}^{(r)} = H_{p^r}$. We define $\phi : H_{p^r} \to H_{p^r}^{(1)}$ by $x + y\sqrt{\delta_r} \mapsto x^p + y^p\sqrt{\delta_r^p}$. Let $\Phi : G_{p^r} \to G_{p^r}$, $\begin{pmatrix} a & b \\ c & d \end{pmatrix} \mapsto \begin{pmatrix} a^p & b^p \\ c^p & d^p \end{pmatrix}$. Then the pair (Φ, ϕ) is a modular embedding from H_{p^r} into $H_{p^r}^{(1)}$. Let $\pi' : \mathbb{F}_{p^r}(\sqrt{\delta_r^p})^\times \to \mathbb{C}^\times$ be a multiplicative character, and $\widetilde{K}_{p^r}$ the fixed subgroup of $\sqrt{\delta_r^p}$ in G_{p^r}. We know that $\widetilde{K}_{p^r} = \{\begin{pmatrix} a & b\delta_r^p \\ b & a \end{pmatrix} \mid a, b \in \mathbb{F}_{p^r}, a^2 - b^2\delta_r^p \neq 0\}$. For $\widetilde{K}_{p^r}$ and π', consider the Eisenstein sum

$$\widetilde{E}_{\pi', r} = \frac{1}{|\widetilde{K}_{p^r}|} \sum_{k' \in \widetilde{K}_{p^r}} \frac{J_{\pi'}(k', \sqrt{\delta_r^p})}{J_{\pi'}(k', z)},$$

which is a modular form for $\widetilde{K}_{p^r}$ with the multiplier system $m'(k, z) = J_{\pi'}(k, z)/J_{\pi'}(k, \sqrt{\delta_r^p})$. Let us pullback it by (Φ, ϕ). Then we have

$$\phi^* \widetilde{E}_{\pi', r}(z) = |K_{p^r}|^{p-1} E_{\pi, r}(z)^p,$$

where $\pi : \mathbb{F}_{p^r}(\sqrt{\delta_r})^\times \to \mathbb{C}^\times$ is the multiplicative character obtained by composing π' with the map $\mathbb{F}_{p^r}(\sqrt{\delta_r})^\times \to \mathbb{F}_{p^r}(\sqrt{\delta_r^p})^\times$, $x \mapsto x^p$. The function $E_{\pi, r}$ is the Eisenstein sum for K_{p^r} with the multiplier system $m(k, z) = J_\pi(k, z)/J_\pi(k, \sqrt{\delta_r})$.

5   Hilbert Modular Forms

In classical case, Hilbert modular forms are modular forms of several variables defined on the product of the Poincaré upper half plane $\mathbb{H}$. Let n be a natural number greater than one. We take a totally real number field K of degree n. The field K has n embeddings $K \hookrightarrow \mathbb{R}$, $x \mapsto x^{(i)}$ $(i = 1, \ldots, n)$. Let O_K be the ring of integers of K. We write $SL_2(O_K) = \{\begin{pmatrix} a & b \\ c & d \end{pmatrix} \in SL_2(\mathbb{R}) \mid a, b, c, d \in O_K\}$.

The group $SL_2(O_K)$ acts on $\mathbb{H}^n$ as follows: for $\gamma = \begin{pmatrix} a & b \\ c & d \end{pmatrix} \in SL_2(O_K)$, put $\gamma^{(i)} = \begin{pmatrix} a^{(i)} & b^{(i)} \\ c^{(i)} & d^{(i)} \end{pmatrix}$ $(i = 1, \ldots, n)$. For $z = (z_1, \ldots, z_n) \in \mathbb{H}^n$, we define

$$\gamma z = \left(\frac{a^{(1)}z_1 + b^{(1)}}{c^{(1)}z_1 + d^{(1)}}, \ldots, \frac{a^{(n)}z_n + b^{(n)}}{c^{(n)}z_n + d^{(n)}} \right).$$

Let k be a nonnegative integer. Then the holomorphic function $f : \mathbb{H}^n \to \mathbb{C}$ is called *Hilbert modular form* of weight k for $SL_2(O_K)$ if for each $\gamma = \begin{pmatrix} a & b \\ c & d \end{pmatrix} \in SL_2(O_K)$,

$$f(\gamma z) = \prod_{i=1}^{n} j(\gamma^{(i)}, z_i)^k f(z)$$

holds. Here we have used the notation $j(\gamma^{(i)}, z_i) = c^{(i)} z_i + d^{(i)}$.

Now we are going to define Hilbert modular forms on $H_{p^r}^r$. The field $\mathbb{F}_{p^r}$ has r automorphisms over $\mathbb{F}_p$ defined by $x \mapsto x^{(i)} = x^{p^i}$ $(i = 1, \ldots, r)$. The group $G_{p^r} = GL(2, \mathbb{F}_{p^r})$ acts on $H_{p^r}^r$ as follows: for $\gamma = \begin{pmatrix} a & b \\ c & d \end{pmatrix} \in G_{p^r}$, put $\gamma^{(i)} = \begin{pmatrix} a^{(i)} & b^{(i)} \\ c^{(i)} & d^{(i)} \end{pmatrix}$ $(i = 1, \ldots, n)$. For $z = (z_1, \ldots, z_n) \in H_{p^r}^r$, we define

$$\gamma z = \left(\frac{a^{(1)} z_1 + b^{(1)}}{c^{(1)} z_1 + d^{(1)}}, \ldots, \frac{a^{(n)} z_n + b^{(n)}}{c^{(n)} z_n + d^{(n)}} \right).$$

Definition 2. Take a subgroup Γ of G_{p^r}. Let $m : \Gamma \times H_{p^r} \to \mathbb{C}^\times$ be a multiplier system. If the function $f : H_{p^r}^r \to \mathbb{C}$ satisfies the following condition, then we call f *Hilbert modular form* for Γ with the multiplier system m:

$$f(\gamma z) = \left(\prod_{i=1}^{r} m(\gamma^{(i)}, z_i) \right) f(z) \quad (\gamma \in \Gamma).$$

We write $M^H(\Gamma, m)$ for the space of Hilbert modular forms for Γ, m. It should be noted that we can define Hilbert modular form on $H_{p^r}^{(1)} \times \cdots \times H_{p^r}^{(r)}$ in the same way as the above.

The following is an example for Hilbert modular forms. We utilize the notation in Section two. Let $\pi : \mathbb{F}_{p^r}(\sqrt{\delta_r})^\times \to \mathbb{C}^\times$ be a multiplicative character, and $m : K_{p^r} \times H_{p^r} \to \mathbb{C}^\times$ the multiplier system for K_{p^r} defined by $m(k, z) = J_\pi(k, z) / J_\pi(k, \sqrt{\delta_r})$. We put

$$E_{\pi,r}^H(z) = \frac{1}{|K_{p^r}|} \sum_{k \in K_{p^r}} \prod_{i=1}^{r} \frac{J_\pi(k^{(i)}, \sqrt{\delta_r})}{J_\pi(k^{(i)}, z_i)}.$$

We call $E_{\pi,r}^H(z)$ the *Hilbert-Eisenstein sum* associated to K_{p^r}, π. This sum is a Hilbert modular form:

Theorem 2. $E_{\pi,r}^H \in M^H(\Gamma, m)$.

The proof can be done as that of Theorem 1. Therefore we omit it.

In the last section we have defined modular embeddings between two finite upper half planes. Now let us introduce another type of modular embeddings. Under the notation of the last section, let $\psi : H_{p^r} \to H_{p^r}^{(1)} \times \cdots \times H_{p^r}^{(r)}$ be the map defined by

$$x + y\sqrt{\delta_1} \mapsto (x^p + y^p \sqrt{\delta_r^p}, \ldots, x^{p^r} + y^{p^r} \sqrt{\delta_r^{p^r}}),$$

and $\Psi : G_{p^r} \to G_{p^r}$ the identity map. By definition of the action of G_{p^r} on $H_{p^r}^r$, we have

$$\psi(\gamma z) = \Psi(\gamma)\psi(z) \qquad (\gamma \in G_{p^r}).$$

We call the pair (Ψ, ψ) the *diagonal embedding* from H_{p^r} into $H_{p^r}^{(1)} \times \cdots \times H_{p^r}^{(r)}$. This embedding is analogous to the diagonal embedding $\mathbb{H} \to \mathbb{H}^n$ in the classical case. Thanks to the diagonal embedding, we can pull back each Hilbert modular form on $H_{p^r}^{(1)} \times \cdots \times H_{p^r}^{(r)}$ to get a modular form of one variable. This follows from a similar discussion to that done in the last section.

Remark 1. As the related topics, one can define mixed modular forms and modular embeddings between the products of finite upper half planes.

6   Concluding Remarks

It is clear from the above discussion that there is a shortage of good examples to modular forms on finite upper half planes. Hence on one hand, we need to find examples to modular forms other than Eisenstein sums. On the other hand, for further study we should develop the theory of modular forms observing classical modular forms.

References

1. Angel, J., Celniker, N., Poulos, S., Terras, A., Trimble, C., Velasquez, E.: Special functions on finite upper half planes. Contemporary Math. 138, 1–26 (1992)
2. Freitag, E.: Hilbert Modular Forms. Springer-Verlag, Heidelberg (1990)
3. Harish-Chandra.: Eisenstein series over finite fields, in Collected Papers, vol. 4, pp. 8–21. Springer-Verlag, Berlin Heidelberg (1984)
4. Lee, M.H.: Mixed Automorphic Forms, Torus Bundles, and Jacobi Forms. Lecture Notes in Mathematics, vol. 1845. Springer, Heidelberg (2004)
5. Terras, A.: Fourier Analysis on Finite Groups and Applications. London Mathematical Society Student Texts, vol. 43. Cambridge Univ. Press, Cambridge (1999)

A Coprocessor for the Final Exponentiation of the η_T Pairing in Characteristic Three*

Jean-Luc Beuchat[1], Nicolas Brisebarre[2,3], Masaaki Shirase[4], Tsuyoshi Takagi[4], and Eiji Okamoto[1]

[1] Laboratory of Cryptography and Information Security, University of Tsukuba, 1-1-1 Tennodai, Tsukuba, Ibaraki, 305-8573, Japan
[2] LaMUSE, Université J. Monnet, 23, rue du Dr P. Michelon, F-42023 Saint-Étienne Cedex 02, France
[3] LIP/Arénaire (CNRS-ENS Lyon-INRIA-UCBL), ENS Lyon, 46 Allée d'Italie, F-69364 Lyon Cedex 07, France
[4] Future University-Hakodate, School of Systems Information Science, 116-2 Kamedanakano-cho, Hakodate, Hokkaido, 041-8655, Japan

Abstract. Since the introduction of pairings over (hyper)elliptic curves in constructive cryptographic applications, an ever increasing number of protocols based on pairings have appeared in the literature. Software implementations being rather slow, the study of hardware architectures became an active research area. Beuchat *et al.* proposed for instance a coprocessor which computes the characteristic three η_T pairing, from which the Tate pairing can easily be derived, in $33\,\mu s$ on a Cyclone II FPGA. However, a final exponentiation is required to ensure a unique output value and the authors proposed to supplement their η_T pairing accelerator with a coprocessor for exponentiation. Thus, the challenge consists in designing the smallest possible piece of hardware able to perform this task in less than $33\,\mu s$ on a Cyclone II device. In this paper, we propose a novel arithmetic operator implementing addition, cubing, and multiplication over $\mathbb{F}_{3^{97}}$ and show that a coprocessor based on a single such operator meets this timing constraint.

Keywords: η_T pairing, characteristic three, final exponentiation, hardware accelerator, FPGA.

1 Introduction

The first introduction of Weil and Tate pairings in cryptography was due to Menezes *et al.* [20] and Frey and Rück [11] who used them to attack the discrete logarithm problem on some classes of elliptic curves defined over finite fields. More recently, several cryptographic schemes based on those pairings have been proposed: identity-based encryption [6], short signature [8], and efficient broadcast encryption [7] to mention but a few.

---

* This work was supported by the New Energy and Industrial Technology Development Organization (NEDO), Japan.

C. Carlet and B. Sunar (Eds.): WAIFI 2007, LNCS 4547, pp. 25–39, 2007.
© Springer-Verlag Berlin Heidelberg 2007

This article aims at computing the η_T pairing in characteristic three in the case of supersingular elliptic curves over $\mathbb{F}_{3^m}$. These curves are necessarily of the form $E^b : y^2 = x^3 - x + b$, with $b \in \{-1, 1\}$. According to [3], curves over fields of characteristic three often offer the best possible ratio between security level and space requirements. Note that the η_T pairing easily relates to the Tate pairing [2]. In the following, we assume that $m = 97$ and $\mathbb{F}_{3^{97}}$ is given as $\mathbb{F}_3[x]/(x^{97} + x^{12} + 2)$. This choice is currently a good trade-off between security and computation time.

After previous works by Miller [21], Barreto *et al.* [3] and Galbraith *et al.* [12], an efficient algorithm for the characteristic three was proposed by Duursma and Lee [10]. That work was then extended by Kwon [19]. The introduction of the η_T pairing by Barreto *et al.* [2] led to a reduction by a factor two of the number of iterations compared to the approach by Duursma and Lee. Algorithm 1 summarizes the scheme proposed by Barreto *et al.* and uses the following notation: let $\ell > 0$ be an integer relatively prime to 3^m (i.e. to 3). The set $E^b(\mathbb{F}_{3^m})[\ell]$ groups all the points $P \in E^b(\mathbb{F}_{3^m})$ such that $\ell P = \mathcal{O}$, where $\mathcal{O}$ is the point at infinity. Let σ and $\rho \in \mathbb{F}_{3^{6m}}$ which satisfy $\sigma^2 = -1$ and $\rho^3 = \rho + b$. They help to define the distortion map introduced in [3].

Algorithm 1. Computation of η_T pairing in characteristic three [2]

Input: $P = (x_p, y_p)$ and $Q = (x_q, y_q) \in E^b(\mathbb{F}_{3^m})[\ell]$. The algorithm requires R_0 and $R_1 \in \mathbb{F}_{3^{6m}}$, as well as $r_0 \in \mathbb{F}_{3^m}$ for intermediate computations.
Output: $\eta_T(P, Q)$
1: **if** $b = 1$ **then**
2:     $y_p \leftarrow -y_p$;
3: **end if**
4: $r_0 \leftarrow x_p + x_q + b$;
5: $R_0 \leftarrow -y_p r_0 + y_q \sigma + y_p \rho$;
6: **for** $i = 0$ to $(m-1)/2$ **do**
7:     $r_0 \leftarrow x_p + x_q + b$;
8:     $R_1 \leftarrow -r_0^2 + y_p y_q \sigma - r_0 \rho - \rho^2$;
9:     $R_0 \leftarrow R_0 R_1$;
10:     $x_p \leftarrow x_p^{1/3}$; $y_p \leftarrow y_p^{1/3}$;
11:     $x_q \leftarrow x_q^3$; $y_q \leftarrow y_q^3$;
12: **end for**
13: Return R_0;

Algorithm 1 has the drawback of using inverse Frobenius maps (i.e. cube root in characteristic three). In [5], Beuchat *et al.* proposed a modified η_T pairing algorithm in characteristic three that does not require any cube root. However, to ensure a unique output value for the η_T pairing, we have to compute R_0^W, where $W = (3^{3m} - 1)(3^m + 1)(3^m + 1 - b3^{(m+1)/2})$ here. This operation, often referred to as *final exponentiation*, requires among other things a single inversion over $\mathbb{F}_{3^m}$ and multiplications over $\mathbb{F}_{3^{6m}}$ (note that pairing calculation in characteristic two involves an inversion over $\mathbb{F}_{2^m}$ for final exponentiation). Pairing accelerators described in the literature follow two distinct strategies:

- Several researchers designed coprocessors for arithmetic over $\mathbb{F}_{3^m}$ (or $\mathbb{F}_{2^m}$) implementing both pairing calculation and final exponentiation [17,22,23,25]. This last operation is intrinsically sequential and there is unfortunately no parallelism at all when comes the time of inversion. It is therefore crucial to embed a fast inverter to avoid impacting the overall performance of the system. Reference [18] introduces for instance an efficient architecture for Extended Euclidean Algorithm (EEA) based inversion. To our best knowledge, the fastest coprocessor designed according to this philosophy computes $\eta_T(P,Q)^W$ over the field $\mathbb{F}_{3^{97}}$ in $179\,\mu s$ ($114\,\mu s$ for the pairing calculation and $65\,\mu s$ for the final exponentiation) on a Virtex-II Pro 100 Field-Programmable Gate Array (FPGA) [22].
- Consider the computation of the η_T pairing (Algorithm 1) and note that two coefficients of R_1 are null and another one is equal to -1. This observation allowed Beuchat *et al.* to design an optimized multiplier over $\mathbb{F}_{3^{97}}$ which is at the heart of a pairing accelerator computing $\eta_T(P,Q)$ in $33\,\mu s$ [5]. It is worth noticing that the computation of the pairing requires 4849 clock cycles. Since Fermat's little theorem makes it possible to carry out inversion over $\mathbb{F}_{3^m}$ by means of multiplications and cubings, $\eta_T(P,Q)^W$ could be computed on such an accelerator. It seems however more attractive to supplement it with dedicated hardware for final exponentiation.

The challenge consists in designing the smallest possible processor able to compute a final exponentiation in less than $33\,\mu s$ on a Cyclone II device. Our architecture is based on an innovative algorithm introduced by Shirase, Takagi, and Okamoto in [24]. We summarize this scheme in Section 2 and describe a novel arithmetic operator performing addition, subtraction, multiplication, and cubing over $\mathbb{F}_{3^{97}}$ (Section 3). We show that a coprocessor based on a single such processing element allows us to meet our timing constraint. Section 4 provides the reader with a comparison against previously published solutions.

2   Computation of the Final Exponentiation

Algorithm 2 describes a traditional way to perform final exponentiation [5]. Ronan *et al.* took for instance advantage of such a scheme to design their η_T pairing accelerator [22]. Shirase *et al.* proposed a novel algorithm based on the following remark [24]: let $X \in \mathbb{F}_{3^{6m}}$, then $X^{3^{3m}-1}$ belong to the torus $T_2(\mathbb{F}_{3^{3m}})$, a set introduced in [14]. Then they showed that the arithmetic in T_2 is cheaper, hence a significant gain in term of number of operations compared to Algorithm 2 (see Table 1). Algorithm 6 describes this final exponentiation scheme. It uses Algorithms 3 and 5.

Algorithm 3 involves an inversion over $\mathbb{F}_{3^{3m}}$. The tower field representation allows us to substitute this operation with 12 multiplications, 11 additions, and an inversion over $\mathbb{F}_{3^m}$ (see Appendix B for details). In order to keep the

Algorithm 2. Raising $\eta_T(P,Q)$ to the W-th power $(b = 1)$ [5]

Input: $\eta_T(P,Q) \in \mathbb{F}_{3^{6m}}$. Thirteen variables u_i, $0 \le i \le 6$, and v_i, $0 \le i \le 5$ belonging
    to $\mathbb{F}_{3^{6m}}$ store intermediate results.
Output: $\eta_T(P,Q)^W \in \mathbb{F}_{3^{6m}}$
1: $u_0 \leftarrow \eta_T(P,Q)$;
2: **for** $i = 1$ to 5 **do**
3:     $u_i \leftarrow u_{i-1}^{3^{3m}}$;
4: **end for**
5: $u_1 \leftarrow u_1^2$;
6: $u_4 \leftarrow u_4^2$;
7: $v_0 \leftarrow \eta_T(P,Q)^{3^{(m+1)/2}}$;
8: **for** $i = 1$ to 4 **do**
9:     $v_i \leftarrow v_{i-1}^{3^{3m}}$;
10: **end for**
11: $u_6 \leftarrow v_0 \cdot v_1 \cdot u_3 \cdot u_4 \cdot u_5$;
12: $v_5 \leftarrow u_0 \cdot u_1 \cdot u_2 \cdot v_3 \cdot v_4$;
13: Return $u_0 \leftarrow u_6/v_5$;

Algorithm 3. Computation of $X^{3^{3m}-1}$

Input: $X = x_0 + x_1\sigma + x_2\rho + x_3\sigma\rho + x_4\rho^2 + x_5\sigma\rho^2 \in \mathbb{F}_{3^{6m}}^*$.
Output: $X^{3^{3m}-1} \in T_2(\mathbb{F}_{3^{3m}})$
1: $\tau_0 \leftarrow (x_0 + x_2\rho + x_4\rho^2)^2$;
2: $\tau_1 \leftarrow (x_1 + x_3\rho + x_5\rho^2)^2$;
3: $\tau_2 \leftarrow (x_0 + x_2\rho + x_4\rho^2)(x_1 + x_3\rho + x_5\rho^2)$;
4: $Y \leftarrow \dfrac{(\tau_0 - \tau_1) + \tau_2\sigma}{\tau_0 + \tau_1}$;
5: Return Y;

---

circuit area as small as possible, we suggest to perform inversion according to Fermat's little theorem and Itoh and Tsujii's work [16]. Since $m = 97$, inversion requires 9 multiplications and 96 cubings over $\mathbb{F}_{3^{97}}$ (Algorithm 4, see Appendix A for a proof of correctness). Therefore, final exponentiation requires 87 multiplications, 390 cubings, and 477 additions over $\mathbb{F}_{3^{97}}$ (see Appendix B for details about the number of operations over $\mathbb{F}_{3^{97}}$ involved in the final exponentiation). Array multipliers processing D coefficients of an operand at each clock cycle are often at the heart of pairing accelerators (see Section 3.2). In [5], authors suggest to consider $D = 3$ coefficients and multiplication over $\mathbb{F}_{3^{97}}$ involves $\lceil \frac{97}{3} \rceil = 33$ clock cycles. Since addition and cubing are rather straightforward operations, they are carried out in a single clock cycle. Therefore, considering such parameters, final exponentiation requires $477 + 390 + 33 \cdot 87 = 3738$ clock cycles. Note that additional clock cycles are necessary to load and store intermediate results. However, this overhead should be smaller than 10% and a coprocessor embedding a multiplier, an adder/subtracter, as well as a cubing unit should perform this task in less than 4200 clock cycles. It is therefore possible to supplement the η_T pairing

---

Algorithm 4. Inversion over $\mathbb{F}_{3^{97}}$

Input: $a \in \mathbb{F}_{3^{97}}$
Output: $a^{-1} \in \mathbb{F}_{3^{97}}$
  1: $y_0 \leftarrow a$;
  2: **for** $i = 0$ to 5 **do**
  3:    $z_i \leftarrow y_i^{3^{2^i}}$;
  4:    $y_{i+1} \leftarrow y_i z_i$;
  5: **end for**
  6: $z_6 \leftarrow y_6^{3^{32}}$;
  7: $y_7 \leftarrow y_5 z_6$;
  8: $y_8 \leftarrow y_7^2$;
  9: $y_9 \leftarrow y_8^3$;
10: Return $y_0 y_9$;

Algorithm 5. Computation of X^{3^m+1} in the torus $T_2(\mathbb{F}_{3^{3m}})$

Input: $X \in T_2(\mathbb{F}_{3^{3m}})$
Output: $X^{3^m+1} \in T_2(\mathbb{F}_{3^{3m}})$
  1: $z_0 \leftarrow x_0 x_4, z_1 \leftarrow x_1 x_5, z_2 \leftarrow x_2 x_4, z_3 \leftarrow x_3 x_5$;
  2: $z_4 \leftarrow (x_0 + x_1)(x_4 - x_5)$;
  3: $z_5 \leftarrow x_1 x_2, z_6 \leftarrow x_0 x_3$;
  4: $z_7 \leftarrow (x_0 + x_1)(x_2 + x_3)$;
  5: $z_8 \leftarrow (x_2 + x_3)(x_4 - x_5)$;
  6: $y_0 \leftarrow 1 + z_0 + z_1 - bz_2 - bz_3$;
  7: $y_1 \leftarrow z_1 + z_4 + bz_5 - z_0 - bz_6$;
  8: $y_2 \leftarrow z_7 - z_2 - z_3 - z_5 - z_6$;
  9: $y_3 \leftarrow z_3 + z_8 + bz_0 - z_2 - bz_1 - bz_4$;
10: $y_4 \leftarrow bz_2 + bz_3 + bz_7 - bz_5 - bz_6$;
11: $y_5 \leftarrow bz_3 + bz_8 - bz_2$;
12: Return $Y = (y_0 + y_2 \rho + y_4 \rho^2) + (y_1 + y_3 \rho + y_5 \rho^2)\sigma$;

---

accelerator described in [5] (4849 clock cycles) with such a simple processing unit.

3 Hardware Implementation

This section describes the implementation of Algorithm 6 on a Cyclone II EP2C35F672C6 FPGA whose smallest unit of configurable logic is called Logic Element (LE). Each LE includes a 4-input Look-Up Table (LUT), carry logic, and a programmable register. A Cyclone II EP2C35F672C6 device contains for instance 33216 LEs. Readers who are not familiar with Cyclone II devices should refer to [1] for further details. After studying addition, multiplication, and cubing over $\mathbb{F}_{3^m}$, we propose a novel arithmetic operator able to perform these three operations and describe the architecture of a final exponentiation coprocessor based on such a processing element.

Algorithm 6. Final exponentiation of η_T pairing [24]

Input: $X = x_0 + x_1\sigma + x_2\rho + x_3\sigma\rho + x_4\rho^2 + x_5\sigma\rho^2 \in \mathbb{F}_{3^{6m}}^*$.
Output: $X^{(3^{3m}-1)(3^m+1)(3^m+1-b3^{(m+1)/2})}$

1: $Y \leftarrow X^{3^{3m}-1}$ (Algorithm 3);
2: $Y \leftarrow Y^{3^m+1}$ (Algorithm 5);
3: $Z \leftarrow Y$;
4: **for** $i = 0$ to $(m-1)/2$ **do**
5:    $Z \leftarrow Z^3$;
6: **end for**
7: $Y \leftarrow Y^{3^m+1}$ (Algorithm 5);
8: **if** $b = 1$ **then**
9:    Return $Y \cdot (z_0 - z_1\sigma + z_2\rho - z_3\sigma\rho + z_4\rho^2 - z_5\sigma\rho^2)$;
10: **else**
11:    Return YZ;
12: **end if**

Table 1. Comparison of final exponentiation algorithms (number of operations)

Algorithm	Additions over $\mathbb{F}_{3^{97}}$	Cubings over $\mathbb{F}_{3^{97}}$	Multiplications over $\mathbb{F}_{3^{97}}$
Algorithm 2	1022	390	243
Algorithm 6	477	390	87

3.1 Addition and Subtraction over $\mathbb{F}_{3^m}$

Since they are performed component-wise, addition and subtraction over $\mathbb{F}_{3^m}$ are rather straightforward operations. Each element of $\mathbb{F}_3$ is encoded by two bit and addition modulo three on a Cyclone II FPGA requires two 4-input LUTs. Negation over $\mathbb{F}_3$ is performed by multiplying an operand by two. Note that the computation of the y_i's in Algorithm 5 involves the addition of up to six operands. This motivates the design of the accumulator illustrated on Figure 1a.

3.2 Multiplication over $\mathbb{F}_{3^m}$

Three families of algorithms allow one to compute $a(x)b(x) \mod f(x)$. In parallel-serial schemes, a single coefficient of the multiplier $a(x)$ is processed at each step. This leads to small operands performing a multiplication in m steps. Parallel multipliers compute a degree-$(2m-2)$ polynomial and carry out a final modular reduction. They achieve a higher throughput at the price of a larger circuit area. By processing D coefficients of an operand at each clock cycle, array multipliers, introduced by Song and Parhi in [26], offer a good trade-off between computation time and circuit area and are at the heart of several pairing coprocessors (see for instance [5,13,17,22,23,25]). Among the many array multipliers described in the literature (see for instance [15,25]), the one proposed by Shu, Kwon, and Gaj [25] (Algorithm 7) is a good candidate for FPGA implementation when $f(x)$

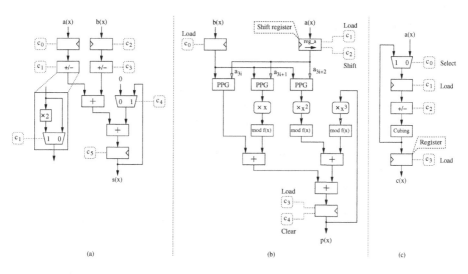

(a)   (b)   (c)

Fig. 1. Arithmetic operators over $\mathbb{F}_{3^m}$. (a) Addition/subtraction of two operands and accumulation. (b) Multiplication [25] ($D = 3$ coefficients of $a(x)$ processed at each clock cycle). (c) Cubing.

Algorithm 7. Multiplication over $\mathbb{F}_{3^m}$ [25]

Input: A degree-m monic polynomial $f(x) = x^m + f_{m-1}x^{m-1} + \ldots + f_1x + f_0$ and two degree-$(m-1)$ polynomials $a(x)$ and $b(x)$. A parameter D which defines the number of coefficients of $a(x)$ processed at each clock cycle. The algorithm requires a degree-$(m-1)$ polynomial $t(x)$ for intermediate computations.

Output: $p(x) = a(x)b(x) \bmod f(x)$

1: $p(x) \leftarrow 0$;
2: **for** i from $\lceil m/D \rceil - 1$ downto 0 **do**
3: $\quad t(x) \leftarrow \sum_{j=0}^{D-1} \left(a_{Di+j}x^j b(x) \right) \bmod f(x)$;
4: $\quad p(x) \leftarrow t(x) + (x^D p(x) \bmod f(x))$;
5: **end for**

---

is a trinomial [4]. Figure 1b illustrates the architecture of an operator processing $D = 3$ coefficients at each clock cycle. It mainly consists of three Partial Product Generators (PPG), three modulo $f(x)$ reduction units, a multioperand adder, and registers to store operands and intermediate results. Five bits make it possible to control this operator.

In the following, we will focus on multiplication over $\mathbb{F}_{3^{97}}$ and assume that $D = 3$ (i.e. multiplication requires 33 clock cycles). With such parameters, the first iteration of Algorithm 7 is defined as follows: $t(x) \leftarrow a_{96}b(x) + (a_{97}xb(x)) \bmod f(x) + (a_{98}x^2b(x)) \bmod f(x)$. To ensure a correct result, we have to guarantee that $a_{97} = a_{98} = 0$. Therefore, the shift register stores a degree-98 polynomial whose two most significant coefficients are set to zero.

3.3 Cubing over $\mathbb{F}_{3^m}$

Since we set $f(x) = x^{97} + x^{12} + 2$, cubing over $\mathbb{F}_{3^m}$ is a pretty simple arithmetic operation: a GP/PARI program provides us with a closed formula:

$$
\begin{aligned}
b_0 &= a_{93} + a_{89} + a_0, & b_1 &= a_{65} + 2a_{61}, & b_2 &= a_{33}, \\
b_3 &= a_{94} + a_{90} + a_1, & \ldots &= \ldots, & b_{94} &= a_{96} + a_{92} + a_{88}, & (1) \\
b_{95} &= a_{64} + 2a_{60}, & b_{96} &= a_{32}.
\end{aligned}
$$

The most complex operation involved in cubing is therefore the addition of three elements belonging to $\mathbb{F}_3$. Recall that inversion over $\mathbb{F}_{3^{97}}$ involves successive cubing operations. Since storing intermediate results in memory would be too time consuming, our cubing unit should include a feedback mechanism to efficiently implement Algorithm 4. Furthermore, cubing over $\mathbb{F}_{3^{6m}}$ requires the computation of $-y_i^3$, where $y_i \in \mathbb{F}_{3^m}$ (see Appendix B for details). These considerations suggest the design of the operator depicted by Figure 1c.

Place-and-Route Results. These three arithmetic operators were captured in the VHDL language and prototyped on an Altera Cyclone II EP2C35F672C6 device. Both synthesis and place-and-route steps were performed with Quartus II 6.0 Web Edition (Table 2). A naive solution would then consist in connecting the outputs of these operators to the memory blocks by means of a three-input multiplexer controlled by two bits. Such an arithmetic and logic unit (ALU) requires 3308 Logic Elements (LEs) and final exponentiation can be carried out within 4082 clock cycles, thus meeting our timing constraint. Cubings only occur in inversion (Algorithm 4) and in the computation of Z (step 5 of Algorithm 6). Due to the sequential nature of these algorithms, both multiplier and adder remain idle at that time. The same observation can be made for additions and multiplications: most of the time, only a single arithmetic operator is processing data. Is it therefore possible to save hardware resources by designing an operator able to perform addition, multiplication, and cubing over $\mathbb{F}_{3^{97}}$?

Table 2. Arithmetic operators over $\mathbb{F}_{3^{97}}$ on a Cyclone II FPGA

	Addition/ subtraction	Multiplication (D = 3)	Cubing	ALU
Area [LEs]	970	1375	668	3308
Control [bits]	6	5	4	17

3.4 An Operator for Multiplication, Addition, and Cubing over $\mathbb{F}_{3^{97}}$

Consider again the closed formula for cubing over $\mathbb{F}_3[x]/(x^{97} + x^{12} + 2)$ (Equation (1)). We can for instance write $b_1 = a_{65} + a_{61} + a_{61}$ and $b_2 = a_{33} + 0 + 0$. Let us define $c_0(x)$, $c_1(x)$, and $c_2(x) \in \mathbb{F}_{3^{97}}$ such that:

$$c_0(x) = a_{93} + a_{65}x + a_{33}x^2 + \ldots + a_{88}x^{94} + a_{64}x^{95} + a_{32}x^{96},$$
$$c_1(x) = a_{89} + a_{61}x + 0 \cdot x^2 + \ldots + a_{92}x^{94} + a_{60}x^{95} + 0 \cdot x^{96}, \qquad (2)$$
$$c_2(x) = a_0 + a_{61}x + 0 \cdot x^2 + \ldots + a_{96}x^{94} + a_{60}x^{95} + 0 \cdot x^{96}.$$

Then, $a(x)^3 = c_0(x) + c_1(x) + c_2(x)$ and cubing requires the addition of three operands as well as some wiring to compute the $c_i(x)$'s. Remember now that our array multiplier (Figure 1b) embeds a three-operand adder and an accumulator, which also makes possible the implementation of addition and cubing. Furthermore, since negation over $\mathbb{F}_{3^m}$ consists in multiplying the operand by two, PPGs can perform this task.

These considerations suggest the design of a three-input arithmetic operator for addition, accumulation, cubing, and multiplication over $\mathbb{F}_{3^{97}}$ (Figure 2). In order to compute the product $a(x)b(x) \bmod f(x)$, it suffices to load $a(x)$ in register R0, and $b(x)$ in registers R1 and R2. Addition and cubing are slightly more complex and we will consider a toy example to illustrate how our operator works. Let us assume we have to compute $-a(x) + b(x)$ and $a(x)^3$, where $a(x)$, $b(x) \in \mathbb{F}_{3^{97}}$. We respectively load $a(x)$ and $b(x)$ in registers R2 and R1 and define a control word stored in R0 so that $d0_{3i} = 2$, $d0_{3i+1} = 1$, and $d0_{3i+2} = 0$. We will thus compute $(2a(x) + b(x) + 0 \cdot a(x)) \bmod f(x) = (-a(x) + b(x)) \bmod f(x)$. For cubing, we load $a(x)$ in both registers R1 and R2. If $d0_{3i} = d0_{3i+1} = d0_{3i+2} = 1$, then our operator implements Equation (2) and returns $a(x)^3$. Thus, register R0 stores either an operand of a multiplication or a control word for up to 33 successive additions and cubings (recall that this shift register stores a degree-98 polynomial and that three coefficients are processed at each clock cycle). Place-and-route results indicate that this processing element requires 2676 LEs instead of 3308 LEs with the naive approach. Furthermore, this architecture allows one to reduce the number of control bits from 17 (see Table 2) to 11.

3.5   Architecture of the Coprocessor

Figure 3 describes the architecture of our coprocessor which embeds a single arithmetic unit performing addition, accumulation, cubing, or multiplication over $\mathbb{F}_{3^{97}}$. Intermediate results (194 bits) and control words for additions and cubings (198 bits) are stored in 64 registers implemented by a dual-port RAM (13 Cyclone II M4K memory blocks). An element of $\mathbb{F}_{3^{6m}}$ returned by the η_T pairing accelerator is sequentially loaded in the RAM. Then, a simple Finite State Machine and a ROM generate all control signals required to perform the final exponentiation according to Algorithm 6. Each instruction stored in the ROM consists of four fields: a control word which specifies the functionality of the processing element, addresses and write enable signals for both ports of the RAM, and a counter which indicates how many times the instruction must be repeated. Inversion over $\mathbb{F}_{3^{97}}$ involves for instance consecutive cubings (Algorithm 4). This approach allows one to execute them with a single instruction.

The implementation of Algorithm 6 on this coprocessor requires 658 instructions which are executed within 4082 clock cycles. Ten control words, stored in

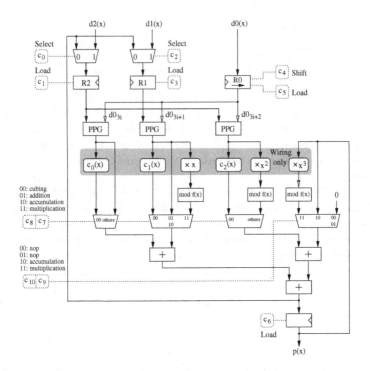

Fig. 2. Addition, accumulation, cubing, and multiplication over $\mathbb{F}_{3^{97}}$

the dual-port RAM, manage all additions and cubings involved in the computation of the final exponentiation.

4   Results and Comparisons

Our final exponentiation coprocessor was implemented on an Altera Cyclone II EP2C35F672C6 FPGA. According to place-and-route tools, this architecture requires 2787 LEs and 21 M4K memory blocks. Since the maximum frequency is 159 MHz, an exponentiation is computed within $26\,\mu s$ and our timing constraint is fully met. It is worth noticing that the inversion over $\mathbb{F}_{3^{97}}$ based on the EEA described in [18] occupies 3422 LEs [27] and needs $2m = 194$ clock cycles. Our approach based on Fermat's little theorem (Algorithm 4) performs the same operation in 394 clock cycles. Therefore, introducing specific hardware for inversion would double the circuit area while reducing the calculation time by only 5%.

To our best knowledge, the only η_T pairing accelerator in characteristic three implementing final exponentiation was proposed by Ronan *et al.* in [22]. In order to easily study the trade-off between calculation time and circuit area, they wrote a C program which automatically generates a VHDL description of a processor and its control according to the number of multipliers to be included and D. The

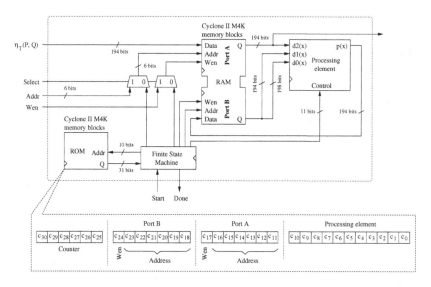

Fig. 3. Architecture of the coprocessor for final exponentiation

ALU also embeds an adder, a subtracter, a cubing unit, and an inversion unit. The most attractive architecture contains three multipliers processing $D = 8$ coefficients at each clock cycle. It computes $\eta_T(P, Q)$ in 114 μs and requires 65 μs to perform final exponentiation according to Algorithm 2 on a Xilinx Virtex-II Pro 100 FPGA (clock frequency: 70.4 MHz). This architecture requires 10000 slices of a Virtex-II Pro FPGA. Each slice of this FPGA family features two 4-input LUTs, carry logic, wide function multiplexers, and two storage elements. Let us assume that Xilinx design tools try to utilize both LUTs of a slice as often as possible (i.e. area optimization). Under this hypothesis, we consider that a slice is roughly equivalent to two LEs and our coprocessor is seven times smaller than the one described in [22].

Recall that Algorithm 6 allows one to divide by 2.8 the number of multiplications over $\mathbb{F}_{3^{97}}$ (Table 1). Therefore, our coprocessor would compute final exponentiation according to Algorithm 2 in around $26 \times 2.8 = 72.8\,\mu$s.

The η_T pairing accelerator described in [5] returns $\eta_T(P, Q)$ in 33 μs using 14895 LEs and 13 memory blocks. We can therefore estimate the total area of a coprocessor computing $\eta_T(P, Q)^W$ to 18000 LEs and 34 M4K memory blocks. Thus, with roughly the same amount of configurable logic, we should achieve five times faster η_T pairing calculation than Ronan *et al.*

5   Concluding Remarks

We proposed a novel arithmetic operator performing addition, accumulation, cubing, and addition over $\mathbb{F}_{3^{97}}$ and designed a coprocessor able to compute the final exponentiation of the η_T pairing in 26 μs on a Cyclone II FPGA. Since the calculation time of the η_T pairing accelerator described in [5] is $33\,\mu s$, we can pipeline both architectures without impacting the overall performance of the system and our approach allows one to divide by five the calculation time of $\eta_T(P,Q)^W$ compared to the best implementation reported in the open literature [22]. Since different FPGA families are involved, it is unfortunately difficult to provide the reader with a fair area comparison. A rough estimate indicates that our coprocessor requires the same hardware resources.

Another important result is that hardware for inversion is not necessary for the calculation of the η_T pairing on a characteristic three elliptic curve over $\mathbb{F}_{3^{97}}$: our final exponentiation coprocessor meets our timing constraint with an algorithm based on Fermat's little theorem. Furthermore, the architecture proposed in [22] computes $\eta_T(P,Q)^W$ in 15113 clock cycles. Since an inverter based on the EEA saves only 200 clock cycles and that no other operation can be performed in parallel, we believe it is not interesting to include dedicated hardware for this operation.

The approach introduced in this paper to design our arithmetic operator offers several further research topics we plan to study in the future. It would for instance be interesting to implement the computation of both pairing and final exponentiation with the coprocessor described in this paper. Such an architecture could for instance be attractive for ASIC implementations. Another open question is if our operator is able to carry out other functions (e.g. cube root) or if this design methodology works for other irreducible polynomials and finite fields. Finally, note that our processor always performs the same operation: at each clock cycle, the content of the shift register is updated (load or shift operation), and a sum of three partial products is computed. Pairing operations could therefore be split into atomic blocks (side-channel atomicity [9]) and such architectures could prevent simple side-channel attacks.

References

1. Altera.: Cyclone II Device Handbook (2006), Available from Altera's web site (http://altera.com)
2. Barreto, P.S.L.M., Galbraith, S.D., Ó hÉigeartaigh, C., Scott, M.: Efficient pairing computation on supersingular abelian varieties. Designs, Codes and Cryptography 42(3), 239–271 (2007)
3. Barreto, P.S.L.M., Kim, H.Y., Lynn, B., Scott, M.: Efficient algorithms for pairing-based cryptosystems. In: Yung, M. (ed.) CRYPTO 2002. LNCS, vol. 2442, pp. 354–368. Springer, Heidelberg (2002)
4. Beuchat, J.-L., Miyoshi, T., Oyama, Y., Okamoto, E.: Multiplication over $\mathbb{F}_{p^m}$ on FPGA: A survey. In: Diniz, P.C., Marques, E., Bertels, K., Fernandes, M.M., Cardoso, J.M.P. (eds.) Reconfigurable Computing: Architectures, Tools and Applications – Proceedings of ARC 2007. LNCS, vol. 4419, pp. 214–225. Springer, Heidelberg (2007)
5. Beuchat, J.-L., Shirase, M., Takagi, T., Okamoto, E.: An algorithm for the η_T pairing calculation in characteristic three and its hardware implementation. In: Proceedings of the 18th IEEE Symposium on Computer Arithmetic (To appear, 2007)

6. Boneh, D., Franklin, M.: Identity-based encryption from the Weil pairing. In: Kilian, J. (ed.) CRYPTO 2001. LNCS, vol. 2139, pp. 213–229. Springer, Heidelberg (2001)

7. Boneh, D., Gentry, C., Waters, B.: Collusion resistant broadcast encryption with short ciphertexts and private keys. In: Shoup, V. (ed.) CRYPTO 2005. LNCS, vol. 3621, pp. 258–275. Springer, Heidelberg (2005)

8. Boneh, D., Lynn, B., Shacham, H.: Short signatures from the Weil pairing. In: Boyd, C. (ed.) ASIACRYPT 2001. LNCS, vol. 2248, pp. 514–532. Springer, Heidelberg (2001)

9. Chevallier-Mames, B., Ciet, M., Joye, M.: Low-cost solutions for preventing simple side-channel analysis: Side-channel atomicity. IEEE Transactions on Computers 53(6), 760–768 (2004)

10. Duursma, I., Lee, H.S.: Tate pairing implementation for hyperelliptic curves $y^2 = x^p - x + d$. In: Laih, C.-S. (ed.) ASIACRYPT 2003. LNCS, vol. 2894, pp. 111–123. Springer, Heidelberg (2003)

11. Frey, G., Rück, H.-G.: A remark concerning m-divisibility and the discrete logarithm in the divisor class group of curves. Math. Comp. 62(206), 865–874 (1994)

12. Galbraith, S.D., Harrison, K., Soldera, D.: Implementing the Tate pairing. In: Fieker, C., Kohel, D.R. (eds.) Algorithmic Number Theory – ANTS V. LNCS, vol. 2369, pp. 324–337. Springer, Heidelberg (2002)

13. Grabher, P., Page, D.: Hardware acceleration of the Tate Pairing in characteristic three. In: Rao, J.R., Sunar, B. (eds.) CHES 2005. LNCS, vol. 3659, pp. 398–411. Springer, Heidelberg (2005)

14. Granger, R., Page, D., Stam, M.: On small characteristic algebraic tori in pairing-based cryptography. LMS Journal of Computation and Mathematics 9, 64–85 (2006), Available from http://www.lms.ac.uk/jcm/9/lms2004-025/

15. Guajardo, J., Güneysu, T., Kumar, S., Paar, C., Pelzl, J.: Efficient hardware implementation of finite fields with applications to cryptography. Acta Applicandae Mathematicae 93 (1–3), 75–118 (2006)

16. Itoh, T., Tsujii, S.: A fast algorithm for computing multiplicative inverses in $GF(2^m)$ using normal bases. Information and Computation 78, 171–177 (1988)

17. Kerins, T., Marnane, W.P., Popovici, E.M., Barreto, P.S.L.M.: Efficient hardware for the Tate Pairing calculation in characteristic three. In: Rao, J.R., Sunar, B. (eds.) CHES 2005. LNCS, vol. 3659, pp. 412–426. Springer, Heidelberg (2005)

18. Kerins, T., Popovici, E., Marnane, W.: Algorithms and architectures for use in FPGA implementations of identity based encryption schemes. In: Becker, J., Platzner, M., Vernalde, S. (eds.) FPL 2004. LNCS, vol. 3203, pp. 74–83. Springer, Heidelberg (2004)

19. Kwon, S.: Efficient Tate pairing computation for supersingular elliptic curves over binary fields. Cryptology ePrint Archive, Report 2004/303 (2004)

20. Menezes, A., Okamoto, T., Vanstone, S.A.: Reducing elliptic curves logarithms to logarithms in a finite field. IEEE Transactions on Information Theory 39(5), 1639–1646 (1993)

21. Miller, V.S.: Short programs for functions on curves. (1986) Unpublished manuscript available at http://crypto.stanford.edu/miller/miller.pdf

22. Ronan, R., O'h'Eigeartaigh, C., Murphy, C., Kerins, T., Barreto, P.S.L.M.: Hardware implementation of the η_T pairing in characteristic 3. Cryptology ePrint Archive, Report 2006/371 (2006)

23. Ronan, R., O'h'Eigeartaigh, C., Murphy, C., Scott, M., Kerins, T., Marnane, W.P.: An embedded processor for a pairing-based cryptosystem. In: Proceedings of the Third International Conference on Information Technology: New Generations (ITNG'06), IEEE Computer Society Press, Los Alamitos (2006)
24. Shirase, M., Takagi, T., Okamoto, E.: Some efficient algorithms for the final exponentiation of η_T pairing. In: 3rd Information Security Practice and Experience Conference – ISPEC 2007. LNCS, Springer, Heidelberg (2007)
25. Shu, C., Kwon, S., Gaj, K.: FPGA accelerated Tate pairing based cryptosystem over binary fields. In: Proceedings of 2006 IEEE International Conference on Field Programmable Technology (FPT 2006), pp. 173–180. IEEE Computer Society Press, Los Alamitos (2006)
26. Song, L., Parhi, K.K.: Low energy digit-serial/parallel finite field multipliers. Journal of VLSI Signal Processing 19(2), 149–166 (1998)
27. Vithanage, A.: Personal communication

A   Proof of Correctness of Algorithm 4

Let $a \in \mathbb{F}_{3^{97}}$. According to Fermat's little theorem, $a^{-1} = a^{3^{97}-2}$. Note that the ternary representation of $3^{97}-2$ is $(\underbrace{22\ldots22}_{96\times}1)_3$. In order to prove the correctness of Algorithm 4, it suffices to show that $y_9 = a^k$, where $k = (\underbrace{22\ldots22}_{96\times}0)_3$:

$$z_0 = y_0^3 = a^{(10)_3}, \qquad y_1 = a^{(11)_3}, \qquad z_1 = y_1^{3^2} = a^{(1100)_3},$$

$$y_2 = a^{(1111)_3}, \qquad z_2 = y_2^{3^4} = a^{(11110000)_3}, \qquad y_3 = a^{(11111111)_3},$$

$$z_3 = y_3^{3^8} = a^{(\overbrace{1\ldots1}^{8\times}\overbrace{0\ldots0}^{8\times})_3}, \qquad y_4 = a^{(\overbrace{1\ldots1}^{16\times})_3}, \qquad z_4 = y_4^{3^{16}} = a^{(\overbrace{1\ldots1}^{16\times}\overbrace{0\ldots0}^{16\times})_3},$$

$$y_5 = a^{(\overbrace{1\ldots1}^{32\times})_3}, \qquad z_5 = y_5^{3^{32}} = a^{(\overbrace{1\ldots1}^{32\times}\overbrace{0\ldots0}^{32\times})_3}, \qquad y_6 = a^{(\overbrace{1\ldots1}^{64\times})_3},$$

$$z_6 = y_6^{3^{32}} = a^{(\overbrace{1\ldots1}^{64\times}\overbrace{0\ldots0}^{32\times})_3}, \qquad y_7 = y_5 z_6 = a^{(\overbrace{1\ldots1}^{96\times})_3}, \qquad y_8 = y_7^2 = a^{(\overbrace{2\ldots2}^{96\times})_3},$$

$$y_9 = y_8^3 = a^{(\overbrace{2\ldots2}^{96\times}0)_3}$$

Then, Algorithm 4 returns $y_0 y_9 = a^{(\overbrace{22\ldots221}^{96\times})_3} = a^{3^{97}-2}$.

B   Arithmetic over $\mathbb{F}_{3^{2m}}$, $\mathbb{F}_{3^{3m}}$, and $\mathbb{F}_{3^{6m}}$

This Appendix summarizes classical algorithms for arithmetic over $\mathbb{F}_{3^{2m}}$, $\mathbb{F}_{3^{3m}}$, and $\mathbb{F}_{3^{6m}}$. Proofs of correctness of such algorithms are for instance provided in [17]. In order to compute the number of operations over $\mathbb{F}_{3^m}$, we assume that the ALU is able to compute $a_i a_j$, $\pm a_i \pm a_j$ and $\pm a_i^3$, where a_i and $a_j \in \mathbb{F}_{3^m}$. We consider the case where the elliptic curve is given by $y^2 = x^3 - x + 1$ (i.e. $b = 1$ and $\rho^3 = \rho + 1$).

Multiplication over $\mathbb{F}_{3^{2m}}$. Let $A = a_0 + a_1\sigma$ and $B = b_0 + b_1\sigma$, where $a_0, a_1, b_0,$ and $b_1 \in \mathbb{F}_{3^m}$. The product $AB = (a_0b_0 - a_1b_1) + ((a_0 + a_1)(b_0 + b_1) - a_0b_0 - a_1b_1)\sigma$ requires 3 multiplications and 5 additions (or subtractions) over $\mathbb{F}_{3^m}$.

Multiplication over $\mathbb{F}_{3^{3m}}$. Assume that $A = a_0 + a_1\rho + a_2\rho^2$ and $B = b_0 + b_1\rho + b_2\rho^2$, where $a_i, b_i \in \mathbb{F}_{3^m}$, $0 \leq i \leq 2$. The product $C = AB$ is then given by:

$$\begin{bmatrix} c_0 \\ c_1 \\ c_2 \end{bmatrix} = \begin{bmatrix} (a_1 + a_2)(b_1 + b_2) + a_0b_0 - a_1b_1 - a_2b_2 \\ (a_0 + a_1)(b_0 + b_1) + (a_1 + a_2)(b_1 + b_2) - a_0b_0 + a_1b_1 \\ (a_0 + a_2)(b_0 + b_2) + a_1b_1 - a_0b_0 \end{bmatrix}.$$

This operation requires 6 multiplications and 14 additions (or subtractions) over $\mathbb{F}_{3^m}$.

Inversion over $\mathbb{F}_{3^{6m}}$. Let $A = a_0 + a_1\rho + a_2\rho^2$, where $a_i \in \mathbb{F}_{3^m}$, $0 \leq i \leq 2$. The inverse C of A is then given by:

$$\begin{bmatrix} c_0 \\ c_1 \\ c_2 \end{bmatrix} = d^{-1} \begin{bmatrix} a_0^2 - (a_1^2 - a_2^2) - a_2(a_0 + a_1) \\ -a_0a_1 + a_2^2 \\ a_1^2 - a_2^2 - a_0a_2 \end{bmatrix},$$

where $d = a_0^2(a_0 - a_2) + a_1^2(-a_0 + a_1) + a_2^2(-(-a_0 + a_1) + a_2)$. This operation involves 12 multiplications, 11 additions (or subtractions), and 1 inversion over $\mathbb{F}_{3^m}$.

Multiplication over $\mathbb{F}_{3^{6m}}$. Let $A = \underbrace{a_0 + a_1\sigma}_{\tilde{a}_0} + \underbrace{(a_2 + a_3\sigma)}_{\tilde{a}_1}\rho + \underbrace{(a_4\rho^2 + a_5\sigma)}_{\tilde{a}_2}\rho^2$

and $B = \underbrace{b_0 + b_1\sigma}_{\tilde{b}_0} + \underbrace{(b_2 + b_3\sigma)}_{\tilde{b}_1}\rho + \underbrace{(b_4\rho^2 + b_5\sigma)}_{\tilde{b}_2}\rho^2$. The product $C = AB$ is then given by (6 multiplications and 14 additions over $\mathbb{F}_{3^{2m}}$):

$$\begin{bmatrix} \tilde{c}_0 \\ \tilde{c}_1 \\ \tilde{c}_2 \end{bmatrix} = \begin{bmatrix} (\tilde{a}_1 + \tilde{a}_2)(\tilde{b}_1 + \tilde{b}_2) + \tilde{a}_0\tilde{b}_0 - \tilde{a}_1\tilde{b}_1 - \tilde{a}_2\tilde{b}_2 \\ (\tilde{a}_0 + \tilde{a}_1)(\tilde{b}_0 + \tilde{b}_1) + (\tilde{a}_1 + \tilde{a}_2)(\tilde{b}_1 + \tilde{b}_2) - \tilde{a}_0\tilde{b}_0 + \tilde{a}_1\tilde{b}_1 \\ (\tilde{a}_0 + \tilde{a}_2)(\tilde{b}_0 + \tilde{b}_2) + \tilde{a}_1\tilde{b}_1 - \tilde{a}_0\tilde{b}_0 \end{bmatrix}.$$

Thus, multiplication over $\mathbb{F}_{3^{6m}}$ requires 18 multiplications and 58 additions (or subtractions) over $\mathbb{F}_{3^m}$.

VLSI Implementation of a Functional Unit to Accelerate ECC and AES on 32-Bit Processors*

Stefan Tillich and Johann Großschädl

Graz University of Technology,
Institute for Applied Information Processing and Communications,
Inffeldgasse 16a, A–8010 Graz, Austria
{Stefan.Tillich,Johann.Groszschaedl}@iaik.tugraz.at

Abstract. Embedded systems require efficient yet flexible implementations of cryptographic primitives with a minimal impact on the overall cost of a device. In this paper we present the design of a functional unit (FU) for accelerating the execution of cryptographic software on 32-bit processors. The FU is basically a multiply-accumulate (MAC) unit able to perform multiplications and MAC operations on integers and binary polynomials. Polynomial arithmetic is a performance-critical building block of numerous cryptosystems using binary extension fields, including public-key primitives based on elliptic curves (e.g. ECDSA), symmetric ciphers (e.g. AES or Twofish), and hash functions (e.g. Whirlpool). We integrated the FU into the Leon2 SPARC V8 core and prototyped the extended processor in an FPGA. All operations provided by the FU are accessible to the programmer through custom instructions. Our results show that the FU allows to accelerate the execution of 128-bit AES by up to 78% compared to a conventional software implementation using only native SPARC V8 instructions. Moreover, the custom instructions reduce the code size by up to 87.4%. The FU increases the silicon area of the Leon2 core by just 8,352 gates and has almost no impact on its cycle time.

1 Introduction

The usual way to accelerate cryptographic operations in embedded devices like smart cards is to offload the computationally heavy parts of an algorithm from the main processor to a dedicated hardware accelerator such as a cryptographic co-processor. However, cryptographic hardware has all the restrictions inherent in any pure hardware implementation, most notably limited flexibility and poor scalability in relation to software. The term scalability refers to the ability to process operands of arbitrary size. Typical RSA hardware implementations, for example, only support operands up to a certain size, e.g. 1024 bits, and can not be used when the need for processing longer operands arises. The term flexibility

---

* The work described in this paper was supported by the Austrian Science Fund under grant P16952-NO4 ("Instruction Set Extensions for Public-Key Cryptography") and by the European Commission under grant FP6-IST-033563 (project SMEPP).

C. Carlet and B. Sunar (Eds.): WAIFI 2007, LNCS 4547, pp. 40–54, 2007.

means the possibility to replace a cryptographic algorithm (e.g. DES) by another one from the same class (e.g. AES) without the need to redesign a system. While cryptographic software can be relatively easily upgraded and/or "patched," an algorithm cast in silicon is fixed and can not be changed without replacing the whole chip. However, the importance of *algorithm agility* becomes evident in light of the recently discovered vulnerabilities in the SHA-1 hash algorithm. SHA-1 is widely used in security protocols like SSL or IPSec and constitutes an integral part of the security concepts specified by the Trusted Computing Group (TCG) [20]. A full break of SHA-1 would be a disaster for TCG-compliant systems since almost all trusted platform modules (TPMs) implement SHA-1 in hardware and lack hash algorithm agility.

In recent years, a new approach for implementing cryptography in embedded systems has emerged that combines the performance and energy-efficiency of hardware with the scalability and algorithm agility of software [10]. This approach is based on the idea of extending an embedded processor by dedicated custom instructions and/or architectural features to allow for efficient execution of cryptographic algorithms. *Instruction set extensions* are well established in the domain of multimedia and digital signal processing. Today, almost every major processor architecture features multimedia extensions; well-known examples are Intel's MMX and SSE technology, AMD's 3DNow, and the AltiVec extensions to the PowerPC architecture. All these extensions boost the performance of multimedia workloads at the expense of a slight increase in silicon area.

The idea of extending a processor's instruction set with the goal to accelerate performance-critical operations is applicable to cryptography as well. Software implementations of cryptographic algorithms often spend the majority of their execution time in a few performance-critical code sections. Typical examples of such code sections are the inner loops of long integer arithmetic operations needed in public-key cryptography [8]. Other examples are certain transformations used in block ciphers (e.g. SubBytes or MixColumns in AES), which can be expensive in terms of computation time when memory constraints or the threat of cache attacks prevent an implementation via lookup tables. Speeding up these code sections through custom instructions can, therefore, result in a significant performance gain. Besides execution time also the code size is reduced since a custom instruction typically replaces several "native" instructions.

The custom instructions can be executed in an application-specific functional unit (FU) or a conventional FU—such as the arithmetic/logic unit (ALU) or the multiplier—augmented with application-specific functionality. A typical example for the latter category is an integer multiplier able to execute not only the standard multiply instructions, but also custom instructions for long integer arithmetic [7]. Functional units are tightly coupled to the processor core and directly controlled by the instruction stream. The operands processed in FUs are read from the general-purpose registers and the result is written back to the register file. Hardware acceleration through custom instructions is cost-effective because tightly coupled FUs can utilize all resources already available in a processor, e.g. the registers and control logic. On the other hand, loosely-coupled hardware accelerators like co-processors have separate registers, datapaths, and state machines for their control. In addition, the interface between processor and co-processor costs silicon area and may also introduce a severe performance bottleneck due to communication and synchronization overhead [9].

In summary, instruction set extensions are a good compromise between the performance and efficiency of cryptographic hardware and the scalability and algorithm agility of software. Application-specific FUs require less silicon area than co-processors, but allow to achieve significantly better performance than "conventional" software implementations [10]. Recent research has demonstrated that instruction set extensions can even outperform a crypto co-processor while demanding only a fraction of the silicon area [18].

1.1   Contributions of This Work

In this paper we introduce the design and implementation of a functional unit (FU) to accelerate the execution of both public-key and secret-key cryptography on embedded processors. The FU is basically a multiply/accumulate (MAC) unit consisting of a (32×16)-bit multiplier and a 72-bit accumulator. It is capable to process signed and unsigned integers as well as binary polynomials, i.e. the FU contains a so-called *unified multiplier*[1] [15]. Besides integer and polynomial multiplication and multiply/accumulate operations, the FU can also perform the reduction of binary polynomials modulo an irreducible polynomial of degree $m = 8$, such as needed for AES en/decryption [3,5]. The rich functionality provided by the FU facilitates efficient software implementation of a broad range of cryptosystems, including the "traditional" public-key schemes involving long integer arithmetic (e.g. RSA, Diffie-Hellman), elliptic curve cryptography (ECC) [8] over both prime fields and binary extension fields, as well as the Advanced Encryption Standard (AES) [13].

A number of unified multiplier architectures for public-key cryptography, in particular ECC, have been published in the past [15,6]. However, the FU presented in this paper extends previous work in two important aspects. First, our FU supports not only ECC but also the AES, in particular the MixColumns and InvMixColumns operations. Second, we integrated the FU into the SPARC V8-compliant Leon2 softcore [4] and prototyped the extended processor in an FPGA, which allowed us, on the one hand, to evaluate the hardware cost and critical path delay of the extended processor and, on the other hand, to analyze the impact of the FU on performance and code size of AES software. All execution times reported in this paper were measured on "working silicon" in form of an FPGA prototype.

The main component of our FU is a (32×16)-bit unified multiplier for signed/unsigned integers and binary polynomials. We used the unified multiplier architecture for ECC described in [6] as starting point for our implementation. The main contribution of this paper is the integration of support for the AES

---

[1] The term unified means that the multiplier uses the same datapath for both integers and binary polynomials.

MixColumns and InvMixColumns operations, which require besides polynomial multiplication also the reduction modulo an irreducible polynomial of degree $m = 8$. Hence, we focus in the remainder of this paper on the implementation of the polynomial modular reduction and refer to [6] for details concerning the original multiplier for ECC. To the best of our knowledge, the FU introduced in this paper is the first approach for integrating AES support into a unified multiplier for integers and binary polynomials.

Although the focus of this paper is directed towards the AES, we point out that the presented concepts can also be applied to other block ciphers requiring polynomial arithmetic, e.g. Twofish, or to hash functions like Whirlpool, which has a similar structure as AES.

2 Arithmetic in Binary Extension Fields

The finite field $\mathbb{F}_q$ of order $q = p^m$ with p prime can be represented in a number of ways, whereby all these representations are isomorphic. The elements of fields of order 2^m are commonly represented as polynomials of degree up to $m-1$ with coefficients in the set $\{0, 1\}$. These fields are called *binary extension fields* and a concrete instance of $\mathbb{F}_{2^m}$ is generated by choosing an irreducible polynomial of degree m over $\mathbb{F}_2$ as reduction polynomial. The arithmetic operations in $\mathbb{F}_{2^m}$ are defined as polynomial operations with a reduction modulo the irreducible polynomial. Binary extension fields have the advantage that addition has no carry propagation. This feature allows efficient implementation of arithmetic in these fields in hardware. Addition can be done with a bitwise exclusive OR (XOR) and multiplication with the simple shift-and-XOR method followed by reduction modulo the irreducible polynomial.

Binary extension fields play an important role in cryptography as they constitute a basic building block of both public-key and secret-key algorithms. For example, the NIST recommends to use binary fields as underlying algebraic structure for the implementation of elliptic curve cryptography (ECC) [8]. The degree m of the fields used in ECC is rather large, typically in the range between 160 and 500. The multiplication of elements of such large fields is very costly on 32-bit processors, even if a custom instruction for multiplying binary polynomials is available. On the other hand, the reduction of the product of two field elements modulo an irreducible polynomial $f(x)$ is fairly fast (in relation to multiplication) and can be accomplished with a few shift and XOR operations if $f(x)$ has few non-zero coefficients, e.g. if $f(x)$ is a trinomial [8].

Contrary to ECC schemes, the binary fields used in secret-key systems like block ciphers are typically very small. For example, AES and Twofish rely on the field $\mathbb{F}_{2^8}$. A multiplication of two binary polynomials of degree ≤ 7 can be easily performed in one clock cycle with the help of a custom instruction like gf2mul [17]. However, the reduction of the product modulo an irreducible polynomial $f(x)$ of degree 8 is relatively slow when done in software, i.e. requires much longer than one cycle. Therefore, it is desirable to provide hardware support for the reduction operation modulo irreducible polynomials of small degree.

3   Implementation Options for AES

The Advanced Encryption Standard (AES) is a block cipher with a fixed block size of 128 bits and a variable key size of 128, 192, or 256 bits [3]. In November 2001, the NIST officially introduced the AES as successor of the Data Encryption Standard (DES). An encryption with AES consists of an initial key addition, a sequence of round transformations, and a (slightly different) final round transformation. The round transformation for encryption is composed of the following four steps: AddRoundKey, SubBytes, ShiftRows, and MixColumns. Decryption is performed in a similar fashion as encryption, but uses the inverse operations (i.e. InvSubBytes, InvShiftRows, and InvMixColumns).

The binary extension field $GF(2^8)$ plays a central role in the AES algorithm [3]. Multiplication in $GF(2^8)$ is part of the MixColumns operation and inversion in $GF(2^8)$ is carried out in the SubBytes operation. The MixColumns/InvMixColumns operation is, in general, one of the most time-consuming parts of the AES [5]. Software implementations on 32-bit platforms try to speed up this operation either by using an alternate data representation [1] or by employing large lookup tables [3]. However, the use of large tables is disadvantageous for embedded systems since they occupy scarce memory resources, increase cache pollution, and may open up potential vulnerabilities to cache-based side channel attacks [14].

The MixColumns transformation of AES can be defined as multiplication in an extension field of degree 4 over $\mathbb{F}_{2^8}$ [3]. Elements of this field are polynomials of degree ≤ 3 with coefficients in $\mathbb{F}_{2^8}$. The coefficient field $\mathbb{F}_{2^8}$ is generated by the irreducible polynomial $f(x) = x^8 + x^4 + x^3 + x + 1$ (0x11B in hexadecimal notation). For the extension field $\mathbb{F}_{2^8}[t]/(g(t))$ the irreducible polynomial $g(t)$ is $\{1\}t^4 + \{1\}$ with $\{1\} \in \mathbb{F}_{2^8}$. The multiplier operand for MixColumns and InvMixColumns is fixed and its coefficients in $\mathbb{F}_{2^8}$ have a degree of ≤ 3. A multiplication in this extension field over $\mathbb{F}_{2^8}$ can be performed in three steps:

1. Multiplication of binary polynomials.
2. Reduction of product-polynomials modulo $f(x)$.
3. Reduction of a polynomial over $\mathbb{F}_{2^8}$ modulo $g(t)$.

3.1   Instruction Set Extensions

Previous work on instruction set extensions for AES was aimed at both increasing performance as well as minimizing memory requirements. Nadehara et al. [12] designed custom instructions that calculate the result of the (Inv)MixColumns operation in a dedicated functional unit (FU). Bertoni et al. [2] proposed custom instructions to speed up AES software following the approach of [1]. Lim and Benaissa [11] implemented a subword-parallel ALU for binary polynomials that supports AES and ECC over $GF(2^m)$. The work of Tillich et al. focused on reducing memory requirements [19] as well as optimizing performance [18] with the help of custom instructions and dedicated functional units for AES. In addition, they also investigated the potential for speeding up AES using instruction set extensions for ECC [17]. Their results show that three custom instructions originally designed for ECC (gf2mul, gf2mac, and shacr) allow to accelerate AES by up to 25%.

4   Design of a Unified Multiplier with AES Support

Our base architecture is the unified multiply-accumulate (MAC) unit presented in [6]. It is capable of performing unsigned and signed integer multiplication as well as multiplication of binary polynomials. Our original implementation of the MAC unit has been optimized for the SPARC V8 Leon2 processor and consists of two stages. The first stage contains a unified (32×16)-bit multiplier that requires two cycles to produce the result of a (32×32)-bit multiplication. The second stage features a 72-bit unified carry-propagation adder, which adds the product to the accumulator.

Of the three steps described in Section 3, binary polynomial multiplication is already provided by the original multiplier from [6]. The special structure of the reduction polynomial $g(t)$ for step 3 allows a very simple reduction: The higher word (i.e. 32 bits) of the multiplication result after step 2 (with reduced coefficients) is added to the lower word. This operation can be implemented in the second stage (i.e. the accumulator) of the unified MAC unit without much overhead. The only remaining operation to perform is the reduction modulo $f(x)$ (step 2). In the following we introduce the basic ideas for integrating this operation into the unified multiplier presented in [6].

4.1   Basic Unified Multiplier Architecture

The white blocks in Figure 1 show the structure of our baseline multiplier. All grey blocks are added for AES MixColumns support and will be described in detail in Section 5. The multiplier proposed in [6] employs unified radix-4 partial product generators (PPGs) for unsigned and signed integers as well as binary polynomials. In integer mode, the partial products are generated according to the modified Booth recoding technique, i.e. three bits of the multiplier B are examined at a time. On the other hand, the output of each PPG in polynomial mode depends on exactly two bits of B. A total of $\lfloor n/2 \rfloor + 1$ partial products are generated for an n-bit multiplier B if performing an unsigned multiplication, but only $\lfloor n/2 \rfloor$ partial products in the case of signed multiplication or when binary polynomials are multiplied.

The unified MAC unit described in [6] uses *dual-field adders (DFAs)* arranged in an array structure to sum up the partial products. However, we decided to implement the multiplier in form of a Wallace tree to minimize the critical path delay. Another difference between our unified MAC unit for the SPARC V8 Leon2 core and the design from [6] is that our unit adds the multiplication result to the accumulator in a separate stage. Therefore, our unified (32×16)-bit multiplier has to sum up only the 9 partial products generated by the modified Booth recoder. This is done in a Wallace-tree structure with four summation stages using dual-field adders. The first three stages use unified carry-save adders

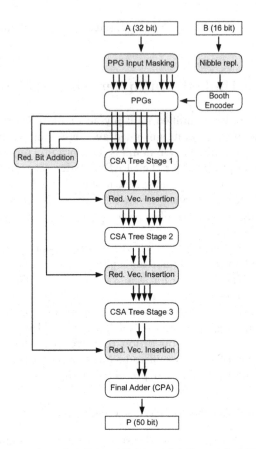

Fig. 1. Proposed unified (32×16)-bit multiplier with AES support

(CSAs) with either (3:2) or (4:2) compressors. The result of each adder is in a redundant form, split up into a carry-vector and a sum-vector. This redundant representation allows for addition without carry-propagation and minimizes the contribution of these summation stages to the overall circuit delay. The fourth and last stage consists of a unified carry-propagate adder (CPA), which produces the final result in non-redundant representation.

4.2 Concepts for Support of AES MixColumns Multiplication

Two observations are important to be able to integrate AES MixColumns support into the basic unified multiplier:

1. For AES MixColuns/InvMixColumns the coefficients of the constant multiplier B have a degree of ≤ 3. At least half of the PPGs will, therefore, have both input multiplier bits at 0 and will produce a partial product of 0 in polynomial mode.

2. As binary polynomials have no carry propagation in addition, the carry-vectors of the carry-save summation stages will always be 0 in polynomial mode.

When two polynomials over $\mathbb{F}_{2^8}$ are multiplied with the unified multiplier in polynomial mode, the result will be incorrect. The coefficients of the polynomial over $\mathbb{F}_{2^8}$ will exceed the maximum degree of 7, i.e. they will be in non-reduced form. The coefficient bits of degree > 7 are added to the bits of the next-higher coefficient in the partial product generators and in the subsequent summation stage. But in order to perform a reduction of the coefficients to non-redundant form (degree ≤ 7), it is necessary to have access to the excessive bits of each coefficient. In the following we will denote these excessive bits as *reduction bits*. The reduction bits indicate whether the irreducible polynomial $f(x)$ must be added to the respective coefficient with a specific offset in order to reduce the degree of the coefficient.

The reduction bits can be isolated in separate partial products. A modification of the PPGs can be prevented by making use of the "idle" PPGs to process the highest three bits of every coefficient of the multiplicand A. This is achieved with the following modifications:

- The "normal" (i.e. not "idle") PPGs are supplied with multiplicand A where only the lowest 5 bits of each coefficients are present (A AND 0x1F1F1F1F).
- Multiplicand A for the "idle" PPGs contains only the highest 3 bits of every coefficient (A AND 0xE0E0E0E0) and is shifted to the right by 4 bits.
- The multiplier B has the lower nibble (4 bits) of each byte replicated in the respective higher nibble (e.g. 0x0C0D $\rightarrow$ 0xCCDD).

These modifications entail a different generation of partial products but still result in the same multiplication result after the summation tree. This is because processing of the multiplicand A is spread across all PPGs (which is done by the masking of A). The "idle" PPGs are activated through replication of the nibbles of the multiplier B. Moreover, the "idle" PPGs produce partial products with a higher weight than intended, which is compensated by the right-shift of the input multiplicand A for these PPGs. Figure 2 and 3 illustrate the partial product generation for a multiplication of a polynomial over $\mathbb{F}_{2^8}$ of degree 1 (16-bit multiplicand A) with a polynomial of degree 0 (8-bit multiplier B). Note that partial product 1 in Figure 2 is split into the partial products 1 and 3 in Figure 3. The same occurs for partial product 2, which is split into the partial products 2 and 4. The PPG-scheme in Figure 3 yields partial products which directly contain the reduction bits.

To determine whether the reduction polynomial needs to be added to a coefficient of the multiplication result with a specific offset, it is necessary to combine (add) reduction bits with the same weight from different partial products. In order to minimize delay, these few additional XOR gates are placed in parallel to the summation tree stages. The resulting reduction bits determine the value of the so-called reduction vectors, which are injected via the carry-vectors of the summation tree and which reduce the coefficients to non-redundant form. More

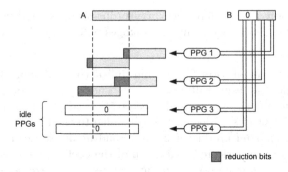

Fig. 2. Multiplication of polynomials over $\mathbb{F}_{2^8}$ with a radix-4 multiplier for binary polynomials

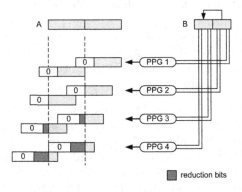

Fig. 3. Multiplication of polynomials over $\mathbb{F}_{2^8}$ with the modified PPG-scheme for AES support

specifically, if a reduction bit is set, then a portion of a carry-vector (with the correct offset) is forced to the value of the reduction polynomial $f(x)$ (0x11B), otherwise it is left 0. Reduction vectors for different coefficients can be injected in the same carry-vector, as long as they do not overlap and the carry-vector is long enough. Thus, by making use of the "idle" PPGs and the carry-vectors of the summation tree, the multiplier can be extended to support AES MixColumns multiplication.

5   Implementation Details

The general concepts for integrating AES MixColumns support into the unified multiplier of [6] are described in Section 4.2. Figure 1 shows our modified multiplier with all additional components. *PPG Input Masking* and *Nibble Replication* make sure that the partial products are generated in a redundant fashion where the reduction bits are subsequently accessible. *Reduction Bit Addition* adds up reduction bits of coefficients of partial products with the same weight. *Reduction Vector Insertion* conditionally injects reduction polynomials for the coefficients with different offsets, depending on the reduction bits. The result P will be a polynomial over $\mathbb{F}_{2^8}$ of degree 4 with fully reduced coefficients. In the following we briefly describe the implementation of the additional components.

PPG Input Masking. The AES MixColumns mode is controlled with the signal *ff_mix*. This signal selects the input multiplier A for the PPGs either as unmodified or masked (and shifted) as described in Section 4.2.

Multiplier Nibble Replication. In our implementation the multiplier B is set by the processor in dependance on the required operation (AES MixColumns or InvMixColumns). Nibble replication is therefore performed outside of our multiplier. If it is to be done within the multiplier, it just requires an additional multiplexor for the multiplier B controlled by *ff_mix*.

Reduction Bit Addition. Reduction bits of the same weight are XORed in parallel to the summation tree stages. For the (32×16)-bit case, the resulting reduction bits have contributions from one, two, or four partial products.

Reduction Vector Insertion. For each reduction polynomial, the *ff_mix* and the corresponding reduction bit are combined with a logical AND. The result is used to conditionally inject the reduction polynomial over a logical OR with the required bit-lines of a carry-vector. Reduction bits which have contributions from more partial products are used in later stages of the summation tree than reduction bits which depend on less partial products.

6 Experimental Results

We integrated our functional unit into the SPARC V8-compatible Leon2 core [4] and prototyped the extended processor in an FPGA[2]. For performing AES MixColumns and InvMixColumns, four custom instructions were defined: Two of these instructions (mcmuls, imcmuls) can be used for the MixColumns and InvMixColumns transformation only, while the other two (mcmacs, imcmacs) include an addition of the transformation result to the accumulator. The latter two instructions write their result only to the accumulator registers and not to the general-purpose register file. They require two clock cycles to produce the result[3]. If the subsequent instruction does not need the multiplication result or access to the multiply-accumulate unit, then it can be processed in parallel to the multiply instruction, resulting in one cycle per instruction. In addition, our new custom instructions assemble the 32-bit multiplicand for AES multiplication from the two source register operands of the instruction (the 16 higher bits of the first register and the 16 lower bits of the second register), in order to facilitate the AES ShiftRows/InvShiftRows transformation.

---

[2] The HDL source code of the extended processor, denoted Leon2-CIS, is available for download from the ISEC project page at http://www.iaik.tugraz.at/isec.

[3] Although the multiply-accumulate unit takes three cycles for the calculation, subsequent instructions can access the result after two cycles without a pipeline stall due to the implementation characteristics of the accumulator registers.

6.1 Silicon Area and Critical Path

The impact of our modifications on the critical path of the multiplier is very small. One additional multiplexor delay is required to select the input for the PPGs. The reduction bits are added in parallel to the summation tree, which should not extend the critical path. For injection of the reduction vectors, there is one additional OR-delay for the 2nd, 3rd and 4th summation tree stage, i.e. in the worst case three OR-delays altogether.

We synthesized the original unified multiplier from [6] (unimul32x16) and our proposed unified multiplier with AES support (unimul_mix32x16) using a 0.13 μm standard-cell library in order to estimate the overhead in silicon area and the impact on the critical path delay. These results were compared with the conventional (32 × 16)-bit integer multiplier that is part of the Leon2 soft-core (intmul32x16). We also made comparisons including the enclosing unified multiply accumulate units (unimac32x16, unimac_mix32x16) and the five-stage processor pipeline, denoted as integer unit (IU). The results are summarized in Table 1.

Table 1. Area and delay of the functional units and the extended Leon2 core

FU/Component	Minimal Delay		Typical Delay	
	Area (GE)	Delay (ns)	Area (GE)	Delay (ns)
intmul32x16	7,308	2.05	5,402	2.50
unimul32x16	9,660	2.15	7,413	2.50
unimul_mix32x16	9,988	2.21	8,418	2.50
unimac32x16	14,728	2.53	12,037	3.00
unimac_mix32x16	16,145	2.56	12,914	3.00
Leon2 IU (intmul32x16)	27,250	2.59	17,867	4.97
Leon2 IU (unimac32x16)	38,705	2.77	24,927	5.00
Leon2 IU (unimac_mix32x16)	39,306	2.85	26,219	4.99

All results in Table 1 are given for the minimal and for a typical critical path delay. The former give an estimate of the maximum frequency with which the processor can be clocked, while the latter allow to assess the increase in silicon area due to our proposed modifications. Taking a Leon2 processor with a unified MAC unit for ECC (unimac32x16) as reference, our modifications for AES support increase the critical path by about 5% and the silicon area by less than 1.3 kGates. The overall size of the FU with support for ECC and AES is approximately 12.9 kGates when synthesized for a delay of 3 ns. However, it must be considered that the "original" (32 × 16)-bit integer multiplier of the Leon2 core has an area of about 5.4 kGates. Therefore, the extensions for ECC and AES increase the size of the multiplier by just 7.5 kGates and the overall size of the Leon2 core by approximately 8.35 kGates.

6.2   AES Performance

In order to estimate the achievable speedup with our proposed FU, we proto-typed the extended Leon2 on an FPGA board. We evaluated AES encryption and decryption functions with 128-bit keys (AES-128) both for precomputed key schedule and on-the-fly key expansion. The number of cycles was determined with an integrated cycle counter using the timing code of the well-known AES software implementation of Brian Gladman [5]. Note that the AES decryption function with on-the-fly key expansion is supplied with the last round-key. The code size for each implementation is also listed, which encompasses all required functions as well as any necessary constants (e.g. S-box lookup table).

Table 2. AES-128 encryption and decryption: Performance and code size

Implementation	Key exp.	Performance		Code size	
	Cycles	Cycles	Speedup	Bytes	Rel. change
Encryption, Precomputed Key Schedule					
No extensions (pure SW)	739	1,637	1.00	2,168	0.0%
mcmuls (C)	498	1,011	1.62	1,240	−42.8%
sbox4s & mcmuls (ASM)	316	260	6.30	460	−78.8%
Decryption, Precomputed Key Schedule					
No extensions (pure SW)	739	1,955	1.00	2,520	0.0%
mcmuls (C)	316	1,299	1.51	1,572	−37.6%
sbox4s & mcmuls (ASM)	465	259	7.55	520	−79.4%
Encryption, On-the-fly Key Expansion					
No extensions (pure SW)		2,239	1.00	1,636	0.0%
mcmuls (C)		1,258	1.78	1,228	−21.3%
sbox4s & mcmuls (ASM)		296	7.56	308	−81.2%
Decryption, On-the-fly Key Expansion					
No extensions (pure SW)		2,434	1.00	2,504	0.0%
mcmuls (C)		1,596	1.53	1,616	−35.5%
sbox4s & mcmuls (ASM)		305	7.98	316	−87.4%

Table 2 specifies the number of clock cycles per encryption/decryption and the code size for implementations using precomputed key schedule as well as on-the-fly key expansion. Our baseline implementation is a C function which uses only native SPARC V8 instructions. The mcmuls implementation refers to a function written in C where MixColumns or InvMixColumns is realized using our proposed functional unit. The sbox4s & mcmuls implementation is written in assembly and uses our multiplier as well as an additional custom instruction for performing the S-box substitution. This instruction for the S-box requires less than 2 kGates. It is described and performance-evaluated in [18] along with other custom instructions dedicated to AES.

The C implementations can be sped up with the proposed custom instructions by a factor of up to 1.78. However, our extensions are designed to deliver maximal performance in combination with the custom instruction for S-box substitution described in [18]. By combining these extensions, a 128-bit AES encryption can be done in less than 300 clock cycles, which corresponds to a speed-up factor of between 6.3 (pre-computed key schedule) and 7.98 (on-the-fly key expansion) compared to the baseline implementation. Moreover, the custom instructions for AES reduce the code size by up to 87.4%.

The AES performance can be further improved by reducing the latency of the multiply-accumulate unit. With a (32×32)-bit multiplier and integration of the accumulation into the summation tree (as proposed in [6]), an instruction for MixColumns/InvMixColumns could be executed in a single cycle and could also include the subsequent AddRoundKey transformation. With such an instruction a complete AES round could be executed in only 12 clock cycles, and a complete AES-128 encryption or decryption in about 160 cycles (including all loads and stores of the data and key).

6.3   Comparison with Designs Using an AES Coprocessor

Hodjat et al. [9] and Schaumont et al. [16] attached an AES coprocessor to the Leon2 core and analyzed the effects on performance and hardware cost. The implementation reported by Hodjat et al. used a dedicated coprocessor interface to connect the AES hardware with the Leon2 core. Schaumont et al. transferred data to/from the coprocessor via memory-mapped I/O. Both systems were pro-totyped on a Xilinx Virtex-II FPGA on which the "pure" Leon2 core consumes approximately 4,856 LUTs, leaving some 5,400 LUTs for the implementation of the AES coprocessor. Table 3 summarizes the execution time of a 128-bit encryption and the additional hardware cost due to the AES coprocessor. For comparison, the corresponding performance and area figures of the extensions proposed in this paper are also specified.

Table 3. Performance and cost of AES coprocessor vs. instruction set extensions

Reference	Implementation	Performance	HW cost
Hodjat [9]	Coprocessor (COP interface)	704 cycles	4,900 LUTs
Schaumont [16]	Coprocessor (mem. mapped)	1,494 cycles	3,474 LUTs
This work	ISE for MixColumns	1,011/1,299 cycles	3,194 LUTs
This work	ISE for MixColumns + S-box	260 cycles	3,695 LUTs

Hodjat et al's AES coprocessor uses about 4,900 LUTs (i.e. requires more resources than the Leon2 core) and is able to encrypt a 128-bit block of data in 11 clock cycles. However, loading the data and key into the coprocessor, performing the AES encryption itself, and returning the result back to the software routine takes 704 cycles altogether [9, page 492]. Schaumont et al's coprocessor with the memory-mapped interface requires less hardware and is slower than the implementation of Hodjat et al. The performance of our AES extensions lies between the two coprocessor systems. As mentioned in Section 6.2, the custom instruction for S-box substitution from [18] would allow to reduce the execution time of 128-bit AES encryption to 260 cycles, which is significantly faster than the coprocessor systems. The additional hardware cost of the FU is comparable to that of the two co-processors. However, contrary to AES coprocessors, the FU presented in this paper supports not only the AES, but also ECC over both prime fields and binary extension fields.

7 Summary of Results and Conclusions

In this paper we introduced a functional unit (FU) for increasing the performance of embedded processors when executing cryptographic algorithms. The main component of the FU is a unified multiply-accumulate (MAC) capable to perform integer and polynomial multiplication as well as reduction modulo an irreducible polynomial of degree 8. Due to its rich functionality and high degree of flexibility, the FU facilitates efficient implementation of a wide range of cryptosystems, including ECC and AES. When integrated into the Leon2 SPARC V8 processor, the FU allows to execute a 128-bit AES encryption with precomputed key schedule in about 1,000 clock cycles. Hardware support for the S-box operation further reduces the execution time to 260 cycles, which is more than six times faster than a conventional software implementation on the Leon2 processor. The hardware cost of the AES extensions is roughly 1,300 gates and the additional area for the support of ECC and AES amounts to just 8,352 gates altogether. These results confirm that the functional unit presented in this paper is a flexible and cost-effective alternative to a cryptographic co-processor.

References

1. Bertoni, G., Breveglieri, L., Fragneto, P., Macchetti, M., Marchesin, S.: Efficient software implementation of AES on 32-bit platforms. In: Kaliski Jr., B.S., Koç, Ç.K., Paar, C. (eds.) CHES 2002. LNCS, vol. 2523, pp. 159–171. Springer Verlag, Heidelberg (2003)
2. Bertoni, G., Breveglieri, L., Farina, R., Regazzoni, F.: Speeding up AES by extending a 32-bit processor instruction set. In: Proceedings of the 17th IEEE International Conference on Application-Specific Systems, Architectures and Processors (ASAP 2006), pp. 275–282. IEEE Computer Society Press, Los Alamitos (2006)
3. Daemen, J., Rijmen, V.: The Design of Rijndael: AES – The Advanced Encryption Standard. Springer Verlag, Heidelberg (2002)
4. Gaisler, J.: The LEON-2 Processor User's Manual (Version 1.0.10) (2003) Available for download at http://www.gaisler.com/doc/leon2-1.0.10.pdf
5. Gladman, B.: Implementations of AES (Rijndael) in C/C++ and assembler. Available for download at http://fp.gladman.plus.com/cryptography_technology/rijndael/index.htm.
6. Großschädl, J., Kamendje, G.-A.: Low-power design of a functional unit for arithmetic in finite fields $GF(p)$ and $GF(2^m)$. In: Chae, K.-J., Yung, M. (eds.) Information Security Applications — WISA 2003. LNCS, vol. 2908, pp. 227–243. Springer Verlag, Heidelberg (2003)
7. Großschädl, J., Tillich, S., Szekely, A., Wurm, M.: Cryptography instruction set extensions to the SPARC V8 architecture. Preprint, submitted for publication

8. Hankerson, D.R., Menezes, A.J., Vanstone, S.A.: Guide to Elliptic Curve Cryptography. Springer Verlag, Heidelberg (2004)
9. Hodjat, A., Verbauwhede, I.: Interfacing a high speed crypto accelerator to an embedded CPU. In: Proceedings of the 38th Asilomar Conference on Signals, Systems, and Computers, vol. 1, pp. 488–492. IEEE, New York (2004)
10. Koufopavlou, O., Selimis, G., Sklavos, N., Kitsos, P.: Cryptography: Circuits and systems approach. In: Proceedings of the 5th IEEE Symposium on Signal Processing and Information Technology (ISSPIT 2005), December 2005, pp. 918–923. IEEE, New York (2005)
11. Lim, W.M., Benaissa, M.: Subword parallel $GF(2^m)$ ALU: An implementation for a cryptographic processor. In: Proceedings of the 17th IEEE Workshop on Signal Processing Systems (SIPS 2003), pp. 63–68. IEEE, New York (2003)
12. Nadehara, K., Ikekawa, M., Kuroda, I.: Extended instructions for the AES cryptography and their efficient implementation. In: Proceedings of the 18th IEEE Workshop on Signal Processing Systems (SIPS 2004), pp. 152–157. IEEE, New York (2004)
13. National Institute of Standards and Technology. FIPS-197: Advanced Encryption Standard (November 2001) Available online at http://www.itl.nist.gov/fipspubs/
14. Page, D.: Theoretical Use of Cache Memory as a Cryptanalytic Side-Channel. Technical Report CSTR-02-003, Department of Computer Science, University of Bristol (June 2002)
15. Savaş, E., Tenca, A.F., Koç, Ç.K.: A scalable and unified multiplier architecture for finite fields $GF(p)$ and $GF(2^m)$. In: Paar, C., Koç, Ç.K. (eds.) CHES 2000. LNCS, vol. 1965, pp. 277–292. Springer Verlag, Heidelberg (2000)
16. Schaumont, P., Sakiyama, K., Hodjat, A., Verbauwhede, I.: Embedded software integration for coarse-grain reconfigurable systems. In: Proceedings of the 18th International Parallel and Distributed Processing Symposium (IPDPS 2004), pp. 137–142. IEEE Computer Society Press, Los Alamitos (2004)
17. Tillich, S., Großschädl, J.: Accelerating AES using instruction set extensions for elliptic curve cryptography. In: Gervasi, O., Gavrilova, M., Kumar, V., Laganà, A., Lee, H.P., Mun, Y., Taniar, D., Tan, C.J.K. (eds.) Computational Science and Its Applications — ICCSA 2005. LNCS, vol. 3481, pp. 665–675. Springer, Heidelberg (2005)
18. Tillich, S., Großschädl, J.: Instruction set extensions for efficient AES implementation on 32-bit processors. In: Goubin, L., Matsui, M. (eds.) CHES 2006. LNCS, vol. 4249, pp. 270–284. Springer, Heidelberg (2006)
19. Tillich, S., Großschädl, J., Szekely, A.: An instruction set extension for fast and memory-efficient AES implementation. In: Dittmann, J., Katzenbeisser, S., Uhl, A. (eds.) CMS 2005. LNCS, vol. 3677, pp. 11–21. Springer, Heidelberg (2005)
20. Trusted Computing Group. TCG Specification Architecture Overview (Revision 1.2) (April 2004), Available for download at https://www.trustedcomputinggroup.org/groups/TCG_1_0_Architecture_Overview.pdf

Efficient Multiplication Using
Type 2 Optimal Normal Bases

Joachim von zur Gathen[1], Amin Shokrollahi[2], and Jamshid Shokrollahi[3,*]

[1] B-IT, Dahlmannstr. 2, Universität Bonn, 53113 Bonn, Germany
gathen@bit.uni-bonn.de
[2] ALGO, Station 14, Batiment BC, EPFL, 1015 Lausanne, Switzerland
amin.shokrollahi@epfl.ch
[3] B-IT, Dahlmannstr. 2, Universität Bonn, 53113 Bonn, Germany
current address: System Security Group, Ruhr-Universität Bochum, D-44780
Bochum, Germany
jamshid@crypto.rub.de

Abstract. In this paper we propose a new structure for multiplication using optimal normal bases of type 2. The multiplier uses an efficient linear transformation to convert the normal basis representations of elements of $\mathbb{F}_{q^n}$ to suitable polynomials of degree at most n over $\mathbb{F}_q$. These polynomials are multiplied using any method which is suitable for the implementation platform, then the product is converted back to the normal basis using the inverse of the above transformation. The efficiency of the transformation arises from a special factorization of its matrix into sparse matrices. This factorization — which resembles the FFT factorization of the DFT matrix — allows to compute the transformation and its inverse using $O(n \log n)$ operations in $\mathbb{F}_q$, rather than $O(n^2)$ operations needed for a general change of basis. Using this technique we can reduce the asymptotic cost of multiplication in optimal normal bases of type 2 from $2\mathsf{M}(n) + O(n)$ reported by Gao et al. (2000) to $\mathsf{M}(n) + O(n \log n)$ operations in $\mathbb{F}_q$, where $\mathsf{M}(n)$ is the number of $\mathbb{F}_q$-operations to multiply two polynomials of degree $n - 1$ over $\mathbb{F}_q$. We show that this cost is also smaller than other proposed multipliers for $n > 160$, values which are used in elliptic curve cryptography.

Keywords: Finite field arithmetic, optimal normal bases, asymptotically fast algorithms.

1 Introduction

The normal basis representation of finite fields enables easy computation of the qth power of elements. Assuming q to be a prime power, a basis of the form $(\alpha, \alpha^q, \cdots, \alpha^{q^{n-1}})$ for $\mathbb{F}_{q^n}$, as a vector space over $\mathbb{F}_q$, is called a normal basis generated by the normal element $\alpha \in \mathbb{F}_{q^n}$. In this basis the qth power of an

* Partially funded by the German Research Foundation (DFG) under project RU 477/8.

C. Carlet and B. Sunar (Eds.): WAIFI 2007, LNCS 4547, pp. 55–68, 2007.

element can be computed by means of a single cyclic shift. This property makes such bases attractive for parallel exponentiation in finite fields (see Nöcker 2001).

Naive multiplication in these bases is more expensive than in polynomial bases, especially when using linear algebra (cf. Mullin et al. (1989)). Hence substantial effort has gone into reducing the multiplication cost. In this paper a new method for multiplication in normal bases of type 2 is suggested. It uses an area efficient circuit to convert the normal basis representation to polynomials and vice versa. Any method can be used to multiply the resulting polynomials. Although this structure has small area, its propagation delay is longer than other methods and, when used in cryptography, is mostly suitable for applications where the area is limited, like in RFIDs.

One popular normal basis multiplier is the Massey-Omura multiplier presented by Omura & Massey. The space and time costs of this multiplier increase with the number of nonzero coefficients in the matrix representation of the endomorphism $x \rightarrow \alpha x$ over $\mathbb{F}_{q^n}$, where α generates the normal basis. Mullin et al. (1989) show that this number is at least $2n - 1$ which can be achieved for optimal normal bases. Gao & Lenstra (1992) specify exactly the finite fields for which optimal normal bases exist. They are related to Gauss periods, and can be grouped into optimal normal bases of type 1 and 2.

For security reasons only prime extension degrees are used in cryptography, whereas the extension degrees of the finite fields containing an optimal normal basis of type 1 are always composite numbers. Cryptography standards often suggest finite fields for which the type of normal bases are small (see for example the Digital Signature Standard (2000)) to enable designers to deploy normal bases. Applications in cryptography have stimulated research about efficient multiplication using optimal normal bases of type 2. The best space complexity results for the type 2 multipliers are n^2 and $3n(n - 1)/2$ gates of types AND and XOR, respectively reported by Sunar & Koç (2001) and Reyhani-Masoleh & Hasan (2002). Their suggested circuits are obtained by suitably modifying the Massey-Omura multiplier. A classical polynomial basis multiplier, however, requires only n^2 and $(n - 1)^2$ gates of types AND and XOR respectively for the polynomial multiplication, followed by a modular reduction. The latter is done using a small circuit of size of $(r - 1)n$, where r is the number of nonzero coefficients in the polynomial which is used to create the polynomial basis. It is conjectured by von zur Gathen & Nöcker (2004) that there are usually irreducible trinomials or pentanomials of degree n. The above costs and the fact that there are asymptotically fast methods for polynomial arithmetic suggest the use of polynomial multipliers for normal bases to make good use of both representations. The proposed multiplier in this paper works for normal bases but its space complexity is similar to that of polynomial multipliers. Using classical polynomial multiplication methods, it requires $2n^2 + 16n \log_2 n$ gates in $\mathbb{F}_{2^n}$. With the Karatsuba algorithm, we can decrease the space asymptotically even further down to $O(n^{\log_2 3})$. The usefulness of this approach in hardware has first been demonstrated in Grabbe et al. (2003). The proposed structure can be employed to compute the Tate-pairing in characteristic three, for example.

Applications of optimal normal bases of type 2 for pairing-based cryptography have been proposed by Granger et al. (2005).

The connection between polynomial and normal bases, together with its application in achieving high performance multiplication in normal bases, has been investigated by Gao et al. (1995, 2000). The present work can be viewed as a conceptual continuation of the approach in those papers. They describe how multiplication in normal basis representation by Gauss periods for $\mathbb{F}_{q^n}$ can be reduced to multiplication of two $2n$-coefficient polynomials, which because of the existing symmetries can be done by two multiplications of n-coefficient polynomials: here any method, including asymptotically fast ones, can be deployed.

The multiplier of this work is based on a similar approach. For optimal normal bases of type 2 we present an efficient transformation which changes the representations from the normal basis to suitable polynomials. These polynomials are multiplied using any method, such as the classical or the Karatsuba multiplier. Using the inverse transformation circuit and an additional small circuit the result is converted back into the normal basis representation. The heart of this method is a factorization of the transformation matrix between the two representations into a small product of sparse matrices. The circuit requires roughly $O(n \log n)$ operations in $\mathbb{F}_q$ and resembles the circuit used for computing the Fast Fourier Transformation (FFT). The analogy to the FFT circuit goes even further: as with the FFT, the inverse of the transformation has a very similar circuit. It should be noted that a general basis conversion requires $O(n^2)$ operations, as also reported by Kaliski & Liskov (1999). Recently Fan & Hasan (2006) found a new multiplier for normal bases with asymptotically low cost of $O(n^{\log_2 3})$, which uses fast multiplication methods by Toeplitz matrices. One advantage of our multiplier is the ability of working with any polynomial multiplication. Hence using the Cantor multiplier, we can achieve a cost of $O(n(\log n)^2 (\log \log n)^3)$.

This paper begins with a review of Gauss periods and normal bases of type 2. Then the structure of the multiplier is introduced and the costs of each part of the multiplier are computed. The last section compares the results with the literature.

In this Extended Abstract most of the proofs and also some possible improvements for fields of characteristic 2 are omitted for lack of space. The only exception is Lemma 1 which is a central part of this paper. But we have tried to intuitively describe why the theorems are correct. The proofs can be found in the full paper, or the work of Shokrollahi (2006).

2   Permuted Normal Basis

It is well known (see Wassermann (1990), Gao et al. (2000), Sunar & Koç (2001)) that a type 2 optimal normal basis for $\mathbb{F}_{q^n}$ over $\mathbb{F}_q$ is of the form

$$\mathcal{N} = (\beta + \beta^{-1}, \beta^q + \beta^{-q}, \cdots, \beta^{q^{n-1}} + \beta^{-q^{n-1}}), \tag{1}$$

where β is a $2n + 1$st primitive root of unity in $\mathbb{F}_{q^{2n}}$, and that the basis

$$\mathcal{N}' = (\beta + \beta^{-1}, \beta^2 + \beta^{-2}, \cdots, \beta^n + \beta^{-n}),$$

which we call the permuted normal basis, is a permutation of $\mathcal{N}$. Hence the normal basis representation of an element $a = \sum_{k=0}^{n-1} a_k^{(\mathcal{N})}(\beta^{q^k} + \beta^{-q^k}) \in \mathbb{F}_{q^n}$ can be written as

$$a = \sum_{l=1}^{n} a_l^{(\mathcal{N}')}(\beta^l + \beta^{-l}), \tag{2}$$

where $(a_l^{(\mathcal{N}')})_{1 \le l \le n}$ is a permutation of $(a_k^{(\mathcal{N})})_{0 \le k < n}$, called the permuted normal representation of a. The $a_k^{(\mathcal{N})}$ and $a_l^{(\mathcal{N}')}$ are elements of $\mathbb{F}_q$.

3   Multiplier Structure

The structure of the multiplier is described in Figure 1. To multiply two elements $a, b \in \mathbb{F}_{q^n}$ given in the basis (1) we first convert their representations to the permuted form as

$$a = \sum_{i=1}^{n} a_i^{(\mathcal{N}')}(\beta^i + \beta^{-i}), \text{ and } b = \sum_{i=1}^{n} b_i^{(\mathcal{N}')}(\beta^i + \beta^{-i}).$$

By inserting a zero at the beginning of the representation vectors and applying a linear mapping π_{n+1}, which we define in Section 4, from $\mathbb{F}_q^{n+1}$ to $\mathbb{F}_q[x]^{\le n}$, the vectors of these representations are converted to polynomials ϕ_a and ϕ_b of degree at most n, such that their values at $\beta + \beta^{-1}$ are a and b, respectively. Then ϕ_a and ϕ_b are multiplied using an appropriate method with respect to the polynomial degrees and implementation platform. Obviously, the value of the resulting polynomial ϕ_c at $\beta + \beta^{-1}$ is the product $c = a \cdot b$. The degree of ϕ_c is at most $2n$, and the evaluation is a linear combination of $(\beta + \beta^{-1})^i$ for $0 \le i \le 2n$. Using another linear mapping ν_{2n+1} from $\mathbb{F}_q[x]^{\le 2n}$ to $\mathbb{F}_q^{2n+1}$, namely the inverse of π_{2n+1}, ϕ_c is converted to a linear combination of the vectors 1 and $\beta^i + \beta^{-i}$ for $1 \le i \le 2n$. This is then converted to the permuted normal basis using another linear mapping τ_n.

The linear mapping ν_{2n+1} takes a polynomial in $\mathbb{F}_q[x]^{\le 2n}$, evaluates it at $\beta + \beta^{-1}$, and represents the result as a linear combination of 1 and $\beta^i + \beta^{-i}$, for $1 \le i \le 2n$. Since the above vectors are linearly dependent there are several choices for ν_{2n+1}. One way to compute the resulting linear combination is to expand $(\beta + \beta^{-1})^j$, for $1 \le j \le 2n$, as a linear combination of $\beta^i + \beta^{-i}$, for $1 \le i \le 2n$. The coefficients of these expansions are closely connected to the binomial coefficients, that is, the entries of the Pascal triangle. The matrix representation of ν_{2n+1} has a structure similar to the Pascal triangle reduced modulo p, the characteristic of $\mathbb{F}_q$. This infinite triangle has a fractal structure, which has attracted a lot of attention and has been given various names in the literature, among them "Sierpinski triangle" or "Sierpinski gasket" (see Wikipedia) for $p = 2$. The central result of this paper is in Section 5, where we find a special factorization for the matrix representation of ν_{2n+1} in an appropriate basis which allows the mapping to be computed in $O(n \log n)$ operations. This cost is also sufficient for π_{n+1}.

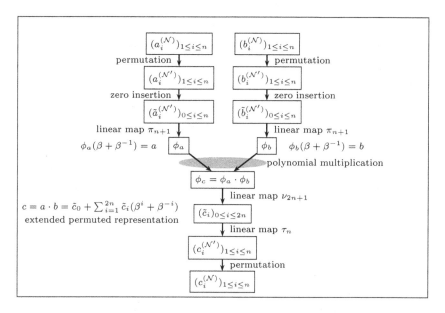

Fig. 1. Overview of our multiplier structure to multiply two elements $a, b \in \mathbb{F}_{q^n}$ in the representation $(*_i^{(\mathcal{N})})_{1 \leq i \leq n}$ with respect to the normal basis $\mathcal{N}$

4 Polynomials from Normal Bases

In this paper we always represent the characteristic of $\mathbb{F}_q$ by p.

The most important parts of our multiplier are the converters between polynomial and permuted normal representations. Since the elements $(\beta + \beta^{-1})^i$, for $0 \leq i \leq n$, and also 1 and $\beta^i + \beta^{-i}$, for $1 \leq i \leq 2n$, are linearly dependent, there are different possibilities for the maps π_{n+1} and ν_{2n+1} from Section 3. We define our selection via matrices $P_{n+1} \in \mathbb{F}_p^{(n+1) \times (n+1)}$ and $L_{2n+1} \in \mathbb{F}_p^{(2n+1) \times (2n+1)}$. These matrices have special factorizations which allow to multiply them by vectors of appropriate length using $O(n \log n)$ operations in $\mathbb{F}_q$.

To construct the polynomial representation for a and b, their permuted representations are preceded by zero and P_{n+1} is multiplied by the resulting vectors. The structure of the inverse of P_{n+1}, which we denote by L_{n+1}, is easier to describe. Hence we define a candidate for L_{n+1}. This matrix can be used to convert from polynomial to the extended permuted normal representation, i.e., it satisfies

$$(1, \beta + \beta^{-1}, \beta^2 + \beta^{-2}, \cdots, \beta^n + \beta^{-n})L_{n+1} =$$
$$(1, \beta + \beta^{-1}, (\beta + \beta^{-1})^2, \cdots, (\beta + \beta^{-1})^n).$$

Furthermore L_{n+1} is invertible. Then we study its structure and exhibit a factorization into sparse factors in Section 5, which is also used to find a factorization for P_n.

Definition 1. *For integers i, j let $l_{i,j} \in \mathbb{F}_p$ be such that $(x + x^{-1})^j = \sum_{i \in \mathbb{Z}} l_{i,j} x^i$ in $\mathbb{F}_p[x]$, and $L_n = (l_{i,j})_{0 \leq i,j < n} \in \mathbb{F}_p^{n \times n}$.*

Obviously $l_{i,j} = 0$ for $|i| > |j|$.

Example 1. Let $q = 9$, i.e., $p = 3$. For $0 \leq j < 9$, the powers $(x + x^{-1})^j$ and hence the matrix L_9 are:

$$
\begin{array}{c||l}
j & (x + x^{-1})^j \\
\hline
0 & 1 \\
1 & x + x^{-1} \\
2 & x^2 + 2 + x^{-2} \\
3 & x^3 + x^{-3} \\
4 & x^4 + x^2 + x^{-2} + x^{-4} \\
5 & x^5 + 2x^3 + x + x^{-1} + 2x^{-3} + x^{-5} \\
6 & x^6 + 2 + x^{-6} \\
7 & x^7 + x^5 + 2x + 2x^{-1} + x^{-5} + x^{-7} \\
8 & x^8 + 2x^6 + x^4 + 2x^2 + 1 + 2x^{-2} + x^{-4} + 2x^{-6} + x^{-8}
\end{array}
\qquad
L_9 =
\begin{pmatrix}
1 & 0 & 2 & 0 & 0 & 0 & 2 & 0 & 1 \\
0 & 1 & 0 & 0 & 0 & 1 & 0 & 2 & 0 \\
0 & 0 & 1 & 0 & 1 & 0 & 0 & 0 & 2 \\
0 & 0 & 0 & 1 & 0 & 2 & 0 & 0 & 0 \\
0 & 0 & 0 & 0 & 1 & 0 & 0 & 0 & 1 \\
0 & 0 & 0 & 0 & 0 & 1 & 0 & 1 & 0 \\
0 & 0 & 0 & 0 & 0 & 0 & 1 & 0 & 2 \\
0 & 0 & 0 & 0 & 0 & 0 & 0 & 1 & 0 \\
0 & 0 & 0 & 0 & 0 & 0 & 0 & 0 & 1
\end{pmatrix}
$$

Theorem 1. *The matrix L_n of Definition 1 satisfies*

$$
\begin{aligned}
&(1, \beta + \beta^{-1}, \beta^2 + \beta^{-2}, \cdots, \beta^{n-1} + \beta^{-n+1}) L_n = \\
&(1, \beta + \beta^{-1}, (\beta + \beta^{-1})^2, \cdots, (\beta + \beta^{-1})^{n-1}),
\end{aligned}
\tag{3}
$$

is upper triangular with 1 on the diagonal, hence nonsingular, and its entries satisfy the relation:

$$
(L_n)_{i,j} = \begin{cases} 0 & \text{if } i > j \text{ or } j - i \text{ is odd, and} \\ \binom{j}{(j-i)/2} & \text{otherwise.} \end{cases}
$$

Definition 2. *We denote the inverse of L_n by $P_n = (p_{i,j})_{0 \leq i,j < n} \in \mathbb{F}_p^{n \times n}$.*

5   Factorizations of the Conversion Matrices

The cost of computing the isomorphisms π_n and ν_n of Section 3 depends on the structure of the corresponding matrices. As in the last section, it is easier to initially study the structure of L_n and use this information to analyze P_n. The former study will be simplified by assuming n to be a power of p, say $n = p^r$, and extending the results to general n later. This simplification enables a recursive representation of L_{p^r} which is exhibited in Lemma 1. This recursive structure is then used in Theorem 3 to find a factorization of L_{p^r} into sparse matrices. To describe the recursive structure of L_{p^r} we define three matrices of *reflection, shifting,* and *factorization* denoted by Θ_n, Ψ_n, and B_r, respectively.

Definition 3. *The entries of the* reflection *and* shifting *matrices* $\Theta_n = (\theta_{i,j})_{0 \leq i,j < n} \in \mathbb{F}_p^{n \times n}$ *and* $\Psi_n = (\psi_{i,j})_{0 \leq i,j < n} \in \mathbb{F}_p^{n \times n}$, *respectively, are defined by the relations:*

$$
\theta_{i,j} = \begin{cases} 1 \text{ if } i + j = n, \\ 0 \text{ otherwise,} \end{cases}
\qquad
\psi_{i,j} = \begin{cases} 1 \text{ if } j - i = 1, \\ 0 \text{ otherwise.} \end{cases}
$$

a   b

Fig. 2. (a) The matrix Θ_5 and (b) the matrix Ψ_5

As an example, Θ_5 and Ψ_5, are shown in Figure 2, where the coefficients equal to 0 and 1 are represented by empty and filled boxes, respectively. Left multiplication by Θ_n reflects a matrix horizontally and shifts the result by one row downwards. Right multiplication by Ψ_n shifts a matrix by one position upwards.

Definition 4. *The factorization matrix B_r is:*

$$B_r = L_p \otimes I_{p^{r-1}} + (\Psi_p L_p) \otimes \Theta_{p^{r-1}} \in \mathbb{F}_p^{p^r \times p^r},$$

where $\otimes$ is the Kronecker or tensor product operator.

Using Definitions 1 and 4 it is easy to prove the following theorem which gives more information about the structure of B_r and can be helpful for constructing this matrix. The matrices B_3 and L_{27} are shown in Figure 3.

Theorem 2. *The matrix B_r can be split into $p \times p$ blocks $B^{(i_1, j_1)} \in \mathbb{F}_p^{p^{r-1} \times p^{r-1}}$ such that $B_r = (B^{(i_1,j_1)})_{0 \leq i_1, j_1 < p}$ and*

$$B^{(i_1,j_1)} = \begin{cases} \text{the zero block} & \text{if } i_1 > j_1, \\ \binom{j_1}{(j_1-i_1)/2} I_{p^{r-1}} & \text{if } i_1 \leq j_1 \text{ and } j_1 - i_1 \text{ is even, and} \\ \binom{j_1}{(j_1-i_1-1)/2} \Theta_{p^{r-1}} & \text{otherwise.} \end{cases}$$

Lemma 1. *For $r \geq 1$, we have*

$$L_{p^r} = B_r(I_p \otimes L_{p^{r-1}}). \tag{4}$$

Proof. For $0 \leq i, j < p^r$ we compute $(L_{p^r})_{i,j}$ by writing

$$i = i_1 p^{r-1} + i_0, \, j = j_1 p^{r-1} + j_0, \tag{5}$$

with $0 \leq i_1, j_1 < p$ and $0 \leq i_0, j_0 < p^{r-1}$. Since $p \cdot x = 0$, we have

$$(x + x^{-1})^j = (x + x^{-1})^{j_1 p^{r-1}} (x + x^{-1})^{j_0} = (x^{p^{r-1}} + x^{-p^{r-1}})^{j_1} (x + x^{-1})^{j_0} =$$

$$\left(\sum_{k_1 \in \mathbb{Z}} l_{k_1, j_1} x^{k_1 p^{r-1}} \right) \left(\sum_{k_0 \in \mathbb{Z}} l_{k_0, j_0} x^{k_0} \right) = \sum_{k_0, k_1 \in \mathbb{Z}} l_{k_1, j_1} l_{k_0, j_0} x^{k_1 p^{r-1} + k_0} \tag{6}$$

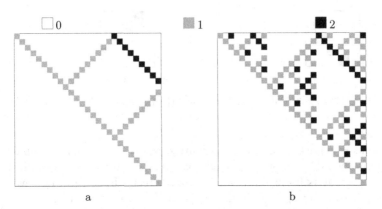

Fig. 3. (a) the matrix B_3 and (b) the matrix L_{27} for $p = 3$

where $l_{k,j}$ is as Definition 1 and is zero for $|k| > |j|$. For the coefficient of $x^i = x^{i_1 p^{r-1} + i_0}$, which is $(L_{p^r})_{i,j}$, we have:

$$k_1 p^{r-1} + k_0 = i_1 p^{r-1} + i_0 \implies k_0 \equiv i_0 \mod p^{r-1} \implies k_0 = i_0 + t p^{r-1} \text{ and } k_1 = i_1 - t \text{ for some } t \in \mathbb{Z}. \tag{7}$$

In the above equation except for $t = -1, 0$ we have $|i_0 + t p^{r-1}| \geq |p^{r-1}| > |j_0|$ which means $l_{i_0 + t p^{r-1}, j_0} = 0$, and hence

$$(L_{p^r})_{i,j} = l_{i_1, j_1} l_{i_0, j_0} + l_{i_1 + 1, j_1} l_{i_0 - p^{r-1}, j_0}, \tag{8}$$

in which $l_{i_1, j_1} = (L_p)_{i_1, j_1}$, $l_{i_0, j_0} = (L_{p^{r-1}})_{i_0, j_0}$, and $l_{i_1 + 1, j_1} = (\Psi_p L_p)_{i_1, j_1}$ according to the definition of Ψ_p. The value of $l_{i_0 - p^{r-1}, j_0}$ can be replaced by $l_{p^{r-1} - i_0, j_0}$ because of the symmetry of the binomial coefficients. The latter can again be replaced by $(\Theta_{p^{r-1}} L_{p^{r-1}})_{i_0, j_0}$, since for $0 < i_0 < p^{r-1}$ the only nonzero entry in the i_0th row of $\Theta_{p^{r-1}}$ is in the $(p^{r-1} - i_0)$th column and hence $(\Theta_{p^{r-1}} L_{p^{r-1}})_{i_0, j_0}$ is the entry in the $(p^{r-1} - i_0)$th row and j_0th column of $L_{p^{r-1}}$. For $i_0 = 0$ the entry $(\Theta_{p^{r-1}} L_{p^{r-1}})_{i_0, j_0}$ is zero since there is no nonzero entry in the i_0th row of $\Theta_{p^{r-1}}$, and l_{p^{r-1}, j_0} is also zero since $j_0 < p^{r-1}$. Substituting all of these into (8) we have

$$(L_{p^r})_{i,j} = (L_p)_{i_1, j_1} (L_{p^{r-1}})_{i_0, j_0} + (\Psi_p L_p)_{i_1, j_1} (\Theta_{p^{r-1}} L_{p^{r-1}})_{i_0, j_0} \tag{9}$$

which together with (5) shows that:

$$L_{p^r} = L_p \otimes L_{p^{r-1}} + (\Psi_p L_p) \otimes (\Theta_{p^{r-1}} L_{p^{r-1}}). \tag{10}$$

It is straightforward, using Definition 4, to show that (10) is equivalent to (4). □

This recursive relation resembles that for the DFT matrix in Chapter 1 of van Loan (1992) and enables us to find a matrix factorization for L_{p^r} in Theorem 3. Using this factorization the map of a vector under the isomorphism ν_n can be computed using $O(n \log n)$ operations as will be shown later in Section 6.

Theorem 3. *For $r \geq 1$, we have*

$$L_{p^r} = (I_1 \otimes B_r)(I_p \otimes B_{r-1}) \cdots (I_{p^{r-2}} \otimes B_2)(I_{p^{r-1}} \otimes B_1). \tag{11}$$

In order to multiply L_{p^r} by a vector, we successively multiply the matrices in the factorization (11) by that vector. In the next section we count the number of operations required for the computations of the mappings π_n and ν_n

6   Cost of Computing ν_n and π_n

Multiplication by L_{p^r} consists of several multiplications by B_k for different values of k. Hence it is better to start the study by counting the required operations for multiplying B_k by a vector in $\mathbb{F}_q^{p^k}$. The number of nonzero entries in the matrices L_p, $\psi_p L_p$, $I_{p^{k-1}}$, and $\Theta_{p^{k-1}}$ are dominated by $p^2/4$, $p^2/4$, p^{k-1}, and p^{k-1}, respectively. Hence using Definition 4, we expect the number of operations to multiply B_k by a vector in $\mathbb{F}_q^{p^k}$ be a polynomial in p, dominated by $p^{k+1}/2$. A more accurate expression for this cost is given in Lemma 2.

Definition 5. *We define $\mu_{add}(k)$ and $\mu_{mult}(k)$ to be the number of additions and multiplications, respectively, in $\mathbb{F}_q$ to multiply B_k by a vector in $\mathbb{F}_q^{p^k}$. We define further δ by the relation*

$$\delta = \begin{cases} 1 \ \textit{if } p = 2, \\ 0 \ \textit{otherwise.} \end{cases}$$

Lemma 2. *For $k \geq 1$, we have*

$$\mu_{add}(k) \leq (p-1)(2p^k - p - 1)/4 - \delta/4,$$
$$\mu_{mult}(k) \leq (1 - \delta) \cdot \mu_{add}(k).$$

For this estimate we use information about the structural zeros in B_{k-1} according to Theorem 2, but ignore the fact that some binomial coefficients might vanish modulo p. As an example since B_1, for $p = 2$, is the identity matrix both $\mu_{add}(1)$ and $\mu_{mult}(1)$ are zero.

Using Lemma 2 and Theorem 3 we are now in the position to estimate the cost of multiplication by L_{p^r}.

Lemma 3. *Multiplying L_{p^r} by a vector in $\mathbb{F}_q^{p^r}$ for $r \geq 1$ requires at most $\eta(r)$ additions, where*

$$\eta(r) = r(p-1)p^r/2 - (p+1)(p^r - 1)/4 - \delta(2^r - 1)/4.$$

The number of multiplications is not larger than the number of additions.

The following theorem is an application of Lemma 3, using $r = \lceil \log_p(n+1) \rceil$.

Theorem 4. *Multiplication of L_n by a vector in $\mathbb{F}_q^n$ can be done using $O(n \log n)$ operations in $\mathbb{F}_q$.*

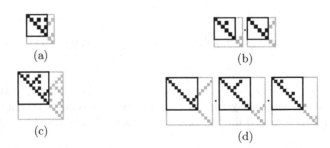

Fig. 4. (a) The matrices P_6 (black) and P_8 (gray), (b) their factorizations, (c) the matrices L_{11} (black) and L_{16} (gray), and (d) their factorizations for $p = 2$

We observe that each B_r is nonsingular since it is upper triangular and all of the entries on the main diagonal are 1. Using (11), the matrix P_{p^r} can be factored as:

$$P_{p^r} = L_{p^r}^{-1} = (I_{p^{r-1}} \otimes B_1^{-1})(I_{p^{r-2}} \otimes B_2^{-1}) \cdots (I_p \otimes B_{r-1}^{-1})(I_1 \otimes B_r^{-1}). \quad (12)$$

Studying the structure of B_r^{-1} in terms of $\Theta_{p^{r-1}}$ and $I_{p^{r-1}}$ reveals that B_r^{-1} does not have more nonzero entries than B_r. We omit the complete proof here for sake of brevity and refer the reader to the full paper or Section 4.6 of Shokrollahi (2006).

Theorem 5. *Multiplication of P_n from Definition 2 by a vector in $\mathbb{F}_q^n$ can be done using $O(n \log n)$ operations in $\mathbb{F}_q$.*

The matrices L_{11} and P_6 when $p = 2$ and their factorizations are shown in Figure 4. We conclude this section with the following theorem. Although its result is not concerned with normal basis multiplication directly, it emphasizes the most important property of our multiplier. Namely a specific change of basis in $\mathbb{F}_{q^n}$ which can be done using $O(n \log n)$ instead of $O(n^2)$ operations, which is the cost of general basis conversion in $\mathbb{F}_{q^n}$.

Theorem 6. *Let $\mathcal{N}$ be a type 2 normal basis of $\mathbb{F}_{q^n}$ over $\mathbb{F}_q$ generated by the normal element $\beta + \beta^{-1}$ and*

$$\mathcal{P} = (1, \beta + \beta^{-1}, \cdots, (\beta + \beta^{-1})^{n-1})$$

be the polynomial basis generated by the minimal polynomial of $\beta + \beta^{-1}$. Then the change of representation between the two bases $\mathcal{N}$ and $\mathcal{P}$ can be done using $O(n \log n)$ operations in $\mathbb{F}_q$.

7   Other Costs

There are two other operations in our multiplier, namely polynomial multiplication and conversion from the extended permuted representation to the normal basis representation.

The polynomial multiplication method can be selected arbitrarily among all available methods depending on the polynomial lengths and the implementation environments. Another cost is the number of bit operations to convert from extended permuted to the permuted representation. By multiplying the polynomials of length $n + 1$, the product which is of length $2n + 1$ is converted to a linear combination of $\beta^i + \beta^{-i}$ for $0 \leq i \leq 2n$. These values should be converted to the permuted representation, i.e., $\beta^i + \beta^{-i}$ for $1 \leq i \leq n$. This conversion is done using the fact that β is a $2n + 1$st root of unity. The cost for the case of odd prime numbers is given in the next theorem.

Theorem 7. *Let q be odd. Conversion from the extended permuted representation of the product in Figure 1 into the permuted basis can be done using at most $2n$ additions and n scalar multiplications in $\mathbb{F}_q$.*

When $p = 2$, the constant term vanishes because of Lucas' theorem, and the above task requires only n additions. Using the material presented herein we can summarize the costs of our multiplier in the following theorem. Since we can use any suitable polynomial multiplier, the cost depends on the polynomial multiplication method used.

Theorem 8. *Let $\mathbb{F}_{q^n}$ be a finite field of characteristic p, which contains an optimal normal basis of type 2 over $\mathbb{F}_q$. Multiplication in this normal basis can be done using at most*

$$n + 2(1 - \delta)n + 2\eta(r_1) + \eta(r_2) + \mathsf{M}(n + 1)$$

operations in $\mathbb{F}_q$, where δ is defined in Definition 5, η in Lemma 3, $\mathsf{M}(n)$ is the number of $\mathbb{F}_q$-operations to multiply two polynomials of degree $n - 1$, $r_1 = \lceil \log_p(n + 1) \rceil$, and $r_2 = \lceil \log_p(2n + 1) \rceil$. For sufficiently large n the above expression is at most

$$\mathsf{M}(n + 1) + 3n + 2(2n + 1)p^2 \log_p(2n + 1).$$

8   Comparison

Our multiplier is especially efficient when the extension degree n is much larger than the size of the ground field q. One practical application of this kind is cryptography in fields of characteristic 2. In this section we compare this multiplier with some other structures from the literature for this task. The field extensions which are discussed here are prime numbers n such that $\mathbb{F}_{2^n}$ contains an optimal normal basis of type 2.

The first structures which we study here are the circuits of Sunar & Koç (2001) and Reyhani-Masoleh & Hasan (2002). Both of these circuits require $n(5n-1)/2$ gates and we group them together as *classical*. The second circuit is from Gao et al. (2000). The idea behind this multiplier is to consider the representation

$$a_1(\beta + \beta^{-1}) + \cdots + a_n(\beta^n + \beta^{-n})$$

as the sum of two polynomials

$$a_1\beta + \cdots + a_n\beta^n \text{ and } a_n\beta^{-n} + \cdots + a_1\beta^{-1}.$$

To multiply two elements four polynomials of degree n should be multiplied together. However, because of the symmetry only two multiplications are necessary which also yield the other two products by mirroring the coefficients. The cost of a multiplication using this circuit is $2\mathsf{M}(n) + 2n$, where $\mathsf{M}(n)$ is the cost of multiplying two polynomials of length n. We may use for $\mathsf{M}(n)$ the cost of the multiplier by von zur Gathen & Shokrollahi (2005) which is not larger than $\lceil 7.6n^{\log_2 3}\rceil$ in our range.

To have a rough measure of hardware cost, we compare the circuits with respect to both area and area-time (AT). By time we mean the depth of the circuit implementation of a parallel multiplier in terms of the number of AND and XOR gates. The propagation delay of the classical multiplier is $1 + \lceil\log_2 n\rceil$ gates. The propagation delay of the multiplier of this chapter consists of two parts: the first one belongs to the conversion circuits which is $2 + 2\lceil\log_2 n\rceil$ and the other part corresponds to the polynomial multiplier. We compute the propagation delay of each polynomial multiplier for that special case. The propagation delay of the multiplier of Gao et al. (2000) is two plus the delay of each polynomial multiplier which must again be calculated for each special case.

The area and AT parameters of these three circuits are compared with each other and the results are shown in Figure 5. Please note that the costs of our designs are exact values from Theorem 8. In these diagrams polynomial multiplication is done using the methods of von zur Gathen and Shokrollahi (2005). As it can be seen the area of the proposed multiplier is always better than the other two structures. But the AT parameter is larger for small finite fields. This shows that, as we have mentioned, this method is appropriate for applications where only small area is available, or where the finite fields are large. Economical applications where small FPGAs should be used or RFID technology, are situations of

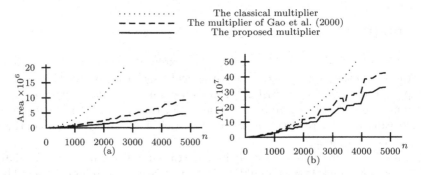

Fig. 5. Comparing the (a) area (as the number of two-input gates) and (b) the AT parameter (as the product of the number of two-input gates and the delay of a single gate) of three multipliers for binary finite fields $\mathbb{F}_{2^n}$ such that n is a prime smaller than 5000 and $\mathbb{F}_{2^n}$ contains an optimal normal basis of type 2

this sort. The AT parameter of the proposed multiplier is $O(n \log^3 n (\log \log n)^3)$, whereas that of the classical multiplier is $O(n^2 \log n)$.

Another method which should be compared to ours is the method of Fan & Hasan (2006). This work has been introduced to us by one of the referees and we did not have enough time for an exact comparison. This method requires roughly $13n^{1.6}$ gates, whereas the Karatsuba method for polynomials of length n needs $9n^{1.6}$ gates. Hence we approximate the number of operations for this method to be $13/9\mathsf{M}(n)$ for the Karatsuba method. The delay of this method equals the delay of the Karatsuba method. We guess the number of operations for this method to be larger than ours for the given bound, but the area-time parameter must be better. We again emphasize that their methods is comparable to the Karatsuba method, whereas ours can use any asymptotically fast method like that of Cantor with a cost of $O(n \log^3 n (\log \log n)^2)$.

9   Conclusion

This work presents a new algorithm for multiplication in finite fields using optimal normal bases of type 2 which reduces the asymptotic number of operations from $2\mathsf{M}(n) + O(n)$ reported by Gao et al. (2000) to $\mathsf{M}(n) + O(n \log n)$. The efficiency of this multiplier arises from a fast transformation between normal bases and suitable polynomial representation. This transformation can be done by $O(n \log n)$ operations instead of generic $O(n^2)$ operations for the general case. This algorithm is especially attractive for hardware implementations where area resources are limited as also shown in comparisons with other methods from the literature.

Acknowledgement

We thank anonymous referees for their helpful suggestions and also for pointing to the work of Fan & Hasan (2006).

References

1. U.S. Department of Commerce / National Institute of Standards and Technology: Digital Signature Standard (DSS) Federal Information Processings Standards Publication 186-2 (2000)
2. Fan, H., Hasan, M.A.: Subquadratic multiplication using optimal normal bases. Technical Report cacr2006-26, University of Waterloo, Waterloo (2006)
3. Gao, S., von zur Gathen, J., Panario, D., Shoup, V.: Algorithms for exponentiation in finite fields. Journal of Symbolic Computation 29, 879–889 (2000)
4. Gao von, S., von zur Gathen, J., Panario, D.: Gauss periods and fast exponentiation in finite fields. In: Baeza-Yates, R.A., Poblete, P.V., Goles, E. (eds.) LATIN 1995. LNCS, vol. 911, pp. 311–322. Springer, Heidelberg (1995)
5. Gao, S., Lenstra Jr., H.W.: Optimal normal bases. Designs, Codes, and Cryptography 2, 315–323 (1992)

6. von zur Gathen, J., Nöcker, M.: Polynomial and normal bases for finite fields. Journal of Cryptology 18, 313–335 (2005)
7. von zur Gathen, J., Shokrollahi, J.: Efficient FPGA-based Karatsuba multipliers for polynomials over $\mathbb{F}_2$. In: Preneel, B., Tavares, S. (eds.) SAC 2005. LNCS, vol. 3897, pp. 359–369. Springer, Heidelberg (2006)
8. Grabbe, C., Bednara, M., Shokrollahi, J., Teich, J., von zur Gathen, J.: FPGA designs of parallel high performance $GF(2^{233})$ multipliers. In: Proc. of the IEEE International Symposium on Circuits and Systems (ISCAS-03), Bangkok, Thailand, vol. II, pp. 268–271 (2003)
9. Granger, R., Page, D., Stam, M.: Hardware and software normal basis arithmetic for pairing-based cryptogaphy in characteristic three. IEEE Transactions on Computers 54, 852–860 (2005)
10. Kaliski, B.S., Liskov, M.: Efficient Finite Field Basis Conversion Involving Dual Bases. In: Koç, Ç.K., Paar, C. (eds.) CHES 1999. LNCS, vol. 1717, pp. 135–143. Springer, Heidelberg (1999)
11. van Loan, C.: Computational Frameworks for the Fast Fourier Transform. Society for Industrial and Applied Mathematics (SIAM), Philadelphia (1992)
12. Mullin, R.C., Onyszchuk, I.M., Vanstone, S.A., Wilson, R.M.: Optimal normal bases in GF(p^n). Discrete Applied Mathematics 22, 149–161 (1989)
13. Nöcker, M.: Data structures for parallel exponentiation in finite fields. Doktorarbeit, Universität Paderborn, Germany (2001)
14. Omura, J.K., Massey, J.L.: Computational method and apparatus for finite field arithmetic. United States Patent vol. 4, pp. 587,627 (1986) (Date of Patent: May 6, 1986)
15. Reyhani-Masoleh, A., Hasan, M.A.: A new construction of Massey-Omura parallel multiplier over $GF(2^m)$. IEEE Transactions on Computers 51, 511–520 (2002)
16. Shokrollahi, J.: Efficient Implementation of Elliptic Curve Cryptography on FPGAs. PhD thesis, Bonn University, Bonn (2006) http://hss.ulb.uni-bonn.de/diss_online/math_nat_fak/2007/shokrollahi_jamshid/index.htm.
17. Sunar, B., Koç, Ç.K.: An efficient optimal normal basis type II multiplier. IEEE Transactions on Computers 50, 83–87 (2001)
18. Wassermann, A.: Konstruktion von Normalbasen. Bayreuther Math. Schriften 31, 155–164 (1990)
19. Wikipedia: Sierpinski triangle. (2006), Webpage http://en.wikipedia.org/wiki/Sierpinski_triangle

Effects of Optimizations for Software Implementations of Small Binary Field Arithmetic

Roberto Avanzi[1] and Nicolas Thériault[2]

[1] Fakultät für Mathematik, Ruhr-Universität Bochum and
the Horst Görtz Institut für IT-Sicherheit, Germany
roberto.avanzi@rub.de
[2] Instituto de Matemática y Física, Universidad de Talca, Chile
ntheriau@inst-mat.utalca.cl

Abstract. We describe an implementation of binary field arithmetic written in the C programming language. Even though the implementation targets 32-bit CPUs, the results can be applied also to CPUs with different granularity.

We begin with separate routines for each operand size *in words* to minimize performance penalties that have a bigger relative impact for shorter operands – such as those used to implement modern curve based cryptography. We then proceed to use techniques specific to operand size *in bits* for several field sizes.

This results in an implementation of field arithmetic where the curve representing field multiplication performance closely resembles the theoretical quadratic bit-complexity that can be expected for small inputs.

This has important practical consequences: For instance, it will allow us to compare the performance of the arithmetic on curves of different genera and defined over fields of different sizes without worrying about penalties introduced by field arithmetic and concentrating on the curve arithmetic itself. Moreover, the cost of field inversion is very low, making the use of affine coordinates in curve arithmetic more interesting. These applications will be mentioned.

Keywords. Binary fields, efficient implementation, curve-based cryptography.

1 Introduction

The performance of binary field arithmetic is crucial in several contexts. In many cases, the most important operation to optimize is the multiplication, which is based on the multiplication of polynomials over $\mathbb{F}_2$. A lot of work has been devoted to improve the speed of multiplication of binary polynomials of very large degree, for example to break polynomial factorization records. For other important applications such as the implementation of elliptic curve (EC) and hyperelliptic curve (HEC) cryptography the focus is instead on relatively small

C. Carlet and B. Sunar (Eds.): WAIFI 2007, LNCS 4547, pp. 69–84, 2007.
© Springer-Verlag Berlin Heidelberg 2007

fields, of roughly 40 to 300 bits. Only in special circumstances will larger fields be contemplated.

One of the most celebrated advantages of using higher genus curves in place of EC is the fact that to achieve the same level of security of the latter they need to be defined over smaller fields. For hardware designers this is a bonus since they can use smaller multiplication and inversion circuits, and thus higher genus curves hold their own advantage, such as a low area-time product [22].

However, most software implementations of finite fields penalize smaller fields more than larger ones: [1] shows that HEC in odd characteristic are often heavily penalized with respect to EC because of many types of overheads in the field arithmetic whose impact increases as field sizes get smaller. This is also the case in characteristic 2, but the nature of the worst overheads and the techniques used to address them differ.

The techniques described in this paper are thus aimed at small fields such as those used in curve-based cryptography. They are also devised to improve the use of basic field operations in other practical contexts, such as polynomial factorization algorithms.

The paper is structured as follows: Section 2 describes the types of overheads we must handle. We then look at the field multiplication (§ 3.1), sequential multiplications (§ 3.2), squaring (§ 3.3), modular inversion (§ 3.5), and modular reduction (§ 3.4). We conclude in Section 4 with performance results, including timings of EC and HEC arithmetic (that will be discussed in detail in a forthcoming paper) and their consequences.

2   Types of Overheads

We begin by introducing some notation. A field $\mathbb{F}_{2^n}$ is represented using a polynomial basis as the quotient ring $\mathbb{F}_2[t]/(p(t))$, where $p(t)$ is an irreducible polynomial of degree n. This is the most usual representation of small fields and their elements for software implementations (cf. [2, Ch. 11]). Field elements are represented by binary polynomials of degree less than n. Multiplication (resp. squaring) in $\mathbb{F}_{2^n}$, is performed by first multiplying (resp. squaring) the polynomial(s) representing the input(s), and then reducing the result modulo $p(t)$. Let γ be the number of bits in a computer word. A field elements a occupies $s = \lceil n/\gamma \rceil$ words $A_0, A_1, \ldots, A_{s-1}$. The γ least significant bits of the polynomial representing a are contained in the words A_0, the next γ bits in A_1 and so on.

In our implementation, we address the following types of overheads:

1. *Software loops to process operands, pipelining, and branch mispredictions.*

   Algorithm 1 is the standard method for performing multiplication of two small polynomials over GF(2). The issue of expensive branch mispredictions at loop boundaries is addressed by full loop unrolling for all input sizes. This is common in the long integer arithmetic underlying the implementation of prime fields [1]. Loop unrolling is also useful in even characteristic, but it is

---

Algorithm 1. Field multiplication [13]

INPUT: $A = (A_{s-1}, \ldots, A_0)$, $B = (B_{s-1}, \ldots, B_0)$, and a comb size w
OUTPUT: $R = (R_{2s-1}, \ldots, R_0) = A \times B$

1. **for** $j = 0$ **to** $2^w - 1$ **do**

2. $P_j(t) \leftarrow (j_{w-1}t^{w-1} + \cdots + j_1 t + j_0)A(t)$ where $j = (j_{w-1} \ldots j_2 j_1 j_0)_2$.
        [Here the polynomial $P_j(t)$ is at most $s + 1$ words long]

3. **for** $i = 0, \ldots, 2s - 1$ **do** $R_i \leftarrow 0$

4. **for** $j = \lceil \gamma/w \rceil - 1$ **downto** 0 **do**

5.   **for** $i = 0$ **to** $s - 1$ **do**

6. $u \leftarrow (B_i \gg jw) \bmod t^w$ [mask out w bits at time]

7.   **for** $k = 0$ **to** s **do**

8. $R_{k+i} \leftarrow R_{k+i} \oplus P_u[k]$

9.   If $j \neq 0$ **then** $R \leftarrow t^w R$ [bit shift]

10. **return** R

---

used differently: see § 3.1 for more details. In all cases, loop unrolling also helps the compiler to produce code which exploits the processor pipelines more efficiently.

2. *Architecture granularity.*

Algorithm 1 requires a precomputation phase to compute the multiples of the first input by all binary polynomials of degree less than w (Steps 1 and 2). The complexity of this stage is exponential in w, and the optimal value of w to minimize the total number of operations is an integer of the form $O(\log \log n)$ (Theorem D, § 4.6.3 in [9] also applies to binary multiplication). The constants vary depending on the architecture however, and in most cases the optimal w can only be found by trial-and-error.

For large ranges of the input sizes, the value of w will remain constant. The complexity of multiplication in these ranges is therefore roughly quadratic in s, as a generic implementation does not distinguish between fields of $s\gamma + 1$-bits and $(s + 1)\gamma$-bits. This *granularity* introduces irregular performance penalties. To see its effect in curve cryptography, let us consider an example.

For the same security level, we can use an EC over a field of 223 bits, or a genus 2 curve over a field of roughly 112 bits, using respectively 7 and 4 words per field element. The expected number of field multiplications for a group operation increases roughly quadratically with the genus[1],but in this scenario the cost of a multiplication in the smaller field (which requires 4

---

[1] This is an average assuming a windowed scalar multiplication method, where the most common group operation is doublings. By choosing curves of a special type, group doublings can be implemented with quadratically many field multiplications in the genus, even though group additions have cubic complexity. This approximation works nicely for curve of genus up to 4 (and possibly more), see for example [11,14,15,5,3,23] and [2, Ch. 13, 14].

words store each element) is about $1/3$ of the cost of a multiplication in the larger field (where elements are 7 words long), as $4^2 = 16$ is 32.6% of $7^2 = 49$. As a result, the granularity would penalize the HEC of genus two by a factor of 1.32.

Little can be done to defeat granularity problems in the prime field case [1], but more options are available in even characteristic (cf. § 3.1).

3. Multiplications of a single field element by a vector of field elements.

Once again, let us consider curve-based cryptography. In this context, occurrences of several multiplications by the same element become more common as the genus increases. It is possible to speed up these multiplications appreciably by treating them as vector multiplications rather than sets of independent multiplications. A similar situation occurs when multiplying polynomials of low degree, where vector multiplications can be faster than Karatsuba methods. The technique for doing this, described in § 3.2, is not new, but so far its impact has not been thoroughly evaluated in the literature. Similar optimization technique do not seem to be possible in the prime field case.

Another important difference with the prime field case lies in the relative cost of modular reduction compared to multiprecision multiplication.

For prime fields, this relative cost increases as the operand size decreases, and in odd characteristic HEC implementations more time is spent doing modular reductions than in EC implementations. To decrease the total cost of reductions, one can delay modular reductions when computing the sum of several products. The idea is described in [17], and [1] shows that this approach also works with the Montgomery representation of modular integers.

In even characteristic, modular reduction is much cheaper, and the additional memory traffic caused by delaying reductions can even reduce performance. After doing some operation counts and practical testing, we opted not to allow delayed reductions.

This in turn raises the issue of the number of function calls, since calling the reduction code after each multiplication increases this number.

4. Function call overheads.

Function call overheads also play a different role in even characteristic. In fact, the code for implementing arithmetic routines is bigger than in the prime field case. Inlining it would usually cause the code size to increase dramatically, creating a large penalty in performance. As a result, field additions are inlined in the main code of our programs (such as explicit formulæ for curve arithmetic) but multiplications are not. However, we inline the reduction code *into* the special multiplication and squaring routines for each field: the reduction code is compact and it is inlined only a few times, thus keeping code size manageable while saving function calls.

3   Implementation Techniques

3.1   Field Multiplication and Architecture Granularity

The starting point for our implementation of field multiplication is Algorithm 1, by López and Dahab [13]. It is based on comb exponentiation, that is usually attributed to Lim and Lee [12], but that in fact is due to Pippenger [16]. Recall that s is the length of the inputs in computer words and γ the number of bits per word. There are a few obvious optimizations of Algorithm 1.

If the inputs are at most $s\gamma - w + 1$ bits long, all the $P_j(t)$'s fit in s words. In this case, the counter k of the loop at Step 7 will only need to go from 0 to $s - 1$.

If operands are between $s\gamma - w + 2$ and $s\gamma$ bits long, proceed as follows; Zero the $w - 1$ most significant bits of A for the computations in Steps 2 and 3, thus obtaining polynomials $P_j(t)$ that fit in s words; Perform the computation as in the case of shorter operands, with $s - 1$ as the upper bound in Step 7; Finally, add the multiples of $B(t)$ corresponding to the $w - 1$ most significant bits of A to R before returning the result. This leads to a much faster implementation since several memory writes in Step 2 and memory reads in Step 8 are traded for a minimal amount of memory operations later to "adjust" the final result.

More optimizations can be applied if the field size is known in advance, such as partial or full loop unrolling. Also, some operands (containing the most significant bits of R) are known to be zero in the first repeats of the loop bodies, hence parts of the computation can be explicitly omitted during the first iterations. This can be done in the full unrolling or just by splitting a loop in two or more parts whereby one part contain less instructions that the following one.

Steps 7 and 8 could be skipped when $u = 0$ but inserting an "If $u \neq 0$" statement before Step 7. de facto slows down the algorithm (because of frequent branch mispredictions), so it is more efficient to allow these "multiplications by zero".

We considered different comb sizes and different loop splitting and unrolling techniques for several input sizes ranging from 43 to 283 bits – the exact choice being determined by the application described in § 4.2.

As we mentioned in the previous section, the optimal choice for the comb size w will depend both on the field size and the architecture of the processor. *It should be noted that the choice of the comb size and the point at which Karatsuba multiplication becomes interesting (see below) are the only optimizations that are processor-dependent (given a fixed word size), whereas the other techniques presented in this paper are not affected by the processor's architecture.* because it did change with the processor

Special treatment is reserved to polynomials whose size is just a few bits more than a multiple k of the architecture granularity. Two such cases are polynomials of 67 and 97 bits, where the values of k are 64 and 96, respectively. We perform scalar multiplication of the polynomials consisting of the lower k bits first, and then handle the remaining most significant bits one by one. In other words, write $A = A' + t^k \cdot A''$ and $B = B' + t^k \cdot B''$ with the degrees of A', B' smaller than k, and perform $A \cdot B = A' \cdot B' + t^k \cdot A'' \cdot B' + t^k \cdot A' \cdot B'' + t^{2k} \cdot A'' \cdot B''$. In some cases, a little regrouping such as $A \cdot B = A' \cdot B' + t^k \cdot A'' \cdot B' + t^k \cdot B'' \cdot A$ is slightly more efficient. The resulting code is 10 to 15% faster than if we applied the previous approaches to the one-word-longer multiplication.

Multiplication of polynomials whose degree is high enough is also done using Karatubsa's technique [7] of reducing the multiplication of two polynomials to three multiplications of half size polynomials. After some testing, we observed that Karatsuba multiplication performs slightly better than comb-based multiplication for $s \geq 6$ on the PowerPC, and for $s \geq 7$ on the Intel Core architecture (the half-size multiplications are performed with the comb method), but not for smaller sizes.

The performance of multiplication routines can be seen in Tables 1 and 2 and Figures 1 and 2. They will be discussed in more detail in Section 4, but we can already observe that the cost of multiplication grows quite smoothly with the field size, and in fact it approached a curve of quadratic bit complexity (as we might expect from theory) much better than a coarser approach that works at the word level.

3.2 Sequential Multiplications

In certain situations, for instance in explicit formulæ for curves of genus two to four, we find sets of multiplications with a common multiplicand. This usually occurs as a result of polynomial arithmetic.

A natural approach to reduce average multiplication times is to keep the precomputations (Steps 1 and 2 of Algorithm 1) associated to the common multiplicand and re-use them in the next multiplications. However, this would require extra variables in the implementation of the explicit formulæ and demand additional memory bookkeeping. We thus opted for a slightly different approach: we wrote routines that perform the precomputations once and then repeat Steps 3 to 10 of Algorithm 1 for each multiplication. We call this type of operation *sequential multiplication* (the more common terminology of scalar multiplication having another signification in curve based cryptography...).

An important observation is that the optimal comb size used in the multiplication (on a given architecture) may vary depending on the number of multiplications that are performed together. For example, for the field of 73 bits on the PowerPC, a comb of size 3 is optimal for the single multiplication and for sequences of two multiplications, but for 3 to 5 multiplications the optimal comb of size is 4. For the field of 89 bits, the optimal comb size for single multiplications is again 3, but it already increases to 4 for the double multiplication.

If a comb method is used for 6-word fields on the PowerPC and Core architectures, then 4 is the optimal comb size for single multiplications and 5 is optimal for groups of at least 3 multiplications. However, on the PowerPC, Karatsuba is used not only for the single multiplications, but also for the sequential ones, where a sequential multiplication of s-word operands is turned into three sequential multiplications of $s/2$-word operands for $s \geq 6$. On the Core CPU, Karatsuba's approach becomes more efficient for sequential multiplications only when $s \geq 8$.

To keep function call overheads low, the sequential multiplication procedures for at least 3 multiplications use input and output *vectors* of elements which are adjacent in memory. This also speeds up memory accesses, taking better advantage of the structure of modern caches.

Static precomputations have already been used for multiplications by a constant parameter (coming from the curve or the field) – for example, [4] suggests this in the context of square root extraction. In [8], King uses a similar approach to reduce the number of precomputed tables in the projective formulæ for elliptic curves, however he does not estimate the costs of several multiplications performed by this method in comparison to the cost of one multiplication, nor does he adapt the comb size to the number of multiplications performed.

3.3   Polynomial Squaring

Squaring is a linear operation in even characteristic: if $f(t) = \sum_{i=0}^{n} e_i t^i$ where $e_i \in \mathbb{F}_2$, then $\left(f(t)\right)^2 = \sum_{i=0}^{n-1} e_i t^{2i}$. That is, the result is obtained by inserting a zero bit between every two adjacent bits of the input. To efficiently implement this process, a 512-byte table is precomputed for converting 8-bits polynomials into their expanded 16-bits counterparts [18]. In practice, this technique is faster than King's method [8] (which otherwise has the advantage of requiring less memory).

3.4   Modular Reduction

We implemented two sets of routines for modular reduction.

The first consists of just one generic routine that reduces an arbitrary polynomial over $\mathbb{F}_2$ modulo another arbitrary polynomial over $\mathbb{F}_2$. This code is in fact rather efficient, and a reduction by means of this routine can often take less than 20% of the time of a multiplication. The approach is similar to the one taken in SUN's ECC contributions to OpenSSL [20] or in NTL [19]. Let the reduction polynomial be $p(t) = \sum_{i=0}^{k-1} t^{n_i}$ with $n_0 > n_1 > ... > n_{k-1} = 0$. Its degree is n_0 and its *sediment* is $\sum_{i=1}^{k-1} t^{n_i}$. If an irreducible trinomial ($k = 3$) exists, we use it, otherwise we use a pentanomial ($k = 5$).

The second set of routines uses fixed reduction polynomials, and is therefore specific for each polynomial degree. The code is very compact. Here we sometimes prefer reduction eptanomials ($k = 7$) to pentanomials when the reduction is faster due to the form of the polynomial, for instance when the sediment has lower degree and/or it factors nicely.

As an example, for degree 59 we have two good irreducible polynomials: $p_1(t) = t^{59} + (t+1)(t^5 + t^3 + 1)$ and $p_2(t) = t^{59} + t^7 + t^4 + t^2 + 1$. The first C code fragment takes a polynomial of degree up to 116 ($= 2 \cdot 58$) stored in variables r3 (most significant word), r2, r1 and r0 (least significant word), and reduces it modulo $p_1(t)$, leaving the result in r1 and r0:

```
#define bf_mod_59_6_5_4_3_1_0(r3,r2,r1,r0) do {                    \
   r3  = ((r3) <<  5) ^ ((r3) <<  6);                              \
   r1 ^= (r3) ^ ((r3) << 3) ^ ((r3) << 5);                        \
   r3  = ((r2) <<  5) ^ ((r2) <<  6);                              \
   r0 ^= (r3) ^ ((r3) << 3) ^ ((r3) << 5);                        \
   r3  = ((r2) >> 22) ^ ((r2) >> 21);                             \
   r1 ^= (r3) ^ ((r3) >> 2) ^ ((r3) >> 5);                        \
   r2  = (r1) >> 27; r2 ^= (r2) << 1;                             \
   r1 &= 0x07ffffff; r0 ^= (r2) ^ ((r2) << 3) ^ ((r2) << 5);     \
} while (0)
```

The C code to reduce the same input modulo $p_2(t)$ is

```
#define bf_mod_59_7_4_2_0(r3,r2,r1,r0) do {                                      \
   r1 ^= ((r3) <<  5) ^ ((r3) <<  7) ^ ((r3) <<  9) ^ ((r3) << 12);  \
   r2 ^=               ((r3) >> 25) ^ ((r3) >> 23) ^ ((r3) >> 20);  \
   r0 ^= ((r2) <<  5) ^ ((r2) <<  7) ^ ((r2) <<  9) ^ ((r2) << 12);  \
   r1 ^= ((r2) >> 27) ^ ((r2) >> 25) ^ ((r2) >> 23) ^ ((r2) >> 20);  \
   r2  = (r1) >> 27;   r1 &= 0x07ffffff;                              \
   r0 ^= (r2) ^ ((r2) << 2) ^ ((r2) << 4) ^ ((r2) << 7);            \
} while (0)
```

We found the first reduction routine slightly more efficient. A similar choice occurs at degree 107 (the eptanomial being $t^{107} + (t^6 + t^2 + 1)(t + 1)$), and the idea of factoring the "lower degree part" of the reduction polynomial is also used for degree 109 (the polynomial is $t^{109} + (t^6 + 1)(t + 1)$).

These considerations were applied to the degrees 43, 47, 53, 59, 67, 73, 79, 83, 89, 97, 101, 107, 109, 113, 127, 131, 137, 149, 157, 163, 179, 199, 211, 233, 239, 251, 269 and 283. For degrees 47, 79, 89, 97, 113, 127, 137, 199, 233 and 239 we used a trinomial. For degrees 59 and 107 we opted for eptanomials (cf. remarks above), and in all other cases pentanomials were used. The time for the modular reduction is kept between 3 and 5% of the time required for a multiplication if we use a trinomial, and between 6 and 10% in the other cases. Reduction modulo a trinomial is about twice fast as polynomial squaring because the latter requires more memory accesses.

3.5   Modular Inversion

There are many algorithms for computing the inverse of a polynomial $a(t)$ modulo another polynomial $p(t)$, where both polynomials are defined over the field $\mathbb{F}_2$. In [6] three methods are compared:

- The Extended Euclidean Algorithm (EEA), where partial quotients are approximated by powers of t, and no polynomial division is required.
- The Almost Inverse Algorithm (AIA), a variant of the binary extended GCD algorithm that computes a polynomial $b(t)$ together with an integer ℓ such that $b(t)a(t) \equiv t^\ell \bmod p(t)$. The final result must then be adjusted.
- The Modified Almost Inverse Algorithm (MAIA), which is a variant of the binary extended GCD algorithm which returns the correct inverse as a result.

We refer to [6] for details. In agreement with [6], we find that EEA performs consistently better than the two other methods in our context.

For inputs of up to 8 words, we always keep all words (limbs) of all multiprecision operands in separate integer variables explicitly, not in indexed arrays. This allows the compiler to allocate a register for each of these integer variables if enough registers are provided by the architecture (such as on RISC processors). Furthermore, it does not penalize architectures with fewer registers: the contents of many variables containing individual words are spilled on the stack, but this data would still be stored in memory if we used arrays.

Another advantage of the EEA is that it offers good control on the bit lengths of the intermediate variables. We can therefore split the main loop in several copies optimised for the different sizes of the intermediate operands, with n sections of code for inputs of n words. Since some intermediate values grow in size as other values get shorter, we can reduce the local usage of registers, allowing an increase in the size of the inputs before the compiler starts to produce code that spills some data to memory. See Section 4 for inversion performance.

4   Performance Results, Comparisons, and Conclusions

We compiled and ran our code on several architectures. Due to space constraints we present here two sets of timings taken on very different CPUs. The first set was obtained on a 1.5 GHz PowerPC G4 (Motorola 7447) CPU, a 32-bits RISC architecture with 32 general purpose integer registers, no level 3 cache support and slow memory bus. The second set was taken on a 1.83 GHz Intel Core 2 Duo (running 32-bit code on one core), an architecture with fewer registers, a better cache system and faster memory bus.

In our C code, we declare all limbs of all intermediate operands as single 32-bit word variables, and operate on them with logic and shift operations. The code compiles on any 32-bits architecture supported by the gnu compiler collection (we used gcc 4.0.1, Apple branch, build 5367, under Mac OS X 10.4.9 on both architectures). We did not use assembler code – which would be necessary to get satisfactory performance in long integer arithmetic – since near-optimal performance can be attained for binary field arithmetic by carefully crafting the C code. This is a known fact: the binary arithmetic of NTL [19], which is very well known for its performance, is written in C (in fact C++), as are SUN's ECC contributions to OpenSSL [20].

Tables 1 and 2 contain the timings of the fundamental operations in the fields that we considered. The operations are: single multiplication (Mul), squaring (Sqr), multiplication of 2 to 5 different field elements by a fixed one (columns from Mul2 to Mul5) and inversion (Inv). Modular reduction is always included. We give the absolute times in microseconds and the relative costs compared to a single multiplication. We also give the timings of our best generic routines for field multiplications (more or less on-par with the NTL libraries) together with the speedup factor gained by the field-specific routines.

Table 1. Timings of field operations in μsec and ratios (1.5 GHz Powerpc G4)

Field Size Bits	Timings of optimized library													Standard Mult.	
	Absolute timings							Timings relative to multiplication						Time	Speed-up
	Mul	Sqr	Mul2	Mul3	Mul4	Mul5	Inv	Sqr	Mul2	Mul3	Mul4	Mul5	Inv		
43	.100	.017	.169	.227	.287	.344	.450	.169	1.693	2.271	2.874	3.451	4.505	.331	3.310
47	.087	.014	.142	.192	.243	.296	.483	.164	1.640	2.225	2.795	3.421	5.511	.338	3.885
53	.098	.018	.167	.218	.276	.337	.512	.183	1.707	2.224	2.817	3.434	5.224	.280	2.857
59	.121	.021	.210	.272	.333	.394	.531	.172	1.740	2.257	2.762	3.267	4.402	.336	2.776
67	.168	.025	.276	.348	.436	.516	.820	.148	1.647	2.073	2.598	3.077	4.886	.444	2.627
73	.190	.024	.329	.466	.572	.674	.857	.128	1.732	2.453	3.011	3.547	4.508	.521	2.742
79	.193	.019	.321	.452	.549	.649	.899	.099	1.662	2.342	2.848	3.364	4.658	.448	2.321
83	.213	.050	.387	.518	.640	.760	.922	.236	1.818	2.430	3.003	3.569	4.329	.521	2.446
89	.254	.025	.420	.538	.667	.796	.962	.100	1.653	2.119	2.627	3.136	3.790	.443	1.744
97	.311	.025	.455	.612	.761	.913	1.589	.081	1.462	1.967	2.445	2.933	5.105	.602	1.936
101	.353	.057	.552	.722	.903	1.083	1.621	.161	1.564	2.046	2.557	3.068	4.592	.695	1.969
107	.358	.058	.559	.760	.954	1.149	1.659	.162	1.560	2.122	2.665	3.209	4.633	.852	2.380
109	.371	.029	.567	.799	.998	1.201	1.692	.079	1.528	2.155	2.693	3.241	4.564	.695	1.873
113	.373	.029	.580	.783	.981	1.185	1.702	.078	1.554	2.099	2.629	3.177	4.561	.593	1.590
127	.415	.053	.674	.957	1.201	1.447	1.832	.128	1.625	2.306	2.894	3.486	4.415	.574	1.383
131	.460	.067	.763	1.046	1.329	1.605	3.664	.147	1.659	2.275	2.890	3.489	7.968	1.042	2.265
137	.625	.062	1.084	1.485	1.907	2.325	3.733	.100	1.734	2.375	3.050	3.718	5.969	.926	1.539
149	.677	.090	1.191	1.680	2.170	2.656	3.908	.133	1.760	2.482	3.206	3.924	5.773	.817	1.207
157	.774	.090	1.469	1.951	2.441	2.910	4.055	.116	1.899	2.522	3.155	3.762	5.243	1.027	1.327
163	.815	.085	1.342	1.902	2.455	3.009	5.336	.105	1.647	2.334	3.012	3.692	6.547	1.229	1.508
179	1.116	.124	2.117	2.916	3.623	4.508	5.552	.111	1.896	2.613	3.246	4.038	4.974	1.229	1.101
199	1.195	.091	2.085	2.716	3.423	4.126	12.114	.076	1.745	2.273	2.865	3.454	10.140	1.390	1.163
211	1.225	.145	2.169	2.937	3.695	4.435	12.559	.119	1.771	2.398	3.016	3.621	10.254	1.531	1.250
233	1.380	.114	2.347	3.206	4.013	4.883	14.613	.079	1.701	2.323	2.907	3.537	10.596	1.594	1.155
239	1.528	.187	2.630	3.637	4.604	5.638	14.828	.122	1.722	2.381	3.014	3.691	9.706	1.596	1.060
251	1.675	.228	2.978	4.157	5.297	6.498	15.157	.137	1.778	2.482	3.163	3.879	9.050	1.741	1.040
269	2.035	.210	3.790	5.146	6.431	7.853	20.495	.103	1.863	2.529	3.161	3.860	10.073	2.438	1.197
283	2.148	.229	3.943	5.338	6.817	8.134	20.845	.107	1.835	2.485	3.174	3.787	9.704	2.447	1.139

Table 2. Timings of field operations in μsec and ratios (1.83 GHz Intel Core 2 Duo)

Field Size Bits	Timings of optimized library													Standard Mult.	
	Absolute timings							Timings relative to multiplication						Time	Speed-up
	Mul	Sqr	Mul2	Mul3	Mul4	Mul5	Inv	Sqr	Mul2	Mul3	Mul4	Mul5	Inv		
43	.087	.017	.137	.182	.228	.260	1.381	.195	1.575	2.091	2.621	2.989	15.830	.253	2.908
47	.064	.012	.108	.153	.194	.239	1.502	.191	1.688	2.393	3.028	3.742	23.484	.211	3.297
53	.072	.025	.131	.185	.240	.294	1.623	.350	1.820	2.573	3.350	4.107	22.635	.253	3.514
59	.117	.027	.201	.279	.361	.444	1.740	.228	1.723	2.396	3.097	3.813	14.935	.285	2.436
67	.174	.027	.227	.313	.410	.502	2.386	.154	1.302	1.797	2.354	2.881	13.693	.371	2.132
73	.174	.026	.321	.473	.545	.667	2.512	.151	1.840	2.716	3.128	3.827	14.413	.377	2.167
79	.171	.025	.257	.332	.425	.511	2.649	.146	1.506	1.947	2.487	2.994	15.520	.374	2.187
83	.184	.041	.285	.350	.458	.554	2.727	.224	1.553	1.908	2.492	3.018	14.857	.371	2.016
89	.207	.028	.318	.426	.526	.619	2.813	.135	1.536	2.106	2.589	3.039	13.589	.345	1.665
97	.223	.032	.379	.531	.687	.838	3.677	.142	1.701	2.388	3.085	3.765	16.523	.405	1.816
101	.240	.039	.431	.594	.767	.945	3.760	.164	1.796	2.472	3.194	3.934	15.659	.507	2.113
107	.240	.038	.427	.620	.813	1.001	3.895	.157	1.779	2.582	3.385	4.171	16.223	.566	2.358
109	.274	.035	.505	.702	.918	1.131	3.961	.128	1.840	2.559	3.345	4.121	14.433	.473	2.111
113	.255	.047	.444	.637	.828	1.022	4.008	.183	1.741	2.496	3.245	4.005	15.708	.414	1.624
127	.305	.038	.563	.836	1.084	1.335	4.375	.124	1.848	2.745	3.560	4.385	14.365	.488	1.467
131	.366	.064	.679	.961	1.269	1.581	5.278	.174	1.855	2.624	3.466	4.316	14.413	.660	1.803
137	.384	.049	.706	.933	1.219	1.485	5.396	.129	1.837	2.426	3.172	3.863	14.039	.515	1.341
149	.413	.072	.740	1.081	1.420	1.760	5.676	.174	1.791	2.617	3.437	4.261	13.741	.647	1.567
157	.447	.072	.848	1.170	1.508	1.844	5.887	.161	1.900	2.621	3.378	4.129	13.185	.660	1.477
163	.495	.069	.942	1.358	1.787	2.221	6.855	.139	1.904	2.744	3.611	4.488	13.855	.798	1.612
179	.598	.085	1.101	1.513	1.951	2.390	7.301	.143	1.843	2.533	3.266	4.000	12.218	.798	1.334
199	.674	.067	1.211	1.975	2.128	2.604	9.542	.099	1.797	2.930	3.157	3.864	14.157	.957	1.420
211	.868	.089	1.580	2.256	2.926	3.572	10.591	.102	1.821	2.600	3.372	4.117	12.206	1.112	1.281
233	.945	.086	1.759	2.532	3.256	4.005	12.246	.091	1.860	2.678	3.444	4.236	12.954	1.047	1.080
239	1.061	.187	2.014	2.880	3.748	4.584	12.378	.176	1.898	2.715	3.533	4.321	11.668	1.131	1.066
251	1.120	.206	2.075	3.008	3.884	4.809	12.641	.184	1.853	2.686	3.468	4.294	11.287	1.260	1.116
269	1.224	.127	2.209	3.128	4.050	4.939	14.924	.104	1.805	2.556	3.309	4.036	12.195	1.667	1.362
283	1.325	.141	2.282	3.206	4.135	5.070	15.284	.106	1.722	2.420	3.120	3.826	11.535	1.625	1.316

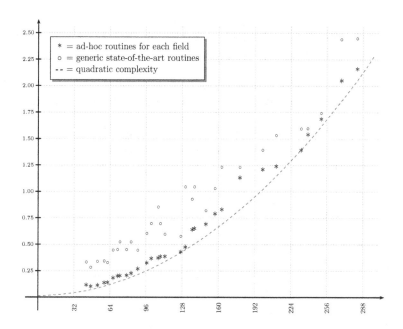

Fig. 1. Field multiplication performance on the 1.5 GHz PowerPC G4 (μsec)

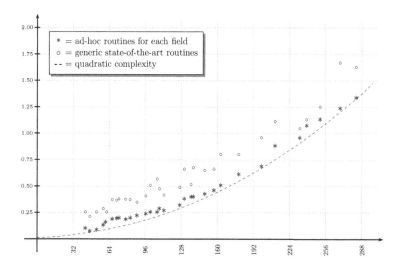

Fig. 2. Field multiplication performance on the 1.83 GHz Intel Core 2 Duo (μsec)

Figures 1 and 2 represent the results of the single multiplications. The timings of our per-field ad-hoc routines are compared to those of our generic implementation. The parabola arcs represent interpolations of the form $c_1 \cdot n^2 + c_0$ bit operations using our best performance values.

The way we implemented inversion (§ 3.5) allows us to get extra performance from CPUs with several registers. For several input sizes, most of the computation can take place only between registers on these processors, without accessing external memory except to load the inputs and to store the final result – thus, a slow memory bus or a high level one cache latency are not big problems. This is reflected in the exceptionally low I/M ratio on the PowerPC: for inputs of up to 6 words all the operands of the EEA fit in the registers. For longer inputs the registers no longer suffice, the compiler must store some partial data in the main memory (as confirmed by disassembly of the compiled code), and a "bump" in inversion performance occurs.

The multiplication routines make extensive use of tables of precomputations, hence they are more penalized by slow memory bus or high level one cache latency than by register paucity. This is reflected in the fact the Intel Core 2 Duo offers better multiplication performance than the PowerPC, especially for larger fields, that use larger tables. Some of the deficiencies of older CISC architectures, such as the small register set, are now mitigated by register renaming and wide-execution units, but their impact is still noticeable. This can be seen, for example, by the higher I/M ratio on the Intel Core 2 Duo.

The choice of trinomials, pentanomials or eptanomials is reflected in the timings. The ratio Sqr/Mul is higher when pentanomials or eptanomials are used because the reduction is significantly slower than it would have been if an irreducible trinomial of the same degree had existed. The reduction has a smaller impact on field multiplication than on field squaring. In the case of generic routines, the variations due to the choice of polynomials are bigger, and it is easy to recognize that the zigzagging multiplication performance curve follows the shape of a staircase over which pentanomial-induced "wedges" are placed. Apart from this "zigzagging", it is difficult to find more patterns in the performance gap between specialized routines and generic ones. It seems to remain (on average) similar for small and larger fields, with maybe only a small linear component in the bit size. This suggests that most of the costs we eliminated were roughly constant, and therefore had much bigger impact on the performance of smaller fields.

4.1   Comparisons with Other Literature

It can be difficult to compare our results with the literature, since in most cases only a few fields were implemented and benchmarked, and usually only the larger ones (for EC) or the smaller ones (for HEC), but not both.

Our goal was to get performance curves that would resemble the parabola arcs expected from theory instead of the usual "broken staircases" (which can only be shown with the implementation of several fields), and at the same time matching or improving on the best results for individual fields that are scattered in the literature. We wanted to show that in even characteristic the granularity of the architecture does not play the same crucial role as in the prime field case. when, say comparing performances of different types of curves over fields of various sizes. For this reason we did not use vector extensions like MMX,

SSE, or Altivec: these only "change the granularity", but results would have been similar.

In [15], two sets of implementations are reported for binary fields of 32, 40 and 47 bits, one on an ARM 7 and the other on a 1.8 GHz Pentium 4. For the canonical trio of fields of 163, 233 and 283 bits, a few more papers [21,6,4] are available. These timings can be seen in Table 3. The best timings in [13] essentially agree with those in [6] for the Pentium II and are slighly slower than those in [4] for the Sparc.

In [8], the timing for the multiplications must be adjusted by a factor of 10 so they correspond to the overall timings of the elliptic curve operations. This is most likely due to a simple error in notation. The results as they are stated, suitably scaled, mean that King's code would perform a multiplication about 8 times faster than we and [6] do, but a whole scalar multiplication 20% slower.

Scaling the 47-bit field performance on the 80 Mhz ARM 7, we see that our routines are about 6 times more efficient, even though in this case a factor at least 2 is due to the processor architecture. The 1.8 Pentium 4 performance from [15] would scale to about 0.1008 μsec at 3.0 GHz – our code performs a 47-bit finite field multiplication in 0.079 μsec on a 3.0 GHz Pentium 4. Comparable scaled performance of our code with those from [6,4], two reference papers, can be noted (the timings of their gcc-compiled non-mmx versions are given).

Table 3. Timings in μsecs of field operations in other papers

Field	[15]: ARM 7, 80 MHz			[15]: Intel P4, 1.8 GHz		
Size	Mult.	Inv.	Ratio	Mult.	Inv.	Ratio
32	2.6	26.8	10.16	0.168	1.650	9.82
40	7.3	49.2	6.73	0.413	2.519	6.04
47	7.3	77.5	10.5	0.402	3.752	9.33

Field	[6]: Intel Pentium II 400 Mhz, gcc			[4]: Intel Pentium 3 800 Mhz, gcc			[4]: Sparc 500 Mhz, gcc			[21]: 900 Mhz Sparc	[21]: 1 Ghz Intel
Size	Mult.	Inv.	Ratio	Mult.	Inv.	Ratio	Mult.	Inv.	Ratio	Mult.	Mult.
163	3.00	30.99	10.33	1.8	12.0	6.67	1.9	16.8	8.85	2.9	2.3
233	5.07	53.22	10.49	3.0	21.9	7.3	4.0	36.8	9.2	3.1	3.2
283	6.23	70.32	11.29							5.2	4.9

4.2 Application to Curve-Based Cryptography

An implementation of EC and of HEC of genus up to four has been written. The aim was to compare the performance of curves of different genera offering the same security level. Therefore, under consideration of the best attacks for each type of curve (see [2, Chs. 19, 20, 21 and 23]), we tried to find suitable quadruplets of fields to use as fields of definition for EC and HEC of genera 2, 3 and 4 at each of the chosen security levels. Due to the irregularity of the distribution of the primes, we had to choose security levels (expressed in bit-equivalents for EC) that permitted us to find matches for at least three curves, with tolerances of at most 2% (in bits) of security level between the curves in each level. The details of this implementation, which include several new explicit formulæ, will be presented elsewhere.

Figure 3 shows the timings for each curve type and security level on the 1.5 GHz PowerPC G4. The scalar multiplication algorithm used is a windowed method. Precomputation times are included in the timings. For EC, the best choice of coordinate system proved to be mixed affine/López-Dahab coordinates. For higher genus HEC, affine coordinates are used.

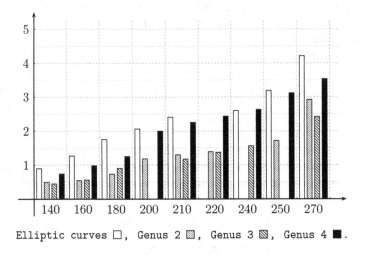

Elliptic curves □, Genus 2 ▨, Genus 3 ▧, Genus 4 ■.

Fig. 3. Scalar multiplication timings (μsec) for curves of different genera at various security levels, expressed in bit-equivalents for elliptic curves

The performance of curves of genus four is comparable to that of EC, sometimes better. Curves of genus two and three have similar performance, with genus three winning in some cases despite the use of the Lange-Stevens genus two doubling [11], and both perform better than curves of genus one and four.

We mention the fact that for curves of genus 3 and 4 the gain obtained by using sequential multiplications is significant, often around 15% and 20% respectively. On the other hand, there are no gains for EC in affine coordinates and a very small gain using López-Dahab coordinates (less than 1%.) For curves of genus two, the improvement is around 2%. In other words, without sequential multiplications the relative performance of genus 3 and 4 curves would have been much worse.

In recent literature there have been a few remarkable implementations of low genus HEC, such as [23,15], where a 1.8 GHz Pentium 4 was used. A comparison of the results must be taken with a pinch of salt, but our results seem to represent an improvement in the performance of HEC. In [23,15] genus 2 and 3 HEC roughly match EC, while we show significant gains. Their implementation of genus 4 HEC is about 4 times slower than EC, but our results show that performance is close.

4.3 Conclusions and Perspectives

Our implementation "smooths" the performance of binary fields as a function of the extension degree: jumps corresponding to crossing granularity boundaries and erratic behavior depending on modular reduction mostly disappear. The shape of the graph of multiplication timings is quite close to the parabola arc given by the theoretical quadratic bit-complexity. For small fields, the performance gain due to our optimizations can be close to a factor 4 for multiplication and even around 10 for inversion, with respect to state-of-the-art generic libraries.

Sequences of several multiplications with a common multiplicand can be implemented faster by reusing the precomputations. We assess the resulting gains. These vary wildly but we can still see that 2, 3, 4 and 5 multiplications with a common multiplicand can be performed at roughly the cost of 1.75, 2.35, 3 and 3.65 single multiplications respectively.

This prompts the development of new explicit formulae for arithmetic on elliptic and hyperelliptic curves that take into account these routines and ratios. In fact, we can already report on an implementation of curve-based cryptographic primitives that depends in a significant way on our optimized routines.

Acknowledgements. Parts of this work were done while the second author was at the Department of Combinatorics and Optimization, University of Waterloo, Canada, and at the Fields Institute, Toronto, Canada, and while the first author was visiting the same two institutions.

References

1. Avanzi, R.: Aspects of hyperelliptic curves over large prime fields in software implementations. In: Joye, M., Quisquater, J.-J. (eds.) CHES 2004. LNCS, vol. 3156, pp. 148–162. Springer, Heidelberg (2004)
2. Avanzi, R., Cohen, H., Doche, C., Frey, G., Lange, T., Nguyen, K., Vercauteren, F.: The Handbook of Elliptic and Hyperelliptic Curve Cryptography. CRC Press, Boca Raton (2005)
3. Fan, X., Wollinger, T., Wang, Y.: Efficient Doubling on Genus 3 Curves over Binary Fields. IACR ePrint 2005/228
4. Fong, K., Hankerson, D., López, J., Menezes, A.: Field Inversion and Point Halving Revisited. IEEE Trans. Computers 53(8), 1047–1059 (2004)
5. Guyot, C., Kaveh, K., Patankar, V.M.: Explicit algorithm for the arithmetic on the hyperelliptic Jacobians of genus 3. J. Ramanujan Math. Soc. 19(2), 75–115 (2004)
6. Hankerson, D., López-Hernandez, J., Menezes, A.: Software Implementation of Elliptic Curve Cryprography over Binary Fields. In: Paar, C., Koç, Ç.K. (eds.) CHES 2000. LNCS, vol. 1965, pp. 1–24. Springer, Heidelberg (2000)
7. Karatsuba, A., Ofman, Y.: Multiplication of Multidigit Numbers on Automata. Soviet Physics - Doklady 7, 595–596 (1963)
8. King, B.: An Improved Implementation of Elliptic Curves over $GF(2^n)$ when Using Projective Point Arithmetic. In: Vaudenay, S., Youssef, A.M. (eds.) SAC 2001. LNCS, vol. 2259, pp. 134–150. Springer, Heidelberg (2001)

9. Knuth, D.: The Art of Computer Programming, 3rd edn. Seminumerical Algorithms, vol. 2. Addison Wesley Longman, Redwood City (1998)

10. Lange, T.: Efficient Arithmetic on Genus 2 Hyperelliptic Curves over Finite Fields via Explicit Formulae. Cryptology ePrint Archive, Report 2002/121

11. Lange, T., Stevens, M.: Efficient doubling for genus two curves over binary fields. In: Handschuh, H., Hasan, M.A. (eds.) SAC 2004. LNCS, vol. 3357, pp. 170–181. Springer, Heidelberg (2005)

12. Lim, C., Lee, P.: More flexible exponentiation with precomputation. In: Desmedt, Y.G. (ed.) CRYPTO 1994. LNCS, vol. 839, pp. 95–107. Springer, Heidelberg (1994)

13. López, J., Dahab, R.: High-speed software multiplication in $\mathbb{F}_{2^m}$. In: Roy, B., Okamoto, E. (eds.) INDOCRYPT 2000. LNCS, vol. 1977, pp. 203–212. Springer, Heidelberg (2000)

14. Pelzl, J., Wollinger, T., Guajardo, J., Paar, C.: Hyperelliptic curve cryptosystems: closing the perfomance gap to elliptic curves (Update). IACR ePrint (2003)/026

15. Pelzl, J., Wollinger, T., Paar, C.: Low cost Security: Explicit Formulae for Genus 4 Hyperelliptic Curves. In: Matsui, M., Zuccherato, R.J. (eds.) SAC 2003. LNCS, vol. 3006, pp. 1–16. Springer, Heidelberg (2004)

16. Pippenger, N.: On the evaluation of powers and related problems (preliminary version). 17th Annual Symp. on Foundations of Comp. Sci, pp. 258–263. IEEE Computer Society, Los Alamitos (1976)

17. Schönhage, A., Grotefeld, A.F.W., Vetter, E.: Fast Algorithms–A Multitape Turing Machine Implementation. BI Wissenschafts-Verlag, Mannheim (1994)

18. Schroeppel, R., Orman, H., O'Malley, S., Spatscheck, O.: Fast key exchange with elliptic curve systems. In: Coppersmith, D. (ed.) CRYPTO 1995. LNCS, vol. 963, pp. 43–56. Springer, Heidelberg (1995)

19. Shoup, V.: NTL: A Library for doing number theory. URL: http://shoup.net/ntl/

20. Sun Corporation's Elliptic Curve Cryptography contributions to OpenSSL. Available at http://research.sun.com/projects/crypto/

21. Weimerskirch, A., Stebila, D., Shantz, S.C.: Generic GF(2^m) Arithmetic in Software and its Application to ECC. In: Safavi-Naini, R., Seberry, J. (eds.) ACISP 2003. LNCS, vol. 2727, pp. 79–92. Springer, Heidelberg (2003)

22. Wollinger, T.: Software and Hardware Implementation of Hyperelliptic Curve Cryptosystems. Ph.D. Thesis, Ruhr-Universität Bochum, Germany (2004)

23. Wollinger, T., Pelzl, J., Paar, C.: Cantor versus Harley: Optimization and Analysis of Explicit Formulae for Hyperelliptic Curve Cryptosystems. To appear in IEEE Transactions on Computers

Disclaimer: *The information in this document reflects only the authors' views, is provided as is and no guarantee or warranty is given that the information is fit for any particular purpose. The user thereof uses the information at its sole risk and liability.*

Software Implementation of Arithmetic in $\mathbb{F}_{3^m}$

Omran Ahmadi[1], Darrel Hankerson[2], and Alfred Menezes[3]

[1] Dept. of Electrical and Computer Engineering, University of Toronto
oahmadid@comm.utoronto.ca
[2] Dept. of Mathematics and Statistics, Auburn University
hankedr@auburn.edu
[3] Dept. of Combinatorics & Optimization, University of Waterloo
ajmeneze@uwaterloo.ca

Abstract. Fast arithmetic for characteristic three finite fields $\mathbb{F}_{3^m}$ is desirable in pairing-based cryptography because there is a suitable family of elliptic curves over $\mathbb{F}_{3^m}$ having embedding degree 6. In this paper we present some structure results for Gaussian normal bases of $\mathbb{F}_{3^m}$, and use the results to devise faster multiplication algorithms. We carefully compare multiplication in $\mathbb{F}_{3^m}$ using polynomial bases and Gaussian normal bases. Finally, we compare the speed of encryption and decryption for the Boneh-Franklin and Sakai-Kasahara identity-based encryption schemes at the 128-bit security level, in the case where supersingular elliptic curves with embedding degrees 2, 4 and 6 are employed.

1 Introduction

Pairing-based cryptographic protocols are realized using algebraic curves of low-embedding degree. Several families of low-embedding degree elliptic curves have been considered, including supersingular curves with embedding degrees 2, 4, and 6, and ordinary curves with embedding degrees 2 [29], 6 [23], and 12 [5]. The family of supersingular elliptic curves with embedding degree 6 is defined over characteristic three finite fields $\mathbb{F}_{3^m}$. Consequently, the software and hardware implementation of arithmetic in these fields has been intensively studied in recent years [16,25,13,20,14].

The elements of $\mathbb{F}_{3^m}$ can be represented using a polynomial basis or a normal basis. We present some new structure results for Gaussian normal bases of $\mathbb{F}_{3^m}$, and use these results to devise faster multiplication algorithms. Our implementation on a Pentium 4 machine shows that our fastest algorithm for normal basis multiplication in $\mathbb{F}_{3^{239}}$ is about 50% faster than standard Ning-Yin multiplication [24], and about 4.4 times faster than the Ning-Yin implementation reported by Granger, Page and Stam [14]. Our experiments also suggest that the comb method for polynomial basis multiplication [22] (perhaps combined with shallow-depth Karatsuba-like techniques) is faster than Karatsuba multiplication. In particular, our implementation for polynomial basis multiplication in $\mathbb{F}_{3^{239}}$ is about 4.6 times faster than that reported in [14]. We conclude, as in [14], that polynomial bases are preferred over normal bases for the software implementation of characteristic three field arithmetic.

C. Carlet and B. Sunar (Eds.): WAIFI 2007, LNCS 4547, pp. 85–102, 2007.

A recent IETF draft standard [8] for identity-based encryption (IBE) mandates use of supersingular elliptic curves over prime fields — these curves have embedding degree 2. We compare the speed of encryption and decryption for the Boneh-Franklin [7] and Sakai-Kasahara [27] identity-based encryption schemes, when the underlying elliptic curves are supersingular and defined over a prime field (embedding degree 2), a characteristic two finite field (embedding degree 4), and a characteristic three finite field (embedding degree 6). We focus our attention on 1536-bit prime fields $\mathbb{F}_p$, the characteristic two field $\mathbb{F}_{2^{1223}}$, and the characteristic three field $\mathbb{F}_{3^{509}}$. Each of these choices achieves the 128-bit security level in the sense that the best attacks known on the discrete logarithm problem in the extension fields $\mathbb{F}_{p^2}$, $\mathbb{F}_{2^{4 \cdot 1223}}$, and $\mathbb{F}_{3^{6 \cdot 509}}$ have running time approximately 2^{128} [21].

We acknowledge that the Barreto-Naehrig (BN) ordinary elliptic curves [5] over 256-bit fields $\mathbb{F}_p$ with embedding degree 12 are ideally suited for pairing applications at the 128-security level, and can be expected to yield faster implementations than supersingular elliptic curves especially when the Ate pairing algorithm [18] is employed. However, some people are reluctant to use the BN curves because recent work by Schirokauer [28] has raised the possibility that the special number field sieve may be effective for computing discrete logarithms in $\mathbb{F}_{p^{12}}$. Furthermore, in some cases expensive hashing or the absence of an efficiently-computable isomorphism ψ (cf. [12]) may be a concern. Thus it is worthwhile to consider the relative merits of supersingular elliptic curves in pairing-based cryptography.

The remainder of the paper is organized as follows. Methods for performing arithmetic in $\mathbb{F}_{3^m}$ using a polynomial basis representation are reviewed in §2. Our structure results for Gaussian normal bases are developed in §3. In §4, we present our implementation results for $\mathbb{F}_{3^m}$. Estimates for Boneh-Franklin and Sakai-Kasahara IBE are given in §5. Summary conclusions appear in §6.

2   Polynomial Bases for $\mathbb{F}_{3^m}$

Elements $a \in \mathbb{F}_{3^m}$ can be regarded as polynomials $a = a_{m-1}x^{m-1} + \cdots + a_0$ where $a_i \in \mathbb{F}_3$ and arithmetic is performed modulo an irreducible polynomial f of degree m. We associate a with the vector of coefficients $(a_{m-1}, \ldots, a_0)$.

Harrison, Page, and Smart [16] considered two coefficient representations suitable for implementation. Their "Type II" representation is closer to common techniques used for binary fields. Each coefficient $a_i \in \mathbb{F}_3$ is represented uniquely in $\{0, 1, -1\}$ using a pair (a_i^0, a_i^1) of bits, where $a_i = a_i^0 - a_i^1$ and not both bits are 1. Elements a are represented by vectors $a^j = (a_{m-1}^j, \ldots, a_0^j)$, $j \in \{0, 1\}$. Addition $c = a + b$ is

$$t \leftarrow (a^0 \vee b^1) \oplus (a^1 \vee b^0), \quad c^0 \leftarrow (a^1 \vee b^1) \oplus t, \quad c^1 \leftarrow (a^0 \vee b^0) \oplus t.$$

The seven operations involve only bitwise "or" ($\vee$) and "exclusive-or" ($\oplus$), and it is easy to order the instructions to cooperate with processor pipelining. Negation is $-a = (a^1, a^0)$.

The analysis and experimental data in [16] strongly favour the Type II approach. This also has the advantage that representations are unique, and common techniques employed for multiplication in binary fields have direct analogues. Hence, as in [14], we consider only this representation.

2.1 Field Multiplication

Techniques for binary field multiplication that extend directly to the representation $a = (a^0, a^1)$ include table lookup and "comb" methods [22], possibly combined with Karatsuba-like techniques to reduce the number of word-level multiplications at the expense of more additions.

Briefly, a common case of the comb [22] calculates ab with a single table of precomputation containing ub for polynomials u of degree less than w for some small w (e.g., $w = 4$). The words of a are then "combed" w bits at a time to select the appropriate precomputed value to add at the desired location of the accumulator.

In binary fields, our experience and analysis suggests the comb method will be among the fastest on common processors. We also found this to be the case for characteristic three finite fields, contrary to the findings in [14] where the Karatsuba-Ofman style approach was the fastest choice. Indeed, the times in Table 1 (see §4) show that a comb method is dramatically faster on the processor used in [14] (an Intel Pentium 4).

The comb method, while not difficult to implement, requires attention to processor and compiler characteristics [15, Chapter 5]. It appears necessary to code a multiplication for each field size (in number of words to hold an element). With fields of interest in methods for elliptic curves, Karatsuba-Ofman can be used at shallow depth to reduce the multiplication to a few comb sizes, so code expansion can be controlled.

2.2 Cubing and Cube Roots

Since $a^3 = (\sum a_i x^i)^3 = \sum a_i x^{3i}$, cubing is an inexpensive operation when performed using an expansion table to "thin" the coefficients followed by a reduction. This is analogous to squaring in binary fields. The cost of cube roots depends on the reduction polynomial. Since the operation is linear, write

$$a^{1/3} = \sum_{i=0}^{\lceil m/3 \rceil - 1} a_{3i} x^i + x^{1/3} \left(\sum_{i=0}^{\lceil m/3 \rceil - 1} a_{3i+1} x^i \right) + x^{2/3} \left(\sum_{i=0}^{\lceil m/3 \rceil - 1} a_{3i+2} x^i \right).$$

If $x^{1/3}$ and $x^{2/3}$ are precomputed, then $a^{1/3}$ can be found using a table-lookup method to extract coefficients from a followed by two multiplications where each has an operand that has all nonzero coefficients in the lower third of the vector representation. If the reduction polynomial can be chosen so that $x^{1/3}$ has only a few nonzero terms, then roots are especially inexpensive.

Example 1. Consider $\mathbb{F}_3[x]/(f)$ where $f(x) = x^m + f_k x^k + f_0$ is irreducible. If $m \equiv k \pmod 3$, then $x^{1/3}$ has $4 - (m \bmod 3)$ nonzero terms [3]. For $m = 239$ (of interest in [14], for example), the polynomial $f(x) = x^m - x^k + 1$ for $k = 5$ is irreducible and has $m \equiv k \equiv 2$ so that $x^{1/3} = -x^{80} + x^2$ has just two terms. However, [14] selected $f(x) = x^{239} + x^{24} - 1$, and [1] shows that $x^{1/3}$ will have $\lceil \frac{2m-1}{3k} \rceil + \lceil \frac{2m-1-k}{3k} \rceil + \lceil \frac{2m-1-2k}{3k} \rceil + j = 20 + j$ nonzero terms where $0 \le j \le 3$ (in fact, there are 23 nonzero terms in $x^{1/3}$ and 9 nonzero terms in $x^{2/3}$).

There are m where no irreducible trinomial yields sparse $x^{1/3}$; an almost worst case is illustrated by $m = 163$ where $x^{1/3}$ has $m - 1$ nonzero terms. However, there is an irreducible tetranomial where $x^{1/3}$ has five terms, and an irreducible pentanomial where $x^{1/3}$ has 3 terms. The case $m = 509$ also has no trinomial giving a low-weight $x^{1/3}$. There is a tetranomial giving a 17-term $x^{1/3}$ and a pentanomial where $x^{1/3}$ has 5 terms. To minimize the combined hamming weight of $x^{1/3}$ and $x^{2/3}$, a pentanomial can be chosen (e.g., $x^{509} + x^{294} - x^{215} + x^{79} - 1$) where $x^{1/3}$ has 6 terms and $x^{2/3}$ has 3 terms.

2.3 Inversion

Euclidean algorithm variants for inversion in binary fields adapt fairly directly to the representation $a = (a^0, a^1)$. As an alternative, the inverse can be found by exponentiation. Although [14] remark that "one cannot use Itoh-Tsujii type methods to reduce the cost", in fact such methods apply. To see this, note that

$$a^{3^k - 1} = \begin{cases} a^2 (a^{3^{k-1}-1})^3, & k \text{ odd,} \\ (a^{3^{k/2}-1})^{3^{k/2}+1} = c \cdot c^{3^{k/2}}, & k \text{ even,} \end{cases}$$

where $c = a^{3^{k/2}-1}$. Then $a^{-1} = a^{3^m - 2} = a(a^{3^{m-1}-1})^3$. Since a^2 can be calculated once, the cost of inversion by a recursive approach is $\lfloor \log(m-1) \rfloor + \text{wt}(m-1) + 1$ field multiplications (where wt is Hamming weight) along with many cubings. The technique is most applicable when $m - 1$ has low weight and cubings are extremely cheap (as in normal basis representations). In a polynomial basis, it is expected to be more expensive than inversion based on the Euclidean algorithm, although it has the advantage that it requires very little additional code over multiplication (and is thus especially suitable for hardware).

3 Normal Bases

Let α generate a normal basis N for $\mathbb{F}_{q^m}$ over $\mathbb{F}_q$, and let $\alpha_i = \alpha^{q^i}$ for $0 \le i \le m - 1$. Let $\alpha_i \alpha_j = \sum_{k=0}^{m-1} t_{ij}^{(k)} \alpha_k$, where $t_{ij}^{(k)} \in \mathbb{F}_q$. For $a \in \mathbb{F}_{q^m}$ let $a_i \in \mathbb{F}_q$ be defined by $a = \sum_{i=0}^{m-1} a_i \alpha_i$ and let $A = (a_0, \ldots, a_{m-1})$. Then $c = ab$ is given by $c_k = \sum_{i,j} a_i b_j t_{ij}^{(k)} = A T_k B'$ where the collection of matrices $\{T_k = (t_{ij}^{(k)})\}$ is known as a multiplication table for $\mathbb{F}_{q^m}$ over $\mathbb{F}_q$. It is known that $t_{ij}^{(k)} = t_{i-k,j-k}^{(0)}$ and hence $c_k = A^{(k)} T_0 B^{(k)'}$ where $A^{(k)}$ denotes the left cyclic shift of the vector A by k positions.

Let $\alpha\alpha_i = \sum_{j=0}^{m-1} t_{ij}\alpha_j$, for $0 \le i \le m-1$. Then $t_{ij}^{(k)} = t_{i-j,k-j}$, and $T = (t_{ij})$ and T_0 have the same number of nonzero entries, known as the *complexity* C_N of the normal basis N. It is known that $C_N \ge 2m-1$, and N is said to be *optimal* if $C_N = 2m-1$.

3.1 Gauss Periods

Let k, m be such that $r = mk+1$ is prime and $\gcd(r, q) = 1$. Let β be a primitive r-th root of unity in an extension of $\mathbb{F}_q$, and let γ be a primitive k-th root of unity in $\mathbb{Z}_r$. The element $\alpha = \sum_{j=0}^{k-1} \beta^{\gamma^j}$ is a *Gauss period of type* (m, k) for $\mathbb{F}_q$. In fact, $\alpha \in \mathbb{F}_{q^m}$ and is normal if and only if $\langle q^i \gamma^j \rangle = \mathbb{Z}_r^*$. In this case, every element of $\mathbb{Z}_r^*$ can be written uniquely as $q^i \gamma^j$ where $0 \le i \le m-1$ and $0 \le j \le k-1$ [2]. In the following, $\mathrm{Tr} : a \mapsto \sum_{i=0}^{m-1} a^{3^i}$ is the trace function of $\mathbb{F}_{3^m}$ over $\mathbb{F}_3$.

Lemma 1. *If α is a Gauss period and is normal, then* $\mathrm{Tr}(\alpha) = -1$.

Proof. In the notation of this section,

$$\mathrm{Tr}(\alpha) = \sum_{i=0}^{m-1} \left(\sum_{j=0}^{k-1} \beta^{\gamma^j} \right)^{q^i} = \sum_{i=0}^{m-1}\sum_{j=0}^{k-1} \beta^{\gamma^j q^i} = \sum_{\ell=1}^{r-1} \beta^\ell$$

since $\langle q^i \gamma^j \rangle = \mathbb{Z}_r^*$. The last sum is $(\beta^r - 1)/(\beta - 1) - 1 = -1$. □

Since $\mathrm{Tr}(\alpha) = \sum_{i=0}^{m-1} \alpha_i = -1$, the normal basis representation of the identity element in $\mathbb{F}_{3^m}$ is the vector all of whose entries are -1.

3.2 Complexity and Structure for T When $q = 3$

Fix $q = 3$ and assume that α is a Gauss period of type (m, k) and that α is normal. Note that m odd implies that k is even. Since our main interest is prime $m > 2$, we shall henceforth assume k is even. The normal basis N generated by α is called a *Gaussian normal basis (GNB)* for $\mathbb{F}_{3^m}$, and is said to be of *type* k.

We are interested in the complexity of the multiplication and in the "structure" of the multiplication matrix T, in particular, the number of entries that are -1. As a direct consequence, we will obtain results for T that are of practical interest, including a decomposition that accelerates the multiplication significantly when k is 2 or 4.

We define the complexity C_i of the ith row of the matrix T to be its number of nonzero entries. Notice that by definition $C_N = \sum_{i=0}^{m-1} C_i$. Now we may write

$$\sum_{j=0}^{m-1} t_{ij}\alpha_j = \alpha\alpha_i = \alpha\alpha^{3^i} = \sum_{s=0}^{k-1}\sum_{\ell=0}^{k-1} \beta^{\gamma^\ell(1+\gamma^s 3^i)} = \sum_{s=0}^{k-1} f(i, s), \tag{1}$$

where $f(i, s)$ is defined to be $\sum_{\ell=0}^{k-1} \beta^{\gamma^\ell(1+\gamma^s 3^i)}$. Since α is a Gauss period which is normal, there is a unique pair (i, s) such that $1 + \gamma^s 3^i = 0$, namely $(i = 0,$

$s = k/2$), and then $f(0, k/2) = k$. If $(i, s) \neq (0, k/2)$, then $1 + \gamma^s 3^i = \gamma^t 3^j$ for some t and j, and then $f(i, s) = \alpha_j$. Hence, if $i \geq 1$, then $C_i \leq k$. These observations lead to well-known upper bounds on C_N.

Theorem 1 ([6, Theorem 5.5]). $C_N \leq (m-1)k + m = (k+1)m - k$ and $C_N \leq (m-1)k + k - 1 = mk - 1$ if $k \equiv 0 \pmod 3$. If $k = 2$, then $C_N = 3m - 2$.

The next result establishes some lower bounds on C_N. Further, the number of -1 entries in the multiplication matrix is given for some cases; this number is of practical interest because it can affect implementation optimizations.

Theorem 2. *Suppose that α is a Gauss period of type (m, k) for $\mathbb{F}_3$, k is an even number, and α is normal. Let f be defined by (1). Then a lower bound on the complexity of the Gaussian normal basis generated by α is*

$$C_N \geq \begin{cases} mk - 1 - (k/2 - 1)(k - 1), & \text{if } k \equiv 0 \pmod 3, \\ (k+1)m - k - 1 - (k/2 - 1)(k - 2), & \text{if } k \equiv 1 \pmod 3, \\ (k+1)m - k - (k/2 - 1)(k - 1), & \text{if } k \equiv 2 \pmod 3. \end{cases}$$

Furthermore, the lower bound is achieved if and only if

(i) *there are no $i \geq 1$ and distinct s_1, s_2, s_3, s_4 such that $f(i, s_1) = \cdots = f(i, s_4)$;*

(ii) *if $k \equiv 0 \pmod 3$, then there are no distinct s_1, s_2, s_3, s_4 such that $f(0, s_1) = \cdots = f(0, s_4)$; and*

(iii) *if $k \not\equiv 0 \pmod 3$, then there are no distinct s_1, s_2, s_3 such that $f(0, s_1) = f(0, s_2) = f(0, s_3)$.*

Also if there are no i, $0 \leq i \leq m - 1$, and distinct s_1, s_2, s_3 such that $f(i, s_1) = f(i, s_2) = f(i, s_3)$, then the number of -1 entries in the matrix T is

$$\begin{cases} (k/2 - 1)(k - 1), & \text{if } k \equiv 0 \pmod 3, \\ m + (k/2 - 1)(k - 1) - k + 1, & \text{if } k \equiv 1 \pmod 3, \\ (k/2 - 1)(k - 2) + 1, & \text{if } k \equiv 2 \pmod 3. \end{cases}$$

Proof. For $i = 0, \ldots, m - 1$, let E_i denote the number of triples (i, s', s'') such that $s' \neq s''$ and $f(i, s') = f(i, s'')$. We have $C_i \geq k - E_i/2$ for $i \geq 1$. This is because each $f(i, s)$ is a basis element for every s and $i \geq 1$. Moreover, if there exist distinct $s_1, \ldots, s_\ell$ such that $f(i, s_1) = \cdots = f(i, s_\ell)$ and $\ell \geq 2$, then the complexity of row i is decreased by 1 if $\ell = 2$ and at most by ℓ if $\ell \geq 3$ while E_i is increased by $\ell(\ell - 1)$. Also it is easy to see that $C_i = k - E_i/2$ if and only if there are no distinct s_1, s_2, s_3, s_4 such that $f(i, s_1) = \cdots = f(i, s_4)$. In the following we obtain some inequalities involving C_0 which, together with the inequalities obtained for C_i, $i = 1, \ldots, m - 1$, will allow us to establish a lower bound for C_N.

Let $i = 0$. Now, if $1 \leq s < k/2$ and $f(0, s) = \alpha_j$, then $1 + \gamma^s = \gamma^t 3^j$ for some t. Hence $1 + \gamma^{k-s} = 1 + \gamma^{-s} = \gamma^{-s}(\gamma^s + 1) = \gamma^{t-s} 3^j$, and so $f(0, k - s) = f(0, s)$. Furthermore, $f(0, k/2) = k$ and $f(0, 0)$ is a basis element, whence $f(0, 0) \neq f(0, k/2)$. We have the following three cases:

(A) If $k \equiv 0 \pmod{3}$, then we claim that $C_0 + E_0/2 \geq k-1$. As we mentioned above the sum (1) for $i = 0$ produces

$$f(0,1) = f(0, k-1), \ \ldots, \ f(0, k/2-1) = f(0, k/2+1), \ f(0,0), \ f(0, k/2) = k.$$

If $f(0,0), f(0,1), \ldots, f(0, k/2-1)$ are pairwise distinct, then we have $C_0 + E_0/2 = k - 1$, and if they are not pairwise distinct then C_0 will decrease while there wll be an increase in E_0. It is easy to see that the increment in $E_0/2$ will be greater than or equal to the decrement in C_0. From this the claim follows. Also it is easy to see that $C_0 + E_0/2 = k - 1$ if and only if there are no distinct s_1, s_2, s_3, s_4 such that $f(0, s_1) = \cdots = f(0, s_4)$.

(B) If $k \equiv 1 \pmod{3}$, then $\mathrm{Tr}(\alpha) = -1 \equiv -k \pmod{3}$. From the fact that the trace is the sum of the basis elements and an argument similar to above we obtain $C_0 + E_0/2 \geq m + k/2 - 2$. Furthermore we see that $C_0 + E_0/2 = m + k/2 - 2$ if and only if there are no distinct s_1, s_2, s_3 such that $f(0, s_1) = f(0, s_2) = f(0, s_3)$.

(C) If $k \equiv 2 \pmod{3}$, then $\mathrm{Tr}(\alpha) = -1 \equiv k \pmod{3}$. Similar arguments as above lead to $C_0 + E_0/2 \geq m$. Again it is easily seen that $C_0 + E_0/2 = m$ if and only if there are no distinct s_1, s_2, s_3 such that $f(0, s_1) = f(0, s_2) = f(0, s_3)$.

Using the inequalities we have obtained for $C_0, C_1, \ldots, C_{m-1}$, it suffices to compute $E_0 + E_1 + \cdots + E_{m-1}$ in order to obtain a lower bound for $C_N = \sum_{i=0}^{m-1} C_i$. This is done in the following through a double counting argument.

A triple (i, s', s'') for E_i exists if and only if there is some j with

$$1 + \gamma^{s'} 3^i \equiv \gamma^j (1 + \gamma^{s''} 3^i) \pmod{r} \quad \text{or} \quad \gamma^{s''} 3^i (\gamma^{s'-s''} - \gamma^j) \equiv \gamma^j - 1 \pmod{r}.$$

Now $\gamma^j - 1$ and $\gamma^{s'-s''} - \gamma^j$ cannot both be zero because otherwise $s' = s''$. For a given j and $s' - s''$, there is either no solution or exactly one solution (i, s''). Solutions are obtained for $0 < j < k$ and $s' - s'' \notin \{0, j\}$, giving $(k-1)(k-2)$ triples (i, s', s'').

The claim about the count of -1 entries follows from the fact that $\sum_{i=0}^{m-1} E_i = (k-1)(k-2)$ and by examining the first row of the matrix T. $\qquad\square$

Corollary 1. If $k = 4$, then $C_N = 5m - 7$.

Proof. We verify that conditions (i) and (iii) of Theorem 2 are satisfied. Since $k = 4$, we have $f(0,1) = f(0,3)$ and $E_0 + \cdots + E_{m-1} = 6$. Thus there are no $i \geq 1$ and distinct s_1, s_2, s_3 such that $f(i, s_1) = f(i, s_2) = f(i, s_3)$. Suppose now that $f(0,0) = f(0,1)$. Then $1 + \gamma^0 = \gamma^\ell (1 + \gamma)$ for some ℓ. Squaring both sides gives $4 = \pm 2\gamma$, and hence $2 = \pm\gamma$. Squaring again yields $4 = \gamma^2 = -1$ which is impossible if $m > 1$. Hence $f(0,0) \neq f(0,1)$ and the result follows. $\qquad\square$

The lower bound of Theorem 2 is not always met with equality when $k \geq 6$. A computer search found that the values (m, k) for which $k \leq 26$ is even, $m \in [k, 1000]$ is prime, a type k GNB for $\mathbb{F}_{3^m}$ exists, but the lower bound of Theorem 2 is not met with equality are $(17, 14)$, $(53, 20)$, $(31, 22)$, and $(103, 24)$.

The proof of Theorem 2 yields "structure" results concerning the matrix T_0 that can lead to significant computational savings when k is 2 or 4. The basic idea is that T_0 can be written as $P+Q$ where the total number of nonzero entries is essentially unchanged, but the multiplication $A^{(\ell)}PB^{(\ell)'}$ is independent of ℓ. The complexity of the multiplication is then essentially the number of nonzero terms in Q.

This type of decomposition was shown in $\mathbb{F}_{2^m}$ for optimal normal bases of type 1 by Hasan, Wang, and Bhargava [17]. For their case, the corresponding T_0 has $2m-1$ nonzero entries, and Q has $m-1$ nonzero entries. Exploiting the decomposition gives significant speed (and possibly storage) improvements [26,10]. However, this decomposition is for type 1 bases, and so m is necessarily even.

The following decomposition result is obtained for characteristic 3. For $k=2$, the multiplication complexity is essentially reduced from $3m$ to $2m$. This result can be applied, for example, to the Ning-Yin multiplication presented in [14]. For $k=4$, the multiplication complexity is essentially reduced from $5m$ to $4m$.

Theorem 3. *Let T_0 correspond to a GNB of type k for $\mathbb{F}_{3^m}$ with $m > k$. If $k=2$, then $T_0 = I + Q$ where Q has $2m-1$ nonzero entries, each of which is 1. If $k=4$, then $T_0 = -I + Q$ where Q has $4m-4$ nonzero entries, three of which are -1.*

Proof. The entries of T_0 are obtained from T via $t_{i,j}^{(0)} = t_{i-j,-j}$ and hence the diagonal entries of T_0 are $t_{j,j}^{(0)} = t_{0,-j}$.

For $k=2$, we have $f(0,0) = \alpha_\ell$ for some ℓ, and $f(0,1) = 2$. Hence $m-1$ of the $t_{0,-j}$ entries are 1 and one entry is -1. We can thus write $T_0 = I + Q$ where Q receives a "correction term" corresponding to the -1 entry. It is easy to see that Q will then have $2m-1$ nonzero entries.

For $k=4$, in the proof of Corollary 1 we showed that there cannot be an i with $f(i,s)$ constant for three distinct s. Thus from $f(0,1) = f(0,3)$ and $f(0,2) = 4$, we have $C_0 = m-1$ where $m-2$ entries are -1 and one entry is 1. The result then follows in a fashion similar to $k=2$ using the fact that $C_N = 5m-7$. $\square$

A consequence of Theorem 3 is that a GNB of type 2 over $\mathbb{F}_{3^m}$ is essentially optimal in the sense that the complexity of the multiplication is effectively $2m-1$ (since $A^{(\ell)}IB^{(\ell)'} = A \cdot B$ is independent of ℓ and essentially cost-free).

4   Implementation Notes and Timings

In this section, we provide details on the implementation along with timings on a Pentium 4, a common platform chosen for such comparisons. This processor is briefly described as 32-bit with extensions for wider operations in special registers, and has relatively few general-purpose registers. Compared to the preceding generation Pentium III processors, the instruction pipeline is deeper and penalties are larger for branch misprediction [19].

The implementation language was C, and only general-purpose registers were employed. Cooperating with processor characteristics and compiler optimizing peculiarities can be a difficult task, and our efforts in this area were modest.

In particular, the GNU C compiler can be weaker on register allocation strategies, and favours scalars over structures and arrays (even when array indices are known at compile-time). Limited effort was applied to cooperate with such weaknesses, but the timings in Table 1 show that significant differences between compiler performance remain.

Table 1. Timings (in μs) for field arithmetic on a 2.4 GHz Pentium 4[a]

		Polynomial Basis					Type 2 Normal Basis			
add	mult	a^3	$a^{1/3}$	invert by exp	invert by EEA	mult	mult Thm 3	mult ring map[b]	a^3 or $a^{1/3}$	invert by exp
$\mathbb{F}_{3^{239}} = \mathbb{F}_3[x]/(x^{239} + x^{24} - 1)$										
gcc	.05	5.0	.32	1.6^c	137^d	55	21.0	16.2	13.9	.04	195^d
icc	.04	4.2	.30	1.2^c	122^d	46	17.2	14.2	11.5	.02	171^d
GPS[e]	.69	23.0	1.59	19.3		159	60.9			.60	14182
$\mathbb{F}_{3^{509}} = \mathbb{F}_3[x]/(x^{509} - x^{477} + x^{445} + x^{32} - 1)$										
gcc	.09	15.5	.70	2.5^f	575^g	213	98.7	74.5	58.1	.07	1034^g
icc	.07	12.8	.66	1.7^f	508^g	190	74.3	58.8	52.0	.04	829^g
$\mathbb{F}_{2^{1223}} = \mathbb{F}_2[x]/(x^{1223} + x^{255} + 1)$										
gcc	.06	17.9									
icc	.06	15.6									

[a] Compilers are GNU C (gcc) 3.3 and Intel C (icc) 6.0. Timings done under Linux/x86.
[b] Map to $\mathbb{F}_3[x]/((x^{mk+1} - 1)/(x - 1))$ and use a modified comb polynomial multiplication [10]. Fields here have $k = 2$ (the type of the Gaussian normal basis).
[c] Sparse multiplication; $x^{1/3}$ has 23 nonzero terms and $x^{2/3}$ has 9 nonzero terms.
[d] Addition-chain with 12 multiplications.
[e] Timings in [14] are given for a 2.8 GHz Pentium 4 running Linux with gcc 3.3; times here are obtained by scaling linearly to 2.4 GHz.
[f] Sparse multiplication; $x^{1/3}$ has 6 nonzero terms and $x^{2/3}$ has 3 nonzero terms.
[g] Addition-chain with 14 multiplications.

4.1   Field Multiplication

For polynomial multiplication, we employed a "comb" multiplier [22] suitably modified for characteristic 3. We used width $w = 3$, which extracts 3 bits each of a^0 and a^1 at each step. Only 27 (rather than $2^6 = 64$) distinct values are possible due to the representation. A lookup was performed on the 6 bits in order to select from the table of (data-dependent) precomputation. Since half the elements are obtained by simple negation, precomputation is less expensive than it may appear. For fields of sufficient size, a shallow-depth Karatsuba split was used. At smaller field sizes this need not be faster; the results vary by platform, but typically the times are competitive with a "full comb" and the split has the advantage of less code size and dynamic memory consumption. For example, on the test platform, a full comb on eight 32-bit word pairs (e.g., $\mathbb{F}_{3^{239}}$) is 8-18% faster than a depth-1 split (giving three 4-word-pair multiplications).

Normal basis multiplication uses the precomputation strategy of Ning and Yin [24]. The basic idea is to calculate rotations required in a (data-dependent) precomputation phase to reduce costs in the main evaluation. For low-complexity Gaussian normal bases, the multiplication matrix is sparse and of regular structure, and the corresponding algorithm is relatively simple to code. Granger et al. [14] adapt the Ning and Yin algorithm directly. Our implementation apparently differs in the order of evaluation in that our outer loop is on the rows of the multiplication matrix, which reduces the number of lookups. We also give timings for the reduced-complexity version given by Theorem 3.

The "ring mapping" approach is detailed in [10]; only an outline is given here. For a field $\mathbb{F}_{p^m}$ with a type k normal basis, there is a fast mapping ϕ from $\mathbb{F}_{p^m}$ to the ring $\mathbb{F}_p[x]/((x^{mk+1} - 1)/(x - 1))$. The basic idea is to perform normal basis multiplication by mapping into the ring and applying fast polynomial-basis methods and then map back. A downside in this approach is the expansion by a factor k. However, the last $mk/2$ coefficients for elements in $\phi(\mathbb{F}_{p^m})$ are a mirror reflection of the first $mk/2$ [34]. Hence, it suffices to find half the coefficients in the ring product.

Each coefficient in $\phi(\cdot)$ or $\phi^{-1}(\cdot)$ can be obtained with a shift and mask.[1] The comb multiplier is defined for only a subset of the ring. However, the expansion in the mapping means that the method will be significantly more expensive than polynomial multiplication in the field. In particular, precomputation is for ring elements, and an entire ring element is "combed." On the positive side, only half the product in the ring is calculated, and reduction is especially simple.

4.2   Cubing and Cube Roots

For cubing in polynomial representations, we used an 8-to-24-bit lookup table to "thin" coefficients and then reduced the result. This is analogous to the common method for squaring in binary fields, and cubing is similarly inexpensive. Cube roots were obtained by the method described in §2.2, with a 64-word lookup table to extract coefficients. The cost depends on the number of nonzero terms in $x^{1/3}$ and $x^{2/3}$.

For $m = 239$, we used the reduction polynomial $x^{239} + x^{24} - 1$ so that direct comparisons could be made to [14]. This choice gives 23 terms in $x^{1/3}$ and 9 terms in $x^{2/3}$. As noted in Example 1, this is not optimal, but nonetheless leads to fairly fast cube roots via sparse multiplication. For $m = 509$, there is no trinomial giving sparse $x^{1/3}$. We searched for irreducible tetranomial or pentanomial $x^{509} + p(x)$ giving the lowest combined weight for $x^{1/3}$ and $x^{2/3}$ subject to $\deg p \leq 509 - 32$. There are multiple candidate pentanomials, but $x^{509} - x^{477} + x^{445} + x^{32} - 1$ was chosen for the fortuitous spacing between powers that permits optimizations. The choice gives $x^{1/3}$ with 6 terms and $x^{2/3}$ with 3.

---

[1] More precisely, in our representation for characteristic 3, a shift and mask is applied to a pair of words to obtain an output coefficient. For the mapping into the ring, only half the coefficients are found this way—the remainder are obtained at low cost by symmetry.

A possible implementation downside is the high degree of $p(x)$, although this is not a significant issue here.

In a normal basis representation, cubing and cube roots are rotations. In our specific representation, this is rotation of a pair, an inexpensive operation.

4.3 Inversion

Inversion is performed via a Euclidean algorithm variant and also using exponentiation. Although Euclidean algorithm variants can be faster, coding for speed typically involves code expansion (to efficiently track lengths of operands, etc.). The method using exponentiation is short and easy to code once field multiplication is done.

For the Euclidean algorithm approach in characteristic 3, [16] and [14] employ the "binary" Euclidean algorithm. We adapted the "usual" Euclidean algorithm [15, Algorithm 2.48]. Unlike the binary Euclidean algorithm, explicit degree calculations are required. Some processors have hardware support that can aid in these calculations. The Intel Pentium family has "bit scan" instructions to find the left- or right-most 1 bit in a word, and we used an assembly language fragment to exploit this capability. A binary search can be used on processors without such support,[2] and in fact "bit scan" on the Pentium is less effective for our code here than in [11] for characteristic 2, in part because of the difference in characteristic and also that the characteristic 2 code uses more optimization via code expansion.

For inversion via exponentiation, we used Itoh-Tsujii methods (see §2.3). Rather than the direct recursive approach, a few multiplications can sometimes be saved by choosing short addition chains. We used the following chains:

$$\mathbb{F}_{3^{239}} : 1, 2, 3, 6, \ 8, 14, 28, 56, 112, 224, 238$$
$$\mathbb{F}_{3^{509}} : 1, 2, 4, 8, 12, 24, 28, 56, 112, 224, 252, 504, 508$$

These give inversion via 12 and 14 multiplications, respectively, saving a multiplication in each case over the direct recursive approach. (The corresponding inversion code using these chains has low resource requirements.)

4.4 Analysis

Table 1 shows that the times in [14, Table 4] are unnecessarily pessimistic (on this platform) for both polynomial basis and normal basis field arithmetic in characteristic 3. For the example field $\mathbb{F}_{3^{239}}$, multiplication times in a polynomial basis are approximately a factor 5 faster than reported in [14], in part due to algorithm differences.

For normal basis representations, significant improvement can be obtained for the type 2 case exhibited by $\mathbb{F}_{3^{239}}$ by exploiting Theorem 3 to reduce the complexity. Further improvement is obtained by the "ring mapping" approach.

---

[2] Sun recommends using a "population count" (*popc*) instruction to build a seven-instruction bit-scan on SPARC [33]. However, *popc* is implemented via a trap, and bit-scan will be faster via binary search.

Our results are consistent with [14] in the sense that normal basis multiplication is sufficiently expensive relative to multiplication in a polynomial basis to discourage the use of normal basis representations in this environment.

Nonetheless, normal bases continue to be of interest in some environments, and choosing between the Ning-Yin approach and the ring mapping method will depend, in part, on the type k of the Gaussian basis. Type 2 bases are of course advantageous in both methods, but larger type may be the only choice if supersingular curves with low cofactor are demanded. For example, in the range $239 < m < 487$, only $m = 317$ and $m = 353$ give supersingular curves over $\mathbb{F}_{3^m}$ with small cofactor, and the corresponding types are 26 and 14, resp.

The Ning-Yin precomputation requires $4m$ words of data-dependent storage, and this amount is not affected by the type of the basis. The method is especially easy to code, and (roughly speaking) the running time increases linearly with k, although the number and location of -1 entries in the multiplication matrix complicates performance tuning. In contrast, the ring mapping approach has an expansion by a factor k, although symmetry lessens the impact. For the test platform, there will be a threshold k where multiplication will be fastest via conversion to polynomial basis representation (at cost equivalent to a few multiplications provided the conversion matrix is known). Of less practical importance, [14] note the "exceptionally high cost of inversion in normal bases;" however, in fact the methods of Itoh and Tsujii apply and inversion cost is relatively modest.

For curve arithmetic, [16] provide comparisons for a supersingular curve over $\mathbb{F}_{3^{97}}$ against a curve over $\mathbb{F}_{2^{241}}$, which offer similar security in the context of pairing-based cryptography (the corresponding embedding degrees are 6 and 4, respectively). In [16, Table 4], the best times for point multiplication favour the characteristic three case by roughly a factor 2. However, the scalar recodings selected are binary, ternary, and nonary, and this favours the characteristic three curve; in fact, the ternary and nonary methods are not useful for the characteristic two case. Since the nonary method is permitted storage for a few points of precomputation, a more meaningful comparison would involve a similar-storage width-w NAF method in the characteristic two case. In fact, the calculation of the usual width-w NAF adapts directly to the base 3 case, and so we'd recommend that the nonary method be replaced by a method employing a (base 3) width-3 NAF (which uses the same amount of storage).

5   Identity-Based Encryption

In this section we compare the speed of encryption and decryption for the Boneh-Franklin (BF) [7] and Sakai-Kasahara (SK) [27] identity-based encryption schemes at the 128-bit security level.

5.1   Symmetric Pairings

Let E be a supersingular elliptic curve defined over $\mathbb{F}_q$. Let n be a prime divisor of $\#E(\mathbb{F}_q)$, and suppose that $n^2 \nmid \#E(\mathbb{F}_q)$. The embedding degree of E (with respect to n) is the smallest positive integer k such that $n \mid q^k - 1$. Let $P \in E(\mathbb{F}_q)$ be a point of order n, and let μ_n denote the order-n cyclic subgroup of $\mathbb{F}_{q^k}^*$. The (reduced) Tate pairing is a bilinear map $\hat{e} : \langle P \rangle \times \langle P \rangle \to \mu_n$.

The three pairings we consider are described next. We let m, s, c denote the cost of multiplication, squaring, and cubing in the base field $\mathbb{F}_q$, and let M, S, C denote the cost of multiplication, squaring and cubing in the extension field $\mathbb{F}_{q^k}$. Also, we let A, D, T denote the cost of point addition (using mixed affine-projective coordinates), point doubling (using projective coordinates), and point tripling (using projective coordinates) in $E(\mathbb{F}_q)$.

Type I Pairing. Let $q = p$ be a 1536-bit prime such that $p \equiv 3 \pmod{4}$ and $p+1$ has a 256-bit low Hamming weight prime divisor n. Then the elliptic curve

$$E_1/\mathbb{F}_p : Y^2 = X^3 - 3X$$

is supersingular and $n \mid \#E_1(\mathbb{F}_p)$. The embedding degree of E_1 is $k = 2$. The extension field $\mathbb{F}_{p^2}$ is represented as $\mathbb{F}_{p^2} = \mathbb{F}_p[i]/(i^2 + 1)$, and a distortion map is $(x, y) \mapsto (-x, iy)$. We have $m \approx s$, $M \approx 3m$, $S \approx 2m$, $A \approx 3s + 8m \approx 11m$, and $D \approx 4s + 4m \approx 8m$. The Tate pairing, computed using the algorithm described by Scott [29], costs $4s + 8m + S + M$ per bit of n (for the Miller operation) plus $5s + 5m$ per bit of n (for the final exponentiation by $(p^2 - 1)/n$). If one of the two input points is fixed then, as observed in [29], precomputing 768 $\mathbb{F}_p$-elements can reduce the cost of the Miller operation to $m + S + M$ per bit of n.

Type II Pairing. Let $q = 2^{1223}$, $\mathbb{F}_{2^{1223}} = \mathbb{F}_2[x]/(x^{1223} + x^{255} + 1)$, and

$$E_2/\mathbb{F}_{2^{1223}} : Y^2 + Y = X^3 + X.$$

Then E_2 is supersingular, and $\#E_2(\mathbb{F}_{2^{1223}}) = 5n$ where n is a 1221-bit prime. The embedding degree of E_2 is $k = 4$. The extension field $\mathbb{F}_{q^4}$ is represented using tower extensions $\mathbb{F}_{q^2} = \mathbb{F}_q[u]/(u^2 + u + 1)$ and $\mathbb{F}_{q^4} = \mathbb{F}_{q^2}[v]/(v^2 + v + u)$. We have $M \approx 9m$ and $A \approx 9m$, while s, S and D are essentially free. Inversion of an element $\alpha \in \mu_n$ is also essentially free since $\alpha^{-1} = \alpha^{q^2}$. The BGhS [4] algorithm for computing the Tate pairing costs approximately $612 \times 7m$.

Type III Pairing. Let $q = 3^{509}$, $\mathbb{F}_{3^{509}} = \mathbb{F}_3[x]/(x^{509} - x^{477} + x^{445} + x^{32} - 1)$, and

$$E_3/\mathbb{F}_{3^{509}} : Y^2 = X^3 - X + 1.$$

Then E_3 is supersingular, and $\#E_3(\mathbb{F}_{3^{509}}) = 7n$ where n is a 804-bit prime. The embedding degree of E_2 is $k = 6$. The extension field $\mathbb{F}_{q^6}$ is represented using tower extensions $\mathbb{F}_{q^3}[u] = \mathbb{F}_q[u]/(u^3 - u - 1)$ and $\mathbb{F}_{q^6} = \mathbb{F}_{q^3}[v]/(v^2 + 1)$. We have $M \approx 18m$ and $A \approx 9m$, while c, C and T are essentially free. Inversion of an element $\alpha \in \mu_n$ is also essentially free since $\alpha^{-1} = \alpha^{q^3}$. The BGhS [4] algorithm for computing the Tate pairing costs approximately $255 \times 15m$.

5.2   Boneh-Franklin and Sakai-Kasahara IBE

Let $P \in E(\mathbb{F}_q)$ be a point of order n, and let $\hat{e} : \langle P \rangle \times \langle P \rangle \to \mu_n$ be a (symmetric) bilinear pairing. Let $H_1 : \{0,1\}^* \to \langle P \rangle$, $H_2 : \{0,1\}^\lambda \times \{0,1\}^\lambda \to [1, n-1]$, $H_3 : \mu_n \to \{0,1\}^\lambda$, $H_4 : \{0,1\}^\lambda \to \{0,1\}^\lambda$, $H_5 : \{0,1\}^* \to [1, n-1]$ be hash functions. The Key Generator's private key is $t \in_R [1, n-1]$, while its public key is $T = tP$.

Boneh-Franklin IBE. Party A's private key is $d = tQ$, where $Q = H_1(\text{ID}_A)$.

To encrypt a message $m \in \{0,1\}^\lambda$ for A, a party B does the following: Select $\sigma \in_R \{0,1\}^\lambda$, and compute $Q = H_1(\text{ID}_A)$, $r = H_2(\sigma, m)$, $R = rP$, $V = \sigma \oplus H_3(\hat{e}(T, Q)^r)$, and $c = m \oplus H_4(\sigma)$. B sends (R, V, c) to A.

To decrypt, A computes $\sigma = V \oplus H_3(\hat{e}(d, R))$, $m = c \oplus H_4(\sigma)$, and $r = H_2(\sigma, m)$. Finally, A accepts m provided that $R = rP$.

The BF scheme requires a hash function $H_1 : \{0,1\}^* \to \langle P \rangle$. For Type I pairings, H_1 can be implemented by first hashing to a point Q' in $E(\mathbb{F}_p)$, and then multiplying Q' by the cofactor $h = \#E(\mathbb{F}_p)/n$. As noted by Scott [29], the cofactor multiplication can be avoided; thus the essential cost of hashing is a square-root computation in $\mathbb{F}_p$. Square roots in $\mathbb{F}_p$ can be obtained by an exponentiation to the power $(p+1)/4$, an operation which requires about 1806 $\mathbb{F}_p$-multiplications using width-5 sliding windows. The hash function H_1 for the Type II pairing (and the Type III pairing) is relatively inexpensive since square roots (resp. cube roots) can be efficiently computed, and since the cofactor h is small.

The dominant operations in BF encryption are the point multiplication rP where P is fixed, and the computation of $\gamma = \hat{e}(T, Q)^r$ (plus a square-root computation for Type I pairings). For Type I pairings, the fastest way to compute γ is to first evaluate the Tate pairing $\hat{e}(T, Q)$ (where T is fixed), and then perform the exponentiation to the power r (where the base element $\hat{e}(T, Q)$ is unknown in advance). For Type II and III pairings, γ should be computed by first computing rT (where T is fixed), and then evaluating $\hat{e}(rT, Q)$. The dominant operations in BF decryption are the Tate pairing evaluation $\hat{e}(d, R)$ and the point multiplication rP where the points d and P are fixed.

Sakai-Kasahara IBE. Party A's private key is $d = (1/(H_5(\text{ID}_A) + t)P$.

To encrypt a message $m \in \{0,1\}^\lambda$ for A, a party B does the following: Select $\sigma \in_R \{0,1\}^\lambda$, and compute $Q = H_5(\text{ID}_A)P + T$, $r = H_2(\sigma, m)$, $R = rQ$, $V = \sigma \oplus H_3(\hat{e}(P, P)^r)$, and $c = m \oplus H_4(\sigma)$. B sends (R, V, c) to A.

To decrypt, A computes $\sigma = V \oplus H_3(\hat{e}(d, R))$, $m = c \oplus H_4(\sigma)$, and $r = H_2(\sigma, m)$. Finally, A accepts m provided that $R = rQ$.

The dominant operations in SK encryption are the point multiplication $H_5(\text{ID}_A)P$ where the base point P is fixed, the point multiplication rQ where the base point Q is unknown in advance, and the exponentiation $\hat{e}(P, P)^r$ where the base element $\hat{e}(P, P)$ is fixed. The dominant operations in SK decryption are the same as for BF decryption.

5.3   Costs

Table 2 lists the approximate number of $\mathbb{F}_q$-multiplications needed to compute the Tate pairing, and to perform point multiplication and exponentiation in μ_n for the Type I, II and III pairings. We acknowledge that these raw multiplication counts do *not* include the cost of other field operations such as additions, square roots, cube roots, and inversions that are either relatively cheap or few in number. Nonetheless, these multiplication counts are reasonably accurate estimates of the actual running times of the operations listed in Table 2. For example, the timings for our implementation of the Tate pairing, point multiplication, and μ_n-exponentiation for the Type II pairing are 4578, 2163, and 2037 $\mathbb{F}_{2^{1223}}$-multiplications, which are close to the estimated costs of 4284, 1895, and 1895 in Table 2. Similarly, the timings for our implementation of the Tate pairing, point multiplication, and μ_n-exponentiation for the Type III pairing are 4359, 1602, and 2695 $\mathbb{F}_{3^{509}}$-multiplications, which are close to the estimated costs of 3825, 1259, and 2518 in Table 2.

Table 2. Cost (number of $\mathbb{F}_q$-multiplications) of the Tate pairing, point multiplication rP in $E(\mathbb{F}_q)$, and exponentiation α^r in μ_n. The bitlength of n is denoted by ℓ.

Pairing type	$\mathbb{F}_q$	ℓ	Tate pairing	rP P unknown	rP P fixed	Exp in μ_n α uknown	Exp in μ_n α fixed
I	1536-bit p	256	$6912/4096^a$	2602^b	745^c	512^d	199^c
II	$\mathbb{F}_{2^{1223}}$	1221	4284	$1895^b/447^e$	$1832^b/384^e$	$1895^b/447^e$	$1832^b/384^e$
III	$\mathbb{F}_{3^{509}}$	804	3825	$1259^f/498^g$	$1016^f/324^g$	$2518^f/996^g$	$2032^f/648^g$

[a] Applicable when one of the input points is fixed; requires 144 Kbytes for precomputed values.
[b] 5-NAF point multiplication/exponentiation (see Algorithm 3.36 of [15]) with ℓ-bit multiplier r.
[c] Width-5 two-table comb method (see Algorithm 3.45 of [15]) with ℓ-bit multiplier r.
[d] Lucas-method of exponentiation [29].
[e] 5-NAF point multiplication/exponentiation with 256-bit multiplier r.
[f] 4-NAF point multiplication/exponentiation with ℓ-bit multiplier r.
[g] 3-NAF point multiplication/exponentiation with 256-bit multiplier r.

Table 3 lists the multiplication costs of the dominant operations in encryption and decryption for the BF and SK IBE schemes. The costs have been normalized to $\mathbb{F}_p$-multiplications (where p is a 1536-bit prime), using our timings of 12.8μs for multiplication in $\mathbb{F}_{3^{509}}$, 15.6μs for multiplication in $\mathbb{F}_{2^{1223}}$, and 26.5μs for a multiplication in $\mathbb{F}_p$ (the latter obtained using Mike Scott's MIRACL multi-precision library [30]).[3] The first set of (I,II,III) timings in Table 3 use full-length

---

[3] Timings were obtained on a 2.4 GHz Pentium 4 running Linux/x86. Compilers were GNU C (gcc) 3.3 for $\mathbb{F}_p$ and the Intel compiler (icc) 6.0 for the others. The expensive portions in the $\mathbb{F}_p$ multiplcation are written in assembly, and times for these fragments are not affected by compiler selection.

multipliers r for the Type II and III pairings, and do not include any precomputation of the Type I Tate pairing computation. The second set of (I,II,III) timings, on the other hand, use 256-bit multipliers r for the Type II and II pairings, and allow for the 144 Kbytes of precomputed values that accelerate the Tate pairing computation for Type I pairings. Short $2t$-bit multipliers (where t is the security level) instead of ℓ-bit multipliers (where ℓ is the bitlength of n) have been used in some previous papers (e.g., [31] and [32]). The drawback of using short multipliers is that the BF [7] and SK [9] security proofs are no longer applicable.

Table 3. Normalized cost (in terms of $\mathbb{F}_p$-multiplications) of encryption and decryption for BF and SK IBE using the Type I, II and III pairings

	I^a	II^b	III^b	I^c	II^d	III^d
BF encrypt	9975	4679	2829	7159	2974	2161
BF decrypt	7657	3600	2338	4841	2748	2004
SK encrypt	3546	3272	2080	3546	715	710
SK decrypt	7657	3600	2338	4841	2748	2004

[a] No precomputation for Tate pairing computations.
[b] Full length multipliers r.
[c] 144 Kbytes of precomputed values for the Tate pairing computation.
[d] 256-bit multipliers r.

6   Conclusions

We devised faster multiplication algorithms for characteristic three finite fields when elements are represented using a Gaussian normal bases. Despite our structure results and fast implementations, our analysis confirms the conclusions of previous papers that multiplication is faster when a polynomial basis representation is employed. We also compared the relative speed of the BF and SK IBE schemes at the 128-bit security levels when a pairing based on a supersingular

---

MIRACL has optimized code for several operations on the Pentium 4, the processor used in this comparison. In particular, the timing was obtained using general-purpose registers and a multiplier that uses Karatsuba down to a specified operand-size threshold. The threshold t is specified in terms of words and so that the modulus size is $t \cdot 2^n$ words for some n. Code size grows quadratically with t, and t between 8 and 16 is reasonable on this platform. Hence, for 1536-bit primes, we chose $t = 12$.

The Pentium 4 has special-purpose "multi-media" (SSE2) registers that can be employed for field multiplication. Roughly speaking, the basic advantage for prime fields is additional registers that participate in multiplications and accumulation can be on 64 bits, and the advantage for characteristic 2 and 3 is wider operations. Multiplication in MIRACL for prime fields is nearly a factor 2 faster with these registers, and [15] reports similar acceleration for characteristic 2 fields (on a Pentium III via the SSE subset); similar techniques apply to characteristic 3.

elliptic curve is used. The Type III pairing (over $\mathbb{F}_{3^{509}}$) yields the fastest encryption and decryption operations, which are several times faster than with a Type I pairing (over a 1536-bit prime field). Moreover, when using a Type III pairing, SK encryption is about 3 times as fast as BF encryption, while SK and BF decryption have similar running times.

References

1. Ahmadi, O., Hankerson, D., Menezes, A.: Formulas for cube roots in $\mathbb{F}_{3^m}$. Discrete Applied Mathematics 155, 260–270 (2007)
2. Ash, D., Blake, I., Vanstone, S.: Low complexity normal bases. Discrete Applied Mathematics 25, 191–210 (1989)
3. Barreto, P.: A note on efficient computation of cube roots in characteristic 3, Technical Report 2004/305, Cryptology ePrint Archive (2004)
4. Barreto, P., Galbraith, S., hÉigeartaigh, C., Scott, M.: Efficient pairing computation on supersingular abelian varieties. Designs, Codes and Cryptography 42, 239–271 (2007)
5. Barreto, P., Naehrig, M.: Pairing-friendly elliptic curves of prime order. In: Preneel, B., Tavares, S. (eds.) SAC 2005. LNCS, vol. 3897, pp. 319–331. Springer, Heidelberg (2006)
6. Blake, I., Gao, X., Menezes, A., Mullin, R., Vanstone, S., Yaghoobian, T.: Applications of Finite Fields. Kluwer, Dordrecht (1993)
7. Boneh, D., Franklin, M.: Identity-based encryption from the Weil pairing. SIAM Journal on Computing 32, 586–615 (2003)
8. Boyen, X., Martin, L.: Identity-based cryptography standard (IBCS) #1: Supersingular curve implementations of the BF and BB1 cryptosystems, IETF Internet Draft (December 2006)
9. Chen, L., Cheng, Z.: Security proof of Sakai-Kasahara's identity-based encryption scheme. In: Smart, N.P. (ed.) Cryptography and Coding. LNCS, vol. 3796, pp. 442–459. Springer, Heidelberg (2005)
10. Dahab, R., Hankerson, D., Hu, F., Long, M., López, J., Menezes, A.: Software multiplication using Gaussian normal bases. IEEE Transactions on Computers 55, 974–984 (2006)
11. Fong, K., Hankerson, D., López, J., Menezes, A.: Field inversion and point halving revisited. IEEE Transactions on Computers 53, 1047–1059 (2004)
12. Galbraith, S., Paterson, K., Smart, N.: Pairings for cryptographers, Technical Report 2006/165, Cryptology ePrint Archive (2006)
13. Grabher, P., Page, D.: Hardware acceleration of the Tate pairing in characteristic three. In: Rao, J.R., Sunar, B. (eds.) CHES 2005. LNCS, vol. 3659, pp. 398–411. Springer, Heidelberg (2005)
14. Granger, R., Page, D., Stam, M.: Hardware and software normal basis arithmetic for pairing based cryptography in characteristic three. IEEE Transactions on Computers 54, 852–860 (2005)
15. Hankerson, D., Menezes, A., Vanstone, S.: Guide to Elliptic Curve Cryptography. Springer, Heidelberg (2004)
16. Harrison, K., Page, D., Smart, N.: Software implementation of finite fields of characteristic three, for use in pairing-based cryptosystems. LMS Journal of Computation and Mathematics 5, 181–193 (2002)

17. Hasan, M., Wang, M., Bhargava, V.: A modified Massey-Omura parallel multiplier for a class of finite fields. IEEE Transactions on Computers 42, 1278–1280 (1993)
18. Hess, F., Smart, N., Vercauteren, F.: The eta pairing revisited. IEEE Transactions on Information Theory 52, 4595–4602 (2006)
19. Intel Corporation, IA-32 Intel Architecture Software Developer's Manual, Vol. 1: Basic Architecture. Number 245470-007, (2002) available from http://developer.intel.com.
20. Kerins, T., Marnane, W., Popovici, E., Barreto, P.: Efficient hardware for the Tate pairing calculation in characteristic three. In: Rao, J.R., Sunar, B. (eds.) CHES 2005. LNCS, vol. 3659, pp. 412–426. Springer, Heidelberg (2005)
21. Lenstra, A.: Unbelievable security: Matching AES security using public key systems. In: Boyd, C. (ed.) ASIACRYPT 2001. LNCS, vol. 2248, pp. 67–86. Springer, Heidelberg (2001)
22. López, J., Dahab, R.: High-speed software multiplication in $\mathbb{F}_{2^m}$. In: Roy, B., Okamoto, E. (eds.) INDOCRYPT 2000. LNCS, vol. 1977, pp. 203–212. Springer, Heidelberg (2000)
23. Miyaji, A., Nakabayashi, M., Takano, S.: New explicit conditions of elliptic curve traces for FR-reduction. IEICE Transactions on Fundamentals of Electronics, Communications and Computer Sciences E84-A, 1234–1243 (2001)
24. Ning, P., Yin, Y.: Efficient software implementation for finite field multiplication in normal basis. In: Qing, S., Okamoto, T., Zhou, J. (eds.) ICICS 2001. LNCS, vol. 2229, pp. 177–189. Springer, Heidelberg (2001)
25. Page, D., Smart, N.: Hardware implementation of finite fields of characteristic three. In: Kaliski Jr., B.S., Koç, Ç.K., Paar, C. (eds.) CHES 2002. LNCS, vol. 2523, pp. 529–539. Springer, Heidelberg (2003)
26. Reyhani-Masoleh, A.: Efficient algorithms and architectures for field multiplication using Gaussian normal bases. IEEE Transactions on Computers 55, 34–47 (2006)
27. Sakai, R., Kasahara, M.: ID based cryptosystems with pairing on elliptic curve, Technical Report 2003/054, Cryptology ePrint Archive (2003)
28. Schirokauer, O.: The number field sieve for integers of low weight, Technical Report 2006/107, Cryptology ePrint Archive (2006)
29. Scott, M.: Computing the Tate pairing. In: Menezes, A.J. (ed.) CT-RSA 2005. LNCS, vol. 3376, pp. 293–304. Springer, Heidelberg (2005)
30. Scott, M.: MIRACL – Multiprecision Integer and Rational Arithmetic C Library, http://www.computing.dcu.ie/~mike/miracl.html
31. Scott, M.: Implementing cryptographic pairings, preprint (2006)
32. Scott, M., Costigan, N., Abdulwahab, W.: Implementing cryptographic pairings on smartcards. In: Goubin, L., Matsui, M. (eds.) CHES 2006. LNCS, vol. 4249, pp. 134–147. Springer, Heidelberg (2006)
33. Weaver, D., Germond, T. (eds.): The SPARC Architecture Manual (Version 9). Prentice-Hall, Englewood Cliffs (1994)
34. Wu, H., Hasan, A., Blake, I., Gao, S.: Finite field multiplier using redundant representation. IEEE Transactions on Computers 51, 1306–1316 (2002)

Complexity Reduction of Constant Matrix Computations over the Binary Field

Oscar Gustafsson and Mikael Olofsson

Department of Electrical Engineering, Linköping University,
SE-581 83 Linköping, Sweden
oscarg@isy.liu.se, mikael@isy.liu.se

Abstract. In this work an algorithm for realizing a multiplication of a vector by a constant matrix over the binary field with few two-input XOR-gates is proposed. This type of problem occurs in, e.g., Galois field computations, syndrome computation for linear error correcting codes, cyclic redundancy checks (CRCs), linear feedback shift-registers (LFSRs), and implementations of the Advanced Encryption Standard (AES) algorithm. As the proposed algorithm can utilize cancellation of terms it outperforms in general previously proposed algorithms based on sub-expression sharing.

Keywords: binary field, low-complexity, Galois field arithmetic, constant multiplication.

1   Introduction

The binary field $GF(2)$ is the set $\{0, 1\}$ together with addition and multiplication reduced modulo 2, i.e., addition and multiplication correspond to the logical operations XOR and AND, respectively [1,2]. The extension field $GF(2^m)$ is the set of univariate binary polynomials over $GF(2)$ of degree less than m. Arithmetic in $GF(2^m)$ is the corresponding polynomial arithmetic reduced modulo

$$p(x) = \sum_{i=0}^{m} p_i x^i,$$

where $p(x)$ is an irreducible polynomial over $GF(2)$. The reduction modulo $p(x)$ is performed by successive uses of

$$x^m \equiv \sum_{i=0}^{m-1} p_i x^i \mod p(x).$$

If all non-zero elements of the field can be generated by x reduced modulo $p(x)$, i.e., if the set $\{x^i \mod p(x) : 0 \leq i < 2m - 1\}$ contains all non-zero elements in $GF(2^m)$, then $p(x)$ is said to be primitive.

C. Carlet and B. Sunar (Eds.): WAIFI 2007, LNCS 4547, pp. 103–115, 2007.

Consider multiplying two arbitrary elements in $GF(2^m)$, represented in their polynomial basis, $a(x)$ and $b(x)$ where we have

$$a(x) = \sum_{i=0}^{m-1} a_i x^i \qquad \text{and} \qquad b(x) = \sum_{i=0}^{m-1} b_i x^i.$$

The product $c(x)$ is then given by

$$c(x) = \sum_{i=0}^{m-1} c_i x^i = a(x)b(x) = \sum_{i=0}^{m-1} b_i x^i a(x) \bmod p(x). \qquad (1)$$

Now, introduce a matrix X as

$$X = \begin{bmatrix} 0 \dots 0 & p_0 \\ 1 \quad 0 & p_1 \\ \ddots & \vdots \\ 0 \quad 1 & p_{m-1} \end{bmatrix},$$

and the vector representations

$$\bar{a} = [a_0 \dots a_{m-1}]^\mathsf{T}, \quad \bar{b} = [b_0 \dots b_{m-1}]^\mathsf{T}, \quad \text{and} \quad \bar{c} = [c_0 \dots c_{m-1}]^\mathsf{T}.$$

Equation (1) can now be rewritten as $\bar{c} = A\bar{b}$, where

$$A = (\bar{a}, X\bar{a}, \dots, X^{m-1}\bar{a}).$$

Hence, for constant $a(x)$ and $p(x)$, the matrix A is a constant binary matrix. This means that a multiplication by a constant in $GF(2^m)$ is a linear operation over $GF(2)$. Other linear operations in $GF(2^m)$ include e.g. squaring. In implementation of Reed-Solomon error correcting codes, constant computations may be used in the encoder [3], the syndrome computation and Chien search of the decoder [4]. Furthermore, constant computations are also used in, e.g., general Galois field multipliers based on composite fields [5,6].

The same type of problem also occurs in linear error correcting codes [7], parallel realizations of cyclic redundancy checks (CRCs) [8,9] and linear feedback shift-registers (LFSRs), and implementations of the Advanced Encryption Standard (AES) algorithm [10].

The aim of this work is to realize a constant matrix multiplication like A with as few additions (XOR-gates) as possible. This problem has been explicitly studied for computations over the binary field in [3, 4, 7, 10]. However, these previous works only cover a subset of the available algorithms proposed for a similar problem, namely multiple constant multiplication (MCM). The MCM problem is mainly motivated by the design of low-complexity FIR digital filters [11, 12, 13, 14, 15, 16, 17].

In the next section the proposed algorithm is presented. Then, in Section 3, the algorithm is applied to an example matrix to illustrate the behavior. In Section 4 the proposed algorithm is compared to previous algorithms. Finally, in Section 5, some conclusions are drawn.

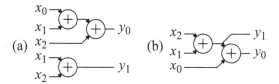

Fig. 1. Realization of the computation in (2) (a) without and (b) with complexity reduction

2   Complexity Reduction

Consider the computation

$$\begin{bmatrix} y_0 \\ y_1 \end{bmatrix} = \begin{bmatrix} 1 & 1 & 1 \\ 0 & 1 & 1 \end{bmatrix} \begin{bmatrix} x_0 \\ x_1 \\ x_2 \end{bmatrix}. \tag{2}$$

A straightforward realization of this computation is shown in Fig. 1(a) which requires three adders/XOR-gates. However, it is also possible to share the computation of $x_1 \oplus x_2$ with a resulting complexity of two XOR-gates as illustrated in Fig. 1(b).

2.1   Previous Approaches

The previously proposed algorithms for the considered problems are all based on sub-expression sharing, i.e., common patterns are found and extracted for realization. In [3, 4, 10] the most common combination of two inputs are realized and used for later iterations. As long as there are common inputs the algorithm continues extracting sub-expressions. The algorithms mainly differ in how they handle the case where several combinations are as frequent. We will refer to this type of algorithm as pairwise iterative matching sub-expression sharing.

In [14] an approach were the largest, i.e., with the most non-zero positions, possible common sub-expression is extracted. This sub-expression is then, given that it includes more than two non-zero positions, added as a new row of the problem. Similarly to the pairwise iterative matching method, this approach continues until there are no more common sub-expressions to extract. While this method has not been explicitly proposed for computations over the binary field, we will use it for comparison. This method is here referred to as maximum length sub-expression sharing.

In [7] a method similar to both the pairwise iterative matching method and the maximum length method were proposed. Here, the most common sub-expression independent of the number of non-zero positions is used.

Other approaches for sub-expression sharing has been proposed for the realization of transposed form FIR filters. Among them are the use of a contention resolution algorithm to select the best sub-expression taking the effect of subsequent sub-expressions into account [15]. In [17] an integer linear programming model was proposed to solve the sub-expression sharing problem. Furthermore, in [16] an exact method for selecting the sub-expressions was proposed.

2.2 Proposed Approach

A completely different approach from sub-expression sharing is the adder graph methodology introduced in [11]. The idea is the following: In iteration i, given an initial set of available rows, V_i, compute all possible rows, C_i, using one operation (here an XOR-gate). One row is here a vector with binary elements corresponding to the sum of the non-zero elements, i.e., the row $\begin{pmatrix} 0 & 1 & 0 & 1 \end{pmatrix}$ corresponds to $x_1 \oplus x_3$. The set of rows remaining is denoted R_i. If any of the possible rows are required, i.e. $R_i \cap C_i \neq \emptyset$, they are moved to the set of available rows used in the next iteration, V_{i+1}. If not, a heuristic decision must be taken to add one possible row to the available rows in order to be able to find the required rows at later stages. The output is in [11] represented as a graph with each node corresponding to a computed value (row) and each edge to an interconnection, hence the use of graph in the name. Note that if the algorithm terminates without using the heuristic part, an optimal solution (one XOR-gate per unique row) is found.

The proposed adder graph algorithm for XOR-gates with heuristic can be described as

1. Initialize the set of available rows, V_1, with the rows from the identity matrix of size N. Initialize the set of required rows, R_1, with the non-trivial[1] rows from the matrix to be realized, A. Remove any duplicate rows. Let $i = 1$.
2. Form a matrix C_i of all elementwise XOR-ed pairs of rows in V_i, i.e.,

$$C_i = \{a \oplus b : a, b \in V_i\}.$$

3. Form a new set of available rows, V_{i+1}, and a new set of required rows, R_{i+1}, according to

$$V_{i+1} = V_i \cup (C_i \cap R_i),$$
$$R_{i+1} = R_i - (C_i \cap R_i) = R_i - C_i,$$

i.e., move rows that are in both C_i and in the set of required rows from the set of required rows to the set of available rows.
4. If $R_{i+1} = \emptyset$, stop.
5. If $(C_i \cap R_i) \neq \emptyset$, increment i and go to 2.
6. Compute the Hamming distance between all rows in R_i and all rows in C_i. For the combination with minimum Hamming distance, add the corresponding row in C_i to R_i to form a new R_{i+1}. If several combinations have the same

---

[1] Trivial rows do not require any XOR-gates to be realized, i.e., their Hamming weight is ≤ 1.

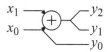

Fig. 2. Realization of the transposed matrix in (3)

minimum Hamming distance, add the row from C_i that is present in most combinations. If several such rows are present in equally many combinations, pick one among them at random. Increment i and go to step 2.

The proposed heuristic in Step 6 generally realizes the row with minimum Hamming weight initially. However, the order in which the intermediate rows are generated does not consider the other rows to be generated later on. This may lead to an increased number of XOR-gates for matrices with more columns than rows.

2.3 Transposition

It is possible to obtain an alternative realization for a matrix by running our algorithm on the transposed matrix, and then transposing the resulting network. Transposition of a network is done by reverting the direction of the signals such that the inputs become the outputs and the outputs become the inputs [18]. Furthermore, each 1-to-2-branch is replaced by a 2-input XOR-gate and each 2-input XOR-gate is replaced by a 1-to-2-branch.

Consider the following example which realizes the transposed matrix from (2). The resulting realization is shown in Fig. 2.

$$\begin{bmatrix} y_0 \\ y_1 \\ y_2 \end{bmatrix} = \begin{bmatrix} 1 & 1 & 1 \\ 0 & 1 & 1 \end{bmatrix}^T \begin{bmatrix} x_0 \\ x_1 \end{bmatrix} = \begin{bmatrix} 1 & 0 \\ 1 & 1 \\ 1 & 1 \end{bmatrix} \begin{bmatrix} x_0 \\ x_1 \end{bmatrix}. \tag{3}$$

If the number of inputs and outputs of the original realization is I and O, respectively and the realization is done using X XOR-gates, then the transposed realization will have

$$X + O - I \qquad \text{XOR-gates.} \tag{4}$$

For the example in (2) and (3) this evaluates to $2 + 2 - 3 = 1$, which can be seen from Fig. 2 to be correct.

Hence, as the proposed algorithm is expected to perform worse on matrices with more columns than rows, it is possible to apply the algorithm to the transposed matrix instead and transpose the resulting realization. In Section 4 we will give empirical support that transposing the network corresponding to the transposed matrix actually can result in a significantly decreased complexity.

3 Example

To clarify the operation of the the proposed algorithm, we illustrate the procedure by finding a solution for the matrix

$$A = \begin{bmatrix} 1\,1\,0\,1\,1\,0 \\ 1\,1\,1\,0\,0\,0 \\ 0\,0\,1\,1\,1\,0 \\ 1\,0\,1\,0\,0\,0 \\ 1\,0\,0\,1\,1\,1 \\ 0\,0\,1\,0\,1\,0 \end{bmatrix}. \tag{5}$$

A straightforward realization of A requires twelve XOR-gates.

We initialize the algorithm according to step 1, i.e., we set $i = 1$ and

$$V_1 = \left\{ \begin{array}{l} (1\,0\,0\,0\,0\,0), (0\,1\,0\,0\,0\,0) \\ (0\,0\,1\,0\,0\,0), (0\,0\,0\,1\,0\,0) \\ (0\,0\,0\,0\,1\,0), (0\,0\,0\,0\,0\,1) \end{array} \right\}, \qquad R_1 = \left\{ \begin{array}{l} (1\,1\,0\,1\,1\,0), (1\,1\,1\,0\,0\,0) \\ (0\,0\,1\,1\,1\,0), (1\,0\,1\,0\,0\,0) \\ (1\,0\,0\,1\,1\,1), (0\,0\,1\,0\,1\,0) \end{array} \right\}.$$

According to steps 2 and 3, we calculate

$$C_1 = \left\{ \begin{array}{l} (1\,1\,0\,0\,0\,0), (1\,0\,1\,0\,0\,0) \\ (1\,0\,0\,1\,0\,0), (1\,0\,0\,0\,1\,0) \\ (1\,0\,0\,0\,0\,1), (0\,1\,1\,0\,0\,0) \\ (0\,1\,0\,1\,0\,0), (0\,1\,0\,0\,1\,0) \\ (0\,1\,0\,0\,0\,1), (0\,0\,1\,1\,0\,0) \\ (0\,0\,1\,0\,1\,0), (0\,0\,1\,0\,0\,1) \\ (0\,0\,0\,1\,1\,0), (0\,0\,0\,1\,0\,1) \\ (0\,0\,0\,0\,1\,1) \end{array} \right\}, \qquad C_1 \cap R_1 = \left\{ \begin{array}{l} (1\,0\,1\,0\,0\,0) \\ (0\,0\,1\,0\,1\,0) \end{array} \right\},$$

and

$$V_2 = \left\{ \begin{array}{l} (1\,0\,0\,0\,0\,0), (0\,1\,0\,0\,0\,0) \\ (0\,0\,1\,0\,0\,0), (0\,0\,0\,1\,0\,0) \\ (0\,0\,0\,0\,1\,0), (0\,0\,0\,0\,0\,1) \\ (1\,0\,1\,0\,0\,0), (0\,0\,1\,0\,1\,0) \end{array} \right\}, \qquad R_2 = \left\{ \begin{array}{l} (1\,1\,0\,1\,1\,0) \\ (1\,1\,1\,0\,0\,0) \\ (0\,0\,1\,1\,1\,0) \\ (1\,0\,0\,1\,1\,1) \end{array} \right\}.$$

R_2 is non-empty. Thus, in step 4 we continue to step 5, where we note that $C_1 \cap R_1$ is non-empty. Thus, we increment i to $i = 2$ and return to step 2.

According to steps 2 and 3, we calculate

$$C_2 = C_1 \cup \left\{ \begin{array}{l} (0\,0\,1\,0\,0\,0), (1\,1\,1\,0\,0\,0) \\ (1\,0\,0\,0\,0\,0), (1\,0\,1\,1\,0\,0) \\ (1\,0\,1\,0\,1\,0), (1\,0\,1\,0\,0\,1) \\ (0\,1\,1\,0\,1\,0), (0\,0\,0\,0\,1\,0) \\ (0\,0\,1\,1\,1\,0), (0\,0\,1\,0\,1\,1) \end{array} \right\}, \qquad C_2 \cap R_2 = \left\{ \begin{array}{l} (1\,1\,1\,0\,0\,0) \\ (0\,0\,1\,1\,1\,0) \end{array} \right\},$$

and

$$V_3 = \left\{ \begin{array}{l} (1\,0\,0\,0\,0\,0), (0\,1\,0\,0\,0\,0) \\ (0\,0\,1\,0\,0\,0), (0\,0\,0\,1\,0\,0) \\ (0\,0\,0\,0\,1\,0), (0\,0\,0\,0\,0\,1) \\ (1\,0\,1\,0\,0\,0), (0\,0\,1\,0\,1\,0) \\ (1\,1\,1\,0\,0\,0), (0\,0\,1\,1\,1\,0) \end{array} \right\}, \qquad R_3 = \left\{ \begin{array}{l} (1\,1\,0\,1\,1\,0) \\ (1\,0\,0\,1\,1\,1) \end{array} \right\}.$$

R_3 is non-empty. Again, in step 4 we continue to step 5, where we note that $C_2 \cap R_2$ is non-empty. Thus, we increment i to $i = 3$ and return to step 2.

According to steps 2 and 3, we calculate

$$C_3 = C_2 \cup \left\{ \begin{array}{l} (1\,1\,1\,1\,0\,0), (1\,1\,1\,0\,1\,0) \\ (1\,1\,1\,0\,0\,1), (0\,1\,0\,0\,0\,0) \\ (1\,1\,0\,0\,1\,0), (1\,0\,1\,1\,1\,0) \\ (0\,1\,1\,1\,1\,0), (0\,0\,1\,1\,1\,1) \\ (1\,0\,0\,1\,1\,0), (1\,1\,0\,1\,1\,0) \end{array} \right\}, \qquad C_3 \cap R_3 = \{(1\,1\,0\,1\,1\,0)\},$$

and

$$V_4 = \left\{ \begin{array}{l} (1\,0\,0\,0\,0\,0), (0\,1\,0\,0\,0\,0) \\ (0\,0\,1\,0\,0\,0), (0\,0\,0\,1\,0\,0) \\ (0\,0\,0\,0\,1\,0), (0\,0\,0\,0\,0\,1) \\ (1\,0\,1\,0\,0\,0), (0\,0\,1\,0\,1\,0) \\ (1\,1\,1\,0\,0\,0), (0\,0\,1\,1\,1\,0) \\ (1\,1\,0\,1\,1\,0) \end{array} \right\}, \qquad R_4 = \{(1\,0\,0\,1\,1\,1)\}.$$

R_4 is non-empty. Yet again, in step 4 we continue to step 5, where we note that $C_3 \cap R_3$ is non-empty. Thus, we increment i to $i = 4$ and return to step 2.

According to steps 2 and 3, we calculate

$$C_4 = C_3 \cup \left\{ \begin{array}{l} (0\,1\,0\,1\,1\,0), (1\,1\,1\,1\,1\,0) \\ (1\,1\,0\,1\,0\,0), (1\,1\,0\,1\,1\,1) \end{array} \right\}, \qquad C_4 \cap R_4 = \emptyset$$

and

$$V_5 = V_4, \qquad R_5 = R_4.$$

R_5 is non-empty. Thus, in step 4 we continue to step 5, where we this time note that $C_4 \cap R_4$ is empty. Thus, we continue with the the heuristic part of the algorithm, i.e., step 6, where we find two rows in C_4 on hamming distance 1 from the only row in R_4, and those are

$$(1\,0\,0\,1\,1\,0) \quad \text{and} \quad (1\,1\,0\,1\,1\,1).$$

Both occur once, so we choose one at random, say the first one. We update R_5 with this row, and now we have

$$R_5 = \{(1\,0\,0\,1\,1\,1), (1\,0\,0\,1\,1\,0)\}.$$

We increment i to $i = 5$ and return to step 2.

According to steps 2 and 3, we calculate

$$C_5 = C_4, \qquad C_5 \cap R_5 = \{(1\,0\,0\,1\,1\,0)\},$$

and

$$V_6 = \left\{ \begin{array}{l} (1\,0\,0\,0\,0\,0), (0\,1\,0\,0\,0\,0) \\ (0\,0\,1\,0\,0\,0), (0\,0\,0\,1\,0\,0) \\ (0\,0\,0\,0\,1\,0), (0\,0\,0\,0\,0\,1) \\ (1\,0\,1\,0\,0\,0), (0\,0\,1\,0\,1\,0) \\ (1\,1\,1\,0\,0\,0), (0\,0\,1\,1\,1\,0) \\ (1\,1\,0\,1\,1\,0), (1\,0\,0\,1\,1\,0) \end{array} \right\}, \qquad R_6 = \{(1\,0\,0\,1\,1\,1)\}.$$

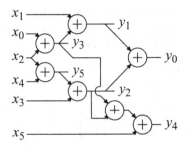

Fig. 3. Realization of the computation in (5) using the proposed approach

R_6 is non-empty. Again, in step 4 we continue to step 5, where we note that $C_5 \cap R_5$ is non-empty. Thus, we increment i to $i = 6$ and return to step 2.

According to steps 2 and 3, we calculate

$$C_6 = C_5 \cup \{(1\,0\,0\,1\,1\,1)\}, \quad C_6 \cap R_6 = \{(1\,0\,0\,1\,1\,1)\},$$

and

$$V_7 = \left\{ \begin{array}{l} (1\,0\,0\,0\,0\,0), (0\,1\,0\,0\,0\,0) \\ (0\,0\,1\,0\,0\,0), (0\,0\,0\,1\,0\,0) \\ (0\,0\,0\,0\,1\,0), (0\,0\,0\,0\,0\,1) \\ (1\,0\,1\,0\,0\,0), (0\,0\,1\,0\,1\,0) \\ (1\,1\,1\,0\,0\,0), (0\,0\,1\,1\,1\,0) \\ (1\,1\,0\,1\,1\,0), (1\,0\,0\,1\,1\,0) \\ (1\,0\,0\,1\,1\,1) \end{array} \right\}, \quad R_7 = \emptyset.$$

R_7 is empty. Finally, in step 4 we stop.

Each row in V_7, except the initial rows from the identity matrix, corresponds to one XOR-gate. The network is built by identifying the rows in V_7 in the order that they were added to the set of available rows. The corresponding implementation is given in Fig. 3.

4   Results

To evaluate the efficiency of the different methods a number of different tests has been performed, both on random matrices as well as constant multiplications in Galois fields.

4.1   Random Matrices

In this test random matrices with varying number of rows and columns were used. For each position the probability of a one is 0.5 independent of adjacent positions. For each row/column combination 100 matrices were used.

In Fig. 4, the average number of XOR-gates using the different algorithms for matrices with ten columns is shown. Clearly, the graph based algorithm

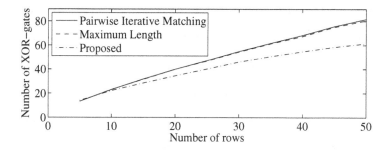

Fig. 4. Average number of XOR-gates for matrices with ten columns and a varying number of rows. *Pairwise Iterative Matching* refers to the method in [3, 4, 10], while *Maximum Length* refers to the method in [14].

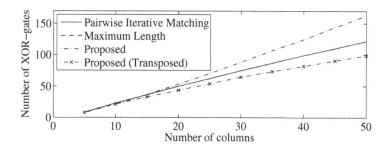

Fig. 5. Average number of XOR-gates for matrices with ten rows and a varying number of columns. *Pairwise Iterative Matching* refers to the method in [3, 4, 10], while *Maximum Length* refers to the method in [14]. The difference between those two methods is not visible in this graph.

outperforms the sub-expression sharing algorithms. Even though the average number of XOR-gates are similar for both sub-expression sharing algorithms, they vary with a few gates for each individual case.

In a similar way, using matrices with ten rows and varying the number of columns, the results in Fig. 5 are obtained. Here, it is obvious that the sub-expression sharing algorithms perform better with increasing number of columns. The reason for this is that the heuristic step for the proposed algorithm, step 6, does not work well enough when the Hamming distance between the possible rows and the required rows is large. Instead, it is for these cases better to find the few sub-expressions that are common between the rows. However, by realizing the transposed matrix and transposing the resulting network, the proposed algorithm shows significantly better results compared to the sub-expression sharing based methods.

Finally, in Fig. 6, the same number of rows and columns are used. Here it is seen that the proposed algorithm is slightly better for sizes between five and 20 rows and columns, while for larger sizes the sub-expression sharing methods are better. This could be explained by the heuristic part as discussed above.

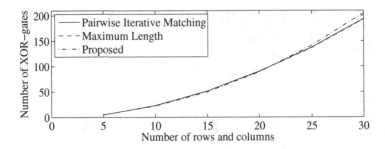

Fig. 6. Average number of XOR-gates for matrices with varying number of rows and columns. *Pairwise Iterative Matching* refers to the method in [3,4,10], while *Maximum Length* refers to the method in [14].

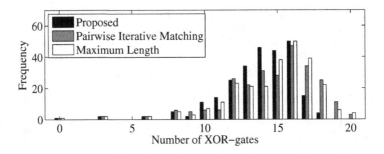

Fig. 7. Frequency of XOR-gate count for all multiplications in $GF(2^8)$ with primitive polynomial $x^8 + x^4 + x^3 + x^2 + 1$. *Pairwise Iterative Matching* refers to the method in [3,4,10], while *Maximum Length* refers to the method in [14].

4.2 Galois Field Arithmetic

During syndrome calculation and Chien search in decoding of Reed-Solomon codes, multiplication by constants can be used. In a fully parallel implementation of the syndrome calculation, one instance of each multiplier constant is required. Hence, the number of XOR-gates required to realize all constant multiplications for a given Galois field is a relevant measure.

For $GF(2^8)$ a commonly used polynomial is $x^8 + x^4 + x^3 + x^2 + 1$. The result of applying the algorithms to all constant multiplications using the polynomial basis is shown in Fig. 7. There are 255 different non-zero constants, and, hence, 255 8 × 8-matrices. The total number of required XOR-gates are 3800, 3776, and 3537 for the pairwise iterative matching, maximum length, and the proposed algorithms, respectively. Using the best solution for each constant multiplier among the three considered algorithms requires 3509 XOR-gates. A straightforward realization requires 6152 XOR-gates.

In a similar way the total number of XOR-gates for all constant multiplications in $GF(2^{16})$ with primitive polynomial $x^{16} + x^{12} + x^3 + x + 1$ is 3681652 for the proposed algorithm. For the sub-expression sharing based ones 3847420 and

Table 1. Number of XOR-gates required for all constant multiplications in $GF(2^8)$. *Pairwise Iterative Matching (PIM)* refers to the method in [3, 4, 10], while *Maximum Length (ML)* refers to the method in [14].

Primitive polynomial	PIM	ML	Proposed	Best[i]	No sharing
$x^8 + x^7 + x^3 + x^2 + 1$	3725	3711	3522	3471	6152
$x^8 + x^5 + x^3 + x + 1$	3738	3742	3513	3473	6152
$x^8 + x^7 + x^5 + x^3 + 1$	3755	3725	3524	3475	6152
$x^8 + x^6 + x^5 + x + 1$	3730	3718	3543	3482	6152
$x^8 + x^6 + x^5 + x^2 + 1$	3776	3752	3533	3492	6152
$x^8 + x^6 + x^3 + x^2 + 1$	3777	3747	3537	3498	6152
$x^8 + x^6 + x^5 + x^4 + 1$	3799	3777	3539	3504	6152
$x^8 + x^5 + x^3 + x^2 + 1$	3781	3764	3553	3505	6152
$x^8 + x^4 + x^3 + x^2 + 1$	3800	3776	3537	3509	6152
$x^8 + x^6 + x^5 + x^3 + 1$	3800	3761	3551	3509	6152
$x^8 + x^6 + x^4 + x^3 + x^2 + x + 1$	3768	3749	3543	3510	6152
$x^8 + x^7 + x^6 + x + 1$	3803	3782	3553	3511	6152
$x^8 + x^7 + x^2 + x + 1$	3795	3802	3565	3513	6152
$x^8 + x^7 + x^6 + x^5 + x^2 + x + 1$	3812	3807	3550	3517	6152
$x^8 + x^7 + x^6 + x^5 + x^4 + x^2 + 1$	3781	3749	3560	3520	6152
$x^8 + x^7 + x^6 + x^3 + x^2 + x + 1$	3820	3796	3576	3541	6152

[i] As a comparison: For each constant multiplication, the best solution among the three considered algorithms is used.

3839031 XOR-gates are required, respectively. Taking the best solution for each constant multiplication requires 3613084 XOR-gates, while a straightforward realization requires 7322438 XOR-gates.

Another aspect is the selection of a primitive polynomial so that the minimum complexity is obtained. In Table 1, the number of XOR-gates for all constant multiplications in $GF(2^8)$ using different primitive polynomials using the different algorithms are shown, along with the minimum number of XOR-gates using the best solution for each constant. It can be seen that the proposed algorithm outperforms the sub-expression sharing case when the total number of XOR-gates is considered and only in rare cases results in a worse XOR-gate count for single multiplications.

In [4] a given multiplication in $GF(2^8)$ was used to describe the algorithm. It was claimed that the 18 XOR-gates required were globally optimal. Applying the proposed algorithm gives 17 XOR-gates (16 and 17 for the sub-expression sharing based). In another example using four constant multipliers with the same input, the result in [4] was 44 XOR-gates. Here, using the proposed algorithm 33 XOR-gates are required (40 and 38 for the sub-expression sharing based).

5   Conclusions

In this work the problem of minimization of the number of binary field additions (XOR-gates) for constant matrix multiplication over the binary field was considered. This is a common type of operation in, e.g., Galois field arithmetic, linear error-correcting codes, CRC computation, LFSRs, and the AES algorithm.

An algorithm for solving the problem was proposed and it was shown by examples both on random matrices as well as matrices originating from Galois field computations that the propose algorithm worked better than earlier proposed algorithms.

The main advantage of the proposed algorithm is the ability to utilize cancellation of terms when realizing the matrix multiplication. This brings an advantage over sub-expression sharing based methods even if exact solutions of these are considered.

References

1. Lidl, R., Niederreiter, H.: Finite Fields. Cambridge University Press, Cambridge (1996)
2. McEliece, R.J.: Finite Fields for Computer Scientists and Engineers. Springer, Heidelberg (1987)
3. Paar, C.: Optimized Arithmetic for Reed-Solomon Encoders. In: Proceedings of IEEE International Symposium on Information Theory, Ulm, Germany, p. 250 (June 1997)
4. Hu, Q., Wang, Z., Zhang, J., Xiao, J.: Low Complexity Parallel Chien Search Architecture for RS Decoder. In: Proceedings of IEEE International Symposium on Circuits and Systems, Kobe, Japan, May 2005, vol. 1, pp. 340–343 (2005)
5. Paar, C.: A New Architecture for a Parallel Finite Field Multiplier with Low Complexity Based on Composite Fields. IEEE Transactions on Computers 45(7), 856–861 (1996)
6. Olofsson, M.: VLSI Aspects on Inversion in Finite Fields. PhD thesis, Linköping University, Linköping, Sweden, No. 731 (February 2002)
7. Chen, Y., Parhi, K.K.: Small Area Parallel Chien Search Architectures for Long BCH Codes. IEEE Transactions on VLSI Systems 12(5), 401–412 (2004)
8. Pei, T.B., Zukowski, C.: High-speed Parallel CRC Circuits in VLSI. IEEE Transactions on Communications 40(4), 653–657 (1992)
9. Cheng, C., Parhi, K.K.: High-speed Parallel CRC Implementation Based on Unfolding, Pipelining, and Retiming. IEEE Transactions on Circuits and Systems II 53(10), 1017–1021 (2006)
10. Zhang, X., Parhi, K.K.: Implementation approaches for the Advanced Encryption Standard algorithm. IEEE Circuits and Systems Magazine 2(4), 24–46 (2002)
11. Bull, D.R., Horrocks, D.H.: Primitive Operator Digital Filters. IEE Proceedings G 138(3), 401–412 (1991)
12. Potkonjak, M., Shrivasta, M.B., Chandrakasan, A.P.: Multiple Constant Multiplication: Efficient and Versatile Framework and Algorithms for Exploring Common Subexpression Elimination. IEEE Transactions on Computer-Aided Design 15(2), 151–161 (1996)
13. Hartley, R.I.: Subexpression Sharing in Filters Using Canonic Signed Digit Multipliers. IEEE Transactions on Circuits and Systems II 43(10), 677–688 (1996)
14. Pasko, R., Schaumont, P., Derudder, V., Vernalde, S., Durackova, D.: A New Algorithm for Elimination of Common Subexpressions. IEEE Transactions on Computer-Aided Design 18(1), 58–68 (1999)

15. Xu, F., Chang, C.H., Jong, C.C.: Contention Resolution Algorithm for Common Subexpression Elimination in Digital Filter Design. IEEE Transactions on Circuits and Systems II 52(10), 695–700 (2005)
16. Flores, P., Monteiro, J., Costa, E.: An Exact Algorithm for the Maximal Sharing of Partial Terms in Multiple Constant Multiplications. In: IEEE/ACM International Conference on Computer-Aided Design, San Jose, CA, November 2005, pp. 13–16 (2005)
17. Gustafsson, O., Wanhammar, L.: ILP Modelling of the Common Subexpression Sharing Problem. In: International Conference on Electronics, Circuits and Systems. Dubrovnik, Croatia, vol. 3, pp. 1171–1174 (September 2002)
18. Bordewijk, J.L.: Inter-reciprocity Applied to Electrical Networks. Applied Scientific Research 6(1), 1–74 (1957)

Towards Optimal Toom-Cook Multiplication for Univariate and Multivariate Polynomials in Characteristic 2 and 0

Marco Bodrato

Centro "Vito Volterra" – Università di Roma Tor Vergata
Via Columbia 2 – 00133 Rome, Italy
waifi2007@bodrato.it

Abstract. Toom-Cook strategy is a well-known method for building algorithms to efficiently multiply dense univariate polynomials. Efficiency of the algorithm depends on the choice of interpolation points and on the exact sequence of operations for evaluation and interpolation. If carefully tuned, it gives the fastest algorithm for a wide range of inputs.

This work smoothly extends the Toom strategy to polynomial rings, with a focus on $GF_2[x]$. Moreover a method is proposed to find the faster Toom multiplication algorithm for any given splitting order. New results found with it, for polynomials in characteristic 2, are presented.

A new extension for multivariate polynomials is also introduced; through a new definition of density leading Toom strategy to be efficient.

Keywords: Polynomial multiplication, multivariate, finite fields, Toom-Cook, Karatsuba, GF2x, binary polynomials, squaring, convolution.

1 Introduction

Starting with the works of Karatsuba[9] and Toom[13], who found methods to lower asymptotic complexity for polynomial multiplication from $O(n^2)$ to $O(n^{1+\epsilon})$ with $0 < \epsilon < 1$, many efforts have been done in finding optimised implementations in arithmetic software packages[5,6,12].

The family of so-called Toom-Cook methods is an infinite set of algorithms. Each of them requires polynomial evaluation of the two operands and a polynomial interpolation problem, with base points not specified *a priori*, giving rise to many possible Toom-k algorithms, even for a fixed size of the operands.

Moreover, to implement one of them, we will need a sequence of many basic operations, which typically are sums and subtractions of arbitrary long operands, multiplication and exact division of long operand by *small* one, optimised, when possible, by bit-shifts.

The exact sequence is important because it determines the real efficiency of the algorithm. It is well known[10] that the recursive application of a single Toom-k algorithm to multiply two polynomials of degree n gives an asymptotic complexity of $O(n^{log_k(2k-1)})$. There is even the well known Schönhage-Strassen method[14,15], which complexity is asymptotically better than any Toom-k:

C. Carlet and B. Sunar (Eds.): WAIFI 2007, LNCS 4547, pp. 116–133, 2007.
© Springer-Verlag Berlin Heidelberg 2007

$O(n \log n \log \log n)$. But the O-notation hides a constant factor which is very important in practice.

All the advanced software libraries actually implement more than one method because the asymptotically better ones, are not practical for small operands. So there can be a wide range of operands where Toom-Cook methods can be the preferred ones. The widely known GMP library[5] uses Toom-2 from around 250 decimal digits, then Toom-3, and finally uses FFT based multiplication over 35,000 digits. Hence the interest for improvement in Toom-k.

On the multivariate side, the problem is much more complex. Even if the combination of Kronecker's trick[11] with FFT multiplication can give asymptotically fast methods, the overhead is often too big to have algorithms useful in practice. The constraint for the polynomials to be dense is most of the time false, for real world multivariate problems. A more flexible definition for density can help.

1.1 Representation of $GF_2[x]$ and Notation

All the algorithms in this paper work smoothly with elements of $GF_2[x]$ stored in compact dense binary form, where each bit represents a coefficient and any degree 7 polynomial fits in one byte.

For compactness and simpler reading, we will somewhere use hexadecimal notation. Every hexadecimal number h corresponds to the element $p \in GF_2[x]$ such that $p(2) = h$. For example $p \in GF_2[x] \leftrightarrow \text{hex}, 1 \leftrightarrow 1, x \leftrightarrow 2, x + 1 \leftrightarrow 3, \ldots, x^3 + x^2 + x + 1 \leftrightarrow F, \ldots, x^8 + x^7 + x^6 \leftrightarrow 1C0, \ldots$.

We will also use the symbols $\ll$ and $\gg$ for bit-shifts. Meaning multiplication and division by power of x, in $GF_2[x]$, or by power of 2 in $\mathbb{Z}[x]$.

2   Toom-Cook Algorithm for Polynomials, Revisited

A general description of the Toom algorithm follows. Starting from two polynomials $u, v \in \mathbb{R}[x]$, on some integral domain $\mathbb{R}$, we want to compute the product $\mathbb{R}[x] \ni w = u \cdot v$. The whole algorithm can be described in five steps.

Splitting: Choose some base $Y = x^b$, and represent u and v by means of two polynomials $\mathfrak{u}(y, z) = \sum_{i=0}^{n-1} u_i z^{n-1-i} y^i$, $\mathfrak{v}(y, z) = \sum_{i=0}^{m-1} v_i z^{m-1-i} y^i$, both homogeneous, with respectively n and m coefficients and degrees $\deg(\mathfrak{u}) = n - 1, \deg(\mathfrak{v}) = m - 1$. Such that $\mathfrak{u}(x^b, 1) = u, \mathfrak{v}(x^b, 1) = v$. The coefficients $u_i, v_i \in \mathbb{R}[x]$ are themselves polynomials and can be chosen to have degree $\forall i, \deg(u_i) < b, \deg(v_i) < b$.

Traditionally the Toom-n algorithm requires balanced operands so that $m = n$, but we can easily generalise to unbalanced ones. We assume commutativity, hence we also assume $n \geq m > 1$.

Evaluation: We want to compute $\mathfrak{w} = \mathfrak{u} \cdot \mathfrak{v}$ which degree is $d = n + m - 2$, so we need $d + 1 = n + m - 1$ evaluation points $P_d = \{(\alpha_0, \beta_0), \ldots, (\alpha_d, \beta_d)\}$ where $\alpha_i, \beta_i \in \mathbb{R}[x]$ can be polynomials. We define $c = \max_i(\deg(\alpha_i), \deg(\beta_i))$.

The evaluation of a single polynomial (for example $\mathfrak{u}$) on the points (α_i, β_i), can be computed with a matrix by vector multiplication. The matrix $E_{d,n}$ is a $(d + 1) \times n$ Vandermonde-like matrix. $\overline{\mathfrak{u}(\alpha, \beta)} = E_{d,n} \overline{u} \implies$

$$
\begin{pmatrix} \mathfrak{u}(\alpha_0,\beta_0) \\ \mathfrak{u}(\alpha_1,\beta_1) \\ \vdots \\ \mathfrak{u}(\alpha_d,\beta_d) \end{pmatrix} = \begin{pmatrix} \beta_0^{n-1} & \alpha_0\cdot\beta_0^{n-2} & \cdots & \alpha_0^{n-2}\cdot\beta_0 & \alpha_1^{n-1} \\ \beta_1^{n-1} & \alpha_1\cdot\beta_1^{n-2} & \cdots & \alpha_1^{n-2}\cdot\beta_1 & \alpha_2^{n-1} \\ \vdots & \vdots & & \vdots & \vdots \\ \beta_k^{n-1} & \alpha_k\cdot\beta_k^{n-2} & \cdots & \alpha_k^{n-2}\cdot\beta_k & \alpha_k^{n-1} \end{pmatrix} \begin{pmatrix} u_0 \\ u_1 \\ \vdots \\ u_{n-1} \end{pmatrix} \tag{1}
$$

Recursive multiplication: We compute $\forall i$, $\mathfrak{w}(\alpha_i,\beta_i) = \mathfrak{u}(\alpha_i,\beta_i)\cdot\mathfrak{v}(\alpha_i,\beta_i)$, with $d+1$ multiplications of polynomials which degree is paragonable to that of $Y = x^b$. Exactly we have $\deg(\mathfrak{u}(\alpha_i,\beta_i)) \le c(n-1)+b$, $\deg(\mathfrak{v}(\alpha_i,\beta_i)) \le c(m-1)+b$, and the results $\deg(\mathfrak{w}(\alpha_i,\beta_i)) \le c(n+m-2)+2b = cd+2b$. We note that c,d,m,n are fixed numbers for a chosen implementation, b instead will grow as the operands grow.

Interpolation: This step depends only on the expected degree of the result d, and on the $d+1$ chosen points (α_i,β_i), no more on n and m separately. We now need the coefficients of the polynomial $\mathfrak{w}(y,z) = \sum_{i=0}^{d} w_i z^{d-i} y^i$. We know the values of $\mathfrak{w}$ evaluated in $d+1$ points, so we face a classical interpolation problem. We need to apply the inverse of A_d, a $(d+1)\times(d+1)$ Vandermonde-like matrix. $\overline{\mathfrak{w}(\alpha,\beta)} = A_d\overline{w} \Longrightarrow$

$$
\begin{pmatrix} w_0 \\ w_1 \\ \vdots \\ w_d \end{pmatrix} = \begin{pmatrix} \beta_0^{d} & \alpha_0\cdot\beta_0^{d-1} & \cdots & \alpha_0^{d-1}\cdot\beta_0 & \alpha_0^{d} \\ \beta_1^{d} & \alpha_1\cdot\beta_1^{d-1} & \cdots & \alpha_1^{d-1}\cdot\beta_1 & \alpha_1^{d} \\ \vdots & \vdots & & \vdots & \vdots \\ \beta_d^{d} & \alpha_d\cdot\beta_d^{d-1} & \cdots & \alpha_k^{d-1}\cdot\beta_d & \alpha_d^{d} \end{pmatrix}^{-1} \begin{pmatrix} \mathfrak{w}(\alpha_0,\beta_0) \\ \mathfrak{w}(\alpha_1,\beta_1) \\ \vdots \\ \mathfrak{w}(\alpha_d,\beta_d) \end{pmatrix} \tag{2}
$$

Recomposition: The desired result can be simply computed with one more evaluation: $w = \mathfrak{w}(x^b,1)$. This step requires at most d *shifts* and sums.

The two critical phases are **evaluation** and **interpolation**. As stated by formulas (1) and (2), both require a matrix by vector multiplication. This two phases can require many sums and subtractions, shifts, and even small multiplications or exact divisions (interpolation only) by small elements in $\mathbb{R}[x]$. The goal of this paper is to find some optimal Evaluation Sequences of operations (called ES from now on) as well as Interpolation Sequences (IS), leading to optimal algorithms.

2.1   References on Collected Ideas

After the first proposals [13,4], many small improvements where introduced beside Toom ideas. Winograd[17] proposed ∞ and fractions for the evaluation points; same results are obtained here with homogenisation. Zimmermann and Quercia[18] proposed to evaluate also on positive and negative powers of $x \in \mathrm{GF}_2[x]$; this idea is extended here using any coprime couple $\alpha_i, \beta_i \in \mathbb{R}[x]$ in the polynomial ring. Bodrato and Zanoni[2], underlined the need to consider unbalanced operands; this idea was inherited by this paper.

3   The Matrices

Two kind of matrices are involved in any Toom-k algorithm, the square invertible matrix A_d and the two, possibly equal, matrices $E_{d,n}, E_{d,m}$ with the same number $d + 1 = 2k - 1$ of rows, but fewer (respectively. $n \leq d$ and $m \leq d$) columns.

3.1 Matrices for the Interpolation Sequence

Since the matrices from equation 2 must be invertible, we are interested in the determinant. Which can be computed from the points in P_d.

Theorem 1. *For the Vandermonde-like matrix A_d generated from the $d + 1$ points in $P_d = \{(\alpha_0, \beta_0), \ldots, (\alpha_d, \beta_d)\}$, the determinant can be computed with:*

$$\det(A_d) = \prod_{0 \leq i < j \leq d} (\alpha_i \beta_j - \alpha_j \beta_i)$$

Proof. It can be easily seen that a matrix with two points with $\beta_i = \beta_j = 0$ is not invertible, and the above formula correctly gives 0.

   If one point, suppose (α_0, β_0), has $\beta_0 = 0$, the matrix will start with the line $(0, \ldots, 0, \alpha_0^d)$. Computing the determinant starting from this row, we will have $\alpha_i^d \det(\widetilde{A}_d)$ where $\widetilde{A}_d$ is the complementary minor. $\widetilde{A}_d$ is a Vandermonde $d \times d$ matrix for the points α_i / β_i, where the i-th line was multiplied by β_i^d.

   Using the classical formula for Vandermonde matrices, we obtain:

$$\det(A_d) = \alpha_0^d \det(\widetilde{A}_d) = \alpha_0^d \prod_{0 < i \leq d} \beta_i^d \prod_{0 < i < j \leq d} (\alpha_i / \beta_i - \alpha_j / \beta_j) = \prod_{0 \leq i < j \leq d} (\alpha_i \beta_j - \alpha_j \beta_i)$$

3.2 The Choice of Evaluation Points

The choice of the evaluation points P_d is one of the most important steps to reach an optimal implementation, and completely determines the matrices $A_d, E_{d,2}, \ldots, E_{d,d}$.

   We will consider two of them as being automatically chosen $(0, 1), (1, 0)$, representing respectively 0 and ∞, and immediately giving $w_0 = u_0 \cdot v_0, w_d = u_{n-1} \cdot v_{m-1}$, and the rows $(1, 0, \ldots), (\ldots, 0, 1)$. An other good choice is the point $(1, 1)$, and (if characteristic $\neq 2$) $(-1, 1)$. We need an invertible matrix A_d, so if we use any point (α_i, β_i), no other multiple point $(\lambda \alpha_i, \lambda \beta_i)$ can be added, or the factor $(\alpha_i \lambda \beta_i - \lambda \alpha_i \beta_i)$ will nullify the determinant.

   Since the dimension of the extra space needed for the *carries* depends on the parameter $c = \max_i (\deg(\alpha_i), \deg(\beta_i))$, we try to keep it as small as possible. That's why in $\mathrm{GF}_2[x]$ we consider only the polynomials with degree at most 1. So we will have $\alpha, \beta \in \{0, 1, x, x + 1\}$, and only 9 possible couples:

$$P_{\mathrm{GF}_2[x]} = \{(0, 1), (1, 0), (1, 1), (x, 1), (1, x), (x+1, 1), (1, x+1), (x+1, x), (x, x+1)\}$$

With this auto-imposed restriction we will be able to analyse Toom-k algorithms up to Toom-5. For any choice of the points the following theorem tells us that any Toom-k in $\mathrm{GF}_2[x]$ with $k > 2$ requires at least one division.

Theorem 2. *Suppose $d > 2$, and the two points $(0,1),(1,0) \in P_d$. Than, for any choice of the other points $d-1$ points in P_d, the determinant of the invertible matrix A_d for a Toom algorithm in $\mathrm{GF}_2[x]$ is not a power of x.*

Proof. From theorem 1 we have:

$$\det(A_d) = \prod_{0<i<d} \alpha_i \beta_i \prod_{0<i<j<d} (\alpha_i \beta_j - \alpha_j \beta_i)$$

By contradiction, if the determinant is a power of x, then any factor of the above formula must be. Then all the α_i and β_i are power of x. Any factor $(\alpha_i \beta_j - \alpha_j \beta_i)$ is then a difference of powers of x, giving 0 (a non invertible matrix) or a non-power of x.

3.3   Matrices for the Evaluation Sequence

The matrices $E_{d,n}$ are non-square, so we can not compute the determinant. But we can compute the rank.

Theorem 3. *If the points P_d give an invertible A_d, then the rank of any $E_{d,n}$ matrix is n.*

Proof. Since the $E_{d,n}$ are sub-matrices of the matrix A_d, modulo some multiplication of rows by non-zero constants, all the n columns are linearly independent, so the rank is n.

4   Optimising Through Graph Search

To study ES and IS, we need at first to fix the operations we admit. We consider 4 basic operations, giving a name for their cost in time:
- add or subtract two elements (cost: add)
- multiply an element by a small constant (cost: Smul)
- exact division by a small constant (cost: Sdiv)
- bit-shift by a small amount (cost: shift)

By *small constant*, we mean an element which fits in a few bytes, hopefully in a register of the target CPU. All the resulting algorithms in this paper use *small constants* needing at most two bytes.

We assume that: sum and subtractions do cost the same, right and left shift do cost the same, multiplication cost and exact division cost do not depend on the constant. Moreover we require some relations on the costs:
- shift < add: it should be faster to compute $a = b \ll 1$ than $a = b + b$
- shift < Smul: it should be faster to compute $a = b \ll 3$ than $a = b \cdot x^3$

In the experiments we also used the empirical relations shift < Smul < add < Sdiv, but we did not assume those to be true in general.

We also assume that any linear combination $l_i \leftarrow \pm c_j \cdot l_j/d_j \pm c_k \cdot l_k/d_k$ is possible without using temporary variables, for any $c_j, d_j, c_k, d_k \in \mathbb{R}[x]$ small constants, even if $i = j$. The cost of this linear combination will be simply computed adding up the cost of single operations, converted to bit-shift whenever possible or skipped when the coefficient is trivial.

4.1 Searching for Evaluation Sequences

The sequences ES and IS will be searched working on their respective matrices. IS can be seen as a sequence of operations on the lines of a matrix, starting from the matrix A_d and reaching the identity matrix. A method to determine the optimal IS with no use of temporaries was already given in [2]; except theorem 2 already shown, all the results from that paper can be directly applied to $GF_2[x]$; the same strategy was used to find optimal IS for this paper.

Here we focus on ES. Also ES can be searched working only on the matrix. Again we require the algorithm not to use any temporary variable.

Any evaluation $u(\alpha_i, \beta_i) = \sum_{j=0}^{n-1} u_j \cdot (\alpha_i^{n-1-j} \cdot \beta_i^j)$ can be directly computed with a cost at most equal to $(n-1) \cdot \mathtt{sum} + n \cdot \mathtt{Smul}$, without any division. So we search for the best ES without divisions.

We will search a sequence of elementary operations on lines starting from the zero matrix and leading to the goal matrix $E_{d,n}$. Computing the evaluations $u(\alpha_i, \beta_i)$ we can always use the coefficients of the polynomial u, the vector $\overline{u} = (\dots, u_j, \dots)$. This values can not be modified. So we use a block matrix, where one block is the identity, and the other is the goal $E_{d,n}$. Moreover, since we always use the two points $(0,1), (1,0)$, and they give two rows already present in the identity matrix, we will cut off two lines and use a smaller $\widetilde{E}_{d,n}$ as the goal.

$$
\begin{pmatrix} \overline{u} \\ \hline u(\alpha_i,\beta_i) \end{pmatrix} = \begin{pmatrix} I \\ \hline E \end{pmatrix} \cdot \overline{u}; \quad \begin{pmatrix} I_n \\ \hline E_{d,n} \end{pmatrix} = \left(\begin{array}{ccc} 1 & 0 & \cdots \\ \vdots & \ddots & \vdots \\ \cdots & 0 & 1 \\ \hline 1 & 0 & \cdots \\ & \widetilde{E}_{d,n} & \\ \cdots & 0 & 1 \end{array} \right) ; \text{ examine } \quad \left(\begin{array}{c|ccc} l_{-1}: & 1 & 0 & \cdots \\ \vdots & \vdots & \ddots & \vdots \\ l_{-n}: & \cdots & 0 & 1 \\ l_1: & & & \\ \vdots & & \widetilde{E}_{d,n} & \\ l_{d-2}: & & & \end{array} \right)
$$

Lines coming from the identity matrix must be left untouched, and are noted with a negative index. Allowed operation are

$$l_i \leftarrow c_j \cdot l_j + c_k \cdot l_k, \quad \text{where } i > 0, k \neq i, \quad c_j, c_k \text{ are null or small constants} \tag{3}$$

Then we look for a sequence starting from the empty matrix $M_0 = (\overline{0})$, reaching the goal matrix $\widetilde{E}_{d,n}$. Every single step changes only one line in the M matrix with a linear combination of lines as formula (3) shows.

$$
\left(\begin{array}{c|ccc} l_{-1}: & 1 & 0 & \cdots \\ \vdots & \vdots & \ddots & \vdots \\ l_{-n}: & \cdots & 0 & 1 \\ l_1: & 0 & \cdots & 0 \\ \vdots & \vdots & 0 & \vdots \\ l_{d-2}: & 0 & \cdots & 0 \end{array} \right) \xrightarrow{l_1 \leftarrow l_{-1}+l_{-2}} \left(\begin{array}{ccc} 1 & 0 & \cdots \\ \vdots & \ddots & \vdots \\ \cdots & 0 & 1 \\ 1 & 1 & 0 \cdots \\ \vdots & 0 & \vdots \\ 0 & \cdots & 0 \end{array} \right) \cdots \begin{pmatrix} I \\ \hline M \end{pmatrix} \cdots \rightsquigarrow \left(\begin{array}{c|ccc} l_{-1}: & 1 & 0 & \cdots \\ \vdots & \vdots & \ddots & \vdots \\ l_{-n}: & \cdots & 0 & 1 \\ l_1: & & & \\ \vdots & & \widetilde{E}_{d,n} & \\ l_{d-2}: & & & \end{array} \right) \tag{4}
$$

4.2   The Graph

Now we have all the ingredients to build a graph, and search for the *shortest path*. The nodes in the graph are all the possible matrices of the form $\left(\frac{I}{M}\right)$ as in (4), and will be labelled by M. From every node, directed arcs represent possible operations as in (3), we only need a limit for the possible coefficients. For the result in this paper we explored combinations with coefficients limited by the biggest coefficient in the goal matrix $E_{d,n}$. In $\mathbb{Z}[x]$ the limit being the absolute value, in $\mathrm{GF}_2[x]$ the degree. The weight of an arc $M \to \widetilde{M}$ is the minimal cost of the operations of the form (3) that lead M to $\widetilde{M}$.

The graphs mentioned above have an infinite number of nodes, so it's essential to use a clever algorithm for the shortest path search. Two possibilities were explored: the Travel Through algorithm described in the previous work[2] on IS and the more standard A* algorithm[7]. The first being slower, but with a smaller memory footprint. While the second is faster but needs too much memory for big matrices.

4.3   Estimate for Evaluation Sequences

Both A* and the Travel Through algorithm need a function to estimate the remaining cost of a path, the estimated cost for a given node M must be smaller or equal to the actual cost of the shortest path from M to the goal $G = \widetilde{E}_{d,n}$.

To build this function we need some preliminary definitions and observations.

Definition 1 (Insertion). *A given arc $l_i \leftarrow c_j \cdot l_j + c_k \cdot l_k$ is an* **insertion** *if and only if $j < 0 \wedge c_j \neq 0$ or $k < 0 \wedge c_k \neq 0$.*

Theorem 4. *If there exist a path of non-insertion arcs from node M to $\widetilde{M}$, then $\mathrm{rank}(M) \geq \mathrm{rank}(\widetilde{M})$.*

Proof. A non-insertion arc, operates inside the matrix M. The resulting line is a linear combination of lines in M, so the rank can not grow.

Theorem 5 (Rank estimate). *The cost of the path from any node M to the goal G is at least $(\mathrm{rank}(G) - \mathrm{rank}(M)) \cdot$ add.*

Proof. By theorem 3, the $\mathrm{rank}(G)$ is maximal and $\mathrm{rank}(G) - \mathrm{rank}(M) \geq 0$. Each step modifies only one line, so the rank will be increased one by one.

Definition 2 (Needed insertions). *Given a matrix M, and a line G_i of the goal matrix, we define $Ni(G_i, M)$ the minimal number of insertions needed to obtain the line G_i from the matrix M.*

If we fix a line M_j of the matrix M, and we note M_{jk}, G_{ik} the k-th elements of the two lines we can compute the minimal needed insertions for a path from M_j to G_i with

$$\widetilde{N}i(G_i, M_j) = n - \max_{\lambda \in \mathrm{GF}_2[x] \setminus \{0\}} (\#\{k : G_{ik} = \lambda M_{jk}\}, 2).$$

Then we can compute the global $Ni(G_i, M) = \min_j(\widetilde{N}i(G_i, M_j))$.

Theorem 6 (Line estimate). *The cost of the path from any node M to the goal G is at least $(\#\{i : Ni(G_i, M) \neq 0\}) \cdot \mathtt{add}$.*

Proof. The function $Ni(G_i, M)$ for a given line G_i gives zero iff G_i is already in M. A combination is needed to change each line which is not yet in the goal.

Theorem 7 (Combined estimate). *To estimate the cost of the path from a node M to the goal G; let $r = \operatorname{rank}(G) - \operatorname{rank}(M)$, $a = \#\{i : Ni(G_i, M) = 1\}$ and $b = \#\{i : Ni(G_i, M) > 1\}$, then*

if $r \leq a$ the cost is at least $(a + b) \cdot \mathtt{sum}$
if $r > a$ the cost is at least $(r + b - \lfloor (r - a)/2 \rfloor) \cdot \mathtt{sum}$

Proof. If $r \leq a$ the cost is that of theorem 6.

If $r > a$ we proceed by induction. For the base case, we note that the formula $f(r, a, b) = (r + b - \lfloor (r - a)/2 \rfloor)$, when $r = a$, gives $f(a, a, b) = a + b$ which is correct.

Then we study how the values a, b, r are modified following an arc from M to an other matrix M'. We can have the new $r' < r$ only if the arc is an insertion, when this happen we have $r' = r - 1$. An insertion can decrease by one the lines counted by a, or move a line from the set counted by b to the set counted by a. If the first condition applies, $b' = b, a' = a - 1 \Rightarrow f(r', a', b') = r - 1 + b - \lfloor (r - 1 - a + 1)/2 \rfloor) = f(r, a, b) - 1$, if the second applies, $b' = b - 1, a' = a + 1 \Rightarrow f(r', a', b') = r - 1 + b - 1 - \lfloor (r - a - 2)/2 \rfloor) = f(r, a, b) - 1$. Otherwise, if $r' = r$, the arc can be a non-insertion, so it can change more than one element, but on a single line, and possibly decrease a or b by 1. In both cases $f(r', a', b') \geq f(r, a, b) - 1$.

The last combined estimate is stronger than the others and is good enough to allow the complete analysis for Toom-4 matrices in $\mathrm{GF}_2[x]$.

5   Results and Algorithms in Characteristic 2

The algorithm described in the following sections were studied to work in $\mathrm{GF}_2[x]$, but can be applied in general for characteristic 2. We skip Toom-2 because it coincides with the well known Karatsuba.

5.1   Toom-2.5 in $\mathrm{GF}_2[x]$

The Toom-2.5 algorithm can be used to multiply two operands whose size is not the same. In particular, one will be divided in 3 parts, the other in 2 parts.

There are many possible choices for the set of points P_3, once inserted the canonical points $(1, 0), (0, 1)$, there is a couple of free points left.

Many pairs of points give a total cost of both ES for $E_{3,2}$ and $E_{3,3}$ equal to $6 \cdot \mathtt{add} + 3 \cdot \mathtt{shift}$ and the evaluation always require 4 multiplications. But only two pairs[1], $(1, 1), (x, 1)$ and $(1, 1), (x + 1, 1)$, reach the minimum cost for the IS: $6 \cdot \mathtt{add} + 2 \cdot \mathtt{shift} + 1 \cdot \mathtt{Sdiv}$.

---

[1] Also their reciprocal $(1, 1), (1, x)$ and $(1, 1), (1, x + 1)$.

We show here both algorithms, they are very similar. The author prefers the first one, involving $x + 1$, because of a slightly better locality.

$$P_3^{x+1} = \{(0,1),(1,1),(x+1,1),(1,0)\} \quad P_3^x = \{(0,1),(1,1),(x,1),(1,0)\}$$

$$E_{3,2}^{x+1} = \begin{pmatrix} 1 & 0 \\ \begin{smallmatrix} 1 & 1 \\ 1 & 3 \end{smallmatrix} \\ 0 & 1 \end{pmatrix}; E_{3,3}^{x+1} = \begin{pmatrix} 1 & 0 & 0 \\ \begin{smallmatrix} 1 & 1 & 1 \\ 1 & 3 & 5 \end{smallmatrix} \\ 0 & 0 & 1 \end{pmatrix}; A_3^{x+1} = \begin{pmatrix} 1 & 0 & 0 & 0 \\ 1 & 1 & 1 & 1 \\ 1 & 3 & 5 & F \\ 0 & 0 & 0 & 1 \end{pmatrix} \quad E_{3,2}^x = \begin{pmatrix} 1 & 0 \\ \begin{smallmatrix} 1 & 1 \\ 1 & 2 \end{smallmatrix} \\ 0 & 1 \end{pmatrix}; E_{3,3}^x = \begin{pmatrix} 1 & 0 & 0 \\ \begin{smallmatrix} 1 & 1 & 1 \\ 1 & 2 & 4 \end{smallmatrix} \\ 0 & 0 & 1 \end{pmatrix}; A_3^x = \begin{pmatrix} 1 & 0 & 0 & 0 \\ 1 & 1 & 1 & 1 \\ 1 & 2 & 4 & 8 \\ 0 & 0 & 0 & 1 \end{pmatrix}$$

```
U  = U2*Y^2 + U1*Y + U0              U  = U2*Y^2 + U1*Y + U0
V  = V1*Y    + V0                    V  = V1*Y    + V0
\\Evaluation:6 add,3 shift;4 mul     \\Evaluation:6 add,3 shift,4 mul
W3 = U2 + U1 + U0; W0 = V1 + V0      W3 = U2 + U1 + U0; W0 = V1 + V0
W1 = W3 * W0                         W1 = W3 * W0
W3 = W3 +(U1 + U2*(x))*(x)           W3 = U0 +(U1 + U2*(x))*(x)
              W0 = W0 + V1*(x)                     W0 = V0 + V1*(x)
W2 = W3 * W0                         W2 = W3 * W0
W3 = U2 * V1    ; W0 = U0 * V0       W3 = U2 * V1    ; W0 = U0 * V0
\\Interpolate:6 add,2 shift,1 Sdiv   \\Interpolate:6 add,2 shift,1 Sdiv
W2 =(W2 + W1)/(x)                    W2 =(W2 + W1)/(x+1)
W1 = W1 + W0                         W1 = W1 + W0
W2 =(W2 + W1)/(x+1)                  W2 =(W2 + W1)/(x)
W2 = W2 + W3*(x)                     W2 = W2 + W3*(x)
W1 = W1 + W2 + W3                    W1 = W1 + W2
                                     W2 = W2 + W3

\\Recomposition                      \\Recomposition
W = W3*Y^3+ W2*Y^2+ W1*Y + W0        W = W3*Y^3+ W2*Y^2+ W1*Y + W0
W == U*V                            W == U*V
```

5.2   Toom-3 in $GF_2[x]$

Toom-3 is by far the best known and widely used variant of Toom-Cook algorithms. But usually only in characteristic 0, for the multiplication in $\mathbb{Z}$ or $\mathbb{Z}[x]$. While writing this paper, only one implementation of balanced Toom-3 in $GF_2[x]$ was found on the net, by Zimmermann[18], based on the NTL library[12] and carefully optimised. It uses the points $P_4^Z = \{(0,1),(1,x),(1,1),(x,1),(1,0)\}$, an ES requiring $6 \cdot$ add $+ 4 \cdot$ shift for each operand, and an IS with cost $11 \cdot$ add $+ 5 \cdot$ shift $+ 2 \cdot$ Sdiv.

We tested all the triplets of points in $P_{GF_2[x]} \backslash (1,0),(0,1)$, and the combination $(1,1),(1,x),(1,x+1)$ together with its reciprocal $(1,1),(x,1),(x+1,1)$ gave the best results.

Toom-3 has two variants, the balanced one, which is the most interesting, because it can be used recursively; and the unbalanced, good when one operand is about twice as big as the other. The two variants share the same IS but has different evaluation matrices. The balanced version uses twice $E_{4,3}$, while the unbalanced uses $E_{4,2}$ for the smallest operand and $E_{4,4}$ for the bigger one.

The set of points used is $P_4 = \{(0,1),(1,1),(1,x),(1,x+1),(1,0)\}$.

$$E_{4,3} = \begin{pmatrix} 1\,0\,0 \\ \boxed{\begin{matrix} 1\,1\,1 \\ 1\,2\,4 \\ 1\,3\,5 \end{matrix}} \\ 0\,0\,1 \end{pmatrix} ; \ A_4 = \begin{pmatrix} 1\,0\,0\,0\,0 \\ 1\,1\,1\,1\,1 \\ 1\,2\,4\,8\,10 \\ 1\,3\,5\,F\,11 \\ 0\,0\,0\,0\,1 \end{pmatrix} \qquad E_{4,2} = \begin{pmatrix} 1\,0 \\ \boxed{\begin{matrix} 1\,1 \\ 1\,2 \\ 1\,3 \end{matrix}} \\ 0\,1 \end{pmatrix} ; \ E_{4,4} = \begin{pmatrix} 1\,0\,0\,0 \\ \boxed{\begin{matrix} 1\,1\,1\,1 \\ 1\,2\,4\,8 \\ 1\,3\,5\,F \end{matrix}} \\ 0\,0\,0\,1 \end{pmatrix}$$

```
U = U2*Y^2 + U1*Y + U0              U = U3*Y^3 + U2*Y^2 + U1*Y + U0
V = V2*Y^2 + V1*Y + V0              V = V1*Y    + V0
\\Evaluation:10 add,4 shift;5 mul   \\Eval:11 add,4 shift,1 Smul;5 mul
W3 = U2+U1+U0    ; W2 = V2+V1+V0    W3 = U3+U2+U1+U0 ;W2 = V1 + V0
W1 = W3 * W2                        W1 = W2 * W3
W0 = U2*x^2+U1*x ; W4 = V2*x^2+V1*x W0 = U3*(x^3)+U2*(x^2)+U1*(x)
W3 = W3 + W0     ; W2 = W2 + W4     W3 = W3 + W0 + (x^2+x)*U3
                                                 W2 = W2 + V1*(x)
W0 = W0 + U0     ; W4 = W4 + V0     W0 = W0 + U0     ;W4 = W2 + V1
W3 = W3 * W2     ; W2 = W0 * W4     W3 = W3 * W2     ;W2 = W0 * W4
W4 = U2 * V2     ; W0 = U0 * V0     W4 = U3 * V1     ;W0 = U0 * V0
```

```
\\Interpolation:9 add,1 shift,1 Smul,2 Sdiv
W3 = W3 + W2
W2 =((W2+ W0 )/(x)+ W3 + W4*(x^3+1)) / (x+1)
W1 = W1 + W0
W3 =(W3 + W1 )/(x*(x+1))
W1 = W1 + W4 + W2
W2 = W2 + W3
\\Recomposition:
W = W4*Y^4+ W3*Y^3+ W2*Y^2+ W1*Y + W0
W == U*V       \\ check
```

The IS needs two exact divisions, one by the small constant element $x+1$ and one by $x \cdot (x+1)$. Since we know these divisions are exact by very small constant, they can be performed in linear time[8]. For a test implementation in NTL on a 32 bits CPU, the following C code for exact division by $x+1$ was implemented. It is inspired by an analogous function by Michel Quercia[18]. Division by $x^2 + x$ was actually implemented by one shift and the same function.

```
static void ExactDivOnePlusX (_ntl_ulong *c, long n) {
  _ntl_ulong t = 0; long i;
  for (i = 0; i < n; i++) {
    t ^= c[i] ; t ^= t << 1; t ^= t << 2;
    t ^= t << 4; t ^= t << 8; t ^= t << 16;
    c[i] = t; t >>= 32-1;
}}
```

The main idea for this function is to multiply each word by the inverse of $x+1$ modulo x^{2^b}, where 2^b is the number of coefficients stored in one word. This requires $b+1$ shifts and sums for each word. Similar functions can be developed for any exact division needed in Toom-3,4,5 IS.

With this function, the new algorithm is about 5% faster than Zimmermann's, and beats the NTL mul starting from 8 words, meaning degree 256. It is also faster than Karatsuba for operands above 11 words, or degree 352.

5.3   Toom-4 in $GF_2[x]$

The complete analysis of the Toom-4 candidate algorithms, requires too much resources. So here we tested only the most promising choice for the 7 points: $P_6 = \{(0,1), (1, x+1), (1, x), (1,1), (x,1), (x+1,1), (1,0)\}$. Here we show only the balanced algorithm, used when the two operands have about the same size, and the matrices.

```
U = U3*Y^3 + U2*Y^2 + U1*Y + U0
V = V3*Y^3 + V2*Y^2 + V1*Y + V0
\\Evaluation: 13*2 add, 7*2 shift, 2*2 Smul, 7 mul
W1 = U3 + U2 + U1 + U0     ; W2 = V3 + V2 + V1 + V0
W3 = W1 * W2
W0 = U1 + x*(U2 + x*U3)    ; W6 = V1+ x*(V2 + x*V3)
W4 =(W0 + U3*(x+1))*x+W1   ; W5 =(W6 + V3*(x+1))*x+W2
W0 = W0*x + U0             ; W6 = W6*x + V0
W5 = W5 * W4               ; W4 = W0 * W6
W0 = U0*x^3+U1*x^2+U2*x    ; W6 = V0*x^3+V1*x^2+V2*x
W1 = W1 + W0 + U0*(x^2+x)  ; W2 = W2 + W6 + V0*(x^2+x)
W0 = W0 + U3              ; W6 = W6 + V3
W1 = W1 * W2              ; W2 = W0 * W6
W6 = U3 * V3             ; W0 = U0 * V0
\\Interpolation: 22 add, 4 shift, 4 Sdiv, 5mul
W1 = W1 + W2 + W0*(x^4+x^2+1)
W5 =(W5 + W4 + W6*(x^4+x^2+1) + W1)/(x^4+x)
W2 = W2 + W6 + W0*x^6
W4 = W4 + W2 + W6*x^6 + W0
W4 =(W4 + W5*(x^5+x))/(x^4+x^2)
W3 = W3 + W0 + W6
W1 = W1 + W3
W2 = W2 + W1*x + W3*x^2
W3 = W3 + W4 + W5
W1 =(W1 + W3*(x^2+x))/(x^4+x)
W5 = W5 + W1
W2 =(W2 + W5*(x^2+x))/(x^4+x^2)
W4 = W4 + W2
\\Recomposition:
W = W6*Y^6 + W5*Y^5 + W4*Y^4+ W3*Y^3+ W2*Y^2+ W1*Y + W0
W == U*V    \\ check
```

$$E_{6,4} = \begin{pmatrix} 1 & 0 & 0 & 0 \\ F & 5 & 3 & 1 \\ 8 & 4 & 2 & 1 \\ 1 & 1 & 1 & 1 \\ 1 & 2 & 4 & 8 \\ 1 & 3 & 5 & F \\ 0 & 0 & 0 & 1 \end{pmatrix}$$

$$A_6 = \begin{pmatrix} 1 & 0 & 0 & 0 & 0 & 0 & 0 \\ 55 & 33 & 11 & F & 5 & 3 & 1 \\ 40 & 20 & 10 & 8 & 4 & 2 & 1 \\ 1 & 1 & 1 & 1 & 1 & 1 & 1 \\ 1 & 2 & 4 & 8 & 10 & 20 & 40 \\ 1 & 3 & 5 & F & 11 & 33 & 55 \\ 0 & 0 & 0 & 0 & 0 & 0 & 1 \end{pmatrix}$$

5.4   Toom-5 in $GF_2[x]$

The complete analysis of Toom-5 is even harder. There is only one possible choice for the evaluating point, with minimal degree:

$$P_{GF_2[x]} = \{(0,1), (x+1,x), (x+1,1), (x,1), (1,1), (1,x), (1,x+1), (x,x+1), (1,0)\}.$$

The resulting algorithms are too big to be transcribed here, the found cost being:

ES: $2 \times (19 \cdot \mathtt{add} + 6 \cdot \mathtt{shift} + 4 \cdot \mathtt{Smul})$

IS: $36 \cdot \mathtt{add} + 9 \cdot \mathtt{shift} + 5 \cdot \mathtt{Smul} + 6 \cdot \mathtt{Sdiv}$

Only 3 different denominators for exact division are needed: $x^3 \cdot (x+1)^3$, $x \cdot (x+1) \cdot (x^2+x+1)^2$ and $x^2 \cdot (x+1)^2 \cdot (x^2+x+1)$. The two matrices:

$$E_{8,5} = \begin{pmatrix} 1 & 0 & 0 & 0 & 0 \\ 11 & 1E & 14 & 18 & 10 \\ 11 & F & 5 & 3 & 1 \\ 10 & 8 & 4 & 2 & 1 \\ 1 & 1 & 1 & 1 & 1 \\ 1 & 2 & 4 & 8 & 10 \\ 1 & 3 & 5 & F & 11 \\ 10 & 18 & 14 & 1E & 11 \\ 0 & 0 & 0 & 0 & 1 \end{pmatrix} \quad A_8 = \begin{pmatrix} 1 & 0 & 0 & 0 & 0 & 0 & 0 & 0 & 0 \\ 101 & 1FE & 154 & 198 & 110 & 1E0 & 140 & 180 & 100 \\ 101 & FF & 55 & 33 & 11 & F & 5 & 3 & 1 \\ 100 & 80 & 40 & 20 & 10 & 8 & 4 & 2 & 1 \\ 1 & 1 & 1 & 1 & 1 & 1 & 1 & 1 & 1 \\ 1 & 2 & 4 & 8 & 10 & 20 & 40 & 80 & 100 \\ 1 & 3 & 5 & F & 11 & 33 & 55 & FF & 101 \\ 100 & 180 & 140 & 1E0 & 110 & 198 & 154 & 1FE & 101 \\ 0 & 0 & 0 & 0 & 0 & 0 & 0 & 0 & 1 \end{pmatrix}$$

6 Bivariate and Multivariate

The same strategy described in section 2 can be extended to multivariate polynomials. In particular it fits perfectly for those polynomials which homogenisation is dense: polynomials dense with respect to *total degree*.

On the opposite side there is the Kronecker substitution[11], which is very efficient for polynomials dense with respect to *maximal degree*.

Definition 3. *We call* triangular polynomial *a polynomial dense with respect to total degree. We mean a polynomial where coefficients for all the possible terms with* sum *of exponents* limited by a constant d *are mostly non-zero. We will call* square polynomial, *those which are dense with respect to* maximal degree.

A couple of examples, $1 + x + y + x^2 + y^2 + xy$ *will be called* triangular, *while* $1 + x + y + x^2 + y^2 + xy + x^2y + xy^2 + x^2y^2$ *is a* square *polynomial.*

6.1 Multivariate Toom-2

Karatsuba's idea was generalised in many ways, one of them can be the extension to multivariate polynomials. If we start from two *triangular polynomials*, u, v, and we want to compute the product $w = u \cdot v$, we can proceed as in section 2.

If we call X_0 the homogenising variable and X_i the other ones, we will have the canonical splitting $u = \sum_i u_i \cdot X_i$. Than we have many smaller polynomials u_i, where u_0 is a *square*, and the others are *triangular*.

All the evaluations and interpolations can be condensed in a one-line formula, valid in any characteristic: $u = \sum_i u_i X_i \wedge v = \sum_i v_i X_i \Rightarrow$

$$w = u \cdot v = \sum_i (u_i \cdot v_i) X_i^2 + \sum_{i<j} ((u_i - u_j) \cdot (v_j - v_i) + u_i v_i + u_j v_j) X_i X_j$$

where any product $u_i v_i$ is computed only once, and recycled for all the $X_i X_j$ coefficients.

Recurrence is not very easy in this algorithm, because all the products involving u_0 and v_0 are *square* product, where the same algorithm can not be used. On squares we can fall back to the Kronecker's trick or use univariate algorithms recursively on any variable.

Another possible formula for the product, is the nearly equivalent

$$\mathfrak{w} = \mathfrak{u} \cdot \mathfrak{v} = \sum_i (u_i \cdot v_i) X_i^2 + \sum_{i<j} \left((u_i + u_j) \cdot (v_i + v_j) - u_i v_i - u_j v_j \right) X_i X_j$$

which is interesting for one reason: if we use this formula for an univariate polynomial, with the identification $X_i = x^i$, we obtain the Karatsuba generalisation by Weimerskirch and Paar[16].

6.2   Bivariate Toom-2.5 in $GF_2[x]$

The smallest interesting example of multivariate Toom, which is not a generalisation of Karatsuba, is the algorithm to multiply a polynomial of degree 2 by one of degree 1. Both with 2 variables. The product has degree 3, so it will have $\binom{3+2}{2} = 10$ coefficients, and we need 10 points.

After homogenising, we have 3 variables and evaluation points need 3 values. With the points $P_3^2 = \{(1,0,0),(1,1,0),(1,x+1,0),(0,1,0),(0,1,1),(0,1,x+1),(0,0,1),(1,0,1),(x+1,0,1),(1,1,1)\}$ we obtain the block-like matrices:

$$A_3^2 = \begin{pmatrix} 1 & 0 & 0 & 0 & 0 & 0 & 0 & 0 & 0 & 0 \\ 1 & 1 & 1 & 1 & 0 & 0 & 0 & 0 & 0 & 0 \\ F & 5 & 3 & 1 & 0 & 0 & 0 & 0 & 0 & 0 \\ 0 & 0 & 0 & 1 & 0 & 0 & 0 & 0 & 0 & 0 \\ 0 & 0 & 0 & 1 & 1 & 1 & 1 & 0 & 0 & 0 \\ 0 & 0 & 0 & F & 5 & 3 & 1 & 0 & 0 & 0 \\ 0 & 0 & 0 & 0 & 0 & 0 & 1 & 0 & 0 & 0 \\ 1 & 0 & 0 & 0 & 0 & 0 & 1 & 1 & 1 & 0 \\ 1 & 0 & 0 & 0 & 0 & 0 & F & 5 & 3 & 0 \\ 1 & 1 & 1 & 1 & 1 & 1 & 1 & 1 & 1 & 1 \end{pmatrix} \; ; E_{3,2}^2 = \begin{pmatrix} 1 & 0 & 0 \\ 1 & 1 & 0 \\ 3 & 1 & 0 \\ 0 & 1 & 0 \\ 0 & 1 & 1 \\ 0 & 3 & 1 \\ 0 & 0 & 1 \\ 1 & 0 & 1 \\ 1 & 0 & 3 \\ 1 & 1 & 1 \end{pmatrix} \; ; E_{3,3}^2 = \begin{pmatrix} 1 & 0 & 0 & 0 & 0 & 0 \\ 1 & 1 & 1 & 0 & 0 & 0 \\ 5 & 3 & 1 & 0 & 0 & 0 \\ 0 & 0 & 1 & 0 & 0 & 0 \\ 0 & 0 & 1 & 1 & 1 & 0 \\ 0 & 0 & 5 & 3 & 1 & 0 \\ 0 & 0 & 0 & 0 & 1 & 0 \\ 1 & 0 & 0 & 0 & 1 & 1 \\ 1 & 0 & 0 & 0 & 5 & 3 \\ 1 & 1 & 1 & 1 & 1 & 1 \end{pmatrix}$$

We can observe that the square invertible matrix A_{2k-2}^n used for the n-variate Toom-k, is not a Vandermonde matrix when $n > 1$, and it has $\binom{2k-2+n}{n}$ lines (and columns). Also the $E_{2k-2,d}^n$ are somehow sparse $\binom{2k-2+n}{n} \times \binom{d-1+n}{n}$ matrices.

Theorems proved in this paper can not be directly extended to those new Vandermonde-blocks matrices, anyway algorithms developed for the univariate case still works. Sparse matrix give graphs much smaller than expected and can be fully analysed. The best bivariate triangular Toom-2.5 found by our software follows.

```
U =  U00*Z^2 + U10*Z*X + U20*X^2\
   + U01*Z*Y + U11*Y*X \
   + U02*Y^2
V =  V00*Z  + V10*X   \
   + V01*Y
\\ Evaluation: 22 add, 9 shift; 10 mul
```

```
W3 = U20+ U10+ U00          ; W0 = V10+ V00        ; W1 = W0 * W3
W3 = W3 +(U10+U20*(x))*(x) ; W0 = W0 + V10*(x) ; W2 = W0 * W3
W3 = U20+ U11+ U02          ; W0 = V10+ V01        ; W4 = W0 * W3
W3 = W3 +(U11+U02*(x))*(x) ; W0 = W0 + V01*(x) ; W5 = W0 * W3
W3 = U02+ U01+ U00          ; W0 = V00+ V01        ; W7 = W0 * W3
W9 = W3 +(U01+U00*(x))*(x) ; W6 = W0 + V00*(x) ; W8 = W6 * W9
W3 = W3 + U20+ U11+ U10    ; W0 = W0 + V10
W9 = W3 * W0; W6 = U02* V01; W0 = U00* V00; W3 = U20* V10
\\ Interpolation: 21 add, 6 shift; 3 Sdiv
W2 =(W2 + W1)/(x)    ; W5 =(W5 + W4)/(x)    ; W8 =(W8 + W7)/(x)
W1 = W1 + W0          ; W4 = W4 + W3          ; W7 = W7 + W6
                     W9 = W9 - W7 - W4 - W1
W2 =(W2 + W1)/(x+1) ; W5 =(W5 + W4)/(x+1) ; W8 =(W8 + W7)/(x+1)
W2 = W2 + W3*(x)    ; W5 = W5 + W6*(x)    ; W8 = W8 + W0*(x)
W1 = W1 + W2 + W3    ; W4 = W4 + W5 + W6    ; W7 = W7 + W8 + W0
\\Recomposition
W = W0*Z^3    + W1*Z^2*X + W2*Z*X^2 + W3*X^3 \
  + W8*Z^2*Y + W9*Z*Y*X + W4*Y*X^2 \
  + W7*Z*Y^2 + W5*Y^2*X \
  + W6*  Y^3
W==U*V
```

Three instances of the $x+1$ version of univariate Toom-2.5 can be recognised in the code, the same trick could be applied using the x version. Only the point $(1, 1, 1)$ requires some extra operations.

6.3 Bivariate Toom-3 in $GF_2[x]$

The first non-Karatsuba multivariate Toom which can be used for recursion is the bivariate triangular Toom-3, with this algorithm we can multiply two *triangular* bivariate polynomials with degree 2 to obtain a triangular result with degree 4. This time we need $\binom{4+2}{2} = 15$ interpolation points. The choice $P_4^2 = \{ (1,0,0), (1, x+1, 0), (1, x, 0),$ $(1, 1, 0), (0, 1, 0), (0, 1, 1),$ $(0, 1, x), (0, 1, x+1), (0, 0, 1),$ $(x, 0, 1), (1, 0, 1), (1, 0, x),$ $(1, 1, x), (1, 1, 1), (x, 1, 1)\}$ gives again block-like matrices. The algorithm requires 15 smaller multiplication, 5 involving *triangular* polynomials, and 10 requiring some *squared polynomial* algorithm. We choose three different sub-matrices, so it's more difficult to recover sub-IS.

$$A_4^2 = \begin{pmatrix}
1 & 0 & 0 & 0 & 0 & 0 & 0 & 0 & 0 & 0 & 0 & 0 & 0 & 0 & 0 \\
11 & F & 5 & 3 & 1 & 0 & 0 & 0 & 0 & 0 & 0 & 0 & 0 & 0 & 0 \\
10 & 8 & 4 & 2 & 1 & 0 & 0 & 0 & 0 & 0 & 0 & 0 & 0 & 0 & 0 \\
1 & 1 & 1 & 1 & 1 & 0 & 0 & 0 & 0 & 0 & 0 & 0 & 0 & 0 & 0 \\
0 & 0 & 0 & 0 & 1 & 0 & 0 & 0 & 0 & 0 & 0 & 0 & 0 & 0 & 0 \\
0 & 0 & 0 & 0 & 1 & 1 & 1 & 1 & 1 & 0 & 0 & 0 & 0 & 0 & 0 \\
0 & 0 & 0 & 0 & 1 & 2 & 4 & 8 & 10 & 0 & 0 & 0 & 0 & 0 & 0 \\
0 & 0 & 0 & 0 & 1 & 3 & 5 & F & 11 & 0 & 0 & 0 & 0 & 0 & 0 \\
0 & 0 & 0 & 0 & 0 & 0 & 0 & 0 & 1 & 0 & 0 & 0 & 0 & 0 & 0 \\
1 & 0 & 0 & 0 & 0 & 0 & 0 & 0 & 10 & 8 & 4 & 2 & 0 & 0 & 0 \\
1 & 0 & 0 & 0 & 0 & 0 & 0 & 0 & 1 & 1 & 1 & 1 & 0 & 0 & 0 \\
10 & 0 & 0 & 0 & 0 & 0 & 0 & 0 & 1 & 2 & 4 & 8 & 0 & 0 & 0 \\
10 & 8 & 4 & 2 & 1 & 1 & 1 & 1 & 1 & 2 & 4 & 8 & 4 & 2 & 2 \\
1 & 1 & 1 & 1 & 1 & 1 & 1 & 1 & 1 & 1 & 1 & 1 & 1 & 1 & 1 \\
1 & 1 & 1 & 1 & 2 & 4 & 8 & 10 & 8 & 4 & 2 & 2 & 2 & 4 \\
\end{pmatrix}$$

```
U = U00*Z^2+U10*Z*X+U20*X^2+U01*Z*Y+U11*X*Y+U02*Y^2
V = V00*Z^2+V10*Z*X+V20*X^2+V01*Z*Y+V11*X*Y+V02*Y^2
```

```
\\Evaluation: 23*2 add, 6*2 shift; 15 mul
W0 = U00+ U10+ U20       ; W4 = V00+ V10+ V20
W12=(U10+ U00*(x))*(x) ; W10=(V10+ V00*(x))*(x)
W2 = W0 + W12            ; W8 = W4 + W10
W3 = W12+ U20            ; W5 = W10+ V20
W1 = W2 * W8 ; W2 = W3 * W5 ; W3 = W0 * W4
W6 = U20+ U11+ U02       ; W7 = V20+ V11+ V02
W8 = U01*(x)             ; W13= V01*(x)
W11= W6 + W12+ W8        ; W10= W10+ W7 + W13
W12= W11* W10; W5 = W6 * W7
W10= U02*(x^2)           ; W11= V02*(x^2)
W9 = W10+ U11*(x)        ; W14= W11 +V11*(x)
W6 = W6 + W9             ; W7 = W7  +W14
W0 = W0 + W9 +W8         ; W4 = W4  +W14+W13
W9 = W9 + U20            ; W14= W14 +V20
W10= W10+ W8 +U00        ; W11= W11 + W13+V00
W8 = W8 + U02+U00*(x^2); W13= W13 + V02+V00*(x^2)
W7 = W6 * W7 ; W6 = W9 * W14; W14= W0 * W4
W0 = U02+U01+U00         ; W4 = V02+V01+V00
W9 = W10* W11; W11= W8 * W13; W10= W0 * W4
W0 = W0+U20+U11+U10      ; W4 = W4+V20+V11+V10
W13= W0 * W4 ; W8 = U02*V02 ; W0 = U00*V00 ; W4 = U20*V20
```

$$
E_{4,3}^2 = \begin{pmatrix}
1 & 0 & 0 & 0 & 0 & 0 \\
5 & 3 & 1 & 0 & 0 & 0 \\
4 & 2 & 1 & 0 & 0 & 0 \\
1 & 1 & 1 & 0 & 0 & 0 \\
0 & 0 & 1 & 0 & 0 & 0 \\
0 & 0 & 1 & 1 & 1 & 0 \\
0 & 0 & 1 & 2 & 4 & 0 \\
0 & 0 & 1 & 3 & 5 & 0 \\
0 & 0 & 0 & 0 & 1 & 0 \\
1 & 0 & 0 & 0 & 4 & 2 \\
1 & 0 & 0 & 0 & 1 & 1 \\
4 & 0 & 0 & 0 & 1 & 2 \\
4 & 2 & 1 & 1 & 1 & 2 \\
1 & 1 & 1 & 1 & 1 & 1 \\
1 & 1 & 1 & 2 & 4 & 2
\end{pmatrix}
$$

```
\\Interpolation: 42 add, 8 shift, 2 Smul, 8 Sdiv
W12= W12+W2             ;W14= W14+W6            ;W13= W13+ W5
W1 = W1 +W2             ;W7 = W7 +W6            ;W9 = W9 + W0 + W8*(x^4)
W2 =(W2 +W4 )/(x) + W1  ;W6 =(W6 +W4 )/(x) + W7 ;W11= W11+ W8 + W0*(x^4)
W2=(W2+W0*(x^3+1))/(x+1);W6=(W6+W8*(x^3+1))/(x+1);W10= W10+ W8 + W0
W3 = W3 +W4             ;W5 = W5 +W4            ; W13= W13+ W10
   W13 = W13+W3         ; W12 = W12+W5      ;  W12=((W12+W11)/x+W13)/(x+1)
   W14 = W14+W3         ;                      W14=((W14+W9 )/x+W13)/(x+1)
W1 =(W1 +W3)/(x*(x+1))  ;W7 =(W7 +W5)/(x*(x+1)) ;W9 =(W9+W11)/(x*(x^2+1))
W3 = W3 +W0 + W2        ;W5 = W5 +W8 + W6       ;W10= W10+ W9
W2 = W2 +W1             ;W6 = W6 +W7            ;W11=(W11/x+W9+W10*x)/(x^2+1)
   W13 = W13+W12+W14                           ;W9 = W9 +W11
\\Recomposition
W = W0 *Z^4    + W1 *Z^3 *X+ W2 *Z^2*X^2 + W3*Z*X^3 + W4*X^4 \
  + W11*Z^3*Y   + W12*Z^2*Y*X+ W13*Z*Y*X^2 + W5*Y*X^3 \
  + W10*Z^2*Y^2+ W14*Z*Y^2*X+ W6 *Y^2*X^2 \
  + W9 *Z *Y^3+ W7    *Y^3*X \
  + W8       *Y^4
W ==U*V
```

7 Conclusions

The paper presented a method to determine an optimal evaluation sequence of
basic operations to be used in Toom multiplications. Joined with the previous
work on inversion sequences[2], this gives a complete framework for the search of
optimal Toom-Cook algorithms. This method shows his immediate effectiveness

giving new algorithms to be used in $GF_2[x]$, and the best known Toom-3 algorithm for $\mathbb{Z}[x]$ and $\mathbb{Z}$.

New generalisation of Toom described in section 2 open the possibility to easily generate simple Toom multiplication algorithms for polynomials on other integral domains[1]. Moreover section 6 generalise to multivariate polynomials, with a natural definition of density. Further work is needed to find general implementations working with any number of variables and any degree.

Note on Algorithms

Algorithms in this paper uses PARI/GP syntax[6], which should be simple enough to translate to any other language, and allow a fast checking within a GP shell. Some more definitions should be typed to have correct results for algorithms in characteristic 2.

```
U0 = u0 * Mod(1,2) ; U1 = u1 * Mod(1,2) ; U2 = u2 * Mod(1,2)
U3 = u3 * Mod(1,2) ; U4 = u4 * Mod(1,2) ; U5 = u5 * Mod(1,2)
V0 = v0 * Mod(1,2) ; V1 = v1 * Mod(1,2) ; V2 = v2 * Mod(1,2)
V3 = v3 * Mod(1,2) ; V4 = v4 * Mod(1,2) ; V5 = v5 * Mod(1,2)
```

Acknowledgements

The author wants to thank Paul Zimmermann for the many stimulating ideas and beautiful code which gave the start to this paper.

References

1. Bodrato, M.: Notes on Low Degree Toom-Cook Multiplication with Small Characteristic, Technical Report, Centro V.Volterra, Università di Roma Tor Vergata (2007)
2. Bodrato, M., Zanoni, A.: Integer and Polynomial Multiplication: Towards Optimal Toom-Cook Matrices. Proceedings of the ISSAC 2007 Conference. ACM press, New York (2007), http://bodrato.it/papers/#ISSAC2007
3. Chung, J., Anwar Hasan, M.: Asymmetric squaring formulae, Technical Report. CACR 2006-24, University of Waterloo (2006)
4. Cook, S.A.: On the Minimum Computation Time of Functions, Thesis, Harvard University, pp. 51–77 (1966)
5. The GNU Multi-Precision Library (GMP) http://gmplib.org/
6. The GP-Pari Computer Algebra System, http://pari.math.u-bordeaux.fr/
7. Hart, P.E., Nilsson, N.J., Raphael, B.: A Formal Basis for the Heuristic Determination of Minimum Cost Paths. IEEE Transactions on Systems Science and Cybernetics SSC4, pp. 100–107 (1968)
8. Jebelean, T.: An algorithm for exact division. Journal of Symbolic Computation 15, 169–180 (1993)
9. Karatsuba, A., Ofman, Y.: Multiplication of multidigit numbers on automata, Soviet Physics-Doklady, 7, 595–596 (1963); translation from Dokl. Akad. Nauk SSSR, 145(2), 293–294, (1962)
10. Knuth, D.E.: The Art of Computer Programming, Chapter 4, Section 3.3, 2nd edn., vol. 2, pp. 278–301. Addison-Wesley, Reading, MA (1981)
11. Kronecker, L.: Grundzüge einer arithmetischen Theorie der algebraischen Grössen. Journal Für die Reine und Angewandte Mathematik, pp. 92–122 (1882)

12. The Number Theory Library (NTL), http://www.shoup.net/ntl/
13. Toom, A.L.: The Complexity of a Scheme of Functional Elements Realizing the Multiplication of Integers. Soviet Mathematics 3, 714–716 (1963)
14. Schönhage, A., Strassen, V.: Schnelle Multiplikation großer Zahlen. Computing 7, 281–292 (1971)
15. Schönhage, A.: Schnelle Multiplikation von Polynomen über Körpern der Charakteristik 2. Acta. Informatica 7, 395–398 (1977)
16. Weimerskirch, A., Paar, C.: Generalizations of the Karatsuba Algorithm for Efficient Implementations, Cryptology ePrint Archive, Report 224 (2006)
17. Winograd, S.: Arithmetic Complexity of Computations, CBMS-NSF Regional Conference Series in Mathematics, vol. 33 (1980)
18. Zimmermann, P., Quercia, M.: Private communication, October 2006, irred-ntl source code (2003)

Appendix Z: Results in $\mathbb{Z}[x]$

Toom-3 in $\mathbb{Z}[x]$

The IS for Toom-3 in $\mathbb{Z}$ was fully examined in the previous work with Zanoni. We give here the ES for the balanced and unbalanced (4x2) flavour. Both saves at least one shift if compared to the currently used ES. Both were tested against the GMP library, giving a small speedup.

```
U = U2*x^2 + U1*x + U0                  U = U3*x^3 + U2*x^2 + U1*x + U0
V = V2*x^2 + V1*x + V0                  V = V1*x + V0
\\Evaluation: 5*2 add, 2 shift; 5mul    \\Eval: 7+3 add, 3 shift; 5mul
W0 = U2 + U0        ; W4 = V2 + V0      W0 = U1 + U3 ; W4 = U0 + U2
W2 = W0 - U1        ; W1 = W4 - V1      W3 = W4 + W0 ; W4 = W4 - W0
W0 = W0 + U1        ; W4 = W4 + V1      W0 = V0 + V1 ; W2 = V0 - V1
W3 = W2 * W1        ; W1 = W0 * W4      W1 = W3 * W0 ; W3 = W4 * W2
W0 =(W0 + U2)<<1-U0; W4 =(W4 + V2)<<1-V0  W4 =((U3<<1+U2)<<1+U1)<<1+U0
W2 = W0 * W4                            W0 = W0 + V1 ; W2 = W4 * W0
W0 = U0 * V0        ; W4 = U2 * V2      W0 = U0 * V0 ; W4 = U3 * V1
```

```
                  \\Interpolation: 8 add, 3 shift, 1 Sdiv
                  W2 =(W2 - W3)/3
                  W3 =(W1 - W3)>>1
                  W1 = W1 - W0
                  W2 =(W2 - W1)>>1 - W4<<1
                  W1 = W1 - W3 - W4
                  W3 = W3 - W2
                  \\Recomposition:
                  W  = W4*x^4+ W2*x^3+ W1*x^2+ W3*x + W0
                  W == U*V
```

Asymmetrical Squaring in $\mathbb{Z}[x]$

Chung and Anwar Hasan proposed new linear systems useful for squaring only; refer to their report[3] for the details. Here we only show results found by our software starting from their matrices, although not optimised for this case.

The report proposed an evaluation sequence and an inversion algorithm for the 5-way squaring method using temporary variables, with cost respectively $14 \cdot \texttt{add} + 4 \cdot \texttt{shift}$ and $18 \cdot \texttt{add} + 7 \cdot \texttt{shift}$. Using our software we where able to find shorter sequences, reaching $12 \cdot \texttt{add} + 5 \cdot \texttt{shift}$ and $16 \cdot \texttt{add} + 3 \cdot \texttt{shift}$.

```
U = U4*Y^4 + U3*Y^3 + U2*Y^2 + U1*Y + U0
\\Evaluation: 12 add, 5 shift; 5 mul, 4 sqr
W0 = U0 - U3 ; W1 = U3 - U1   ; W6 = U1 - U2
W4 = U1 + U2 ; W5 = W6 - U4   ; W3 = W5 + W0<<1
W0 = W0 - W5 ; W6 = W0 + W6<<1; W7 = W6 + W1
W5 = W7 + W1 ; W8 = W5 + W4<<1; W4 = W4 - U4
  W2 = W4 * W3; W4 = W6 * W5; W3 = W7 * W1
W1 = U0 * U1 * 2      ; W7 = U3 * U4 * 2
W5 = W8^2 ; W6 = W0^2 ; W0 = U0^2 ; W8 = U4^2

\\ Interpolation: 16 add, 3 shift.
W6 =(W6 + W5)>>1 ; W5 = W5 - W6
W4 =(W4 + W6)>>1 ; W6 = W6 - W4; W3 = W3 +W5>>1
W5 = W5 - W3 - W1; W4 = W4 - W8 - W0
W2 = W2 - W8 - W1 - W7 + W4 + W5
W3 = W3 - W7    ; W6 = W6 - W2
\\Recomposition:
W  = W8*Y^8 + W7*Y^7 + W6*Y^6 + W5*Y^5   \
   + W4*Y^4 + W3*Y^3 + W2*Y^2 + W1*Y + W0
W == U^2
```

$$
\widetilde{E} = \begin{pmatrix}
1 & 1 & 1 & 1 & 1 \\
1 & -1 & 1 & -1 & 1 \\
1 & 0 & -1 & 0 & 1 \\
0 & 1 & 0 & -1 & 0 \\
1 & 1 & -1 & -1 & 1 \\
1 & -1 & -1 & 1 & 1 \\
0 & 1 & 1 & 0 & -1 \\
2 & 1 & -1 & -2 & -1
\end{pmatrix}
$$

$$
A_s = \begin{pmatrix}
1 & 0 & 0 & 0 & 0 & 0 & 0 & 0 & 0 \\
0 & 1 & 0 & 0 & 0 & 0 & 0 & 0 & 0 \\
1 & -1 & 1 & -1 & 1 & -1 & 1 & -1 & 1 \\
1 & 1 & 1 & 1 & 1 & 1 & 1 & 1 & 1 \\
1 & 0 & -1 & 0 & 1 & 0 & -1 & 0 & 1 \\
0 & 1 & 0 & -1 & 0 & 1 & 0 & -1 & 0 \\
1 & 1 & 0 & -1 & -1 & 0 & 1 & 1 & 0 \\
0 & 0 & 0 & 0 & 0 & 0 & 0 & 1 & 0 \\
0 & 0 & 0 & 0 & 0 & 0 & 0 & 0 & 1
\end{pmatrix}
$$

Appendix T: Toom Three Timing Tests

We include at last some raw graphs of multiplication timings for balanced n bits long univariate operands. Both show relative timings for different operand sizes.

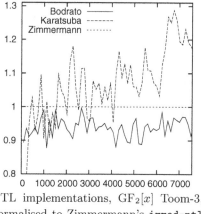

NTL implementations, $GF_2[x]$ Toom-3, normalised to Zimmermann's `irred-ntl` code. Up to degree 7,500.

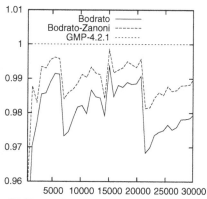

GMP implementations, $\mathbb{Z}$ Toom-3, normalised to GMP-4.2.1 timings. Up to 30,000 bits operands.

A Construction of Differentially 4-Uniform Functions from Commutative Semifields of Characteristic 2

Nobuo Nakagawa[1] and Satoshi Yoshiara[2]

[1] Department of Mathematics, Faculty of Science and Technology, Kinki University,
3-4-1 Kowakae, Higashi Osaka, Osaka 577-8502, Japan
[2] Department of Mathematics, Tokyo Woman's Christian University, Suginami-ku,
Tokyo 167-8585, Japan

Abstract. We construct differentially 4-uniform functions over $GF(2^n)$ through Albert's finite commutative semifields, if n is even. The key observation there is that for some k with $0 \leq k \leq n-1$, the function $f_k(x) := (x^{2^{k+1}} + x)/(x^2 + x)$ is a two to one map on a certain subset $D_k(n)$ of $GF(2^n)$. We conjecture that f_k is two to one on $D_k(n)$ if and only if (n,k) belongs to a certain list. For (n,k) in this list, f_k is proved to be two to one. We also prove that if f_2 is two to one on $D_2(n)$ then $(n,2)$ belongs to the list.

Keywords: Finite field, Almost perfect nonlinear function, Differentially δ-uniformity, Cubic function of a finite semifield, Absolute trace.

1 Introduction

Importance of functions on a finite field with high nonlinearity has been recognized in recent applications to cryptgraphy. On the other hand, such functions are known to have strong connections with finite geometries in the case of odd characteristic, as we shall see below. One of the two purposes of this article is to provide an explicit construction of those functions in the even characteristic case, using the notion in finite geometries. The other is to investigate a class of functions which plays an important role in this construction.

For a prime p and a positive integer n, we use $GF(p^n)$ to denote a finite field of cardinality p^n. Let f be a function on $GF(p^n)$ (from $GF(p^n)$ to itself). For $a \in GF(p^n)^\times := GF(p^n) \setminus \{0\}$ and $b \in GF(p^n)$, we set

$$\delta_f(a,b) := \#\{x \in GF(p^n) \mid f(x+a) - f(x) = b\},$$
$$\delta_f := \max_{b \in GF(p^n)} (\max_{a \in GF(p^n)^\times} \delta_f(a,b)).$$

For a positive integer δ, a function f is called *differentially δ-uniform* if $\delta_f \leq \delta$. Observe that $\delta_f = p^n$ if f is $GF(p)$-linear. Thus functions f with small δ_f are regarded to have high nonlinearity.

C. Carlet and B. Sunar (Eds.): WAIFI 2007, LNCS 4547, pp. 134–146, 2007.
© Springer-Verlag Berlin Heidelberg 2007

The function f is called *planar*, if $\delta_f = 1$. A planar function exists only when p is odd. It is known that the existence of a planar function on $GF(p^n)$ is equivalent to the existence of a certain class of affine planes defined on $GF(p^n) \times GF(p^n)$ [4]. Such an affine plane is coordinatized by an object with some algebraic structure, such as a semifield. Conversely, this structure often provides functions on $GF(p^n)$ with interesting properties. For example, the square mapping on a commutative semifield is a planar function (see section 2).

In the case when $p = 2$, we have $\delta_f \geq 2$ because $f(x + a) - f(x) = f((x + a) + a) - f(x + a)$. The function f on $GF(2^n)$ is called an *APN* (*almost perfect nonlinear*) function if $\delta_f = 2$. Thus APN functions can be thought of as an analogue of planar functions in the even characteristic case. Recently, several quadratic APN functions (inequivalent to any power functions) have been constructed, some of which were given as infinite series [1],[2]. We can construct a certain geometric object from such functions, e.g. [8].

In this paper, we are interestead in the opposite direction: constructing δ-uniform functions for small δ starting from finite geometrical objects. The idea is to consider the "cubic" mapping on a commutative semifield, instead of the square mapping which gives a planar function in the odd characteristic case. After reviewing some standard results about finite commutative semifields (specifically those of characteristic 2 and cubic functions of them) in section 2, a class of differentially 4-uniform functions on $GF(2^{2e})$ is constructed using the cubic functions of the Albert finite commutative semifields (see theorem 2).

This construction is based on the following observation (see lemma 1): the function $f_e(x) := (x^{2^{e+1}} + x)/(x^2 + x)$ defined on $GF(2^{2e}) \setminus GF(2)$ is a two to one map on the subset $GF(2^{2e}) \setminus GF(2^e)$. This naturally leads us to the following question.

Problem. Determine the list of pairs (n, k) of positive integers n and k with $1 \leq k \leq n - 1$ for which $f_k(x) := (x^{2^{k+1}} + x)/(x^2 + x)$ is a two to one map on the subset $D_k(n) := \{x \in GF(2^n) \setminus GF(2) \mid f_k(x) \notin GF(2)\}$ of $GF(2^n)$.

In fact, we found a class of differentially 4-uniform functions indexed by k, under the assumption that f_k is two to one on $D_k(n)$ (see theorem 3). We examined small fields $GF(2^n)$ with $n \leq 16$ using GAP, and obtained the following conjecture.

Conjecture. Assume $\emptyset \neq D_k(n)$. The function f_k above is two to one on $D_k(n)$ if and only if $n \geq 3$ and k is one of the following:

$k = 1$, $k = n - 2$;
$k = (n/2) - 1$ or $k = n/2$ if n is even; $k = (n - 1)/2$ if n is odd.

We also succeeded in solving the problem for the fixed value $k = 2$, exploiting elementary but ingeneous arguments (see proposition 4). The general setting of the problem is discussed in section 4 as well as the proof of proposition 4 in the case when n is even (see proposition 5). Then a proof of proposition 4 for n odd will be given in section 5 (see proposition 6), which is divided into several steps. In the last section, we verify that f_k is in fact two to one on $D_k(n)$ for each of the values k given in the conjecture.

2   Finite Commutative Semifields

A finite algebraic structure $E = (E, +, \circ)$ is called a finite *presemifield*, if it is an abelian group with respect to the operation $+$ and the following properties (i)–(iii) are satisfied for all $x, y, z \in E$ [3,5]:

  (i) $x \circ (y + z) = x \circ y + x \circ z$.
 (ii) $(x + y) \circ z = x \circ z + y \circ z$.
(iii) If $x \circ y = 0$, then $x = 0$ or $y = 0$.

It is called a *semifield* (resp. *commutative*) if it has the identity element with respect to the multiplication $\circ$ (resp. $\circ$ is commutative). With any presemifield, we can associate a semifield by modifying the multiplication.

A finite commutative semifield E is a vector space over $GF(p)$ for a prime p. Therefore $(E, +)$ is isomorphic to the additive group $(GF(p^n), +)$ for some positive integer n. If $p \neq 2$, the *square mapping s* of E defined by

$$f(x) = x \circ x$$

is known to be a planar function on $(E, +)$, regarded as $GF(p^n)$. Explicit shapes of such functions are calculated by K. Minami and the first author for almost all known commutative semifields [7].

Two examples of finite commutative (pre)semifields of size 2^n are known [5,3]. One of them is called *Albert semifields*, where the multiplication is given by

$$x \circ y = xy + \alpha(xy)^\sigma,$$

where $\alpha \notin \{x^{\sigma+1} \mid x \in GF(2^n)\}$ and σ is an automorphism of $GF(2^n)$ which is not a generator. The other is known as *Knuth semifields*, where the multiplication is given by

$$x \circ y = xy + (x\mathrm{Tr}(y) + y\mathrm{Tr}(x))^2,$$

where Tr is the trace mapping of extention $GF(2^n)/K$ with a suitable subfield K of $GF(2^n)$. Sometimes Knuth semifields are described in more general form [5].

Let $E(+, \circ)$ be a finite commutative semifield with $|E| = 2^n$. We consider the *cubic function f* of E, which is defined by

$$f(x) = (x \circ x) \circ x.$$

Then for any $a \in E^\times := E - \{0\}$, we have $f(x + a) + f(x) + f(a) = (x \circ x) \circ a + (a \circ a) \circ x$. The *cubic order* of $a \in E^\times$ is defined to be $\mathrm{co}(a) := \#\{x \in E \mid (x \circ x) \circ a + (a \circ a) \circ x = 0\}$.

It can be verified that the cubic function of E is an APN function if and only if the cubic order $\mathrm{co}(a)$ of any $a \in E^\times$ is 2. We put

$GN(E) =: \{a \mid (a \circ a) \circ x = a \circ (a \circ x) \ (\forall x \in E)\}$ and
$N(E) := \{a \mid (a \circ x) \circ y = a \circ (x \circ y) \ (\forall x, y \in E)\}.$

It is known that $N(E)$ is a finite field with respect to the operations $+$ and $\circ$, which is called the *nucleus* of E. Trivially $N(E) \subset GN(E)$. We note that if $a \in GN(E)$, then $co(a) = 2$, because we have $x = 0$ or $x = a$ from equation $(x \circ x) \circ a + (a \circ a) \circ x = 0$. Thus it is natural to expect that the cubic function of a finite commutative semifield has low differentially uniformity. In the next section, we shall see that this is in fact the case for a special type of Albert semifields.

3 Construction of Differentially 4-Uniform Functions

Let E be the additive group $(GF(2^{2e}), +)$. We define a multiplication $\circ$ in E by

$$x \circ y = xy + \alpha(xy)^{2^e},$$

where α is a generator of $GF(2^{2e})^\times$. We denote by τ the automorphism of $GF(2^{2e})$ of order 2: $x^\tau = x^{2^e}$ $(x \in GF(2^{2e}))$. The subfield of $GF(2^{2e})$ consisting of elements fixed by τ is $GF(2^e)$.

Then E is a commutative semifield, which is a special type of Albert's semifields. The cubic function $f(x) = (x \circ x) \circ x$ of E is calculated to be as follows:

$$f(x) = x^3 + \alpha x^{2^\tau+1} + \alpha x^{3\tau} + \alpha^{\tau+1} x^{\tau+2}.$$

Lemma 1. *The function g defined by $g(x) = (x^{2\tau} + x)/(x^2 + x)$ is a two to one mapping from $GF(2^{2e}) \setminus GF(2^e)$ into $GF(2^{2e})$.*

Proof of Lemma. Choose elements x and y of $GF(2^{2e}) \setminus GF(2^e)$. Assume that $g(x) = g(y)$. Then

$$(x^{2\tau}y^2 + xy^2 + yx^{2\tau}) + (y^{2\tau}x^2 + yx^2 + xy^{2\tau}) = 0 \qquad (1)$$

Sending the both sides of equation (1) by τ, we obtain

$$(x^2y^{2\tau} + x^\tau y^{2\tau} + y^\tau x^2) + (y^2x^{2\tau} + y^\tau x^{2\tau} + x^\tau y^2) = 0. \qquad (2)$$

Adding equations (1) and (2), we have

$$(x + x^\tau)(y + y^\tau)((x + y) + (x + y)^\tau) = 0.$$

Since $x \neq x^\tau$ and $y \neq y^\tau$, we have $(x + y) = (x + y)^\tau$. Thus $c := x + y \in GF(2^e)$. We substitute $y = x + c$ into $g(x) = g(y)$. Then we have

$$\frac{x^{2\tau} + x}{x^2 + x} = \frac{x^{2\tau} + x + c^2 + c}{x^2 + x + c^2 + c}.$$

Thus

$$(c^2 + c)(x^\tau + x)^2 = 0.$$

As $x^\tau + x \neq 0$, we have $c^2 = c$. This proves $y = x$ or $y = x + 1$.

Theorem 2. *The function* f *given by* $f(x) = x^3 + \alpha x^{2\tau+1} + \alpha x^{3\tau} + \alpha^{\tau+1} x^{\tau+2}$
is differentially 4-uniform on $GF(2^{2e})$, *where* α *is a generator of* $GF(2^{2e})^\times$ *and*
$x^\tau = x^{2^e}$ $(x \in GF(2^{2e}))$.

Proof of Theorem. Fix an element $a \in GF(2^{2e})^\times$. Then $f(x+a) + f(x)$ is
calculated to be $(x^2 a + a^2 x) + \alpha(x^{2\tau} a + a^{2\tau} x) + \alpha(x^{2\tau} a^\tau + a^{2\tau} x^\tau) + \alpha^{\tau+1}(x^2 a^\tau + a^2 x^\tau) + (a^3 + \alpha a^{2\tau+1} + \alpha a^{3\tau} + \alpha^{\tau+1} a^{\tau+2})$. Hence it is sufficient to show that the
equation

$$(x^2 a + a^2 x) + \alpha(x^{2\tau} a + a^{2\tau} x) + \alpha(x^{2\tau} a^\tau + a^{2\tau} x^\tau) + \\ \alpha^{\tau+1}(x^2 a^\tau + a^2 x^\tau) = 0 \qquad (3)$$

has at most four solutions. Suppose $x \neq 0$ and $x \neq a$. We denote by $\alpha(3)$ the
equation obtained by applying τ to equation (3) and then multiply the result by
α. Then adding this to equation (3), we have

$$(1 + \alpha^{\tau+1})\{(x^2 a + xa^2) + \alpha(x^{2\tau} a + a^{2\tau} x)\} = 0.$$

As $1 + \alpha^{\tau+1} \neq 0$, it follows that

$$\alpha^{-1} = \frac{a^{2\tau+1}\{(x/a)^{2\tau} + (x/a)\}}{a^3\{(x/a)^2 + (x/a)\}}.$$

Put $t = x/a$. Then $\alpha^{-1} a^{2-2\tau} = (t^{2\tau} + t)/(t^2 + t)$. Suppose that t lies in the
subfiled $GF(2^e)$. Then $\alpha^{-1} = a^{2\tau-2}$ and $(\alpha^{-1})^{\tau+1} = 1$, which is a contradiction.
Therefore t is not contained in $GF(2^e)$. Then it follows from lemma 1 that there
are at most two elements t such that $\alpha^{-1} a^{2-2\tau} = (t^{2\tau} + t)/(t^2 + t)$, say $t = t_0$
and $t = t_0 + 1$. Hence it is proved that equation (3) has at most four solutions
0, a, at_0 and $a(t_0 + 1)$. Theorem 2 is proved.

The inversive function $g(x) = x^{2^{2e}-2}$ is known to be differentially 4-uniform on
$GF(2^{2e})$. It is easily checked that functions in theorem 2 are not extended affine
equivalent to the inversive function.

4   Two to One Property of Some Functions

From now on, we shall examine functions which are natural generalization of
functions appeared in lemma 1. We first fix our notation.

Let $GF(q)$ be the finite field of $q = 2^n$ elements for a positive integer n greater
than 2. For each integer k with $0 \leq k \leq n-1$, we investigate the following
function f_k, which is defined on $GF(2^n) \setminus GF(2)$:

$$f_k(x) := \frac{x^{2^{k+1}} + x}{x^2 + x}. \qquad (4)$$

If $k = n-1$ (resp. 0), the funtion f_{n-1} is the zero map (resp. takes the constant
value 1). Thus we only consider the case where $1 \leq k \leq n-2$.

For $x \in GF(q) \setminus GF(2)$, we have $f_k(x) = 0$ if and only if $x^{2^{k+1}} = x$, that is, x lies in the subfield $GF(2^{(n,k+1)})$ of $GF(q)$, where $(n, k+1)$ denotes the greatest common divisor of n and $k + 1$. Similarly, $f_k(x) = 1$ if and only if x lies in the subfield $GF(2^{(n,k)})$. Thus, setting

$$D_k(n) := GF(q) \setminus (GF(2^{(n,k+1)}) \cup GF(2^{(n,k)})), \tag{5}$$

we have $D_k(n) = \{x \in GF(q) \setminus GF(2) \mid f_k(x) \notin GF(2)\}$. We only consider the case when $D_k(n) \neq \emptyset$.

Observe that the function g appearing in lemma 1 coincides with the function $f_{n/2}$ for $n = 2e$, with the above notation. As we already discussed in the introduction, we are interested in the problem to determine (n, k) for which f_k is a two to one map on $D_k(n)$; namely, $f_k(x) = f_k(y)$ if and only if $x + y \in GF(2)$. In fact, we can construct a class of differentially 4-uniform functions on $GF(2^n)$, if f_k is two to one on $D_k(n)$.

Theorem 3. *Let n and k be positive integers with $1 \leq k \leq n-2$. For a generator α of $GF(2^n)^{\times}$, consider a function $h_{k,\alpha}$ on $GF(2^n)$ given by*

$$h_{k,\alpha}(x) := x^3 + \alpha x^{2^{k+1}+1}.$$

If the function f_k is two to one on $D_k(n)$, then $h_{k,\alpha}$ is differentially 4-uniform.

This theorem can be verified by the similar arguments in the proof of theorem 2. Thus we omit the proof.

The conjectured answer to the above problem is given in introduction. In the subsequent two sections, the authors show that the conjecture is true, if k is fixed to be the smallest nontrivial value 2, but for an arbitrary value n with $k = 2 \leq n - 2$. Namely, the following proposition shall be proved:

Proposition 4. *Let $n \geq 4$. The function f_2 defined by $f_2(x) = (x^8 + x)/(x^2 + x)$ is a two to one map on $D_2(n) = GF(2^n) \setminus (GF(2^{(n,3)}) \cup GF(2^{(n,2)}))$ if and only if $n = 4$, 5 or 6.*

The proof of proposition 4 is divided into two cases according to the parity of n. We end this section with treating the case when n is even.

Proposition 5. *Let n be an even integer with $n \geq 4$. Then the function f_2 defined by $f_2(x) := (x^8 + x)/(x^2 + x)$ for $x \in GF(2^n) \setminus GF(2)$ is two to one on $D_2(n)$ if and only if $n = 4$ or 6.*

Proof of Proposition. It is straightforward to see that f_2 is two to one on $D_2(n)$ for $n = 4$ or 6. We will show the converse. Let $n = 2m$ be an even integer with $m \geq 4$. It is sufficient to show that f_2 is not two to one on $D_2(n)$.

Notice that the assumption $m \geq 4$ implies that the subset $D_2(m) = GF(2^m) \setminus (GF(2^{(2,m)}) \cup GF(2^{(m,3)}))$ is nonempty. We fix $t \in D_2(m)$. Then $f_2(t)$ is an element of $GF(2^m) \setminus GF(2)$. Consider the following polynomial $a(X)$ in $GF(2^m)[X]$, and let $F \cong GF(2^l)$ be the splitting field of $a(X)$:

$$a(X) = X^8 + f_2(t)X^2 + (1 + f_2(t))X.$$

Observe that $\alpha \in F \setminus GF(2)$ is a root of $a(X)$ if and only if $f_2(t) = (\alpha^8 + \alpha)/(\alpha^2 + \alpha) = f_2(\alpha)$. Furthermore $0, 1, t, t + 1$ are roots of $a(X)$ in $GF(2^m)$. Hence, if we find a root α of $a(X)$ in $F \setminus \{0, 1, t, t + 1\}$, then we conclude that f_2 is not two to one on $D_2(l)$. (Remark that $\alpha \notin GF(2^{(2,l)}) \cup GF(2^{(3,l)})$, as $f_2(\alpha) = f_2(t) \notin GF(2)$.)

We shall examine the roots of polynomial $b(X) := a(X)/X(X+1)(X+t)(X+t+1)$ in $GF(2^m)[X]$, which is of degree 4. As $f_2(t) \neq 1$, the derivative $a'(X)$ of $a(X)$ does not have a common root with $a(X)$, whence the multiplicity of every root of $a(X)$ is 1. Thus the roots of $b(X)$ lie in $F \setminus \{0, 1, t, t+1\}$. Since the map $GF(2^m) \ni x \mapsto a(x) \in GF(2^m)$ is $GF(2)$-linear, the roots of $a(X)$ in F form a vector space over $GF(2)$. In particular, if α is a root of $b(X)$ in F, the elements $\alpha + 1$, $\alpha + t$ and $\alpha + t + 1$ are roots of $b(X)$ in F as well.

If we have a root α of $b(X)$ lying in $GF(2^m)$, then α is an element of $D_2(m)$ with $f_2(\alpha) = f_2(t)$. Thus f_2 is not two to one on $D_2(m)$, whence the same holds on $D_2(n)$. (Observe that $D_2(m) = D_2(n) \cap GF(2^m)$, as $n = 2m$.) If we have an irreducible factor $c(X) \in GF(2^m)$ of $b(X)$ of degree 2, take a root α of $c(X)$. Then $GF(2^m)(\alpha) \cong GF(2^n)$. Thus we may assume that α is an element of $GF(2^n) \setminus \{0, 1, t, t+1\}$ with $f_2(\alpha) = f_2(t)$. Then f_2 is not two to one on $D_2(n)$.

In the unique remaining case, the polynomial $b(X)$ of degree 4 is irreducible over $GF(2^m)$. Then $F \cong GF(2^{4m})$ and $b(X) = (X + \alpha)(X + \alpha + 1)(X + \alpha + t)(X + \alpha + t + 1)$ for some $\alpha \in F$. Then there are Galois automorphisms σ and τ in $\mathrm{Gal}(F/GF(2^m))$ satisfying $\alpha^\sigma = \alpha + 1$ and $\alpha^\tau = \alpha + t$. Then we see that α is fixed by both σ^2 and τ^2. As $F = GF(2^m)(\alpha)$, this implies that $\sigma^2 = \tau^2$ is the identity on F. Then, both σ and τ are automorphisms of order 2 in a cylic group $\mathrm{Gal}(F/GF(2^m))$. Thus we have $\sigma = \tau$, whence $\alpha + 1 = \alpha^\sigma = \alpha^\tau = \alpha + t$. However, this implies that $t = 1$, which contradicts the choice of $t \in D_2(m)$. Hence $b(X)$ is not irreducible over $GF(2^m)$.

As we exhausted all the cases, we conlude that f_2 is not two to one on $D_2(n)$, if n is an even integer with $n \geq 8$. This verified the claim.

5   Proof of Proposition 6

In this section, we will treat the remaining case where n is odd. Our aim is to establish the following proposition, whose proof will be divided into several steps.

Proposition 6. *Let n be an odd integer with $n \geq 5$. Then the function f_2 is a two to one map on $D_2(n)$ if and only if $n = 5$.*

Proof of Proposition. It is straightforward to verify that f_2 is two to one on $D_2(5)$ (or by lemma 7(3)). In the following, we will show the only if part of the proposition.

Assume that n is an odd integer with $n \geq 5$ and that the function f_k defined by $f_k(x) := (x^{2^{k+1}} + x)/(x^2 + x)$ for $x \in GF(q) \setminus GF(2)$ is two to one on $D_k(n)$. From Step 2 below, we shall specify k to be 2.

5.1   Step 1

The function t sending $x \in GF(q)$ to $t(x) = x + x^2$ is a two to one map from $D_2(n)$ onto $T_0^\times$. Here $T_0^\times := \{x \in GF(q) \mid Tr(x) = 0\}$ denotes the set of nonzero elements of $GF(q)$ with absolute trace 0, where $Tr(x) = x + x^2 + \cdots x^{2^i} + \cdots + x^{2^{n-1}}$. This notation will be used throughout the note.

Setting $t = x + x^2$, we see $x + x^{2^{k+1}} = t + t^2 + \cdots t^{2^i} + \cdots + t^{2^k}$. Then $f_k(x) = (\sum_{i=0}^{k} t^{2^i})/t = \sum_{i=0}^{k} t^{2^i-1}$. Hence, if we define a map g_k on $T_0^\times$ by $g_k(t) := \sum_{i=0}^{k} t^{2^i-1}$, the function f_k is the composite of t and g_k. Thus f_k is two to one on $D_k(n)$ if and only if the function g_k is injective on $T_0^\times$.

5.2   Step 2

From now on, we consider the special case $k = 2$. As n is an odd integer in this case, we write $n = 2m + 1$. Notice that $Tr(1) = 1$. As $g_2(t) = 1 + t + t^3$, for each element $y \in GF(q)$, we have equality $g_2(y) = g_2(t)$ if and only if $0 = (y + t) + (y^3 + t^3) = (y + t)(y^2 + yt + t^2 + 1)$. This holds exactly when either $y = t$ or y satisfies the following equation with variable z:

$$\left(\frac{z}{t}\right)^2 + \frac{z}{t} + \left(1 + \frac{1}{t}\right)^2 = 0. \tag{6}$$

Note that there exists an element $z \in GF(q)$ satisfying equation (6) if and only if $Tr(1 + (1/t)) = 0$. If this condition is satisfied, we have

$$1 = Tr((1/t)) = \sum_{i=0}^{2m} (1/t)^{2^i}$$

$$= (1/t)^2 + \left(\sum_{i=1}^{m} (1/t)^{2^{2i}}\right) + \left(\sum_{i=1}^{m} (1/t)^{2^{2i+1}}\right)$$

$$= (1/t)^2 + \left(\sum_{i=1}^{m} (1/t)^{2^{2i}}\right) + \left(\sum_{i=1}^{m} (1/t)^{2^{2i}}\right)^2,$$

as $(1/t)^{2^{2m+1}} = (1/t)^{2^n} = (1/t)$. Thus $z_1 := t(\sum_{i=1}^{m} (1/t)^{2^{2i}})$ and $z_2 := t(1 + \sum_{i=1}^{m} (1/t)^{2^{2i}})$ are all solutions of equation (6) lying in $GF(q)$. Hence the injectivity of g_2 is equivalent to the condition that none of z_i $(i = 1, 2)$ lie in $T_0^\times$ for every $t \in T_0^\times$ with $Tr(1/t) = 1$. As $Tr(z_2) = Tr(z_1) + Tr(t) = Tr(z_1)$, this condition is equivalent to the condition that $Tr(z_1) = 1$ for every $t \in GF(q)^\times$ with $Tr(t) = 0$ and $Tr(1/t) = 1$.

Rewriting $w = 1/t$, we verified the following claim:

Claim 1. *For each $w \in GF(q)^\times$, if $Tr(w) = 1$ and $Tr(1/w) = 0$ then we have $Tr((1/w)(\sum_{i=1}^{m} w^{2^{2i}})) = 1$.*

5.3 Step 3

The condition in claim 1 is rephrased as follows. Remark that $1 + s + s^2 \neq 0$ for every $s \in GF(q)$, for otherwise $0 = s^4 + s = s(1 + s)(1 + s + s^2)$ and hence s would lie in $GF(2^{(n,2)}) = GF(2)$.

Claim 2. *For each* $s \in GF(q)^\times$, *if* $Tr(1/(1 + s + s^2)) = 0$ *then* $Tr(s/(1 + s + s^2)) = 0$.

Proof of Claim 2. Choose an element $s \in GF(q)^\times$ with $Tr(1/(1 + s + s^2)) = 0$. Then $w := 1 + s + s^2$ satisfies that $Tr(w) = 1$ and $Tr(1/w) = 0$, whence we can apply claim 1 to this element. Remark that $\sum_{i=1}^{m} w^{2^{2i}} = \sum_{i=1}^{m}(1 + s^{2^{2i}} + s^{2^{2i+1}}) = m + Tr(s) + s^2$, because $s^{2^{2m+1}} = s^{2^n} = s$. Then it follows from claim 1 that

$$0 = 1 + Tr(\frac{m + Tr(s) + s^2}{1 + s + s^2}) = Tr(\frac{m + 1 + Tr(s) + s}{1 + s + s^2})$$

$$= Tr(\frac{m + 1 + Tr(s)}{1 + s + s^2}) + Tr(\frac{s}{1 + s + s^2}).$$

Here, since $m + 1 + Tr(s)$ lies in $GF(2)$, we have

$$Tr(\frac{m + 1 + Tr(s)}{1 + s + s^2}) = (m + 1 + Tr(s))Tr(\frac{1}{1 + s + s^2}) = 0$$

by the assumption on s. Hence we have $0 = Tr(s/(1 + s + s^2))$.   **q.e.d.**

5.4 Step 4

We now count the number of elements $s \in GF(q) \setminus GF(2)$ with $Tr(s/(1 + s + s^2)) = 1$ in two ways. We set

$$A := \{s \in GF(q) \setminus GF(2) \mid Tr(s/(1 + s + s^2)) = 1\}. \tag{7}$$

Before counting $|A|$ in one way, notice that every element $t \in T_0^\times$ is written as $1/(x + x^{-1}) = x/(1 + x^2) = x/(1 + x)^2$ for some $x \in GF(q) \setminus GF(2)$. This fact can be verified as follows. Since

$$\frac{x}{(1 + x)^2} = \frac{1 + x}{(1 + x)^2} + \frac{1}{(1 + x)^2} = \frac{1}{1 + x} + (\frac{1}{1 + x})^2,$$

we have $Tr(1/(x + x^{-1})) = 0$ for all $x \in GF(q) \setminus GF(2)$. Thus the map sending $x \in GF(q) \setminus GF(2)$ to $1/(x + x^{-1})$ is a map into $T_0^\times$. It is easy to see that this map is two to one, where x and x^{-1} correspond to the same element. Then the image of this map consists of $(q - 2)/2 = |T_0^\times|$ elements. Hence we have

$$T_0^\times = \{1/(x + x^{-1}) \mid x \in GF(q) \setminus GF(2)\}.$$

We set $B := \{z \in GF(q) \setminus GF(2) \mid Tr(z + 1) = 0, Tr(1/(z + 1)) = 1\}$. Consider the map ρ sending $s \in A$ to $z = s/(1 + s + s^2)$. Then $z \neq 1$, as

$s \neq 1$. We have $Tr(z+1) = Tr(s/(1+s+s^2)) + 1 = 0$ as $s \in A$. Furthermore, $Tr(1/(z+1)) = Tr((1+s+s^2)/(1+s^2)) = Tr(1) + Tr(s/(1+s^2)) = 1$, by the above remark. Thus the map ρ is into the set B.

To see that the map ρ is surjective onto B, take any $z \in B$. As $Tr(\{1/(z+1)\}+1) = 0$, there is $s \in GF(q) \setminus GF(2)$ with $1/(z+1) = 1 + \{1/(s+s^{-1})\}$ by the above remark. The right hand side of the last equation is calculated to be $(1+s+s^{-1})/(s+s^{-1}) = (1+s+s^2)/(1+s^2)$, from which we have $z = s/(1+s+s^2)$. As $z \in B$, we have $1 = Tr(z)$. This shows that $s \in A$. Now it is easy to see that the map ρ is two to one on A, as $\rho(s) = \rho(1/s) = 1/(1+s+s^{-1})$. Hence we have $|A| = 2|B|$.

It is known that the set $T_{00} := \{x \in GF(q)^\times \mid Tr(x) = 0 = Tr(1/x)\}$ has the cardinality $(q - 3 - a_n)/4$ (e.g. [6, Exercise 6.75]), where

$$a_n = \left(\frac{-1+\sqrt{-7}}{2}\right)^n + \left(\frac{-1-\sqrt{-7}}{2}\right)^n. \tag{8}$$

Since $|B| = {}^\#\{x \in GF(q)^\times \mid Tr(x) = 0, Tr(1/x) = 1\}$ is the cardinality of the compliment to T_{00} in $T_0^\times$, we have $|B| = ((q/2) - 1 - |T_{00}|) = (q - 1 + a_n)/4$. Summarizing, we have

$$|A| = (q - 1 + a_n)/2. \tag{9}$$

5.5   Step 5

Now we start to count $|A|$ in another way. Observe that the arguments in this step except the last paragraph go through for any n, without assuming the injectivity of g_2 on $T_0^\times$.

We set $E(n) := GF(2^n) \setminus GF(2)$ with n odd, and let

$$t_1(s) := Tr(1/(1+s+s^2)) \text{ and } t_2(s) := Tr(s/(1+s+s^2)) \text{ for } s \in E(n).$$

The pair $(t_1(s), t_2(s))$ is one of the four vectors in $GF(2)^2$. For each vector (a, b) of $GF(2)^2$, we define

$$E(n; a, b) := \{s \in E(n) \mid (t_1(s), t_2(s)) = (a, b)\}.$$

Clearly $E(n)$ is the disjoint union of $E(n; a, b)$ for $(a, b) \in GF(2)^2$.

Let G be the subgroup of the group of bijections on $E(n)$ generated by the following two bijections α and ι:

$$x^\alpha := x + 1, \ x^\iota := (1/x) \text{ for } x \in E(n) = GF(q) \setminus GF(2).$$

It is easy to see that G is isomorphic to the symmetric group S_3 of degree 3, where α, ι and $\alpha\iota\alpha$ are involutions and $\alpha\iota$ and $\iota\alpha$ are elements of order 3.

For an element s of $E(n; a, b)$, it is easy to see the following:

$$t_1(s^\alpha) = Tr(1/(1 + (s+1) + (s+1)^2)) = Tr(1/(1+s+s^2)) = a,$$
$$t_2(s^\alpha) = Tr((s+1)/(1+s+s^2))$$
$$= Tr(s/(1+s+s^2)) + Tr(1/(1+s+s^2)) = a + b,$$

$$\begin{aligned}
t_1(s^\iota) &= Tr(1/(1+(1/s)+(1/s)^2)) = Tr(s^2/(1+s+s^2)) \\
&= Tr(1) + Tr(1/(1+s+s^2)) + Tr(s/(1+s+s^2)) \\
&= 1+a+b, \\
t_2(s^\iota) &= Tr((1/s)/(1+(1/s)+(1/s)^2)) \\
&= Tr(s/(1+s+s^2)) = b.
\end{aligned}$$

Thus we have

$$E(n;a,b)^\alpha \subseteq E(n;a,a+b) \text{ and } E(n;a,b)^\iota \subseteq E(n;a+b+1,b).$$

This implies the following inclusions among subsets $E(n;a,b)^g$ for $g \in G = \langle \alpha, \iota \rangle$ and $(a,b) \in GF(2)^2$.

$$E(n;0,0)^{\alpha\iota} \subseteq E(n;1,0), \ E(n;1,0)^{\alpha\iota} \subseteq E(n;1,1),$$
$$E(n;1,1)^{\alpha\iota} \subseteq E(n;0,0) \text{ and } E(n;0,1)^g \subseteq E(n;0,1).$$

As $g \in G$ is a permutation on $E(n)$, we have $|X| = |X^g|$ for a subset X of $E(n)$. Hence we have $|E(n;0,0)| = |E(n;1,0)| = |E(n;1,1)|$ and $E(n;0,1)^g = E(n;0,1)$ for all $g \in G$. As $E(n)$ is the disjoint union of $E(n;a,b)$ for $(a,b) \in GF(2)^2$, we have

$$|E(n;a,b)| = \frac{|E(n)| - |E(n;0,1)|}{3} = \frac{q-2-|E(n;0,1)|}{3}. \tag{10}$$

for any vector (a,b) of $GF(2)^2$ distinct from $(0,1)$.

Observe that so far we did not use the injectivity of g_2. Now we use it. Remark that claim 2 implies $E(n;0,1) = \emptyset$. Thus the set $A = \{s \in E(n) \mid t_2(s) = 1\}$ defind in equation (7) coincides with $E(n;1,1)$. Then, from equation (10) applied to $(a,b) = (1,1)$ we have

$$|A| = (q-2)/3. \tag{11}$$

5.6   Final Step

From equations (9) and (11), if f_2 is a two to one map on $D_2(n)$ for $n \geq 4$ with $(n,2) = 1$, then we have $(q-2)/3 = |A| = (q-1+a_n)/2$, or equivalently

$$-(q+1)/3 = a_n, \tag{12}$$

where a_n is the integer defined in equation (8).

We see that this equation holds only for $n = 5$, as follows. Notice that the absolute value $|a_n|$ is at most $2\sqrt{q}$, as a_n is the sum of two complex numbers $((-1 \pm \sqrt{-7})/2)^n$ with absolute value $(\sqrt{2})^n = \sqrt{q}$. Then we have $((q+1)/3)^2 \leq 4q$ from the above equality (12). This is satisfied only when $q = 2^n < 41$, which implies that $n \leq 5$. As a_n is an integer, the remaining possibilities are $n = 5$ and $n = 3$. However, the latter does not satisfy equation (12), because $a_3 = -a_2 - 2a_1 = 5$. (In fact, a_n is determined by the reccurence relation $a_n = -a_{n-1} - 2a_{n-2}$ with initial terms $a_1 = -1$ and $a_2 = -3$.) Remark that in the case $n = 5$, we in fact have equality $-(q+1)/3 = -11 = a_5$.

This established proposition 6.

6   "If" Part of the Conjecture

We shall verify the "if" part of the conjecture. Remark that $D_{n-1-k}(n) = D_k(n)$ for any k, as $(n, n-1-k) = (n, k+1)$ and $(n, (n-1-k)+1) = (n, k)$. First, it is staightforward to verify the following equation for positive integers n, k with $1 \le k \le n-2$ and $x \in D_k(n)$, where we set $f_0(x) = 1$.

$$f_k(x)/f_{k-1}(x) = \{(1/f_{n-1-k}(x)) + 1\}^{2^k}. \tag{13}$$

Lemma 7. (1) *The function f_k is two to one on $D_k(n)$ for $k = 1$ or $n-2$.*
(2) *For n even, the function f_k is two to one on $D_k(n)$ for $k = (n/2)$ or $(n/2)-1$.*
(3) *For n odd, the function $f_{(n-1)/2}$ is two to one on $D_{(n-1)/2}(n)$.*

Proof (of lemma). (1) As $f_1(x) = (x^{2^2} + x)/(x^2 + x) = x^2 + x + 1$ for $x \in D_1(n)$, it is immediate to see $f_1(x) = f_1(y)$ if and only if $x + y \in GF(2)$. Thus f_1 is two to one on $D_1(n)$. Now by equation (13) with $k = 1$ we have

$$f_1(x) = ((1/f_{n-2}(x)) + 1)^2 \tag{14}$$

for $x \in D_1(n)$, as $f_0(x) = 1$. Suppose that there exists some $z \in D_1(n) \backslash \{y, y+1\}$ with $f_{n-2}(y) = f_{n-2}(z)$. Then the equation (14) applied to $x = y$ and $x = z$ yields

$$f_1(y) = ((1/f_{n-2}(y)) + 1)^2 = ((1/f_{n-2}(z)) + 1)^2 = f_1(z),$$

which contradicts that f_1 is two to one on $D_1(n)$.

(2) Let n be even. It follows from lemma 1 that $f_{n/2}$ is two to one on $D_{n/2}(n)$. Then the two to one property of $f_{(n/2)-1}$ on $D_{(n/2)-1}(n) = D_{(n/2)}$ follows by the same argument as in (1) from the equation

$$f_{(n/2)}(x) = f_{(n/2)-1}(x) \left((1/f_{(n/2)-1}(x)) + 1\right)^{2^{n/2}},$$

which is equation (13) applied to $k = (n/2)$. Note that $k - 1 = n - k - 1$ for $k = n/2$.

(3) Let $n = 2m + 1$ be an odd integer. We denote by ρ and σ respectively the Galois automorphisms defined by $x^\rho = x^2$ and $x^\sigma = x^{2^{(n-1)/2}} = x^{2^m}$ for $x \in GF(q)$. Then we have $x^{\sigma^2 \rho} = x^{2^{2m+1}} = x^{2^n} = x$ for all $x \in GF(q)$, whence $\sigma^2 \rho = id$, or equivalently $\rho = \sigma^{-2}$. Then we have

$$f_m(x) = \frac{x^{2\sigma} + x}{x^2 + x} = \frac{x^{\sigma\rho} + x}{x^\rho + x} = \frac{x^{\sigma^{-1}} + x}{x^{\sigma^{-2}} + x}.$$

Notice that the function f_m is two to one on $D_m(n)$ if and only if the function $1/f_m$ defined by $(1/f_m)(x) = 1/f_m(x)$ is two to one on $D_m(n)$, which is equivalent to the condition that the map $(1/f_m) + 1$ defined by $((1/f_m) + 1)(x) = (1/f_m(x)) + 1$ is two to one on $D_m(n)$. Now we have

$$((1/f_m) + 1)(x) = \frac{x^{\sigma^{-2}} + x}{x^{\sigma^{-1}} + x} + 1 = \frac{x^{\sigma^{-2}} + x^{\sigma^{-1}}}{x^{\sigma^{-1}} + x} = y^{\sigma^{-1}}/y,$$

where we set $y = x^{\sigma^{-1}} + x$. Thus $(1/f_m) + 1$ is the composite of the maps g defined by $g(x) = x^{\sigma^{-1}} + x$ and the map h defined by $h(y) = y^{\sigma^{-1}}/y$. It is easy to see that g is a two to one map from $D_m(n)$ onto $T_0^\times$, the set of nonzero elements of $GF(q)$ of trace 0, where x and $x+1$ correspond to the same element. Furthermore, we see that the map h is injective on $T_0^\times$, because the condition $y^{\sigma^{-1}}/y = z^{\sigma^{-1}}/z$ is equivalent to the condition that $(y/z) = (y/z)^\sigma$, from which we have $y/z = 1$, because ρ generates the group $\mathrm{Gal}(GF(q)/GF(2))$ of odd order and hence σ with $\sigma^2 = \rho^{-1}$ also generates $\mathrm{Gal}(GF(q)/GF(2))$. Thus the composite $(1/f_m) + 1 = h \circ g$ and hence f_m is a two to one map on $D_m(n)$.

References

1. Budaghyan, L., Carlet, C., Leander, G.: A class of quadratic APN binomials inequivalent to power functions (submitted)
2. Budaghyan, L., Carlet, C., Pott, A.: New classes of almost bent and almost perfect nonlinear functions. Trans. Inform. Theory 52(3), 1141–1152 (2006)
3. Cordero, M., Wene, G.P.: A survey of finite semifields. Discrete Math. 208/209, 125–137 (1999)
4. Dembowski, P., Ostrom, T.G.: Planes of order n with collineation groups of order n^2. Math. Z. 103, 239–258 (1968)
5. Kantor, W.: Finite semifields. In: Hulpke, A., Liebler, B., Penttila, T., Seress, A. (eds.) Finite Geometires, Groups, and Computation, Walder de Gruyter, Berlin-New York (2006)
6. Lidle, R., Niederreiter, H.: Finite Fields. In: Encyclopedia of Mathematics and its Applications, 20, Addison-Wesley, Reading, Massachusetts (1983)
7. Minami, K., Nakagawa, N.: On planar functions of elementary abelian p-group type (submitted)
8. Yoshiara, S.: Dimensional dual hyperovals admitting large automorphism groups, manuscript for the proceeding of the symposium on finite groups and related topics, Kyoto University (December 18-21, 2007)

Complete Mapping Polynomials over Finite Field F_{16}*

Yuan Yuan, Yan Tong, and Huanguo Zhang

School of Computer, Wuhan University, Wuhan, Hubei, 430072, China
yuanliuyuan79@163.com, xianquan2006@sohu.com, liss@whu.edu.cn

Abstract. A polynomial $f(x)$ over F_q, the finite field with q elements, is called a complete mapping polynomial if the two mappings $F_q \rightarrow F_q$ respectively defined by $f(x)$ and $f(x) + x$ are one-to-one. In this correspondence, complete mapping polynomials over F_{16} are considered. The nonexistence of the complete mapping polynomial of degree 9 and the existence of the ones of degree 8 and 11 are proved; the result that the reduced degree of complete mapping polynomials over F_{16} are 1, 4, 8, 10, 11, 12, 13 is presented; and by searching with computer, the degree distribution of complete mapping polynomials over the field is given.

Keywords: permutation polynomials, complete mapping polynomials, unique factorization domain.

1 Introduction

F_q is the finite field with q elements, a polynomial $f(x) \in F_q[x]$ is called a *permutation polynomial* on F_q if the mapping defined by f is one-to-one; i.e. the $f(a)$ where $a \in F_q$ are a permutation of the a's. If both $f(x)$ and $f(x) + x$ are permutation polynomials of F_q, then f is called a *complete mapping polynomial* of F_q. By Lagrange's interpolation formula, any mapping of F_q into itself can be represented by a unique polynomial of degree less then q. In fact, for f, $h \in F_q[x]$, we have $f(c) = g(c)$ for all $c \in F_q$ if and only if $f(x) \equiv h(c) \, mod \, (x^q - x)$. The degree of the reduction of $f(x) \, mod \, (x^q - x)$ is called the reduced degree of f, thus, the reduced degree is always less than q.

Although permutation polynomials over finite fields have been a subject of study for many years, complete mapping polynomials over finite fields were introduced in Niederreiter and Robinson [1], and studied in detail in Niederreiter and Robinson [2]. Since then, many results have been given about the subject, but most of them are given from the view of pure theory, and on those polynomials over finite fields of characteristic odd prime.

Now complete mapping polynomials over finite fields F_{2^n}, specially, those ones over the finite fields F_{2^n} with $4|n$, attract more and more attention, for

---

* Supported by National Natural Science Foundation of China (60373087, 60473023, 90104005, 60673071).

C. Carlet and B. Sunar (Eds.): WAIFI 2007, LNCS 4547, pp. 147–158, 2007.

they have useful cryptographic properties and have important applications in block cipher.

Naturally, considerable attention has been given to the case $n = 4$; i.e. the complete mapping polynomials over F_{16}. Over F_{16}, many work have been done on classifying complete mapping polynomials from their reduced degrees, in other words, many work have been done to solve the problem: is there a complete mapping polynomial of degree d with $0 < d < 16$? the results obtained until now are as follows:

1. by Wan'result[3], there is no complete mapping polynomial of degree 14;

2. by Hermite criterion[4], there is no complete mapping polynomials of degree 2, 3, 5;

3. Li [5]proved that there exist complete mapping polynomials of degree 1, 4; and there is no complete mapping polynomial of degree 6, 7 by using transform technique. Unfortunately, her method is not feasible to those polynomials of degree 8, 9, 10, 11, 12, 13. So the above problem over F_{16} has not been solved completely until now.

In this correspondence, we note that $F_{16} \cong \mathbf{Z}[\xi_5]/2\mathbf{Z}[\xi_5]$, where ξ_5 is a primitive 5-th root of unity in the rational number field $\mathbf{Q}$, $\mathbf{Z}[\xi_5]$ is the ring of algebraic integers of the field $\mathbf{Q}(\xi_5)$. By using the property that $\mathbf{Z}[\xi_5]$ is a unique factorization domain and the fact that there is a ring homomorphism from $\mathbf{Z}[\xi_5]$ to F_{16}, we prove that over F_{16} there is no complete mapping polynomial of degree 9; and by discussing the complete mapping binomials and the complete mapping trinomial of the form: $ax^i + bx^j + cx$, $abc \neq 0$ and $15 > i > j > 1$, we prove that there exist complete mapping polynomials of degree 8, 11. We could not prove theoretically there exist complete mapping polynomials of degree 10, 12, 13, which are the work we will do in the future, but it is not difficult for us to find some complete mapping polynomials of degree 10, 12, 13. Hence, the reduced degree of complete mapping polynomial of F_{16} are 1, 4, 8, 10, 11, 12, 13. By searching with computer, we list the degree distribution of the complete mapping polynomials over F_{16}. Thus, we classify the complete mapping polynomials over F_{16} from their reduced degrees, and our result is the theoretical preparation to find the complete mapping polynomials with good cryptographic properties and use them into the design of the block substitution. Moreover, we give a conjecture on the nonexistence of complete mapping polynomial of special form.

2   Preliminaries

At the beginning of the section, we restate some general facts about complete mapping polynomials over F_{2^n} which we freely use throughout this correspondence (see [4]), and after that we present some properties on permutation polynomials over F_{16}.

Definition 1. A polynomial $f(x) \in F_{2^n}[x]$ is called a permutation polynomial, if $f : c \mapsto f(c)$ is a permutation of F_{2^n}.

Lemma 1 (Hermite Criterion). $f(x) \in F_{2^n}[x]$ is a permutation polynomial of F_{2^n} if and only if the following two conditions hold:

    $i)$ $f(x)$ has exactly one root in F_{2^n};

    $ii)$ for each odd integer t with $1 \le t \le 2^n - 2$, the reduction of $f(x)^t \bmod (x^{2^n} - x)$ has degree $\le 2^n - 2$.

Corollary 1. If $d > 1$ is a divisor of $2^n - 1$, then there is no permutation polynomial of F_{2^n} of degree d.

Definition 2. A polynomial $f(x) \in F_{2^n}[x]$ is called a complete mapping polynomial, if both $f(x)$ and $f(x) + x$ are permutation polynomials of F_{2^n}.

Property 1. If $f(x)$ is a permutation (complete mapping) polynomial over F_{2^n}, then so is $f(x) + \gamma$, for any $\gamma \in F_{2^n}$.

Property 2. If $f(x)$ is a complete mapping polynomial over F_{2^n}, then so is $f(x + \gamma)$, for any $\gamma \in F_{2^n}$.

Lemma 2 (ref. [3]). If $n > 1$, then over F_{2^n}, the reduction of any complete mapping polynomial modulo $x^{2^n} - x$ has degree $\le 2^n - 3$.

In the following, we will consider polynomials over the finite field F_{16}. Let ξ_5 be a primitive 5-th root of unity in the rational number field $\mathbf{Q}$, $\mathbf{Z}$ be the integer ring. It is well known(e.g ref[6]) that, as the ring of algebraic integers of the field $\mathbf{Q}(\xi_5)$, $\mathbf{Z}[\xi_5]$ is a unique factorization domain, and the quotient ring $\mathbf{Z}[\xi_5]/2\mathbf{Z}[\xi_5]$ is a finite field, where $2\mathbf{Z}[\xi_5]$ is the principal ideal of $\mathbf{Z}[\xi_5]$ generated by 2. In fact, as finite field, $F_{16} \cong \mathbf{Z}[\xi_5]/2\mathbf{Z}[\xi_5]$. Moreover, 2 is prime in $\mathbf{Z}[\xi_5]$, i.e. for $a, b \in \mathbf{Z}[\xi_5]$, if $2|ab$, then either $2|a$ or $2|b$. For any $a, b, c \in \mathbf{Z}[\xi_5]$, if $c \mid (a - b)$, then we write $a \equiv b(\bmod c)$.

Let $\zeta : \lambda \mapsto \lambda + 2\mathbf{Z}[\xi_5]$ be the natural ring homomorphism from $\mathbf{Z}[\xi_5]$ to the quotient ring $\mathbf{Z}[\xi_5]/2\mathbf{Z}[\xi_5]$, and ϕ be an isomorphism from $\mathbf{Z}[\xi_5]/2\mathbf{Z}[\xi_5]$ to F_{16}, then $\eta = \phi \circ \zeta : \lambda \mapsto \phi(\zeta(\lambda))$, is a ring homomorphism from $\mathbf{Z}[\xi_5]$ to F_{16}. From the definition of η, we have: for any $\beta \in F_{16}$, if $u, v \in \mathbf{Z}[\xi_5]$ such that $\eta(u) = \eta(v) = \beta$, then $u \in v + 2\mathbf{Z}[\xi_5]$, thus there exists a $y \in \mathbf{Z}[\xi_5]$ such that $u = v + 2y$.

We now consider F_{16} as $F_2(\alpha_0)$, where α_0 is a root of the irreducible polynomial $x^4 + x + 1$ over F_2. Obviously, α_0 is a generator of the cyclic multiplicative group $F_{16}^* = F_{16} - \{0\}$, and choose $\lambda_0 \in \mathbf{Z}[\xi_5]$ such that $\eta(\lambda_0) = \alpha_0$, then we have: $\eta(\lambda_0^{15}) = \eta(\lambda_0)^{15} = \alpha_0^{15} = 1, \eta(\lambda_0^i) = \alpha_0^i \ne 1, 0 < i < 15$. Obviously, for $1 \in \mathbf{Z}[\xi_5]$, $\eta(1) = 1$, thus we have $\lambda_0^{15} \in 1 + 2\mathbf{Z}[\xi_5]$, that means $2|(\lambda_0^{15} - 1)$. So we have $\lambda_0^{15} \equiv 1(\bmod 2)$, and $\lambda_0^i \not\equiv 1(\bmod 2), 0 < i < 15$. Specially, let $\alpha = \alpha_0^2$, it is easy to check that α is also a root of $x^4 + x + 1$, thus F_{16} also can be considered as $F_2(\alpha)$, and α is also a generator of F_{16}^*. Let $\lambda = \lambda_0^2 \in \mathbf{Z}[\xi_5]$, then from $2|(\lambda_0^{15} - 1)$, we get that $4|(\lambda^{15} - 1)$. Thus we have $\eta(\lambda) = \eta(\lambda_0^2) = \alpha_0^2 = \alpha$, and in $\mathbf{Z}[\xi_5]$, $\lambda^{15} \equiv 1(\bmod 4), \lambda^i \not\equiv 1(\bmod 2), 0 < i < 15$.

In the following, F_{16} is considered as $F_2(\alpha)$, where α is a root of the irreducible polynomial $x^4 + x + 1$ over F_2, with $\lambda \in \mathbf{Z}[\xi_5]$ satisfying $\eta(\lambda) = \alpha, \lambda^{15} \equiv 1(\bmod 4)$, and $\lambda^i \not\equiv 1(\bmod 2), 0 < i < 15$. Denote $S = \{\lambda^i \mid i = 0, \cdots, 14\} \cup \{0\}$.

Lemma 3. In $\mathbf{Z}[\xi_5]$, $\sum\limits_{x\in S} x^i \equiv \begin{cases} 3\,(\bmod\,4) & 15\mid i \\ 0\,(\bmod\,4) & 15\nmid i \end{cases}$. In particular, when $15\nmid i$ and i is even, $\sum\limits_{x\in S} x^i \equiv 0\,(\bmod\,8)$.

Proof. $\sum\limits_{x\in S} x^i = \sum\limits_{j=0}^{14} (\lambda^j)^i = \begin{cases} 15 & 15\mid i \\ \frac{(\lambda^i)^{15}-1}{\lambda^i-1} & 15\nmid i \end{cases}$. In the case $15\nmid i$, from $4\mid(\lambda^{15}-1)$, we get $4\mid[(\lambda^i)^{15}-1]$ when i is odd, $8\mid[(\lambda^i)^{15}-1]$ when i is even; and $2\nmid(\lambda^i-1)$, then the Lemma follows from the fact that $\mathbf{Z}[\xi_5]$ is a unique factorization domain. □

Proposition 1. If $f(x) = \sum\limits_{i=0}^{14} a_i x^i$ is a permutation polynomial of F_{16}, then

$$\sum\limits_{i+j=15,\,i<j} a_i a_j = 0.$$

Proof. Suppose $f(x) = \sum\limits_{i=0}^{14} a_i x^i$ is a permutation polynomial of F_{16}, then so is $f'(x) = \sum\limits_{i=1}^{14} a_i x^i$ by Property 1. Choose $F(x) = \sum\limits_{i=1}^{14} c_i x^i \in \mathbf{Z}[\xi_5][x]$ such that $\eta(F(x)) = \sum\limits_{i=1}^{14} \eta(c_i)x^i = \sum\limits_{i=1}^{14} a_i x^i = f(x)$; i.e. $\eta(c_i) = a_i$, $1 \le i \le 14$. Then by supposition,

$$\{\eta(F(x)) \mid x \in S\} = \{\eta(\sum\limits_{i=1}^{14} c_i(\lambda^j)^i) \mid 0 \le j \le 14\} \cup \{0\}$$
$$= \{\sum\limits_{i=1}^{14} a_i(\alpha^j)^i \mid 0 \le j \le 14\} \cup \{0\}$$
$$= \{f'(0), f'(1), f'(\alpha), \cdots, f'(\alpha^{14})\} = F_{16}$$

Thus we have $\{\eta(x) \mid x \in S\} = F_{16} = \{\eta(F(x)) \mid x \in S\}$. Hence:

1) $0 \in F_{16}$, then there is only one element $\tilde{x} \in S$ such that $\eta(F(\tilde{x})) = 0$, thus we have $\eta(F(\tilde{x})) = 0 = \eta(0)$, then $\exists \tilde{y} \in \mathbf{Z}[\xi_5]$ such that $F(\tilde{x}) = 0 + 2\tilde{y}$;

2) for each α^i, $0 \le i \le 14$, there is only one element $x_i \in S$ such that $\eta(F(x_i)) = \alpha^i = \eta(\lambda^i)$, thus $\exists y_i \in \mathbf{Z}[\xi_5]$ such that $F(x_i) = \lambda^i + 2y_i$.

So we have $\sum\limits_{x\in S} F(x)^2 = (2\tilde{y})^2 + \sum\limits_{i=0}^{14} (\lambda^i + 2y_i)^2$

$$= \sum\limits_{x\in S} (x + 2y_x)^2, \quad y_x \in \mathbf{Z}[\xi_5]$$

by Lemma 3
$$\equiv \sum\limits_{x\in S} x^2 \,(\bmod\,4)$$
$$\equiv 0 \,(\bmod\,4)$$

on the other hand, $\sum\limits_{x\in S} F(x)^2 = \sum\limits_{x\in S} (\sum\limits_{i=1}^{14} c_i x^i)^2$

$$= \sum\limits_{x\in S}\sum\limits_{i=1}^{14} (c_i^2 x^{2i}) + \sum\limits_{x\in S}\sum\limits_{i<j} 2c_i c_j x^{i+j}$$
$$= \sum\limits_{i=1}^{14} c_i^2 \sum\limits_{x\in S} x^{2i} + \sum\limits_{i<j} 2c_i c_j \sum\limits_{x\in S} x^{i+j}$$

by Lemma 3,
$$\equiv \sum_{i+j=15,\, i<j} 2c_i c_j \sum_{x\in S} x^{15} \,(mod\,4)$$
$$\equiv \sum_{i+j=15,\, i<j} 2c_i c_j \,(mod\,4)$$

thus
$$\sum_{i+j=15,\, i<j} c_i c_j \equiv 0\,(mod\,2)$$

hence
$$\sum_{i+j=15,\, i<j} a_i a_j = \eta\Big(\sum_{i+j=15,\, i<j} c_i c_j\Big) = 0. \qquad \square$$

3   Main Results

In this section, we give our main results and their proofs: in subsection 3.1, the result on polynomials of degree 9 is given and a conjecture is presented; in subsection 3.2, the result on polynomials of degree 8 is given; in subsection 3.3, the result on polynomials of degree 11 is given; we give the conclusions in subsection 3.4.

3.1   Nonexistence of Complete Mapping Polynomial of Degree 9

Let $C_0 = \{0\}$, $C_1 = \{1,2,4,8\}$, $C_3 = \{3,6,9,12\}$, $C_5 = \{5,10\}$, $C_7 = \{7,14, 13,11\}$ be the cyclotomic cosets $mod\,15$ over F_{16}.

Let $f(x) = a_9 x^9 + a_8 x^8 + \ldots + a_1 x$, where $a_9 \neq 0$, if $f(x)$ is a complete mapping polynomial over F_{16}, then by Hermite Criterion, we have that in the reductions of $f(x)^i \,mod\,(x^{16} - x)$ and $[f(x) + x]^i \,mod\,(x^{16} - x)$, $1 \le i \le 14$ and i is odd, coefficients of x^{15} must equal to 0 respectively. In fact, when i, j are in the same cyclotomic coset, $f(x)^i \,mod\,(x^{16} - x)$ and $f(x)^j \,mod\,(x^{16} - x)$, $[f(x)+x]^i \,mod\,(x^{16}-x)$ and $[f(x)+x]^j \,mod\,(x^{16}-x)$ induce the same equations respectively. Thus, we only get 6 equations on the 9 coefficients a_i from the reductions of $f(x)^3 \,mod\,(x^{16}-x)$, $[f(x) + x]^3 \,mod\,(x^{16}-x)$, $f(x)^5 \,mod\,(x^{16}-x)$, $[f(x) + x]^5 \,mod\,(x^{16} - x)$, $f(x)^7 \,mod\,(x^{16}-x)$, $[f(x) + x]^7 \,mod\,(x^{16} - x)$, which are not enough for us to decide whether there is complete mapping polynomial of degree 9. By using the relations between F_{16} and $\mathbf{Z}[\xi_5]$, we get 3 more equations on those coefficients, thus we get 9 equations on 9 coefficients. In fact, among those 9 equations, two are coincident, so we only have 8 equations on the 9 coefficients, but after computing, we still get the following result:

Theorem 1. There is no complete mapping polynomial of degree 9 over F_{16}.

Proof. Let $f(x) = a_9 x^9 + a_8 x^8 + \ldots + a_1 x + a_0$, where $a_9 \neq 0$, then

I. $f(x + a_8 a_9{}^{-1}) = a_9(x + a_8 a_9{}^{-1})^9 + a_8(x + a_8 a_9{}^{-1})^8 + \ldots$
$$+a_1(x + a_8 a_9{}^{-1}) + a_0$$
$$= a_9 x^9 + a_7' x^7 + \ldots + a_1' x + a_0'$$

by Property 2 and Property 1, $f(x)$ is a complete mapping polynomial if and only if $a_9 x^9 + a_7' x^7 + \ldots + a_1' x$ is also a complete mapping polynomial.

Thus, if we can prove there is no complete mapping polynomial of form (1): $f(x) = a_9 x^9 + a_7 x^7 + \ldots + a_1 x$, where $a_9 \neq 0$, then the theorem get proved. We use the proof by contradiction.

II. If $f(x)$ of form (1) is a complete mapping polynomial, then we have
$$f(x)^3 \equiv (a_9a_3{}^2 + a_7a_4{}^2 + a_5{}^3 + a_3a_6{}^2 + a_1a_7{}^2)x^{15} + \varphi(x) \, mod \, (x^{16} - x),$$ where
$\varphi(x)$ is a polynomial of degree ≤ 14;
$$[f(x) + x]^3 \equiv [a_9a_3{}^2 + a_7a_4{}^2 + a_5{}^3 + a_3a_6{}^2 + (a_1+1)a_7{}^2]x^{15} + \varphi'(x) \, mod \, (x^{16} - x),$$
where $\varphi'(x)$ is a polynomial of degree ≤ 14;
    by Hermite Criterion, we have

$$\begin{cases} a_9a_3{}^2 + a_7a_4{}^2 + a_5{}^3 + a_3a_6{}^2 + a_1a_7{}^2 = 0 \\ a_9a_3{}^2 + a_7a_4{}^2 + a_5{}^3 + a_3a_6{}^2 + (a_1 + 1)a_7{}^2 = 0 \end{cases}$$

solve the equations, we obtain $a_7 = 0$ and $a_9a_3{}^2 + a_5{}^3 + a_3a_6{}^2 = 0$.

Thus now, if we can prove there is no complete mapping polynomial of form
(2): $f(x) = a_9x^9 + a_6x^6 + \ldots + a_1x$, where $a_9 \neq 0$, then the theorem get proved.
We use the proof by contradiction again.

III. If $f(x) = a_9x^9 + a_6x^6 + \ldots + a_1x$, where $a_9 \neq 0$, is a complete mapping
polynomial, then by Proposition 1, $a_6 = 0$.

Hence, if we can prove that there is no complete mapping polynomial of form
(3): $f(x) = a_9x^9 + a_5x^5 + \ldots + a_1x$, where $a_9 \neq 0$, then the theorem get proved.
We still use the proof by contradiction.

IV. If $f(x)$ of form (3) is a complete mapping polynomial, then by Hermite
criterion (2), $a_9a_3{}^2 + a_5{}^3 = 0$ must hold. Choose $F(x) = c_9x^9 + c_5x^5 + c_4x^4 + c_3x^3 + c_2x^2 + c_1x \in \mathbf{Z}[\xi_5][x]$ such that $\eta(F(x)) = f(x)$; i.e. $\eta(c_i) = a_i$, $i = 9, 5, 4, 3, 2, 1$.
Obviously, $\eta(1) = 1$, thus $\eta(F(x) + x) = f(x) + x$. By supposition, $f(x)$ and
$f(x) + x$ are permutation polynomials, then similar to the proof of Proposition
1, we have:

$$\{\eta(F(x)) \mid x \in S\} = \{\eta(F(x) + x) \mid x \in S\} = \{\eta(x) \mid x \in S\} = F_{16}$$

1) $\sum_{x \in S} (F(x) + x)^4 = \sum_{x \in S} (x + 2y_x)^4$ where $y_x \in \mathbf{Z}[\xi_5]$

$$\equiv \sum_{x \in S} x^4 \, (mod \, 8)$$

    by Lemma 3 $\equiv 0 \, (mod \, 8)$
on the other hand,
$$\sum_{x \in S} (F(x) + x)^4 = \sum_{x \in S} [F(x)^4 + 4F(x)^3x + 6F(x)^2x^2 + 4x^3F(x) + x^4]$$
compute the reduction of each item $mod \, 8$ respectively:

$$\sum_{x \in S} F(x)^4 \quad \equiv \sum_{x \in S} x^4 (mod \, 8) \equiv 0 (mod \, 8)$$
$$\sum_{x \in S} 4F(x)^3x = \sum_{x \in S} 4(c_9x^9 + c_5x^5 + c_4x^4 + c_3x^3 + c_2x^2 + c_1x)^3 x$$
$$= \sum_{x \in S} 4(c_9{}^2x^{18} + c_5{}^2x^{10} + c_4{}^2x^8 + c_3{}^2x^6 + c_2{}^2x^4 + c_1{}^2x^4 +$$
$$2c_9c_5x^{14} + 2c_9c_4x^{13} + 2c_9c_3x^{12} + 2c_9c_2x^{11} + 2c_9c_1x^{10} +$$
$$2c_5c_4x^9 + 2c_5c_3x^8 + 2c_5c_2x^7 + 2c_5c_1x^6 + 2c_4c_3x^7 + 2c_4c_2x^6 +$$
$$2c_4c_1x^5 + 2c_3c_2x^5 + 2c_3c_1x^4 + 2c_2c_1x^3)$$
$$(c_9x^9 + c_5x^5 + c_4x^4 + c_3x^3 + c_2x^2 + c_1x)x$$

$$\equiv \sum_{x \in S} 4(c_9{}^3 x^{28} + c_9{}^2 c_5 x^{24} + c_9{}^2 c_4 x^{23} + c_9{}^2 c_3 x^{22} + c_9{}^2 c_2 x^{21} + c_9{}^2 c_1 x^{20}$$
$$+ c_5{}^2 c_9 x^{20} + c_5{}^3 x^{16} + \underline{c_5{}^2 c_4 x^{15}} + c_5{}^2 c_3 x^{14} + c_5{}^2 c_2 x^{13} + c_5{}^2 c_1 x^{12}$$
$$+ c_4{}^2 c_9 x^{18} + c_4{}^2 c_5 x^{14} + c_4{}^3 x^{13} + c_4{}^2 c_3 x^{12} + c_4{}^2 c_2 x^{11} + c_4{}^2 c_1 x^{10}$$
$$+ c_3{}^2 c_9 x^{16} + c_3{}^2 c_5 x^{12} + c_3{}^2 c_4 x^{11} + c_4{}^3 x^{10} + c_4{}^2 c_2 x^9 + c_4{}^2 c_1 x^8$$
$$+ c_2{}^2 c_9 x^{14} + c_2{}^2 c_5 x^{10} + c_2{}^2 c_4 x^9 + c_2{}^2 c_3 x^8 + c_2{}^3 x^7 + c_2{}^2 c_1 x^6$$
$$+ c_1{}^2 c_9 x^{12} + c_1{}^2 c_5 x^8 + c_1{}^2 c_4 x^7 + c_1{}^2 c_3 x^6 + c_1 c_2 x^5 + c_1{}^3 x^4)\, (mod\,8)$$
$$\equiv \sum_{x \in S} 4 c_5{}^2 c_4 \,(mod\,8)$$

$$\sum_{x \in S} 6F(x)^2 x^2 = \sum_{x \in S} 6(c_9 x^9 + c_5 x^5 + c_4 x^4 + c_3 x^3 + c_2 x^2 + c_1 x)^2 x^2$$
$$= \sum_{x \in S} 6(c_9{}^2 x^{20} + c_5{}^2 x^{12} + c_4{}^2 x^{10} + c_3{}^2 x^8 + c_2{}^2 x^6 + c_1{}^2 x^4 +$$
$$2 c_9 c_5 x^{16} + 2 c_9 c_4 x^{15} + 2 c_9 c_3 x^{14} + 2 c_9 c_2 x^{13} + 2 c_9 c_1 x^{12} +$$
$$2 c_5 c_4 x^{11} + 2 c_5 c_3 x^{10} + 2 c_5 c_2 x^9 + 2 c_5 c_1 x^8 + 2 c_4 c_3 x^9 + 2 c_4 c_2 x^8$$
$$2 c_4 c_1 x^7 + 2 c_3 c_2 x^7 + 2 c_3 c_1 x^6 + 2 c_2 c_1 x^5)$$
$$\equiv \sum_{x \in S} 4 c_9 c_4 \,(mod\,8),$$
$$\sum_{x \in S} 4 x^3 F(x) \equiv 0 \,(mod\,8)$$
$$\sum_{x \in S} x^4 \equiv 0 \,(mod\,8)$$

then we have $$\sum_{x \in S} (F(x) + x)^4 \equiv 4 c_9 c_4 + 4 c_4 c_5{}^2 \,(mod\,8)$$

hence, we get $c_9 c_4 + c_4 c_5{}^2 \equiv 0 \,(mod\,2).$

thus $a_9 a_4 + a_4 a_5{}^2 = \eta(c_9 c_4 + c_4 c_5{}^2) = 0.$

2) $\sum_{x \in S} (F(x) + x)^6 = \sum_{x \in S} (x + 2y_x)^6 \quad y_x \in \mathbf{Z}[\xi_5]$
$$\equiv \sum_{x \in S} x^6 \,(mod\,4) \equiv 0 \,(mod\,4)$$

on the other hand,
$$\sum_{x \in S} (F(x) + x)^6 = \sum_{x \in S} [F(x)^6 + 6F(x)^5 x + 15F(x)^4 x^2 + 20F(x)^3 x^3 +$$
$$15 x^4 F(x)^2 + 6F(x) x^5 + x^6]$$
$$\equiv \sum_{x \in S} (2 c_9 c_5{}^4 + 2 c_2 a_3{}^4 + 2 c_9 c_2) x^{15} \,(mod\,4)$$
$$\equiv 2 c_9 c_5{}^4 + 2 c_2 a_3{}^4 + 2 c_9 c_2 \,(mod\,4)$$

As above, we obtain $a_9 a_5{}^4 + a_2 a_3{}^4 + a_9 a_2 = 0.$

Now we get 3 equations about the coefficients of $f(x)$ as follows:

$$\begin{cases} a_9 a_3{}^2 + a_5{}^3 = 0 \\ a_9 a_4 + a_4 a_5{}^2 = 0 \\ a_9 a_5{}^4 + a_2 a_3{}^4 + a_9 a_2 = 0 \end{cases}$$

solve the equation system, we get $\begin{cases} a_4 = 0 \\ a_3 = a_5{}^9 a_9{}^7 \\ a_2(1 + a_5{}^6 a_9{}^{12}) = a_5{}^4 \end{cases}$

Hence, if we can prove that there is no complete mapping polynomial of form (4): $f(x) = a_9 x^9 + a_5 x^5 + a_3 x^3 + a_2 x^2 + a_1 x$, where $a_9 \neq 0$, then the theorem get proved. We use the proof by contradiction.

V. If $f(x)$ of form (4) is a complete mapping polynomial, then by Hermite criterion (2), we have

$$\begin{cases} a_3 = a_5{}^9 a_9{}^7 & (1) \\ a_2(1 + a_5{}^6 a_9{}^{12}) = a_5{}^4 & (2) \end{cases}$$

Now we assert that $a_5(1+a_5{}^6a_9{}^{12}) \neq 0$, for otherwise if $1+a_5{}^6a_9{}^{12} = 0$ or $a_5 = 0$, then $a_5 = 0$, $a_2 = 0$ follow from equation (2), hence $a_3 = 0$ by equation (1), which induce that $f(x) = a_9 x^9 + a_1 x$ is a complete mapping polynomial, that is impossible! for when $a_1 \neq 0$, $f(x) = a_9 x^9 + a_1 x$ has more than 2 roots in F_{16}.

Hence, we have

$$\begin{cases} a_9 \neq 0, a_5 \neq 0, 1 + a_5{}^6 a_9{}^{12} \neq 0 \\ a_3 = a_5{}^9 a_9{}^7 & (3) \\ a_2 = a_5{}^4(1 + a_5{}^6 a_9{}^{12})^{-1} = a_5{}^{10} a_9{}^{12} + a_5{}^7 a_9{}^6 & (4) \end{cases}$$

$f(x)^7 \equiv [a_9 a_1{}^6 + a_5 a_3{}^2 a_1{}^4 + a_5 a_2{}^2 a_9{}^4 + a_5 a_2{}^4 a_1{}^2 + a_3 a_2{}^6 + a_3{}^3 a_9{}^4 + a_3{}^2 a_2{}^4 a_1 + a_5{}^2 a_1{}^5 + a_3{}^4 a_1{}^3]x^{15} + \phi(x)(\bmod\, x^{16} - x)$, where $\phi(x)$ is a polynomial of degree ≤ 14;

$[f(x) + x]^7 \equiv [a_9(a_1 + 1)^6 + a_5 a_3{}^2(a_1 + 1)^4 + a_5 a_2{}^2 a_9{}^4 + a_5 a_2{}^4(a_1 + 1)^2 + a_3 a_2{}^6 + a_3{}^3 a_9{}^4 + a_3{}^2 a_2{}^4(a_1+1) + a_5{}^2(a_1 + 1)^5 + a_3{}^4(a_1 + 1)^3]x^{15} + \phi'(x)(\bmod\, x^{16} - x)$, where $\phi'(x)$ is a polynomial of degree ≤ 14;

by Hermite Criterion, we have

$a_9 a_1{}^6 + a_5 a_3{}^2 a_1{}^4 + a_5 a_2{}^2 a_9{}^4 + a_5 a_2{}^4 a_1{}^2 + a_3 a_2{}^6 + a_3{}^3 a_9{}^4 + a_3{}^2 a_2{}^4 a_1 + a_5{}^2 a_1{}^5 + a_3{}^4 a_1{}^3 = 0\,(5)$

$a_9(a_1{}^4 + a_1{}^2 + 1) + a_5 a_3{}^2 + a_5 a_2{}^4 + a_5{}^2(a_1{}^4 + a_1 + 1) + a_3{}^2 a_2{}^4 + a_3{}^4(a_1{}^2 + a_1 + 1) = 0\,(6)$

input (3),(4) to (5) and (6), we get conditions on a_1, a_5 and a_9 as follows:

$$\begin{cases} (a_9 + a_5{}^2)a_1{}^4 + (a_9 + a_5{}^6 a_9{}^{13})a_1{}^2 + (a_5{}^2 + a_5{}^6 a_9{}^{13})a_1 + a_9 + a_5{}^2 + \\ \quad a_5{}^4 a_9{}^{14} + a_5{}^6 a_9{}^{13} + a_5{}^{11} a_9{}^3 + a_5{}^{13} a_9{}^2 + a_5 a_9{}^8 = 0 \\ a_9 a_1{}^6 + a_5{}^2 a_1{}^5 + a_5{}^4 a_9{}^{14} a_1{}^4 + a_5{}^6 a_9{}^{13} a_1{}^3 + (a_5{}^{11} a_9{}^3 + a_5{}^{14} a_9{}^9)a_1{}^2 + \\ \quad (a_5{}^{13} a_9{}^2 + a_5 a_9{}^8)a_1 + a_5{}^9 a_9{}^4 + a_9 + a_5{}^3 a_9{}^7 = 0 \\ 1 + a_5{}^6 a_9{}^{12} \neq 0,\ a_9 \neq 0,\ a_5 \neq 0 \end{cases}$$

When a_9, a_5 respectively run over $F_{16}{}^*$ with $a_5{}^6 a_9{}^{12} \neq 1$, after computing we could not find $a_1 \in F_{16}$ satisfying the above conditions. Hence there is no complete mapping polynomial of form (4) over F_{16}, thus the theorem get proved.   □

Note that in the proof of Theorem 1, the work to prove there is no complete mapping polynomial of form $f(x) = a_9 x^9 + a_8 x^8 + \ldots + a_1 x + a_0$, where $a_9 \neq 0$, is finally reduced to prove there is no complete mapping polynomial of form $f(x) = a_9 x^9 + a_5 x^5 + a_3 x^3 + a_2 x^2 + a_1 x$, where $a_9 \neq 0$. In the case, $f(x) = x(a_9 x^8 + a_5 x^4 + a_3 x^2 + a_2 x + a_1)$, where $a_9 x^8 + a_5 x^4 + a_3 x^2 + a_2 x + a_1$ is a linearized polynomial, a polynomial of the form $\sum_{i=0}^{n-1} a_i x^{2^i}$, $a_i \in F_{2^n}$, is called a linearized polynomial over F_{2^n}. Thus we note the fact that over F_{16}, if $f(x) = xg(x)$, where

$g(x)$ is a linearized polynomial, then $f(x)$ is not a complete mapping polynomial, and we find that the fact still holds over F_8 and there are many examples over other finite fields of characteristic 2, so we have the following conjecture:

Conjecture 1. Over F_{2^n}, there is no complete mapping polynomial of form $xg(x)$, where $g(x)$ is a linearized polynomial over F_{2^n}.

3.2   The Complete Mapping Binomials

In this subsection, we classify the complete mapping binomials of degree less than 16 over F_{16}. From the classification, we obtain that there exist complete mapping polynomials of degree 8.

Lemma 4. The permutation binomials over F_{16} are of the form as follows:

degree	the form of permutation binomials
4	$a(x^4 + bx), ab \neq 0, b \neq \alpha^3, \alpha^6, \alpha^9, \alpha^{12}$
8	$a(x^8 + bx^2), ab \neq 0, b \neq \alpha^3, \alpha^6, \alpha^9, \alpha^{12}$
13	$a(x^{13} + bx^7), ab \neq 0, b \neq \alpha^3, \alpha^6, \alpha^9, \alpha^{12}$
14	$a(x^{14} + bx^{11}), ab \neq 0, b \neq \alpha^3, \alpha^6, \alpha^9, \alpha^{12}$

Proof. Suppose $a(x^i + bx^j)$ is a permutation polynomial, where $ab \neq 0, 0 < j < i < 15$, then we have

(1) $i \nmid 15$, by Corollary 1;

(2) $(i-j, 15) > 1$ and $b \notin (\alpha^{i-j})$, where (α^{i-j}) is the subgroup of F_{16}^* generated by α^{i-j}, otherwise $ax^j(x^{i-j} + b)$ has more than 2 roots in F_{16}, could not be a permutation polynomial;

(3) $i + j \neq 15$, by Proposition 1;

(4) $i, j \in C_1$ or $i, j \in C_7$, for otherwise:

i) if $i \in C_5$, $j \notin C_5$, then in the reduction of $[a(x^i + bx^j)]^3 \bmod (x^{16} - x)$, the coefficient of x^{15} is $a^3 \neq 0$, a contradiction ; the contradiction happens again for the case $j \in C_5$, $i \notin C_5$; but if $i, j \in C_5$, then i must be 10, j must be 5, contradict to (3);

ii) if $i \in C_3$, $j \notin C_3$, by i) we have $j \notin C_5$, if $j \in C_1 \cup C_7$, then in the reduction of $[a(x^i + bx^j)]^5 \bmod (x^{16} - x)$, the coefficient of x^{15} is $a^5 \neq 0$, a contradiction; the contradiction induced again for the case $j \in C_3$, $i \notin C_3$; but if $i \in C_3$, $j \in C_3$, and $i+j \neq 15$, then in the reduction of $[a(x^i + bx^j)]^3 \bmod (x^{16} - x)$, the coefficient of x^{15} is a^3b or a^3b^2, not 0, a contradiction;

iii) if $i \in C_1$, $j \in C_7$, obviously i must be 8, j must be 7, contradict to (3);

iv) if $i \in C_7$, $j \in C_1$, when $i - j = 5$ or 10, by (2) we have $b \notin (\alpha^5)$, hence $b^3 \neq 1$, in this case, the coefficient of x^{15} in the reduction of $[a(x^i + bx^j)]^5 \bmod (x^{16} - x)$ is $a^5(b + b^4) \neq 0$, a contradiction; when $i - j \neq 5, 10$, in the reduction of $[a(x^i + bx^j)]^3 \bmod (x^{16} - x)$, the coefficient of x^{15} is a^3b or a^3b^2, not 0, a contradiction.

From the above four restrictions, and by Hermite Criterion, it is easy for us to get that if $f(x)$ is a permutation polynomial, then $f(x)$ must be of the form as in the above table; and the converse is true. ☐

Theorem 2. The complete mapping binomials over F_{16} are of the form as follows:

degree	the form of complete mapping binomials
4	$a(x^4 + bx), ab \neq 0; b, b + a^{-1} \neq \alpha^3, \alpha^6, \alpha^9, \alpha^{12}$
8	$a(x^8 + bx^2), ab \neq 0; b \neq \alpha^3, \alpha^6, \alpha^9, \alpha^{12};$ and $x^7 + bx + a^{-1}$ has no root in F_{16}

Proof. Suppose $a(x^i + bx^j)$ is a complete mapping polynomial, where $ab \neq 0$, $0 < j < i < 15$, then both $a(x^i + bx^j)$ and $a(x^i + bx^j) + x$ are permutation polynomials, thus by Lemma 4, we have:

(1) $i = 4$, $j = 1$: by Lemma 4, $ab \neq 0$; $b, b + a^{-1} \neq \alpha^3, \alpha^6, \alpha^9, \alpha^{12}$, obviously there exist elements in F_{16} satisfying those conditions, so there exist complete mapping binomials of form $a(x^4 + bx)$.

(2) $i = 8$, $j = 2$: by Hermite criterion, $ab \neq 0$, $b \neq \alpha^3, \alpha^6, \alpha^9, \alpha^{12}$; and $x^7 + bx + a^{-1}$ must have no root in F_{16}. In fact, let $g(x) = x^7 + bx$, then in the reduction of $g(x)^3 \bmod (x^{16} - x)$, the coefficient of x^{15} is $b \neq 0$, so by Hermite criterion, $g(x) = x^7 + bx$ is not a permutation of F_{16}, thus there eixst $a \neq 0$ such that $x^7 + bx + a^{-1}$ has no root in F_{16}. Hence, there exist complete mapping polynomials of form $a(x^8 + bx^2)$.

(3) $i = 13$, $j = 7$: by Hermite criterion, $x^{12} + bx^6 + a^{-1}$ must have no root in F_{16}, with $ab \neq 0, b \neq \alpha^3, \alpha^6, \alpha^9, \alpha^{12}$. Let $y = x^6$, $x^{12} + bx^6 + a^{-1} = y^2 + by + a^{-1}$, obviously, in this case, $y^2 + by$ is a permutation of F_{16}, so $x^{12} + bx^6 + a^{-1} = y^2 + by + a^{-1}$ has one root in F_{16}. Thus, there is no complete mapping polynomial of form $a(x^{13} + bx^7)$. □

3.3   The Complete Mapping Trinomial of Form $ax^i + bx^j + cx$

In the subsection, we classify the complete mapping trinomial of form $ax^i + bx^j + cx$, $abc \neq 0$, and $15 > i > j > 1$. In the process of the classification, we prove that there exist complete mapping polynomials of degree 11 over F_{16}.

Theorem 3. The complete mapping trinomial of form $ax^i + bx^j + cx$, $abc \neq 0$, and $15 > i > j > 1$, must be one of the following: (1) $ax^4 + bx^2 + cx$; (2) $ax^8 + bx^2 + cx$; (3) $ax^8 + bx^4 + cx$; (4) $ax^{11} + bx^7 + cx$.

Proof. Suppose $f(x) = ax^i + bx^j + cx$, $abc \neq 0$, and $15 > i > j > 1$, is a complete mapping polynomial, then four big cases are analyzed as follows:
I. if $i \in C_1$:

(1) if $i = 4$: if $j = 2$, by Li [5], there exist complete mapping polynomials of form $f(x) = ax^4 + bx^2 + cx$; if $j = 3$, then in the reduction of $f(x)^5 \bmod (x^{16} - x)$, the coefficient of x^{15} is $b^5 \neq 0$, a contradiction.

(2) if $i = 8$: if $j = 2$, by Theorem 2, there exist complete mapping polynomials of form $f(x) = ax^8 + bx^2 + cx$; if $j = 4$, by Hermite criterion, $ax^7 + bx^3 + c$, and $ax^7 + bx^3 + c + a^{-1}$ must have no root in F_{16}, it is easy to check that $a = b = 1, c = \alpha^3$ satisfy that conditions, so there exist complete mapping polynomials of form $f(x) = ax^8 + bx^4 + cx$; if $j \in C_3$, then in the reduction of

$f(x)^5 \bmod (x^{16} - x)$, the coefficient of x^{15} is $b^5 \neq 0$, a contradiction; if $j \in C_5$, then in the reduction of $f(x)^3 \bmod (x^{16} - x)$, the coefficient of x^{15} is $b^3 \neq 0$, a contradiction;

II. if $i \in C_5$:

(1) if $i = 5$, by Corollary 2, there is no complete mapping polynomial of degree 5.

(2) if $i = 10$: if $j \in C_1 \cup C_3$, in the reduction of $f(x)^3 \bmod (x^{16} - x)$, the coefficient of x^{15} is $a^3 \neq 0$, a contradiction; if $j = 5$, by Proposition 1, a contradiction; if $j = 7$, by Hermite criterion (2), $a^3 = b^2c$ and $a^3 = b^2c + b^2$ must hold, that means $b = 0$, a contradiction.

III. if $i \in C_3$:

in this case, for there is no complete mapping polynomial of degree 3, 6, and 9, we only need to consider $i = 12$: if $j \in C_1$, then in the reduction of $f(x)^5 \bmod (x^{16} - x)$, the coefficient of x^{15} is $a^5 \neq 0$, a contradiction; if $j \in C_5$, in the reduction of $f(x)^3 \bmod (x^{16} - x)$, the coefficient of x^{15} is $b^3 \neq 0$, a contradiction; if $j \in C_3$, in the reduction of $f(x)^3 \bmod (x^{16} - x)$, the coefficient of x^{15} is a^2b or ab^2, not 0, a contradiction; if $j = 7$, in the reduction of $f(x)^3 \bmod (x^{16} - x)$, the coefficient of x^{15} is $b^2c \neq 0$, a contradiction; if $j = 11$, in the reduction of $f(x)^5 \bmod (x^{16} - x)$ and $[f(x)+x]^5 \bmod (x^{16} - x)$, the coefficients of x^{15} are $a^5 + b^4c$ and $a^5 + b^4(c + 1)$ respectively, could not be 0 at same time, a contradiction;

IV. if $i \in C_7$:

(1) if $i = 7$, there is no complete mapping polynomial of degree 7;

(2) if $i = 11$: if $j = 2$, in the reduction of $f(x)^3 \bmod (x^{16} - x)$, the coefficient of x^{15} is $ab^2 \neq 0$, a contradiction; if $j = 4$, contradict to Proposition 1; if $j = 8$, in the reduction of $f(x)^3 \bmod (x^{16} - x)$, the coefficient of x^{15} is $a^2b \neq 0$, a contradiction; if $j \in C_5$, in the reduction of $f(x)^3 \bmod (x^{16} - x)$, the coefficient of x^{15} is $b^3 \neq 0$, a contradiction; if $j = 3$, in the reduction of $f(x)^5 \bmod (x^{16} - x)$ and $[f(x) + x]^5 \bmod (x^{16} - x)$, the coefficients of x^{15} are $b^5 + a^4c + ac^4$ and $b^5 + a^4(c+1) + a(c^4 + 1)$ respectively, in the reduction of $f(x)^3 \bmod (x^{16} - x)$ and $[f(x)+x]^3 \bmod (x^{16} - x)$, the coefficients of x^{15} are $a^3b^4 + c^3b^4$ and $a^3b^4 + (c+1)^3b^4$ respectively, so the four expressions must equal to 0, which induce $b = 0$, a contradiction; if $j = 9$, in $f(x)^7 \bmod (x^{16} - x)$, the coefficient of x^{15} is $bc^6 \neq 0$, a contradiction; if $j = 6$, by Hermite criterion, $a^4c + c^4a + b^5 = 0$ and $a^3 = 1$ must hold, that means $a(c + c^4) = b^5$, and $a \in (a^5)$ must hold. It is easy to check that for any $\beta \in F_{16}$, $\beta^4 + \beta \in (a^5)$, then choose $a \in (a^5)$ and $a \neq 0$, $c \in F_{16}$ such that $c^4 + c \neq 0$, there exists a $b \neq 0$ with $a(c + c^4) = b^5$. Thus, there exist complete mapping polynomials of form $ax^{11} + bx^7 + cx$; if $j = 7$, in the reduction of $f(x)^3 \bmod (x^{16} - x)$, the coefficient of x^{15} is $b^2c \neq 0$, a contradiction.

(3) if $i = 13$: if $j \in C_5$, in the reduction of $f(x)^3 \bmod (x^{16} - x)$ and $[f(x) + x]^3 \bmod (x^{16} - x)$, the coefficients of x^{15} are $ac^2 + b^3$ and $a(c^2 + 1) + b^3$ respectively, so the two expressions must equal to 0, which induce $a = 0$, a contradiction; if $j \in C_1 \cup C_3$, in the reduction of $f(x)^3 \bmod (x^{16} - x)$, the coefficient of x^{15} is $ac^2 \neq 0$, a contradiction; if $j = 7$, in the reduction of $f(x)^3 \bmod (x^{16} - x)$ and $[f(x)+x]^3 \bmod (x^{16} - x)$, the coefficients of x^{15} are $ac^2 + b^2c$ and $a(c^2 + 1) + b^2(c + 1)$ respectively, so the two expressions must equal to 0, which induce $c = 1$, that

is the case we discussed in the proof of Theorem 2, thus there is no complete mapping polynomial of form $ax^{13} + bx^7 + cx$. $\square$

3.4   Conclusions

Although we could not prove theoretically there exist complete mapping polynomials of degree 10, 12, 13, we find that $\alpha^4 x^{10} + \alpha^{13} x^9 + \alpha^2 x^8 + \alpha^7 x^3 + \alpha^4 x$, $\alpha^8 x^{12} + \alpha^{14} x^9 + \alpha^5 x^6 + \alpha^{11} x^3 + \alpha^5 x$, $\alpha^{14} x^{13} + \alpha^8 x^{10} + \alpha^7 x^7 + \alpha^8 x$ are complete mapping polynomials. Thus, the reduced degree of complete mapping polynomials over F_{16} should be 1, 4, 8, 10, 11, 12, 13. By searching with computer, we get the degree distribution of complete mapping polynomials over F_{16}, listing as follows:

degree	the number of complete mapping polynomials of the degree
1	224
4	6560
8	132480
10	798720
11	933888
12	22179840
13	220692480
sum	244744192

Thus, we classify the complete mapping polynomials over F_{16} from their reduced degrees. From the classification, we even get that: the reduced degree of linearized complete mapping polynomials are 1, 4, and 8; the reduced degree of non-linearized complete mapping polynomials are 8, 10, 11, 12, 13. One of our next work is to find the complete mapping polynomials with good cryptographic properties and use them into the design of the block substitution.

References

1. Niederreiter, H., Robinson, K.H.: Bol loops of order pq. Math. proc. cambridge philos. soc. 89, 241–256 (1981)
2. Niederreiter, H., Robinson, K.H.: Complete mappings of finite fields. J. Austral. Math. Soc. Ser.A 33, 197–212 (1982)
3. Wan, D.: On a problem of Niederreiter and Robinson about finite fields. J.Austral. Math. Soc. Ser. A 41, 336–338 (1986)
4. Lidl, R., Niederreiter, H.: Finite Fields, encyclopedia of mathematics and its application. Addison-Wesley Publishing Company, London (1983)
5. Zhihui Li: The Research on Permutation Theory in Block Cipher System. Ph.D thesis, Northwestern Polytechnical University (2002)
6. Lang, S.: Algebraic number theory, 2nd edn. GTM110. Springer, Berlin Heidelberg New York (1994)

On the Classification of 4 Bit S-Boxes

G. Leander[1,*] and A. Poschmann[2]

[1] GRIM, University Toulon, France
Gregor.Leander@rub.de
[2] Horst-Görtz-Institute for IT-Security, Ruhr-University Bochum, Germany
poschmann@crypto.rub.de

Abstract. In this paper we classify all optimal 4 bit S-boxes. Remark-ably, up to affine equivalence, there are only 16 different optimal S-boxes. This observation can be used to efficiently generate optimal S-boxes ful-filling additional criteria. One result is that an S-box which is optimal against differential and linear attacks is always optimal with respect to algebraic attacks as well. We also classify all optimal S-boxes up to the so called CCZ equivalence. We furthermore generated all S-boxes fulfill-ing the conditions on nonlinearity and uniformity for S-boxes used in the block cipher Serpent. Up to a slightly modified notion of equiva-lence, there are only 14 different S-boxes. Due to this small number it is not surprising that some of the S-boxes of the Serpent cipher are linear equivalent. Another advantage of our characterization is that it eases the highly non-trivial task of choosing good S-boxes for hardware dedicated ciphers a lot.

Keywords. S-box, Vectorial Boolean function, Affine equivalence, Hard-ware Implementation.

1 Introduction

S-boxes play a fundamental role for the security of nearly all modern block ciphers. In the two major design strategies for block ciphers, Feistel networks and Substitution/Permutation networks, the S-boxes form the only non-linear part of a block cipher. Therefore, S-boxes have to be chosen carefully to make the cipher resistant against all kinds of attacks. In particular there are well studied criteria that a good S-box has to fulfill to make the cipher resistant against differential and linear cryptanalyses. There are mainly two ways of generating good S-boxes: (1) picking a random large S-box or (2) generating small S-boxes with good linear and differential properties. The main drawback of picking large S-boxes is, that these S-boxes are much more inefficient to implement, especially in hardware.

Many modern block ciphers use 4 or 8 bit S-boxes. In the AES, for example, an 8 bit S-box is used that provides very good resistance against linear and differential attacks. However, regarding the design of S-boxes there are still some

* Research supported by a DAAD postdoctoral fellowship.

C. Carlet and B. Sunar (Eds.): WAIFI 2007, LNCS 4547, pp. 159–176, 2007.

fundamental questions unsolved. For example, it is not known if the AES S-box is really the optimal choice, it might be true that there exist S-boxes with better resistance against linear and differential attacks. Hence, the AES S-box is a world record S-box, but it is still unclear if this is an optimal result. The problem to find optimal S-boxes is very hard due to the fact that the number of permutations mapping n-bits to n-bits is huge even for very small values of n. Therefore exhaustively checking all permutations to find good S-boxes is no option.

In this paper we focus on 4 bit S-boxes, as used for example in Serpent. One advantage in dimension 4 is that the optimal values for S-boxes with respect to linear and differential cryptanalyses are known. However, the number of 4-bit permutations is still huge: roughly 2^{44}. Furthermore a naive classification of good 4 bit S-boxes is still difficult. However, it is well known that the resistance of S-boxes against most attacks remains unchanged when an invertible affine transformation is applied before and after the S-box. This fairly standard technique allows us to easily classify all optimal 4 bit S-boxes. Surprisingly, up to equivalence, there are only 16 optimal S-boxes and we list them in this paper. This massive reduction enables us to exhaustively check all optimal S-boxes with respect to other criteria, such as algebraic degree or resistance against algebraic attacks and we list some of the results. Most notably an optimal S-box with respect to linear and differential properties is always optimal with respect to algebraic attacks. Furthermore we classify these optimal S-boxes also with respect to the more general CCZ equivalence.

Moreover, this classification simplifies the task to generate optimal S-boxes that:

 – are uniformly distributed among all optimal S-boxes,
 – are not linear equivalent,
 – fulfill additional criteria.

In the second part of this paper we focus on Serpent-type S-boxes. The block cipher Serpent uses 8 S-boxes that were chosen to fulfill additional criteria that are, in general, not invariant under affine transformations. Still it is possible to develop a slightly modified notion of equivalence and again classify these S-boxes. It turns out that, up to equivalence, there are only 14 S-boxes fulfilling the Serpent criteria. Again, using this classification one can easily derive additional properties for these kind of S-boxes. For example we demonstrate that it is not possible to choose a Serpent-type S-box such that all component functions have maximal algebraic degree.

This reduction can also be used for the design of hardware optimized block ciphers. The highly non-trivial task of minimizing the area requirements of the circuit of an S-box in hardware is eased a lot, because only a very small set of S-boxes has to be synthesized.

In [2] it is observed that several of the Serpent-type S-boxes are linear equivalent, although they have been generated in a pseudo random way. Our classification shows that this can be explained as a consequence of the small number of equivalence classes. Instead of 8 different S-boxes our considerations show that Serpent uses only 4 (really) different S-boxes. In other words, it is possible to specify the Serpent cipher, by modifying the linear layer, in such a way that only 4 S-boxes are used. We want to point out, that to the best of our knowledge this does not pose a threat to the security of Serpent. However, due to the small number of linear inequivalent S-boxes, even randomly generated optimal S-boxes might have unexpected and unwished relations. We feel that, when designing a block cipher, one should be aware of this fact and take it into account.

We want to point out an important difference of this work and algorithms that consider the linear equivalence of two given functions (see for example [2]): the task of classifying all functions is not trivial, because the number of functions for which this algorithm has to be applied to might still be too huge.

It should be noted that the results and techniques given here clearly do not work for larger S-boxes. Even for dimension six, a complete classification of all good S-boxes seems elusive. To illustrate the huge amount of computation needed for this classification note that the number of linear inequivalent S-boxes in dimension five is already larger than 2^{61}, see [9].

2   Notation

In this section we introduce the notation used throughout the paper. For two vectors $a, b \in \mathbb{F}_2^n$, we denote by

$$\langle a, b \rangle = \sum_{i=0}^{n-1} a_i b_i$$

the inner product of a and b. The binary weight of a vector a is denoted by $\mathrm{wt}(a)$. For a Boolean function in n variables

$$f : \mathbb{F}_2^n \to \mathbb{F}_2$$

and an element $a \in \mathbb{F}_2^n$ we denote the *Walsh Coefficient* of f at a by

$$f^{\mathcal{W}}(a) = \sum_{x \in \mathbb{F}_2^n} (-1)^{f(x) + \langle a, x \rangle}. \tag{1}$$

The linearity of f is defined as

$$\mathrm{Lin}(f) = \max_{a \in \mathbb{F}_2^n} |f^{\mathcal{W}}(a)|$$

The value $\mathrm{Lin}(f)$ is large if and only if f is close to a linear or affine function, i.e. there exists a linear or affine function which is a good approximation for f. The maximal possible value for $\mathrm{Lin}(f)$ is 2^n and is attained iff f is linear or affine. Moreover, due to Parsevals Equality

$$\sum_{a \in \mathbb{F}_2^n} \left(f^{\mathcal{W}}(a) \right)^2 = 2^{2n}$$

we see that $\text{Lin}(f) \geq 2^{n/2}$. Functions attaining this lower bound are called *bent functions*. Bent functions were introduced by Rothaus [13] and exist if and only if n is even.

This paper deals mainly with S-boxes, i.e. functions with values that are bit strings. Given an S-box mapping n bits to m bits

$$S : \mathbb{F}_2^n \to \mathbb{F}_2^m$$

we denote for any vector $b \in \mathbb{F}_2^m$ the corresponding *component function* S.

$$S_b : \mathbb{F}_2^n \to \mathbb{F}_2$$
$$x \mapsto \langle b, S(x) \rangle$$

The function S_b is the Boolean function derived from S by considering a fixed sum of the output bits determined by $b \in \mathbb{F}_2^m$. In particular, if b is the i-th vector in the canonical base, S_b corresponds to the i-th bit of S. We define the linearity of S as

$$\text{Lin}(S) = \max_{a \in \mathbb{F}_2^n, b \in \mathbb{F}_2^m \setminus \{0\}} |S_b^{\mathcal{W}}(a)|$$

This linearity represents a measure for the resistance against linear cryptanalysis. For even dimension n the smallest known linearity for a permutation is $2^{n/2+1}$. It is a longstanding open problem to find S-boxes with smaller linearity, or to prove that such functions do not exist.

In linear cryptanalysis, introduced by Matsui in [10], one is interested in approximating S with a linear function.

The probability of a linear approximation of a combination of output bits S_b (determined by b) by a linear combination of input bits x (determined by a) can be written as

$$p = \frac{\#\{x | S_b(x) = \langle a, x \rangle\}}{2^n}. \tag{2}$$

Combining equations (1) and (2) leads to

$$p = \frac{1}{2} - \frac{S_b^{\mathcal{W}}(a)}{2^{n+1}}.$$

The *linear probability bias* ε is a correlation measure for this deviation from probability $\frac{1}{2}$ for which it is entirely uncorrelated. We have

$$\varepsilon = \left| p - \frac{1}{2} \right| = \left| \frac{S_b^{\mathcal{W}}(a)}{2^{n+1}} \right|$$

and

$$\varepsilon \leq \left| \frac{\text{Lin}(S)}{2^{n+1}} \right|.$$

Therefore, the smaller the linearity $\text{Lin}(S)$ of a S-box is, the more secure the S-box is against linear cryptanalysis.

The idea of differential cryptanalysis (DC), invented by Biham and Shamir (see [1]), in a nutshell, is to trace how the difference of two encrypted messages m and $m+\delta$ propagates through the different rounds in a block cipher. Basically, if an attacker can guess the output differences with high probability, the cipher will be vulnerable to a differential attack. Thus, a designer of a block cipher has to ensure that, given any nonzero input difference, no fixed output difference occurs with high probability. Since in nearly all block ciphers S-boxes represent the only nonlinear parts, it is particularly important to study the differential properties of these building blocks. To measure the resistance against differential cryptanalysis we define for $a \in \mathbb{F}_2^n$

$$\Delta_{S,a} : \mathbb{F}_2^n \to \mathbb{F}_2^m$$
$$x \mapsto S(x) + S(x + a)$$

and

$$\mathrm{Diff}(S) = \max_{a \neq 0, b \in \mathbb{F}_2^n} |\Delta_{S,a}^{-1}(b)|.$$

Clearly, the value $\mathrm{Diff}(S)$ is related to the maximal probability that any fixed nonzero input difference causes any fixed output difference after applying the S-box. Given an input difference a the value $|\Delta_{S,a}^{-1}(b)|$ is the number of message pairs $(x, x + a)$ with the output difference b.

Clearly it holds for any S-box that $\mathrm{Diff}(S) \geq 2$. Functions attaining this lower bound are called APN functions. However it is unknown if APN permutations exist in even dimension.

3   Optimal 4 Bit S-Boxes

As explained in the introduction, a natural requirement for 4 bit S-boxes is an optimal resistance against linear and differential cryptanalyses. Unlike for higher dimensions the optimal values for $\mathrm{Lin}(S)$ and $\mathrm{Diff}(S)$ are known for dimension $n = 4$. More precisely, for any bijective mapping $S : \mathbb{F}_2^4 \to \mathbb{F}_2^4$ we have $\mathrm{Lin}(S) \geq 8$ and $\mathrm{Diff}(S) \geq 4$. To see that $\mathrm{Lin}(S) \geq 8$ note that, if S is a bijection then all its component functions S_b have even weight and therefore all Walsh coefficients are divisible by 4. Furthermore we must have $\mathrm{Lin}(S) > 4$ since there are no vectorial bent functions from $\mathbb{F}_2^n \to \mathbb{F}_2^n$ for any n as proven by Nyberg in [11]. Knudsen [7] showed that there is no APN Permutation on $\mathbb{F}_2^4$, i.e. no S-boxes with $\mathrm{Diff}(S) = 2$. Therefore, as $\mathrm{Diff}(S)$ is always even, we must have $\mathrm{Diff}(S) \geq 4$. With respect to these observations we call S-boxes attaining these minima *optimal*.

Definition 1. *Let $S : \mathbb{F}_2^4 \to \mathbb{F}_2^4$ be an S-box. If S fulfills the following conditions we call S an optimal S-box*

1. *S is a bijection.*
2. *$\mathrm{Lin}(S) = 8$.*
3. *$\mathrm{Diff}(S) = 4$.*

An example for an optimal S-box is the inverse function, where one identifies the vector space $\mathbb{F}_2^n$ with $\mathbb{F}_{2^n}$, the finite field with 2^n elements, and considers the mapping

$$I : \mathbb{F}_{2^n} \to \mathbb{F}_{2^n}$$
$$I(x) = x^{2^n-2}$$

This mapping fulfils $\mathrm{Lin}(S) = 2^{n/2+1}$ when n is even, as was proven for example in [8]. The proof that $\mathrm{Diff}(S) = 4$ when n is even and $\mathrm{Diff}(S) = 2$ when n is odd is trivial. This type of mapping is also used in the Advanced Encryption Standard (AES). However, in AES we have $n = 8$ and it is not clear that this S-box is optimal in this dimension. It can only be viewed as the world record S-box in a sense that no bijection is known with better resistance against linear and differential cryptanalyses.

When designing a block cipher it is important to know the set of S-boxes to choose from in order to get an optimal resistance against known attacks. Since the number of all permutations on $\mathbb{F}_2^n$ is $2^n!$ and thus huge even for small dimensions, it is crucial to be able to reduce the number of S-boxes which have to be considered. A well known and well suited tool is the notion of linear equivalence.

3.1   Linear Equivalence

It is well known (see for example [4] and [12]) that the values of $\mathrm{Diff}(S)$ and $\mathrm{Lin}(S)$ remain unchanged if we apply affine transformations in the domain or co-domain of S. In particular if we take an optimal S-box in the above sense and transform it in an affine way, we get another optimal S-box. Using such a transformed S-box can also be viewed as changing the linear layer of a block cipher.

This is formalized in the following theorem.

Theorem 1. Let $A, B \in \mathrm{GL}(4, \mathbb{F}_2)$ be two invertible 4×4 matrices and $a, b \in \mathbb{F}_2^4$. Let $S : \mathbb{F}_2^4 \to \mathbb{F}_2^4$ be an optimal S-box. Then the S-box S' with

$$S'(x) = B(S(A(x) + a)) + b$$

is an optimal S-box as well.

This observation can be used to define an equivalence relation on the set of all optimal S-boxes. We call two S-boxes S_1, S_2 equivalent if there exist bijective linear mappings A, B and constants $a, b \in \mathbb{F}_2^4$ such that

$$S'(x) = B(S(A(x) + a)) + b.$$

If two S-boxes S_1 and S_2 are equivalent in the above sense we denote this by $S_1 \sim S_2$. A natural question that arises is in how many equivalence classes the set of all optimal S-boxes is split. As we have already pointed out this reduction to equivalence classes is also important for the practical design of block ciphers, because it simplifies the choice of good S-boxes. We computed the number of equivalence classes using the observations presented in Section 6 and it turns out that this number is very small. There are only 16 different, i.e. non-equivalent, classes for each of them we list a representative in Table 6 and their polynomial representation in Table 7. Each row in Table 6 contains one representative, where we identify vectors $\mathbb{F}_2^n$ with integers in $\{0, \ldots, 15\}$. We list the images of the values in integer ordering, i.e. the first integer represents the image of 0, the second the image of 1, and so on. We summarize this result in the following fact.

Fact 1. *There exist exactly 16 non equivalent optimal S-boxes. Any optimal S-box is equivalent to exactly one S-box given in Table 6.*

Note that G_3 is equivalent to the invers mapping $x \mapsto x^{-1}$.

This massive reduction allows us to exhaustively check all optimal S-boxes up to equivalence with respect to other criteria as explained in the next section.

3.2 Other Criteria

With this reduction to 16 equivalence classes it is now easy to study additional criteria, such as algebraic immunity, improved resistance against linear and differential cryptanalyses, which are again invariant under this equivalence.

Algebraic Degree. Another important criterion for an S-box is to have high algebraic degree. It is well known that every Boolean function $f : \mathbb{F}_2^n \to \mathbb{F}_2$ can be uniquely represented by a multivariate polynomial of degree at most 4, i.e. there exist coefficients $\alpha_u \in \mathbb{F}_2$ such that

$$f(x) = f(x_1, \ldots, x_n) = \sum_{u \in \mathbb{F}_2^n} \alpha_u x_1^{u_1} \cdots x_n^{u_n}$$

The (algebraic) degree of f is defined to be the maximal weight of u such that $\alpha_u \neq 0$. For an S-box we define the algebraic degree as

$$\deg(S) = \max_{b \in \mathbb{F}_2^n, b \neq 0} \deg(S_b)$$

Clearly the degree is invariant under linear equivalence. Moreover it is easy to see that also the multiset

$$\{\deg(S_b) \mid b \in \mathbb{F}_2^n\}$$

is invariant under linear equivalence. It is known that any bijection must have degree smaller than 3. We computed the degree of all 16 representatives and we list the results in Table 1.

It should be noted that a frequently used criterion for good S-boxes is to have high algebraic degree. From this perspective the following fact is of interest

Fact 2. *There exist 8 non linear equivalent optimal S-boxes $S : \mathbb{F}_2^n \to \mathbb{F}_2^n$ such that $\deg(S_b) = 3$ for all $b \in \mathbb{F}_2^n, b \neq 0$.*

Again, one example of such an S-box is the "inverse" S-box, which indeed is equivalent to G_3.

Table 1. Number of $b \in \mathbb{F}_2^4 \setminus \{0\}$ such that $\deg(S_b) = 2, 3$

$S - box$	G_0	G_1	G_2	G_3	G_4	G_5	G_6	G_7	G_8	G_9	G_{10}	G_{11}	G_{12}	G_{13}	G_{14}	G_{15}
$\deg(S_b) = 2$	3	3	3	0	0	0	0	0	3	1	1	0	0	0	1	1
$\deg(S_b) = 3$	12	12	12	15	15	15	15	15	12	14	14	15	15	15	14	14

Linearity. Besides minimizing the linearity $\mathrm{Lin}(S)$ it might also be important to study which Walsh coefficients occur and how often they occur. In particular a reasonable design goal would be to minimize the number of Walsh coefficients with the highest maximal value, as this minimizes the number of linear approximation with maximal probability. We computed the Walsh spectra for all of the 16 S-boxes. In Table 9 we list for each S-box the number of times a certain Walsh coefficient is attained.

Algebraic Relations. Algebraic attacks, invented by Courtois and Pieprzyk (see [5]), are another type of attack which have recently attracted a lot of attention. It still seems to be unclear which conditions exactly enable this attack. However, the main criterion to successfully mount an algebraic attack is the number of linear independent low degree equations that are fulfilled by the input and output values of the S-box, i.e equations of the form $P(x, S(x)) = 0$ for all x. While for large S-boxes or for a huge number of small S-boxes computing the number of such equations needs an enormous computational effort, it is very easy to compute the number of equations for all 16 representative given above. Following [6] we computed the number of quadratic equations, i.e. all equations of the form

$$\sum_{i,j} \alpha_{ij} x_i y_i + \sum_{i \neq j} \beta_{ij} x_i x_j + \sum_{i \neq y} \gamma_{ij} y_i y_j + \sum_i \delta_i x_i + \sum_i \epsilon_i y_j + \nu$$

where $y = S(x)$. Remarkably all 16 representatives, and therefore all optimal S-boxes fulfill exactly 21 linear independent quadratic equations. As explained by Courtois in [6] this is the minimal number of quadratic equations for any mapping from $\mathbb{F}_2^4$ to $\mathbb{F}_2^4$. In this sense the optimal S-boxes are also optimal with respect to algebraic attacks. We summarize these observation in the following fact.

Fact 3. Let $S : \mathbb{F}_2^4 \to \mathbb{F}_2^4$ be an optimal S-box. Then S fulfills 21 quadratic equations, which is the minimum number for S-boxes in dimension 4.

4   Serpent-Type S-Boxes

For the resistance of ciphers against linear and differential cryptanalyses not only the linearity of an S-box might be important, but also where this maximum occurs. Depending on the design strategy, for resistance against DC it might be important, that any one bit input difference causes an output difference of at least two bits (see for example DES or Serpent). Such a condition can be used to increase the minimal number of active S-boxes in two consecutive rounds. A similar requirement for LC is that the probability of a linear approximation using only one input and one output bit is especially low. We formalize these two conditions using the following notation

$$\text{Lin}_1(S) = \max\{|S_b^{\mathcal{W}}(a)| \mid \text{wt}(a) = \text{wt}(b) = 1\}.$$

and

$$\text{Diff}_1(S) = \max_{a \neq 0, b \in \mathbb{F}_2^n} \{|\Delta_{S,a}^{-1}(b)| \ \text{wt}(a) = \text{wt}(b) = 1\}.$$

The S-boxes in the Serpent cipher fulfill the following conditions

Definition 2. *Let* $S : \mathbb{F}_2^4 \to \mathbb{F}_2^4$ *be a S-box. If* S *fulfills the following conditions we call* S *a Serpent-type S-box*

1. S *is optimal*
2. $\text{Diff}_1(S) = 0$, *i.e. any one bit input difference causes at least two bits output difference.*

We generated all S-boxes having 4-bit input and 4-bit output value fulfilling these conditions. The total number of these S-boxes is $2,211,840$ but again this can be reduced using, a slightly modified, notion of linear equivalence.

4.1   Equivalence of Serpent-Type S-Boxes

It is easy to see that the condition $\text{Diff}_1(S) = 0$ is, in general, not invariant under the above defined equivalence relation. However, when we restrict to bit permutations in the domain and the co-domain of a mapping S, instead of allowing arbitrary linear bijections, this gives us a similar tool as before. We have the following Theorem, which is a modified version of Theorem 1.

Theorem 2. *Let* $P_0, P_1 \in \text{GL}(4, \mathbb{F}_2)$ *be two* 4×4 *permutation matrices and* $a, b \in \mathbb{F}_2^4$. *Let* $S : \mathbb{F}_2^4 \to \mathbb{F}_2^4$ *be a Serpent-type S-box. Then the S-boxes* S' *with*

$$S'(x) = P_1(S(P_0(x) + a)) + b$$

is a Serpent-type S-box as well.

Proof. trival.

Clearly this again defines an equivalence relation of the set of S-boxes. If two S-boxes S_1, S_2 are equivalent in the above sense we denote this by

$$S_1 \sim_S S_2.$$

As this notion of equivalence is a special type of the equivalence used in Section 3 we have the implication

$$S_1 \sim_S S_2 \Rightarrow S_1 \sim S_2$$

Again, one might be interested in how many equivalence classes the set of all Serpent-type S-boxes is split. Like before, this number is surprisingly small. There are only 20 different classes and for each class we list a representative in Table 8.

Fact 4. *There exist exactly* 20 *non equivalent Serpent-Type S-boxes. Any Serpent-Type S-box is equivalent to exactly one S-box given in Table 8.*

4.2   Other Criteria

Again this reduction to only 20 representatives allows us to exhaustively check all Serpent-type S-boxes with respect to other criteria. Since Serpent-type S-boxes are in particular optimal S-boxes, Fact 3 immediately applies to Serpent-type S-boxes as well. On the other hand due to the computations given in Table 2, it is impossible to choose a Serpent-type S-box such that all linear combinations of coordinate functions have maximal degree.

Table 2. Number of $b \in \mathbb{F}_2^4$ such that $\deg(S_b) = 2, 3$

$S - box$	R_0	R_1	R_2	R_3	R_4	R_5	R_6	R_7	R_8	R_9
$\deg(S_b) = 2$	3	3	3	1	1	1	3	3	1	3
$\deg(S_b) = 3$	12	12	12	14	14	14	12	12	14	12

$S - box$	R_{10}	R_{11}	R_{12}	R_{13}	R_{14}	R_{15}	R_{16}	R_{17}	R_{18}	R_{19}
$\deg(S_b) = 2$	3	3	3	1	3	1	3	3	3	3
$\deg(S_b) = 3$	12	12	12	14	12	14	12	12	12	12

Fact 5. *For any Serpent-type S-box S there exists an element $b \in \mathbb{F}_2^n$ such that S_b is a quadratic Boolean function.*

In particular this implies that the "inverse" Function is not linear equivalent to a Serpent-type S-box.

Linearity. We computed the Walsh spectra for all of the 20 S-boxes. In Table 10 we list for each S-box the number of times a certain Walsh coefficient is attained.

As already mentioned before, not only the linearity of an S-box is important for the resistance of a cipher against linear cryptanalysis, but also where this maximum occurs. In particular it might be important, depending on the linear layer, that the Walsh coefficients $S_b^{\mathcal{W}}(a)$ with $\mathrm{wt}(a) = \mathrm{wt}(b) = 1$ are especially small. We list the values in Table 3. In particular we see that there exist no Serpent-type S-box such that $\mathrm{Lin}_1(S) = 0$.

Table 3. Linearity of all 20 Classes of Serpent-type S-boxes

$S - box$	R_0	R_1	R_2	R_3	R_4	R_5	R_6	R_7	R_8	R_9	R_{10}	R_{11}	R_{12}	R_{13}	R_{14}	R_{15}	R_{16}	R_{17}	R_{18}	R_{19}
$\mathrm{Lin}_1(S)$	4	4	8	4	4	4	4	4	4	8	8	8	8	4	4	4	4	8	4	4

Fact 6. *Let $S : \mathbb{F}_2^4 \to \mathbb{F}_2^4$ be a Serpent-type S-box. Then $\mathrm{Lin}_1(S) \in \{4, 8\}$.*

This fact demonstrates that the choice made for the block cipher Serpent is indeed optimal with regard to this criterion.

5   Relation Between the Representatives and Inverses

The number of representatives can be further reduced when also the inverses of the S-boxes are considered. This is due to the next Theorem, which is obvious to proof.

Theorem 3. *Let $S : \mathbb{F}_2^4 \to \mathbb{F}_2^4$ be an optimal (resp. a Serpent-type) S-box then its inverse S^{-1} is an optimal (resp. a Serpent-type) S-box as well.*

We have the following relations between the representatives of the optimal S-boxes and their inverses.

$$G_0 \sim G_2^{-1}, \quad G_1 \sim G_1^{-1}, \quad G_2 \sim G_0^{-1}$$
$$G_3 \sim G_3^{-1}, \quad G_4 \sim G_4^{-1}, \quad G_5 \sim G_5^{-1}$$
$$G_6 \sim G_6^{-1}, \quad G_7 \sim G_7^{-1}, \quad G_8 \sim G_8^{-1}$$
$$G_9 \sim G_9^{-1}, \quad G_{10} \sim G_{10}^{-1}, G_{11} \sim G_{11}^{-1}$$
$$G_{12} \sim G_{12}^{-1}, G_{13} \sim G_{13}^{-1}, G_{14} \sim G_{15}^{-1}$$
$$G_{15} \sim G_{14}^{-1}$$

Remarkably, all optimal S-boxes except for G_0, G_2 and G_{14}, G_{15} are linear equivalent to their inverses.

For the Serpent-type S-boxes we have the following relations

$$R_0 \sim_S R_{18}^{-1}, \quad R_1 \sim_S R_6^{-1}, \quad R_2 \sim_S R_{17}^{-1}$$
$$R_3 \sim_S R_5^{-1}, \quad R_4 \sim_S R_{13}^{-1}, \quad R_5 \sim_S R_3^{-1}$$
$$R_6 \sim_S R_1^{-1}, \quad R_7 \sim_S R_{16}^{-1}, \quad R_8 \sim_S R_{15}^{-1}$$
$$R_9 \sim_S R_{10}^{-1}, \quad R_{10} \sim_S R_9^{-1}, \quad R_{11} \sim_S R_{12}^{-1}$$
$$R_{12} \sim_S R_{11}^{-1}, R_{13} \sim_S R_4^{-1}, \quad R_{14} \sim_S R_{19}^{-1}$$
$$R_{15} \sim_S R_8^{-1}, \quad R_{16} \sim_S R_7^{-1}, \quad R_{17} \sim_S R_2^{-1}$$
$$R_{18} \sim_S R_0^{-1}, \quad R_{19} \sim_S R_{14}^{-1}$$

From these relations we see the following Fact.

Fact 7. *No Serpent-type S-box is self-equivalent in a sense that*

$$S \sim_S S^{-1}$$

and in particular no Serpent-type S-box is an involution.

Clearly, as mentioned before, Serpent-type S-boxes are optimal S-boxes and therefore each of the Serpent-type S-boxes R_i must be equivalent to one of the optimal S-boxes G_j. For the sake of completeness we list these relations below.

$$R_0 \sim G_1, \quad R_1 \sim G_1, \quad R_2 \sim G_1$$
$$R_3 \sim G_{10}, \quad R_4 \sim G_9, \quad R_5 \sim G_{10}$$
$$R_6 \sim G_1, \quad R_7 \sim G_2, \quad R_8 \sim G_{15}$$
$$R_9 \sim G_0, \quad R_{10} \sim G_2, \quad R_{11} \sim G_0$$
$$R_{12} \sim G_2, \quad R_{13} \sim G_9, \quad R_{14} \sim G_2$$
$$R_{15} \sim G_{14}, \quad R_{16} \sim G_0, \quad R_{17} \sim G_1$$
$$R_{18} \sim G_1, \quad R_{19} \sim G_0$$

5.1 CCZ Equivalence

The linear equivalence defined above, is a special case of a more general equivalence (see [4]), called Carlet-Charpin-Zinoviev equivalence (CCZ for short). Two functions $F, G : \mathbb{F}_2^n \to \mathbb{F}_2^n$ are called CCZ equivalent if there exist a linear permutation on $\mathbb{F}_2^n \times \mathbb{F}_2^n$ such that the graph of F, i.e. the set $\{(x, F(x)) \mid x \in \mathbb{F}_2^n\}$ is mapped to the graph of G. Note that a permutation is always CCZ equivalent to its inverse (cf. Section 5). CCZ equivalence is in particular interesting, as many cryptographic properties of S-boxes are CCZ invariant. This includes the Walsh spectra, the uniformity and the number of algebraic relations of any degree. Using the algorithms presented in [3] we classified the optimal S-boxes up to CCZ equivalence. There are 6 non CCZ equivalent classes and we list all CCZ equivalence relations between the optimal S-boxes G_i below.

$$G_0 \sim_{ccz} G_1 \sim_{ccz} G_2 \sim_{ccz} G_8 \qquad G_3 \sim_{ccz} G_5$$
$$G_4 \sim_{ccz} G_6 \qquad G_7 \sim_{ccz} G_{11} \sim_{ccz} G_{12}$$
$$G_9 \sim_{ccz} G_{10} \qquad G_{14} \sim_{ccz} G_{15}$$

5.2 The Block Cipher Serpent

As an application of our observations we study the S-boxes in the Serpent cipher, as specified in the AES submission. The 8 different S-boxes in Serpent have been generated in a pseudo random manner from the set of all Serpent-type S-boxes with the additional criterion that $\mathrm{Lin}_1(S) = 4$. We list the S-boxes used in Serpent in Table 4. Up to $\sim_S$-equivalence these S-boxes have been actually chosen from a set of only 14 S-boxes, as 6 of the 20 representatives do not fulfill $\mathrm{Lin}_1(R_i) = 4$. It is therefore no surprise that two of the S-boxes in Serpent are linear equivalent, namely we have

$$S_4 \sim_S S_5$$

Furthermore, if also the more general equivalence and inverses are considered it turns out that the following relations hold

$$S_0 \sim S_1^{-1} \sim G_2 \qquad S_2 \sim S_6 \sim G_1$$
$$S_3 \sim S_7 \sim G_9 \qquad S_4 \sim S_5 \sim G_{14}$$

Table 4. The S-boxes used in the cipher Serpent

S_0	$3, 8, 15, 1, 10, 6, 5, 11, 14, 13, 4, 2, 7, 0, 9, 12$
S_1	$15, 12, 2, 7, 9, 0, 5, 10, 1, 11, 14, 8, 6, 13, 3, 4$
S_2	$8, 6, 7, 9, 3, 12, 10, 15, 13, 1, 14, 4, 0, 11, 5, 2$
S_3	$0, 15, 11, 8, 12, 9, 6, 3, 13, 1, 2, 4, 10, 7, 5, 14$
S_4	$1, 15, 8, 3, 12, 0, 11, 6, 2, 5, 4, 10, 9, 14, 7, 13$
S_5	$15, 5, 2, 11, 4, 10, 9, 12, 0, 3, 14, 8, 13, 6, 7, 1$
S_6	$7, 2, 12, 5, 8, 4, 6, 11, 14, 9, 1, 15, 13, 3, 10, 0$
S_7	$1, 13, 15, 0, 14, 8, 2, 11, 7, 4, 12, 10, 9, 3, 5, 6$

Hence, even though all Serpent S-boxes have been randomly generated, Serpent uses only 4 different S-boxes with respect to linear equivalence and inverses. This can also be viewed as follows: There exists a different specification of the Serpent cipher, which uses a different linear layer, that uses only 4 S-boxes.

6   Implementation Details

In this section we explain some of the shortcuts that have been used to generate the set of representatives G_j for optimal S-boxes. Using these ideas all our computations could be done within a few minutes on a regular PC. The most important speedup was due to the following Lemma.

Lemma 1. *Let $S : \mathbb{F}_2^n \to \mathbb{F}_2^n$ be a bijection. Then there exist bases B, B' of $\mathbb{F}_2^n$ such that*

$$S(B) = B'$$

Proof. We give a proof by induction. For any subset $B \subset \mathbb{F}_2^n$ we denote by $\langle B \rangle$ the linear span of B. Assume that we already constructed two sets $B_i, B_i' \subset \mathbb{F}_2^n$ each consisting of $i < n$ linear independent elements such that $S(B_i) = B_i'$. We have to find an element $x \in \mathbb{F}_2^n \setminus \langle B_i \rangle$ that is mapped into the set $\mathbb{F}_2^n \setminus \langle B_i' \rangle$. There are $2^n - 2^i$ elements in $\mathbb{F}_2^n \setminus \langle B_i \rangle$ but, as S is a permutation, only $2^i - i$ possible images in $\langle B_i' \rangle$. Furthermore, for $i < n$ it holds that $2^n - 2^i > 2^i - i$. Thus, as S is a bijection, at least one element in $\mathbb{F}_2^n \setminus \langle B_i \rangle$ gets mapped to $\mathbb{F}_2^n \setminus \langle B_i' \rangle$.   □

Using this lemma, we can speedup the search for optimal S-boxes. We can restrict to optimal S-boxes fulfilling

$$S(i) = i \quad \text{for } i \in \{0, 1, 2, 4, 8\}$$

as, due to the above lemma, any optimal S-box is equivalent to such an S-box. This observation reduced the search space from $16! \approx 2^{44}$ to only $11! \approx 2^{25}$ permutations that have to be created. We generated all those permutations and tested if they fulfill $\text{Diff}(S) = 4$ and $\text{Lin}(S) = 8$. This resulted in 1396032 optimal S-boxes.

Given this set of optimal S-boxes we created a set of representatives as follows. We started by the first S-box in the set and generated all equivalent S-boxes. Whenever one of these equivalent S-boxes was present in the set, we removed this S-box from the set. Note that there are approximately 2^{15} invertible 4×4 matrices, thus running naively through all invertible matrices A, B and all constants $c, d \in \mathbb{F}_2^4$ to generate all equivalent rows results in generating approximately 2^{38} S-boxes. However, as all S-boxes in the set are chosen such that

$$S(i) = i \quad \text{for } i \in \{0, 1, 2, 4, 8\} \tag{3}$$

then, if we fix A and d, the values for B and c are completely determined. Namely if we have

$$A(S(B(x) + c)) + d = S'(x)$$

for two S-boxes fulfilling (3) it must hold that

$$c = S^{-1}\left(A^{-1}(d)\right)$$

and

$$B(i) = S^{-1}\left(A^{-1}(i+d)\right) + c \quad \text{for } i \in 0,1,2,4,8$$

Using this observation we only had to generate approximately 2^{19} S-boxes.

7   Hardware Implementation

During our investigations we automatically synthesized thousands of S-boxes. Since hardware designers try to avoid using look-up tables, S-boxes are usually realized in combinatorial logic. Therefore, we fed the S-boxes in a combinatorial description into the synthesis tool. We compiled them in an area-efficient way, i.e. we instructed the synthesis tool to minimize the area requirements. Unfortunately logic synthesis tools like Synopsys *Design Compiler* use heuristic algorithms to map VHDL-code to standard-cells. Hence, it is never guaranteed that the resulting gate-level netlist is the smallest possible for a given VHDL-code.

To illustrate these suboptimal results we generated for all representatives R_i of Serpent-type S-boxes all equivalent S-boxes and synthesized them for the AMIS MTC45000 CMOS $0.35\mu m$ standard-cell library. The results are given in Table 5. Table 5 lists for all representatives of Serpent-type S-boxes the minimal and maximal area requirements of equivalent S-boxes.

Table 5. Area Requirement for all 20 Classes of Serpent-type S-boxes

Repr.	R_0	R_1	R_2	R_3	R_4	R_5	R_6	R_7	R_8	R_9
min GE	27.4	25.0	28.0	25.3	27.3	28.0	26.4	25.3	23.7	24.0
max GE	38.7	35.3	37.3	33.3	35.3	35.3	37.3	37	33.3	33.3

Repr.	R_{10}	R_{11}	R_{12}	R_{13}	R_{14}	R_{15}	R_{16}	R_{17}	R_{18}	R_{19}
min GE	26.3	24.4	24.3	28.7	21.3	25.3	24.7	27.0	23.4	25.7
max GE	37	38.7	29.7	36.3	31	34.7	34	39	30.3	39.3

A two bit input XOR-gate typically needs 10 transistors or 2.5 gate equivalences (GE), respectively. If one input bit is fixed, the XOR-gate is either superfluent ($inputbit_a \oplus 0 = inputbit_a$) or can be replaced by an inverter ($inputbit_a \oplus 1 = \neg inputbit_a$), which costs two transistors. Bit-permutations do not need any transistors in hardware. They are realized by wiring, which means that they come for a negligible amount of additional area or even for free. Hence, any two S-boxes which differ only by a permutation of the input bits and an permutation of the output bits should be compiled to the same combinatorial "*core*". Adding a constant before and after the core can be realized with not more than 16 transistors (= 4 GE). Therefore two equivalent Serpent-type S-boxes can be implemented with a difference of at most 4 GE. Hence, S-boxes in the same class should have been compiled to the same core and their size should not differ by more than 4 GEs. However, our figures clearly show that this is unfortunately not the case. We believe that this discrepancy is caused by a suboptimal synthesis due to the heuristic mapping algorithms.

As one can see, the biggest minimal representatives of an S-box class require 28.7 GE, whereas the smallest representatives only require 21.3 GE. Furthermore, the overall biggest representatives require 39.3 GE, which is 84 % more than the smallest we found. This implies an area saving of 46 % when optimal S-boxes are carefully chosen compared to a (worst-case) random selection approach.

One possible next step is to manually synthesize all good candidate S-box with the aim to gain the minimal result. Since this is a cumbersome work, it is impossible for a lot of S-boxes. Our classification of S-boxes into 20 classes greatly reduces the work and helps to find the most area-efficient S-box.

Acknowledgement

The work presented in this paper was supported in part by the European Commission within the STREP UbiSec&Sens of the EU Framework Programme 6 for Research and Development (www.ist-ubisecsens.org). The views and conclusions contained herein are those of the authors and should not be interpreted as necessarily representing the official policies or endorsements, either expressed or implied, of the UbiSecSens project or the European Commission.

References

1. Biham, E., Shamir, A.: Differential cryptanalysis of des-like cryptosystems. In: Menezes, A., Vanstone, S.A. (eds.) CRYPTO 1990. LNCS, vol. 537, pp. 2–21. Springer, Heidelberg (1990)
2. Biryukov, A., De Cannière, C., Braeken, A., Preneel, B.: A toolbox for cryptanalysis: Linear and affine equivalence algorithms. In: Biham, E. (ed.) Advances in Cryptology – EUROCRPYT 2003. LNCS, vol. 2656, pp. 33–50. Springer, Heidelberg (2003)
3. Brinkman, M., Leander, G.: On the classification of apn functions up to dimension five. International Workshop on Coding and Cryptography (2007)
4. Carlet, C., Charpin, P., Zinoviev, V.: Codes, bent functions and permutations suitable for des-like cryptosystems. Des. Codes Cryptography 15(2), 125–156 (1998)
5. Courtois, N., Pieprzyk, J.: Cryptanalysis of block ciphers with overdefined systems of equations. In: Zheng, Y. (ed.) ASIACRYPT 2002. LNCS, vol. 2501, pp. 267–287. Springer, Heidelberg (2002)
6. Courtois, N., Pieprzyk, J.: Cryptanalysis of block ciphers with overdefined systems of equations. Cryptology ePrint Archive, Report 2002/044 (2002), http://eprint.iacr.org/

7. Knudsen, L.: private communication
8. Lachaud, G., Wolfmann, J.: The weights of the orthogonals of the extended quadratic binary goppa codes. IEEE Transactions on Information Theory 36(3), 686 (1990)
9. Lorens, C.S.: Invertible boolean functions. IEEE Trans. Electronic Computers 13(5), 529–541 (1964)
10. Matsui, M.: Linear cryptoanalysis method for des cipher. In: Helleseth, T. (ed.) EUROCRYPT 1993. LNCS, vol. 765, pp. 386–397. Springer, Heidelberg (1993)
11. Nyberg, K.: Perfect nonlinear s-boxes. In: EUROCRYPT 1991. LNCS, vol. 547, pp. 378–386. Springer, Heidelberg (1991)
12. Nyberg, K.: Differentially uniform mappings for cryptography. In: Helleseth, T. (ed.) EUROCRYPT 1993. LNCS, vol. 765, pp. 55–64. Springer, Heidelberg (1994)
13. Rothaus, O.S.: On "bent" functions. J. Comb. Theory, Ser. A 20(3), 300–305 (1976)

A   List of Representatives

Table 6. Representatives for all 16 classes of optimal 4 bit S-boxes (G_3 is equivalent to the invers mapping.)

G_0	0, 1, 2, 13, 4, 7, 15, 6, 8, 11, 12, 9, 3, 14, 10, 5
G_1	0, 1, 2, 13, 4, 7, 15, 6, 8, 11, 14, 3, 5, 9, 10, 12
G_2	0, 1, 2, 13, 4, 7, 15, 6, 8, 11, 14, 3, 10, 12, 5, 9
G_3	0, 1, 2, 13, 4, 7, 15, 6, 8, 12, 5, 3, 10, 14, 11, 9
G_4	0, 1, 2, 13, 4, 7, 15, 6, 8, 12, 9, 11, 10, 14, 5, 3
G_5	0, 1, 2, 13, 4, 7, 15, 6, 8, 12, 11, 9, 10, 14, 3, 5
G_6	0, 1, 2, 13, 4, 7, 15, 6, 8, 12, 11, 9, 10, 14, 5, 3
G_7	0, 1, 2, 13, 4, 7, 15, 6, 8, 12, 14, 11, 10, 9, 3, 5
G_8	0, 1, 2, 13, 4, 7, 15, 6, 8, 14, 9, 5, 10, 11, 3, 12
G_9	0, 1, 2, 13, 4, 7, 15, 6, 8, 14, 11, 3, 5, 9, 10, 12
G_{10}	0, 1, 2, 13, 4, 7, 15, 6, 8, 14, 11, 5, 10, 9, 3, 12
G_{11}	0, 1, 2, 13, 4, 7, 15, 6, 8, 14, 11, 10, 5, 9, 12, 3
G_{12}	0, 1, 2, 13, 4, 7, 15, 6, 8, 14, 11, 10, 9, 3, 12, 5
G_{13}	0, 1, 2, 13, 4, 7, 15, 6, 8, 14, 12, 9, 5, 11, 10, 3
G_{14}	0, 1, 2, 13, 4, 7, 15, 6, 8, 14, 12, 11, 3, 9, 5, 10
G_{15}	0, 1, 2, 13, 4, 7, 15, 6, 8, 14, 12, 11, 9, 3, 10, 5

Table 7. Representatives for all 16 classes of optimal 4 bit S-boxes as Polynomials. g denotes a primitive element in F_{16}^*

G_0	$x^{14} + x^{13} + g^7 x^{12} + g^5 x^{11} + g^5 x^{10}$ $+ g^5 x^9 + x^8 + g^6 x^6 + g^{13} x^5 + g^8 x^4 + g^{13} x^3 + g^{13} x^2$
G_1	$x^{14} + g^5 x^{13} + x^{12} + g x^{11} + g^3 x^{10}$ $+ g^4 x^9 + g^{14} x^8 + g^4 x^7 + g^{12} x^6 + x^5 + g^{13} x^4 + x^3 + g^{13} x^2 + gx$
G_2	$x^{14} + g^6 x^{13} + g^{13} x^{12} + g^{10} x^{10}$ $+ g^6 x^9 + g x^8 + g^2 x^7 + g^{13} x^6 + g^{11} x^5 + g^2 x^4 + g^{13} x^3 + g^2 x^2 + g^6 x$
G_3	$x^{14} + g^{11} x^{13} + g x^{12} + g^3 x^{11}$ $+ g^5 x^9 + g^7 x^8 + g^8 x^7 + g^4 x^6 + g^{11} x^5 + g^2 x^4 + g^4 x^3 + g^{11} x^2$
G_4	$x^{14} + g^{11} x^{13} + g^7 x^{12} + g x^{11}$ $+ g^8 x^{10} + g^{13} x^9 + g^{11} x^8 + g^2 x^6 + g x^5 + g^2 x^4 + g^7 x^3 + g x^2 + g^8 x$
G_5	$x^{14} + g^{13} x^{13} + g^9 x^{12} + g^6 x^{11} + g^{10} x^{10}$ $+ g^7 x^9 + g^{10} x^8 + g^7 x^7 + g^8 x^6 + g^{12} x^5 + g^{12} x^4 + x^3 + g^{11} x^2 + g^5 x$
G_6	$x^{14} + g^4 x^{13} + g^3 x^{12} + g^2 x^{11} + x^{10}$ $+ g^{11} x^9 + g^2 x^8 + g x^7 + g^2 x^6 + g^9 x^5 + g^4 x^4 + g^9 x^3 + g^{12} x^2 + g^{11} x$
G_7	$x^{14} + g x^{13} + g^9 x^{12} + g x^{11} + g^7 x^{10}$ $+ g^6 x^7 + g^{10} x^6 + g x^5 + g^8 x^4 + g^2 x^3 + g^6 x^2 + g^9 x$
G_8	$x^{14} + g x^{13} + x^{12} + g^{10} x^9 + g^{14} x^8$ $+ g^{12} x^7 + g^9 x^5 + g^8 x^4 + g^{13} x^3 + g^{11} x^2 + g^6 x$
G_9	$x^{13} + g^7 x^{12} + g^5 x^{11} + g x^{10}$ $+ g^{11} x^9 + g^{11} x^8 + g^3 x^7 + g^4 x^6 + g^5 x^5 + g x^4 + g^7 x^3 + x^2 + g^6 x$
G_{10}	$x^{13} + g^{13} x^{12} + g^7 x^{11} + g^7 x^{10}$ $+ g^{14} x^9 + g^{10} x^7 + g x^6 + g^5 x^5 + g^7 x^4 + g^{12} x^3 + g^6 x$
G_{11}	$x^{14} + g x^{13} + x^{12} + g^7 x^{11} + g^{13} x^{10}$ $+ g x^9 + g^{11} x^8 + g^{14} x^7 + g^3 x^6 + g^6 x^5 + g x^4 + g^{14} x^3 + g^{14} x^2 + g^9 x$
G_{12}	$x^{14} + g^{10} x^{13} + g x^{12} + g^4 x^{11} + g^{14} x^{10}$ $+ g^4 x^9 + g^5 x^8 + g^2 x^7 + g^9 x^6 + g^4 x^5 + g^8 x^4 + g^{14} x^3 + g^5 x^2 + x$
G_{13}	$x^{14} + g^{12} x^{13} + g^8 x^{12} + g^8 x^{11} + g^{14} x^{10}$ $+ g x^9 + g^8 x^8 + g^{14} x^7 + g^6 x^6 + x^5 + g^{14} x^4 + g^{12} x^3 + g x^2 + g^{14} x$
G_{14}	$x^{14} + g^8 x^{13} + g^{10} x^{12} + g x^{11} + g x^{10}$ $+ g^9 x^9 + x^7 + g^{10} x^6 + g^7 x^5 + g^4 x^4 + g^2 x^3 + g^{12} x^2 + g^{14} x$
G_{15}	$x^{14} + g^6 x^{13} + g^{13} x^{12} + g^5 x^{10} + x^9$ $+ x^8 + x^7 + g^2 x^6 + g^{11} x^5 + g^{10} x^4 + g^4 x^3 + g x^2 + g^3 x$

Table 8. Representatives for all 20 Classes of Serpent-type S-boxes

R_0	0, 3, 5, 6, 7, 10, 11, 12, 13, 4, 14, 9, 8, 1, 2, 15
R_1	0, 3, 5, 8, 6, 9, 10, 7, 11, 12, 14, 2, 1, 15, 13, 4
R_2	0, 3, 5, 8, 6, 9, 11, 2, 13, 4, 14, 1, 10, 15, 7, 12
R_3	0, 3, 5, 8, 6, 10, 15, 4, 14, 13, 9, 2, 1, 7, 12, 11
R_4	0, 3, 5, 8, 6, 12, 11, 7, 9, 14, 10, 13, 15, 2, 1, 4
R_5	0, 3, 5, 8, 6, 12, 11, 7, 10, 4, 9, 14, 15, 1, 2, 13
R_6	0, 3, 5, 8, 6, 12, 11, 7, 10, 13, 9, 14, 15, 1, 2, 4
R_7	0, 3, 5, 8, 6, 12, 11, 7, 13, 10, 14, 4, 1, 15, 2, 9
R_8	0, 3, 5, 8, 6, 12, 15, 1, 10, 4, 9, 14, 13, 11, 2, 7
R_9	0, 3, 5, 8, 6, 12, 15, 2, 14, 9, 11, 7, 13, 10, 4, 1
R_{10}	0, 3, 5, 8, 6, 13, 15, 1, 9, 12, 2, 11, 10, 7, 4, 14
R_{11}	0, 3, 5, 8, 6, 13, 15, 2, 7, 4, 14, 11, 10, 1, 9, 12
R_{12}	0, 3, 5, 8, 6, 13, 15, 2, 12, 9, 10, 4, 11, 14, 1, 7
R_{13}	0, 3, 5, 8, 6, 15, 10, 1, 7, 9, 14, 4, 11, 12, 13, 2
R_{14}	0, 3, 5, 8, 7, 4, 9, 14, 15, 6, 2, 11, 10, 13, 12, 1
R_{15}	0, 3, 5, 8, 7, 9, 11, 14, 10, 13, 15, 4, 12, 2, 6, 1
R_{16}	0, 3, 5, 8, 9, 12, 14, 7, 10, 13, 15, 4, 6, 11, 1, 2
R_{17}	0, 3, 5, 8, 10, 13, 9, 4, 15, 6, 2, 1, 12, 11, 7, 14
R_{18}	0, 3, 5, 8, 11, 12, 6, 15, 14, 9, 2, 7, 4, 10, 13, 1
R_{19}	0, 3, 5, 10, 7, 12, 11, 6, 13, 4, 2, 9, 14, 1, 8, 15

Table 9. Walsh spectra for all 16 Classes of 4 bit S-boxes ($W(a)$ is the number of Walsh coefficients of value a)

S − box	G_0	G_1	G_2	G_3	G_4	G_5	G_6	G_7	G_8	G_9	G_{10}	G_{11}	G_{12}	G_{13}	G_{14}	G_{15}
$W(0)$	108	108	108	90	90	90	90	90	108	96	96	90	90	90	96	96
$W(4)$	60	60	60	76	76	76	76	80	60	68	68	76	72	80	72	72
$W(-4)$	36	36	36	44	44	44	44	40	36	44	44	44	48	40	40	40
$W(8)$	27	27	27	22	22	22	22	20	27	25	25	22	24	20	23	23
$W(-8)$	9	9	9	8	8	8	8	10	9	7	7	8	6	10	9	9

Table 10. Walsh spectra for all 20 Classes of Serpent-type S-boxes ($W(a)$ is the number of Walsh coefficients of value a)

S − box	R_0	R_1	R_2	R_3	R_4	R_5	R_6	R_7	R_8	R_9	R_{10}	R_{11}	R_{12}	R_{13}	R_{14}	R_{15}	R_{16}	R_{17}	R_{18}	R_{19}
$W(0)$	108	108	108	96	96	96	108	108	96	108	108	108	108	96	108	96	108	108	108	108
$W(4)$	60	60	60	72	68	68	60	60	72	60	60	60	60	68	60	72	60	60	60	60
$W(-4)$	36	36	36	40	44	44	36	36	40	36	36	36	36	44	36	40	36	36	36	36
$W(8)$	27	27	27	23	25	25	27	27	23	27	27	27	27	25	27	23	27	27	27	27
$W(-8)$	9	9	9	9	7	7	9	9	9	9	9	9	9	9	7	9	9	9	9	9

The Simplest Method for Constructing APN Polynomials EA-Inequivalent to Power Functions

Lilya Budaghyan

Department of Mathematics, University of Trento, I-38050 Povo (Trento), Italy
`lilia.b@mail.ru`

Abstract. In 2005 Budaghyan, Carlet and Pott constructed the first APN polynomials EA-inequivalent to power functions by applying CCZ-equivalence to the Gold APN functions. It is a natural question whether it is possible to construct APN polynomials EA-inequivalent to power functions by using only EA-equivalence and inverse transformation on a power APN mapping: this would be the simplest method to construct APN polynomials EA-inequivalent to power functions. In the present paper we prove that the answer to this question is positive. By this method we construct a class of APN polynomials EA-inequivalent to power functions. On the other hand it is shown that the APN polynomials constructed by Budaghyan, Carlet and Pott cannot be obtained by the introduced method.

Keywords: Affine equivalence, Almost bent, Almost perfect nonlinear, CCZ-equivalence, Differential uniformity, Nonlinearity, S-box, Vectorial Boolean function.

1 Introduction

A function $F : \mathbf{F}_2^m \to \mathbf{F}_2^m$ is called *almost perfect nonlinear* (APN) if, for every $a \neq 0$ and every b in $\mathbf{F}_2^m$, the equation $F(x) + F(x+a) = b$ admits at most two solutions (it is also called *differentially 2-uniform*). Vectorial Boolean functions used as S-boxes in block ciphers must have low differential uniformity to allow high resistance to the differential cryptanalysis (see [2,30]). In this sense APN functions are optimal. The notion of APN function is closely connected to the notion of almost bent (AB) function. A function $F : \mathbf{F}_2^m \to \mathbf{F}_2^m$ is called AB if the minimum Hamming distance between all the Boolean functions $v \cdot F$, $v \in \mathbf{F}_2^m \setminus \{0\}$, and all affine Boolean functions on $\mathbf{F}_2^m$ is maximal. AB functions exist for m odd only and oppose an optimum resistance to the linear cryptanalysis (see [28,15]). Besides, every AB function is APN [15], and in the m odd case, any quadratic function is APN if and only if it is AB [14].

The APN and AB properties are preserved by some transformations of functions [14,30]. If F is an APN (resp. AB) function, A_1, A_2 are affine permutations and A is affine then the function $F' = A_1 \circ F \circ A_2 + A$ is also APN (resp. AB); the functions F and F' are called extended affine equivalent (*EA-equivalent*). Another case is the *inverse transformation*, that is, the inverse of any APN

C. Carlet and B. Sunar (Eds.): WAIFI 2007, LNCS 4547, pp. 177–188, 2007.
© Springer-Verlag Berlin Heidelberg 2007

(resp. AB) permutation is APN (resp. AB). Until recently, the only known constructions of APN and AB functions were EA-equivalent to power functions $F(x) = x^d$ over finite fields ($\mathbf{F}_{2^m}$ being identified with $\mathbf{F}_2^m$). Table 1 gives all known values of exponents d (up to multiplication by a power of 2 modulo $2^m - 1$, and up to taking the inverse when a function is a permutation) such that the power function x^d over $\mathbf{F}_{2^m}$ is APN. For m odd the Gold, Kasami, Welch and Niho APN functions from Table 1 are also AB (for the proofs of AB property see [11,12,23,24,26,30]).

Table 1. Known APN power functions x^d on $\mathbf{F}_{2^m}$

Functions	Exponents d	Conditions	Proven in
Gold	$2^i + 1$	$\gcd(i, m) = 1$	[23,30]
Kasami	$2^{2i} - 2^i + 1$	$\gcd(i, m) = 1$	[25,26]
Welch	$2^t + 3$	$m = 2t + 1$	[20]
Niho	$2^t + 2^{\frac{t}{2}} - 1$, t even	$m = 2t + 1$	[19]
	$2^t + 2^{\frac{3t+1}{2}} - 1$, t odd		
Inverse	$2^{2t} - 1$	$m = 2t + 1$	[1,30]
Dobbertin	$2^{4t} + 2^{3t} + 2^{2t} + 2^t - 1$	$m = 5t$	[21]

In [14], Carlet, Charpin and Zinoviev introduced an equivalence relation of functions, more recently called CCZ-equivalence, which corresponds to the affine equivalence of the graphs of functions and preserves APN and AB properties. EA-equivalence is a particular case of CCZ-equivalence and any permutation is CCZ-equivalent to its inverse [14]. In [8,9], it is proven that CCZ-equivalence is more general, and applying CCZ-equivalence to the Gold mappings classes of APN functions EA-inequivalent to power functions are constructed. These classes are presented in Table 2. When m is odd, these functions are also AB.

Table 2. Known APN functions EA-inequivalent to power functions on $\mathbf{F}_{2^m}$

Functions	Conditions	Alg. degree
$x^{2^i+1} + (x^{2^i} + x + tr(1) + 1)tr(x^{2^i+1} + x\, tr(1))$	$m \geq 4$ $\gcd(i, m) = 1$	3
$[x + tr_{(m,3)}(x^{2(2^i+1)} + x^{4(2^i+1)})$ $+ tr(x)tr_{(m,3)}(x^{2^i+1} + x^{2^{2i}(2^i+1)})]^{2^i+1}$	m divisible by 6 $\gcd(i, m) = 1$	4
$x^{2^i+1} + tr_{(m,n)}(x^{2^i+1}) + x^{2^i} tr_{(m,n)}(x) + x\, tr_{(m,n)}(x)^{2^i}$ $+ [tr_{(m,n)}(x)^{2^i+1} + tr_{(m,n)}(x^{2^i+1}) + tr_{(m,n)}(x)]^{\frac{1}{2^i+1}}$ $\times (x^{2^i} + tr_{(m,n)}(x)^{2^i} + 1) + [tr_{(m,n)}(x)^{2^i+1}$ $+ tr_{(m,n)}(x^{2^i+1}) + tr_{(m,n)}(x)]^{\frac{2^i}{2^i+1}}(x + tr_{(m,n)}(x))$	$m \neq n$ m divisible by n $\gcd(2i, m) = 1$	$n + 2$

These new results on CCZ-equivalence have solved several problems (see [8,9]) and have also raised some interesting questions. One of these questions is whether the known classes of APN power functions are CCZ-inequivalent. Partly the answer is given in [6]: it is proven that in general the Gold functions are CCZ-inequivalent to the Kasami and Welch functions, and that for different parameters $1 \leq i, j \leq \frac{m-1}{2}$ the Gold functions x^{2^i+1} and x^{2^j+1} are CCZ-inequivalent. Another interesting question is the existence of APN polynomials CCZ-inequivalent to power functions. Different methods for constructing quadratic APN polynomials CCZ-inequivalent to power functions have been proposed in [3,4,17,22,29], and infinite classes of such functions are constructed in [3,4,5,6,7]. In the present paper we consider the natural question whether it is possible to construct APN polynomials EA-inequivalent to power functions by applying only EA-equivalence and the inverse transformation on a power APN function. We prove that the answer is positive and construct a class of AB functions EA-inequivalent to power mappings by applying this method to the Gold AB functions. It should be mentioned that the functions from Table 2 cannot be obtained by this method. It can be illustrated, for instance, by the fact that for $m = 5$ the functions from Table 2 and for m even the Gold functions are EA-inequivalent to permutations [8,9,31], therefore, the inverse transformation cannot be applied in these cases and the method fails.

2   Preliminaries

Let $\mathbf{F}_2^m$ be the m-dimensional vector space over the field $\mathbf{F}_2$. Any function F from $\mathbf{F}_2^m$ to itself can be uniquely represented as a polynomial on m variables with coefficients in $\mathbf{F}_2^m$, whose degree with respect to each coordinate is at most 1:

$$F(x_1, \ldots, x_m) = \sum_{u \in \mathbf{F}_2^m} c(u) \left(\prod_{i=1}^m x_i^{u_i} \right), \qquad c(u) \in \mathbf{F}_2^m.$$

This representation is called the *algebraic normal form* of F and its degree $d^\circ(F)$ the *algebraic degree* of the function F.

Besides, the field $\mathbf{F}_{2^m}$ can be identified with $\mathbf{F}_2^m$ as a vector space. Then, viewed as a function from this field to itself, F has a unique representation as a univariate polynomial over $\mathbf{F}_{2^m}$ of degree smaller than 2^m:

$$F(x) = \sum_{i=0}^{2^m-1} c_i x^i, \qquad c_i \in \mathbf{F}_{2^m}.$$

For any k, $0 \leq k \leq 2^m - 1$, the number $w_2(k)$ of the nonzero coefficients $k_s \in \{0, 1\}$ in the binary expansion $\sum_{s=0}^{m-1} 2^s k_s$ of k is called the *2-weight* of k. The algebraic degree of F is equal to the maximum 2-weight of the exponents i of the polynomial $F(x)$ such that $c_i \neq 0$, that is, $d^\circ(F) = \max_{0 \leq i \leq m-1, c_i \neq 0} w_2(i)$ (see [14]).

A function $F : \mathbf{F}_2^m \to \mathbf{F}_2^m$ is *linear* if and only if $F(x)$ is a linearized polynomial over $\mathbf{F}_{2^m}$, that is,

$$\sum_{i=0}^{m-1} c_i x^{2^i}, \quad c_i \in \mathbf{F}_{2^m}.$$

The sum of a linear function and a constant is called an *affine function*.

Let F be a function from $\mathbf{F}_{2^m}$ to itself and A_1, $A_2 : \mathbf{F}_{2^m} \to \mathbf{F}_{2^m}$ be affine permutations. The functions F and $A_1 \circ F \circ A_2$ are then called *affine equivalent*. Affine equivalent functions have the same algebraic degree (i.e. the algebraic degree is *affine invariant*).

As recalled in the Introduction, we say that the functions F and F' are *extended affine equivalent* if $F' = A_1 \circ F \circ A_2 + A$ for some affine permutations A_1, A_2 and an affine function A. If F is not affine, then F and F' have again the same algebraic degree.

Two mappings F and F' from $\mathbf{F}_{2^m}$ to itself are called Carlet-Charpin-Zinoviev equivalent (*CCZ-equivalent*) if the graphs of F and F', that is, the subsets $G_F = \{(x, F(x)) \mid x \in \mathbf{F}_{2^m}\}$ and $G_{F'} = \{(x, F'(x)) \mid x \in \mathbf{F}_{2^m}\}$ of $\mathbf{F}_{2^m} \times \mathbf{F}_{2^m}$, are affine equivalent. Hence, F and F' are CCZ-equivalent if and only if there exists an affine automorphism $\mathcal{L} = (L_1, L_2)$ of $\mathbf{F}_{2^m} \times \mathbf{F}_{2^m}$ such that

$$y = F(x) \Leftrightarrow L_2(x, y) = F'(L_1(x, y)).$$

Note that since $\mathcal{L}$ is a permutation then the function $L_1(x, F(x))$ has to be a permutation too (see [6]). As shown in [14], EA-equivalence is a particular case of CCZ-equivalence and any permutation is CCZ-equivalent to its inverse.

For a function $F : \mathbf{F}_{2^m} \to \mathbf{F}_{2^m}$ and any elements $a, b \in \mathbf{F}_{2^m}$ we denote

$$\delta_F(a, b) = |\{x \in \mathbf{F}_2^m : F(x + a) + F(x) = b\}|.$$

F is called a *differentially δ-uniform* function if $\max_{a \in \mathbf{F}_{2^m}^*, b \in \mathbf{F}_{2^m}} \delta_F(a, b) \leq \delta$. Note that $\delta \geq 2$ for any function over $\mathbf{F}_{2^m}$. Differentially 2-uniform mappings are called *almost perfect nonlinear*.

For any function $F : \mathbf{F}_{2^m} \to \mathbf{F}_{2^m}$ we denote

$$\lambda_F(a, b) = \sum_{x \in \mathbf{F}_{2^m}} (-1)^{tr(bF(x) + ax)}, \quad a, b \in \mathbf{F}_{2^m},$$

where $tr(x) = x + x^2 + x^4 + \cdots + x^{2^{m-1}}$ is the trace function from $\mathbf{F}_{2^m}$ into $\mathbf{F}_2$. The set $\Lambda_F = \{\lambda_F(a, b) : a, b \in \mathbf{F}_{2^m}, b \neq 0\}$ is called the *Walsh spectrum* of the function F and the multiset $\{|\lambda_F(a, b)| : a, b \in \mathbf{F}_{2^n}, b \neq 0\}$ is called the *extended Walsh spectrum* of F. The value

$$\mathcal{NL}(F) = 2^{m-1} - \frac{1}{2} \max_{a \in \mathbf{F}_{2^m}, b \in \mathbf{F}_{2^m}^*} |\lambda_F(a, b)|$$

equals the *nonlinearity* of the function F. The nonlinearity of any function F satisfies the inequality

$$\mathcal{NL}(F) \leq 2^{m-1} - 2^{\frac{m-1}{2}}$$

([15,32]) and in case of equality F is called *almost bent* or *maximum nonlinear*.

Obviously, AB functions exist only for n odd. It is proven in [15] that every AB function is APN and its Walsh spectrum equals $\{0, \pm 2^{\frac{m+1}{2}}\}$. If m is odd, every APN mapping which is quadratic (that is, whose algebraic degree equals 2) is AB [14], but this is not true for nonquadratic cases: the Dobbertin and the inverse APN functions are not AB (see [12,14]). When m is even, the inverse function x^{2^m-2} is a differentially 4-uniform permutation [30] and has the best known nonlinearity [27], that is $2^{m-1} - 2^{\frac{m}{2}}$ (see [12,18]). This function has been chosen as the basic S-box, with $m = 8$, in the Advanced Encryption Standard (AES), see [16]. A comprehensive survey on APN and AB functions can be found in [13].

It is shown in [14] that, if F and G are CCZ-equivalent, then F is APN (resp. AB) if and only if G is APN (resp. AB). More generally, CCZ-equivalent functions have the same differential uniformity and the same extended Walsh spectrum (see [8]). Further invariants for CCZ-equivalence can be found in [22] (see also [17]) in terms of group algebras.

3   The New Construction

In this section we show that it is possible to construct APN polynomials EA-inequivalent to power functions by applying only EA-equivalence and the inverse transformation on a power APN function. The inverse transformation and EA-equivalence are simple transformations of functions which preserve APN and AB properties. However, applying each of them separately on power mappings it is obviously impossible to construct polynomials EA-inequivalent to power functions. Therefore, our approach for constructing APN polynomials EA-inequivalent to power mappings is the simplest. We shall illustrate this method on the Gold AB functions and in order to do it we need the following result from [8,9].

Proposition 1. ([8,9]) *Let $F : \mathbf{F}_{2^m} \to \mathbf{F}_{2^m}$, $F(x) = L(x^{2^i+1}) + L'(x)$, where $\gcd(i, m) = 1$ and L, L' are linear. Then F is a permutation if and only if, for every $u \neq 0$ in $\mathbf{F}_{2^m}$ and every v such that $tr(v) = tr(1)$, the condition $L(u^{2^i+1}v) \neq L'(u)$ holds.*

Further we use the following notations for any divisor n of m

$$tr_{(m,n)}(x) = x + x^{2^n} + x^{2^{2n}} \ldots + x^{2^{n(m/n-1)}},$$

$$tr_n(x) = x + x^2 + \cdots + x^{2^{n-1}}.$$

Theorem 1. *Let $m \geq 9$ be odd and divisible by 3. Then the function*

$$F'(x) = \left(x^{\frac{1}{2^i+1}} + tr_{(m,3)}(x + x^{2^{2i}}) \right)^{-1},$$

with $1 \leq i \leq m$, $\gcd(i, m) = 1$, is an AB permutation over $\mathbf{F}_{2^m}$. The function F' is EA-inequivalent to the Gold functions and to their inverses, that is, to x^{2^j+1} and $x^{\frac{1}{2^j+1}}$ for any $1 \leq j \leq m$.

Proof. To prove that the function F' is an AB permutation we only need to show that the function $F_1(x) = x^{\frac{1}{2^i+1}} + tr_{(m,3)}(x + x^{2^{2i}})$ is a permutation. Since the function x^{2^i+1} is a permutation when m is odd and $\gcd(i,m)=1$ then F_1 is a permutation if and only if the function $F(x)=F_1(x^{2^i+1})=x+tr_{(m,3)}(x^{2^i+1}+x^{2^{2s}(2^i+1)})$, with $s=i \bmod 3$, is a permutation.

By Proposition 1 the function F is a permutation if for every $v \in \mathbf{F}_{2^m}$ such that $tr(v)=1$ and every $u \in \mathbf{F}_{2^m}^*$ the condition $tr_{(m,3)}(u^{2^i+1}v+(u^{2^i+1}v)^{2^{2s}}) \neq u$ holds. Obviously, if $u \notin \mathbf{F}_{2^3}$ then $tr_{(m,3)}(u^{2^i+1}v+(u^{2^i+1}v)^{2^{2s}}) \neq u$. For any $u \in \mathbf{F}_{2^3}^*$ the condition $tr_{(m,3)}(u^{2^i+1}v+(u^{2^i+1}v)^{2^{2s}}) \neq u$ is equivalent to $u^{2^i+1}tr_{(m,3)}(v)+(u^{2^i+1}tr_{(m,3)}(v))^{2^{2s}} \neq u$. Therefore, F is a permutation if for every $u,w \in \mathbf{F}_{2^3}^*$, $tr_3(w)=1$ the condition $u^{2^i+1}w+(u^{2^i+1}w)^{2^{2s}} \neq u$ is satisfied. Then F is a permutation if $x+x^{2^i+1}+x^{2^{2s}(2^i+1)}$ is a permutation on $\mathbf{F}_{2^3}$ and that was easily checked by a computer.

We have $d^\circ(x^{2^i+1})=2$ and it is proven in [30] that $d^\circ(x^{\frac{1}{2^i+1}})=\frac{m+1}{2}$. We show below that $d^\circ(F')=4$ for $m \geq 9$. Since the function F' has algebraic degree different from 2 and $\frac{m+1}{2}$ then it is EA-inequivalent to the Gold functions and to their inverses.

Since $F'(x)=F_1^{-1}(x)=[F(x^{\frac{1}{2^i+1}})]^{-1}=[F^{-1}(x)]^{2^i+1}$ then to get the representation of the function F' we need the representation of the function F^{-1}. The following computations are helpful to show that $F^{-1}=F \circ F$.

$$tr_{(m,3)}[(x+tr_{(m,3)}(x^{2^i+1}+x^{2^{2s}(2^i+1)}))^{2^i+1}]$$
$$= tr_{(m,3)}(x^{2^i+1})+tr_{(m,3)}(x^{2^s})tr_{(m,3)}(x^{2^i+1}+x^{2^{2s}(2^i+1)})$$
$$+tr_{(m,3)}(x)tr_{(m,3)}(x^{2^i+1}+x^{2^s(2^i+1)})$$
$$+tr_{(m,3)}(x^{2^i+1}+x^{2^{2s}(2^i+1)})tr_{(m,3)}(x^{2^i+1}+x^{2^s(2^i+1)}),$$

since

$$tr_{(m,3)}((x^{2^i+1}+x^{2^{2s}(2^i+1)})^{2^i})=tr_{(m,3)}((x^{2^i+1}+x^{2^{2s}(2^i+1)})^{2^s})$$
$$= tr_{(m,3)}(x^{2^s(2^i+1)}+x^{2^{3s}(2^i+1)})=tr_{(m,3)}(x^{2^s(2^i+1)}+x^{2^i+1}).$$

Then

$$tr_{(m,3)}[(x+tr_{(m,3)}(x^{2^i+1}+x^{2^{2s}(2^i+1)}))^{2^i+1}+(x+tr_{(m,3)}(x^{2^i+1}+x^{2^{2s}(2^i+1)}))^{2^{2s}(2^i+1)}]$$
$$= tr_{(m,3)}(x^{2^i+1}+x^{2^{2s}(2^i+1)})+tr_{(m,3)}(x^{2^s})tr_{(m,3)}(x^{2^i+1}+x^{2^{2s}(2^i+1)})$$
$$+tr_{(m,3)}(x)tr_{(m,3)}(x^{2^{2s}(2^i+1)}+x^{2^s(2^i+1)})+tr_{(m,3)}(x)tr_{(m,3)}(x^{2^i+1}+x^{2^s(2^i+1)})$$
$$+tr_{(m,3)}(x^{2^{2s}})tr_{(m,3)}(x^{2^{2s}(2^i+1)}+x^{(2^i+1)})$$
$$+tr_{(m,3)}(x^{2^i+1}+x^{2^{2s}(2^i+1)})tr_{(m,3)}(x^{2^i+1}+x^{2^s(2^i+1)})$$
$$+tr_{(m,3)}(x^{2^{2s}(2^i+1)}+x^{2^s(2^i+1)})tr_{(m,3)}(x^{2^{2s}(2^i+1)}+x^{(2^i+1)})$$
$$= tr_{(m,3)}(x^{2^i+1}+x^{2^{2s}(2^i+1)})+tr_{(m,3)}(x+x^{2^s}+x^{2^{2s}})tr_{(m,3)}(x^{2^i+1}+x^{2^{2s}(2^i+1)})$$
$$+(tr_{(m,3)}(x^{2^i+1}+x^{2^{2s}(2^i+1)}))^2=tr_{(m,3)}(x^{2^i+1}+x^{2^{2s}(2^i+1)})$$
$$+tr_m(x)tr_{(m,3)}(x^{2^i+1}+x^{2^{2s}(2^i+1)})+(tr_{(m,3)}(x^{2^i+1}+x^{2^{2s}(2^i+1)}))^2$$

and

$$F \circ F(x) = x + tr_m(x)tr_{(m,3)}(x^{2^i+1} + x^{2^{2s}(2^i+1)}) + (tr_{(m,3)}(x^{2^i+1} + x^{2^{2s}(2^i+1)}))^2$$

and, since $tr_m(tr_{(m,3)}(x^{2^i+1} + x^{2^{2s}(2^i+1)})) = 0$,

$$
\begin{aligned}
(F \circ F) \circ F(x) &= x + tr_{(m,3)}(x^{2^i+1} + x^{2^{2s}(2^i+1)}) + tr_m(x)[tr_{(m,3)}(x^{2^i+1} + x^{2^{2s}(2^i+1)}) \\
&\quad + tr_m(x)tr_{(m,3)}(x^{2^i+1} + x^{2^{2s}(2^i+1)}) + (tr_{(m,3)}(x^{2^i+1} + x^{2^{2s}(2^i+1)}))^2] \\
&\quad + [tr_{(m,3)}(x^{2^i+1} + x^{2^{2s}(2^i+1)}) + tr_m(x)tr_{(m,3)}(x^{2^i+1} + x^{2^{2s}(2^i+1)}) \\
&\quad + (tr_{(m,3)}(x^{2^i+1} + x^{2^{2s}(2^i+1)}))^2]^2 = x + tr_{(m,3)}(x^{2^i+1} + x^{2^{2s}(2^i+1)}) \\
&\quad + (tr_{(m,3)}(x^{2^i+1} + x^{2^{2s}(2^i+1)}))^2 + (tr_{(m,3)}(x^{2^i+1} + x^{2^{2s}(2^i+1)}))^4 \\
&= x + tr_3(tr_{(m,3)}(x^{2^i+1} + x^{2^{2s}(2^i+1)})) = x + tr_m(x^{2^i+1} + x^{2^{2s}(2^i+1)})) = x.
\end{aligned}
$$

Therefore,

$$F^{-1}(x) = F \circ F(x) = x + tr_m(x)tr_{(m,3)}(x^{2^i+1} + x^{2^{2s}(2^i+1)}) + (tr_{(m,3)}(x^{2^i+1} + x^{2^{2s}(2^i+1)}))^2.$$

Thus, we have

$$
\begin{aligned}
F'(x) &= [F^{-1}(x)]^{2^i+1} = [x + tr_m(x)tr_{(m,3)}(x^{2^i+1} + x^{2^{2s}(2^i+1)}) + (tr_{(m,3)}(x^{2^i+1} \\
&\quad + x^{2^{2s}(2^i+1)}))^2]^{2^i+1} = x^{2^i+1} + tr_m(x)(tr_{(m,3)}(x^{2^i+1} + x^{2^{2s}(2^i+1)}))^{2^s+1} \\
&\quad + (tr_{(m,3)}(x^{2^i+1} + x^{2^{2s}(2^i+1)}))^{2(2^s+1)} + x^{2^i} tr_m(x)tr_{(m,3)}(x^{2^i+1} + x^{2^{2s}(2^i+1)}) \\
&\quad + x\, tr_m(x)(tr_{(m,3)}(x^{2^i+1} + x^{2^{2s}(2^i+1)}))^{2^s} + x^{2^i} tr_{(m,3)}(x^{2(2^i+1)} + x^{2^{2s+1}(2^i+1)}) \\
&\quad + x\, (tr_{(m,3)}(x^{2(2^i+1)} + x^{2^{2s+1}(2^i+1)}))^{2^s} + tr_m(x)(tr_{(m,3)}(x^{2^i+1} + x^{2^{2s}(2^i+1)}))^{2^s+2} \\
&\quad + tr_m(x)(tr_{(m,3)}(x^{2^i+1} + x^{2^{2s}(2^i+1)}))^{2^{s+1}+1} = x^{2^i+1} + (tr_{(m,3)}(x^{2^i+1} \\
&\quad + x^{2^{2s}(2^i+1)}))^{2(2^s+1)} + x^{2^i} tr_m(x)(tr_{(m,3)}(x^{2^i+1} + x^{2^{2s}(2^i+1)})) \\
&\quad + x\, tr_m(x)tr_{(m,3)}(x^{2^i+1} + x^{2^s(2^i+1)}) + x^{2^i} tr_{(m,3)}(x^{2(2^i+1)} + x^{2^{2s+1}(2^i+1)}) \\
&\quad + x\, tr_{(m,3)}(x^{2(2^i+1)} + x^{2^{s+1}(2^i+1)}) + tr_m(x)[(tr_{(m,3)}(x^{2^i+1} + x^{2^{2s}(2^i+1)}))^{2^s+1} \\
&\quad + (tr_{(m,3)}(x^{2^i+1} + x^{2^{2s}(2^i+1)}))^{2^s+2} + (tr_{(m,3)}(x^{2^i+1} + x^{2^{2s}(2^i+1)}))^{2^{s+1}+1}].
\end{aligned}
$$

The only item in this sum which can give algebraic degree greater than 4 is the last item. We have

$$
\begin{aligned}
&(tr_{(m,3)}(x^{2^i+1} + x^{2^{2s}(2^i+1)}))^{2^s+1} + (tr_{(m,3)}(x^{2^i+1} + x^{2^{2s}(2^i+1)}))^{2^s+2} \\
&\quad + (tr_{(m,3)}(x^{2^i+1} + x^{2^{2s}(2^i+1)}))^{2^{s+1}+1} = (tr_{(m,3)}(x^{2^i+1} + x^{2^{2s}(2^i+1)}))^{2^s+1} \\
&\quad + (tr_{(m,3)}(x^{2^i+1} + x^{2^{2s}(2^i+1)}))^{4(2^s+1)} + (tr_{(m,3)}(x^{2^i+1} + x^{2^{2s}(2^i+1)}))^{2^{2s}},
\end{aligned}
$$

since

$$2^s + 2 = \begin{cases} 4 \text{ if } s = 1 \\ 6 \text{ if } s = 2 \end{cases},$$

$$4(2^s+1) = \begin{cases} 12 = 5 \quad (\mathrm{mod}\ 2^3-1) \text{ if } s=1 \\ 20 = 6 \quad (\mathrm{mod}\ 2^3-1) \text{ if } s=2 \end{cases},$$

$$2^{s+1}+1 = \begin{cases} 5 \qquad\qquad \text{ if } s=1 \\ 9 = 2 \quad (\mathrm{mod}\ 2^3-1) \text{ if } s=2 \end{cases},$$

$$2^{2s} = \begin{cases} 4 \qquad\qquad \text{ if } s=1 \\ 16 = 2 \quad (\mathrm{mod}\ 2^3-1) \text{ if } s=2 \end{cases}.$$

On the other hand,

$$(tr_{(m,3)}(x^{2^i+1} + x^{2^{2s}(2^i+1)}))^{2^s+1} = tr_{(m,3)}(x^{2^i+1} + x^{2^{2s}(2^i+1)})$$
$$\times tr_{(m,3)}(x^{2^i+1} + x^{2^s(2^i+1)}) = tr_{(m,3)}(x^{2^i+1})^2 + (tr_{(m,3)}(x^{2^i+1}))^{2^{2s}+1}$$
$$+ (tr_{(m,3)}(x^{2^i+1}))^{2^s+1} + (tr_{(m,3)}(x^{2^i+1}))^{2^{2s}+2^s}$$

$$= (tr_{(m,3)}(x^{2^i+1}))^6 + (tr_{(m,3)}(x^{2^i+1}))^5 + (tr_{(m,3)}(x^{2^i+1}))^3 + (tr_{(m,3)}(x^{2^i+1}))^2. \quad (1)$$

Using (1) we get

$$(tr_{(m,3)}(x^{2^i+1} + x^{2^{2s}(2^i+1)}))^{2^s+1} + (tr_{(m,3)}(x^{2^i+1} + x^{2^{2s}(2^i+1)}))^{4(2^s+1)}$$
$$+ (tr_{(m,3)}(x^{2^i+1} + x^{2^{2s}(2^i+1)}))^{2^{2s}} = (tr_{(m,3)}(x^{2^i+1}))^6$$
$$+ (tr_{(m,3)}(x^{2^i+1}))^5 + (tr_{(m,3)}(x^{2^i+1}))^3 + (tr_{(m,3)}(x^{2^i+1}))^2$$
$$+ [(tr_{(m,3)}(x^{2^i+1}))^3 + (tr_{(m,3)}(x^{2^i+1}))^6 + (tr_{(m,3)}(x^{2^i+1}))^5$$
$$+ tr_{(m,3)}(x^{2^i+1})] + (tr_{(m,3)}(x^{2^i+1}))^2 + (tr_{(m,3)}(x^{2^i+1}))^4$$

$$= tr_{(m,3)}(x^{2^i+1}) + (tr_{(m,3)}(x^{2^i+1}))^4. \qquad (2)$$

Hence, applying (1) and (2) we get

$$F'(x) = x^{2^i+1} + [(tr_{(m,3)}(x^{2^i+1}))^6 + (tr_{(m,3)}(x^{2^i+1}))^5 + (tr_{(m,3)}(x^{2^i+1}))^3$$
$$+ (tr_{(m,3)}(x^{2^i+1}))^2]^2 + x^{2^i} tr_m(x) tr_{(m,3)}(x^{2^i+1} + x^{2^{2s}(2^i+1)})$$
$$+ x\, tr_m(x) tr_{(m,3)}(x^{2^i+1} + x^{2^s(2^i+1)}) + x^{2^i} tr_{(m,3)}(x^{2(2^i+1)}$$
$$+ x^{2^{2s+1}(2^i+1)}) + x\, tr_{(m,3)}(x^{2(2^i+1)} + x^{2^{s+1}(2^i+1)})$$
$$+ tr_m(x) [tr_{(m,3)}(x^{2^i+1}) + (tr_{(m,3)}(x^{2^i+1}))^4] = x^{2^i+1} + (tr_{(m,3)}(x^{2^i+1}))^6$$
$$+ (tr_{(m,3)}(x^{2^i+1}))^5 + (tr_{(m,3)}(x^{2^i+1}))^3 + (tr_{(m,3)}(x^{2^i+1}))^4$$
$$+ x^{2^i} tr_m(x) tr_{(m,3)}(x^{2^i+1} + x^{2^{2s}(2^i+1)}) + x\, tr_m(x) tr_{(m,3)}(x^{2^i+1} + x^{2^s(2^i+1)})$$
$$+ x^{2^i} tr_{(m,3)}(x^{2(2^i+1)} + x^{2^{2s+1}(2^i+1)}) + x\, tr_{(m,3)}(x^{2(2^i+1)} + x^{2^{s+1}(2^i+1)})$$
$$+ tr_m(x) tr_{(m,3)}(x^{2^i+1} + x^{4(2^i+1)}).$$

Below we consider all items in the sum presenting the function F' which may give the algebraic degree 4:

$$[(tr_{(m,3)}(x^{2^i+1}))^6 + (tr_{(m,3)}(x^{2^i+1}))^5 + (tr_{(m,3)}(x^{2^i+1}))^3]$$

$$+[x^{2^i} tr_m(x)(tr_{(m,3)}(x^{2^i+1} + x^{2^{2s}(2^i+1)})) + x \, tr_m(x)(tr_{(m,3)}(x^{2^i+1} + x^{2^s(2^i+1)}))].$$

For simplicity we take $i = 1$. Obviously, all the items in the second bracket of the algebraic degree 4 have the form $x^{2^j+2^k+2^l+2^r}$, where $r < l < k < j \le m-1$, $r \le 1$. Therefore, if we find an item of algebraic degree 4 in the first bracket of the form $x^{2^j+2^k+2^l+2^r}$, where $2 \le r < l < k < j \le m-1$, which does not cancel, then this item does not vanish in the whole sum.

We have

$$tr_{(m,3)}(x^3) = x^{2+1} + x^{2^4+2^3} + \cdots + x^{2^{m-5}+2^{m-6}} + x^{2^{m-2}+2^{m-3}}$$
$$= \sum_{k=0}^{\frac{m}{3}-1} x^{2^{3k+1}+2^{3k}},$$

$$(tr_{(m,3)}(x^3))^2 = x^{2^2+2} + x^{2^5+2^4} + \cdots + x^{2^{m-4}+2^{m-5}} + x^{2^{m-1}+2^{m-2}}$$
$$= \sum_{k=0}^{\frac{m}{3}-1} x^{2^{3k+2}+2^{3k+1}},$$

$$(tr_{(m,3)}(x^3))^4 = x^{2^3+2^2} + x^{2^6+2^5} + \cdots + x^{2^{m-3}+2^{m-4}} + x^{2^m+2^{m-1}}$$
$$= \sum_{k=0}^{\frac{m}{3}-2} x^{2^{3k+3}+2^{3k+2}} + x^{2^{m-1}+1},$$

$$(tr_{(m,3)}(x^3))^3 = (tr_{(m,3)}(x^3))^2 tr_{(m,3)}(x^3) = \sum_{i,k=0}^{\frac{m}{3}-1} x^{2^{3k+1}+2^{3k}+2^{3i+2}+2^{3i+1}}, \qquad (3)$$

$$(tr_{(m,3)}(x^3))^5 = \sum_{j=0}^{\frac{m}{3}-2}\sum_{k=0}^{\frac{m}{3}-1} x^{2^{3j+3}+2^{3j+2}+2^{3k+1}+2^{3k}} + \sum_{k=0}^{\frac{m}{3}-1} x^{2^{m-1}+1+2^{3k+1}+2^{3k}}, \quad (4)$$

$$(tr_{(m,3)}(x^3))^6 = \sum_{j=0}^{\frac{m}{3}-2}\sum_{k=0}^{\frac{m}{3}-1} x^{2^{3j+3}+2^{3j+2}+2^{3k+2}+2^{3k+1}} + \sum_{k=0}^{\frac{m}{3}-1} x^{2^{m-1}+1+2^{3k+2}+2^{3k+1}}.$$

$$(5)$$

Note that all exponents of weight 4 in (3)-(5) are smaller than 2^m. If $m \ge 9$ then it is obvious that the item $x^{2^6+2^5+2^4+2^3}$ does not vanish in (4) and it definitely differs from all items in (3) and (5).

Hence, the function F' has the algebraic degree 4 when $m \ge 9$ and that completes the proof of the theorem. □

It is proven in [6] that the Gold functions are CCZ-inequivalent to the Welch function for all $m \ge 9$. Therefore, the function F' of Theorem 1 is CCZ-inequivalent to the Welch function. Further, the inverse and the Dobbertin APN functions are not AB (see [12,14]) and, therefore, the AB function F' is

CCZ-inequivalent to them. The algebraic degree of the Kasami function $x^{4^i-2^i+1}$, $2 \leq i \leq \frac{m-1}{2}$, $\gcd(i,m) = 1$, is equal to $i + 1$. Thus, its algebraic degree equals 4 if and only if $i = 3$. Since the function F' is defined only for m divisible by 3 then for $i = 3$ we would have $\gcd(i,m) \neq 1$. On the other hand, if Gold and Kasami functions are CCZ-equivalent then it follows from the proof of Theorem 5 of [6] that the Gold function is EA-equivalent to the inverse of the Kasami function which must be quadratic in this case. Thus, if F' was EA-equivalent to the inverse of a Kasami function then F' would be quadratic. Hence, F' cannot be EA-equivalent to the Kasami functions or to their inverses.

Proposition 2. *The function of Theorem 1 is EA-inequivalent to the Welch, Kasami, inverse, Dobbertin functions and to their inverses.*

For $m = 2t + 1$ the Niho function has the algebraic degree $t + 1$ if t is odd and the algebraic degree $(t + 2)/2$ if t is even. Therefore, its algebraic degree equals 4 if and only if $m = 7, 13$.

Proposition 3. *The function of Theorem 1 is EA-inequivalent to the Niho function.*

We do not have a general proof of EA-inequivalence of F' and the inverse of the Niho function but for $m = 9$ the Niho function coincides with the Welch functions and therefore its inverse cannot be EA-equivalent to the function F'.

Corollary 1. *For $m = 9$ the function of Theorem 1 is EA-inequivalent to any power function.*

When m is odd and divisible by 3 the APN functions from Table 2 have algebraic degrees different from 4. Thus we get the following proposition.

Proposition 4. *The function of Theorem 1 is EA-inequivalent to any APN function from Table 2.*

Acknowledgments

We would like to thank Claude Carlet for many valuable discussions and for detailed and insightful comments on the several drafts of this paper. The main part of this work was carried out while the author was with Otto-von-Guericke University Magdeburg and the research was supported by the State of Saxony Anhalt, Germany; also supported by a postdoctoral fellowship of MIUR-Italy via PRIN 2006.

References

1. Beth, T., Ding, C.: On almost perfect nonlinear permutations. In: Helleseth, T. (ed.) EUROCRYPT 1993. LNCS, vol. 765, pp. 65–76. Springer, Heidelberg (1993)
2. Biham, E., Shamir, A.: Differential Cryptanalysis of DES-like Cryptosystems. Journal of Cryptology 4(1), 3–72 (1991)

3. Budaghyan, L., Carlet, C.: Classes of Quadratic APN Trinomials and Hexanomials and Related Structures. Preprint, available at
   http://eprint.iacr.org/2007/098
4. Budaghyan, L., Carlet, C., Leander, G.: Constructing new APN functions from known ones. Preprint, available at http://eprint.iacr.org/2007/063
5. Budaghyan, L., Carlet, C., Leander, G.: Another class of quadratic APN binomials over $\mathbf{F}_{2^n}$: the case n divisible by 4. In: Proceedings of the Workshop on Coding and Cryptography (2007) (To appear) available at
   http://eprint.iacr.org/2006/428.pdf
6. Budaghyan, L., Carlet, C., Leander, G.: A class of quadratic APN binomials inequivalent to power functions. Submitted to IEEE Trans. Inform. Theory, available at http://eprint.iacr.org/2006/445.pdf
7. Budaghyan, L., Carlet, C., Felke, P., Leander, G.: An infinite class of quadratic APN functions which are not equivalent to power mappings. Proceedings of the IEEE International Symposium on Information Theory 2006, Seattle, USA (July 2006)
8. Budaghyan, L., Carlet, C., Pott, A.: New Classes of Almost Bent and Almost Perfect Nonlinear Functions. IEEE Trans. Inform. Theory 52(3), 1141–1152 (2006)
9. Budaghyan, L., Carlet, C., Pott, A.: New Constructions of Almost Bent and Almost Perfect Nonlinear Functions. In: Charpin, P., Ytrehus, Ø., (eds.) Proceedings of the Workshop on Coding and Cryptography 2005, pp. 306–315 (2005)
10. Canteaut, A., Charpin, P., Dobbertin, H.: A new characterization of almost bent functions. In: Knudsen, L.R. (ed.) FSE 1999. LNCS, vol. 1636, pp. 186–200. Springer, Heidelberg (1999)
11. Canteaut, A., Charpin, P., Dobbertin, H.: Binary m-sequences with three-valued crosscorrelation: A proof of Welch's conjecture. IEEE Trans. Inform. Theory 46(1), 4–8 (2000)
12. Canteaut, A., Charpin, P., Dobbertin, H.: Weight divisibility of cyclic codes, highly nonlinear functions on $\mathbf{F}_{2^m}$, and crosscorrelation of maximum-length sequences. SIAM Journal on Discrete Mathematics 13(1), 105–138 (2000)
13. Carlet, C.: Vectorial (multi-output) Boolean Functions for Cryptography. In: Crama, Y., Hammer, P. (eds.) Chapter of the monography Boolean Methods and Models, Cambridge University Press, to appear soon. Preliminary version available at http://www-rocq.inria.fr/codes/Claude.Carlet/pubs.html
14. Carlet, C., Charpin, P., Zinoviev, V.: Codes, bent functions and permutations suitable for DES-like cryptosystems. Designs, Codes and Cryptography 15(2), 125–156 (1998)
15. Chabaud, F., Vaudenay, S.: Links between differential and linear cryptanalysis. In: De Santis, A. (ed.) EUROCRYPT 1994. LNCS, vol. 950, pp. 356–365. Springer, Heidelberg (1995)
16. Daemen, J., Rijmen, V.: AES proposal: Rijndael (1999),
    http://csrc.nist.gov/encryption/aes/rijndael/Rijndael.pdf
17. Dillon, J.F.: APN Polynomials and Related Codes. Polynomials over Finite Fields and Applications, Banff International Research Station (November 2006)
18. Dobbertin, H.: One-to-One Highly Nonlinear Power Functions on $GF(2^n)$. Appl. Algebra Eng. Commun. Comput. 9(2), 139–152 (1998)
19. Dobbertin, H.: Almost perfect nonlinear power functions over $GF(2^n)$: the Niho case. Inform. and Comput. 151, 57–72 (1999)
20. Dobbertin, H.: Almost perfect nonlinear power functions over $GF(2^n)$: the Welch case. IEEE Trans. Inform. Theory 45, 1271–1275 (1999)

21. Dobbertin, H.: Almost perfect nonlinear power functions over $GF(2^n)$: a new case for n divisible by 5. In: Jungnickel, D., Niederreiter, H. (eds.) Proceedings of Finite Fields and Applications FQ5, Augsburg, Germany, pp. 113–121. Springer, Heidelberg (2000)

22. Edel, Y., Kyureghyan, G., Pott, A.: A new APN function which is not equivalent to a power mapping. IEEE Trans. Inform. Theory 52(2), 744–747 (2006)

23. Gold, R.: Maximal recursive sequences with 3-valued recursive crosscorrelation functions. IEEE Trans. Inform. Theory 14, 154–156 (1968)

24. Hollmann, H., Xiang, Q.: A proof of the Welch and Niho conjectures on crosscorrelations of binary m-sequences. Finite Fields and Their Applications 7, 253–286 (2001)

25. Janwa, H., Wilson, R.: Hyperplane sections of Fermat varieties in P^3 in char. 2 and some applications to cyclic codes. In: Moreno, O., Cohen, G., Mora, T. (eds.) AAECC-10. LNCS, vol. 673, pp. 180–194. Springer, Heidelberg (1993)

26. Kasami, T.: The weight enumerators for several classes of subcodes of the second order binary Reed-Muller codes. Inform. and Control 18, 369–394 (1971)

27. Lachaud, G., Wolfmann, J.: The Weights of the Orthogonals of the Extended Quadratic Binary Goppa Codes. IEEE Trans. Inform. Theory 36, 686–692 (1990)

28. Matsui, M.: Linear cryptanalysis method for DES cipher. In: Helleseth, T. (ed.) EUROCRYPT 1993. LNCS, vol. 765, pp. 386–397. Springer, Heidelberg (1994)

29. Nakagawa, N., Yoshiara, S.: A construction of differentially 4-uniform functions from commutative semifields of characteristic 2. In: Proceedings of WAIFI 2007, LNCS (2007)

30. Nyberg, K.: Differentially uniform mappings for cryptography. In: Helleseth, T. (ed.) EUROCRYPT 1993. LNCS, vol. 765, pp. 55–64. Springer, Heidelberg (1994)

31. Nyberg, K.: S-boxes and Round Functions with Controllable Linearity and Differential Uniformity. In: Preneel, B. (ed.) Fast Software Encryption. LNCS, vol. 1008, pp. 111–130. Springer, Heidelberg (1995)

32. Sidelnikov, V.: On mutual correlation of sequences. Soviet Math. Dokl. 12, 197–201 (1971)

New Point Addition Formulae for ECC Applications

Nicolas Meloni[1,2]

[1] Institut de Mathématiques et de Modélisation de Montpellier,
Univ. Montpellier 2, France
[2] Laboratoire d'Informatique,
de Robotique et de Microélectronique de Montpellier,
CNRS, Univ. Montpellier 2, France
nicolas.meloni@lirmm.fr

Abstract. In this paper we propose a new approach to point scalar multiplication on elliptic curves defined over fields of characteristic greater than 3. It is based on new point addition formulae that suit very well to exponentiation algorithms based on Euclidean addition chains. However finding small chains remains a very difficult problem, so we also develop a specific exponentiation algorithm, based on Zeckendorf representation (i.e. representing the scalar k using Fibonacci numbers instead of powers of 2), which takes advantage of our formulae.

Keywords: elliptic curve, scalar multiplication, exponentiation, Fibonacci, addition chains.

1 Introduction

Since its introduction by Miller and Koblitz [11,9], elliptic curve cryptography (ECC) has received a lot of attention and has subsequently become one of the main standards in public key cryptography. The main operation (in terms of computations) of such systems is the point scalar multiplication, i.e. the computation of the point $[k]P = P + \cdots + P$, where k is an integer and P a point on a curve. It involves hundreds of multiplications on the underlying field which means that some efforts are to be made on optimizing this computation. This is precisely what this paper deals with.

A point scalar multiplication is just a sequence of point additions, being themselves made of several multiplications, squarings and inversions on a finite field. So improvements can be done at the finite field level by developing faster modular multiplication algorithms, at the curve level by improving the point addition and finally at the algorithmic level by proposing exponentiation algorithms adapted to the context of elliptic curves.

In this paper we will contribute to the two last levels. First we will propose new point addition formulae in a specific case. More precisely, if one computes $P_3 = P_1 + P_2$ on a curve then computing $P_3 + P_1$ or $P_3 + P_2$ can be done at a very low computational cost. Then we will compare this approach to existing works

C. Carlet and B. Sunar (Eds.): WAIFI 2007, LNCS 4547, pp. 189–201, 2007.
© Springer-Verlag Berlin Heidelberg 2007

done by Montgomery [13] and generalized by Brier and Joye on one hand [1] and by Lopez and Dahab [10] on the other hand. Those formulae suit very well to Euclidean addition chains which will lead to a very efficient point multiplication algorithm as long as one is able to find a small chain to compute a given integer. This problem being still difficult, we will propose next to represent the scalar k using Fibonacci numbers instead of powers of 2. That is to say writing $k = \sum_{i=1}^{n} F_i$, where F_i is the ith Fibonacci number. This will allow us to propose a "Fibonacci-and-add" algorithm taking advantage of our formula.

2   Elliptic Curve Arithmetic

Definition 1. *An elliptic curve E over a field K denoted by E/K is given by the equation*

$$E : y^2 + a_1 xy + a_3 y = x^3 + a_2 x^2 + a_4 x + a_6$$

where $a_1, a_2, a_3, a_4, a_6 \in K$ are such that, for each point (x, y) on E, the partial derivatives do not vanish simultaneously.

In practice, the equation can be simplified into

$$y^2 = x^3 + ax + b$$

where $a, b \in K$ and $4a^3 + 27b^2 \neq 0$, over field of characteristic greater than 3.

The set of points of E/K is an abelian group. There exist explicit formulae to compute the sum of two points, and several coordinate systems have been proposed to speed up this computation. For a complete overview of those coordinates, one can refer to [3,6]. As an example, in Jacobian coordinates, the curve E (over a field of characteristic greater than 3) is given by $Y^2 = X^3 + a_4 X Z^4 + a_6 Z^6$, the point (X, Y, Z) on E corresponds to the affine point $(\frac{X}{Z^2}, \frac{Y}{Z^3})$ and the formulae are:

Addition:
$P = (X_1, Y_1, Z_1), Q = (X_2, Y_2, Z_2)$ and $P + Q = (X_3, Y_3, Z_3)$

$$A = X_1 Z_1^2, B = X_2 Z_1^2, C = Y_1 Z_2^3, D = Y_2 Z_1^3, E = B - A, F = D - C$$

and

$$X_3 = -E^3 - 2AE^2 + F, Y_3 = -CE^3 + F(AE^2 - X_3), Z_3 = Z_1 Z_2 E$$

Doubling:
$[2]P = (X_3, Y_3, Z_3)$

$$A = 4X_1 Y_1^2, B = 3X_1^2 + a_4 Z_1^4$$

and

$$X_3 = -2A + B^2, Y_3 = -8Y_1^4 + B(A - X_3), Z_3 = 2Y_1 Z_1.$$

The computation cost is 12 multiplications (M) and 4 squarings (S) (8M and 3S if one of the point is given in the form (X,Y,1)) for the addition and 4M and 6S for the doubling.

The computation of $[k]P$ is usually done using Algorithm 1. It requires about $\log_2(k)$ doublings and $w(k)$ additions, where $w(k)$ is the Hamming weight of k. Several methods have been developed to reduce both the cost of a doubling and the number of additions. This can be achieved by using, for example, modified Jacobian coordinates and windowing methods [4]. Other methods include the use of a different number system to represent the scalar k, as the double base number system [5]. Finally over binary fields doubling can be replaced by other endomorphisms such as the Frobenius endomorphism [14] or point halving [7].

Algorithm 1. Double-and-add

Data: $P \in E$ and $k = (k_{l-1}, \ldots, k_0)_2 \in \mathbb{N}$.
Result: $[k]P \in E$.
begin
  | $Q \leftarrow P$
  | **for** $i = l - 2 \ldots 0$ **do**
  | | $Q \leftarrow [2]Q$
  | | **if** $k_i = 1$ **then**
  | | | $Q \leftarrow Q + P$
  | | **end**
  | **end**
end
return Q

---

3 New Point Addition Formulae

Let K be a field of characteristic greater than 3, E/K an elliptic curve, $P_1 = (X_1, Y_1, Z)$ and $P_2 = (X_2, Y_2, Z)$ two points (in Jacobian coordinates) on E sharing the same z-coordinate. Then if we note $P_1 + P_2 = P_3 = (X_3, Y_3, Z_3)$ we have:

$$
\begin{aligned}
X_3 &= (Y_2 Z^3 - Y_1 Z^3)^2 - (X_2 Z^2 - X_1 Z^2)^3 - 2X_1 Z^2 (X_2 Z^2 - X_1 Z^2)^2 \\
&= ((Y_2 - Y_1)^2 - (X_2 - X_1)^3 - 2X_1 (X_2 - X_1)^2) Z^6 \\
&= ((Y_2 - Y_1)^2 - (X_1 + X_2)(X_2 - X_1)^2) Z^6 \\
&= X_3' Z^6 \\
Y_3 &= -Y_1 Z^3 (X_2 Z^2 - X_1 Z^2)^3 \\
&\quad + (Y_2 Z^3 - Y_1 Z^3)(X_1 Z^2 (X_2 Z^2 - X_1 Z^2)^2 - X_3) \\
&= (-Y_1 (X_2 - X_1)^3 + (Y_2 - Y_1)(X_1 (X_2 - X_1)^2 - X_3')) Z^9 \\
&= Y_3' Z^9 \\
Z_3 &= Z^2 (X_2 Z^2 - X_1 Z^2) \\
&= Z(X_2 - X_1) Z^3 \\
&= Z_3' Z^3
\end{aligned}
$$

Thus we have $(X_3, Y_3, Z_3) = (X_3' Z^6, Y_3' Z^9, Z_3' Z^3) \sim (X_3', Y_3', Z_3')$.

So when P_1 and P_2 have the same z-coordinate, $P_1 + P_2$ can be obtained using the following formulae:

Addition:
$P_1 = (X_1, Y_1, Z), P_2 = (X_2, Y_2, Z)$ and $P_1 + P_2 = (X_3', Y_3', Z_3')$

$$A = (X_2 - X_1)^2, \, B = X_1 A, \, C = X_2 A, \, D = (Y_2 - Y_1)^2$$

and

$$X_3' = D - B - C \, ,$$
$$Y_3' = (Y_2 - Y_1)(B - X_3) - Y_1(C - B) \, ,$$
$$Z_3' = Z(X_2 - X_1) \, .$$

This addition involves 5M and 2S.

As they require special conditions, our formulae are logically more efficient than any general or mixed addition formulae. What is more striking is the fact that they are more efficient than any doubling formulae (the best doubling is obtained using modified Jacobian coordinates and requires 4M and 4S).

The comparison with Montgomery's elliptic curves arithmetic is a lot more interesting. At a first sight the approaches look very similar. Indeed on Montgomery's curves the arithmetic is based on the fact that it is easy to compute the x and z-coordinates of $P_1 + P_2$ from the x and z-coordinates of P_1, P_2 and $P_1 - P_2$. The computational cost of this addition is 4M and 2S, which is lower than with our formula, but requires additional computations to recover the y-coordinate. Besides, recovering the y-coordinate requires to perform the point scalar multiplication using the Montgomery ladder algorithm. In the case of Euclidean addition chains exponentiation (treated in the next section) one cannot recover the y-coordinate from Montgomery's formulae. On the other hand we will show that it is possible not to compute the y-coordinate with our formulae. In this case the computational cost of our formulae is reduced to 4M+2S.

Finally notice that not every elliptic curves are Montgomery's curves (Brier and Joye generalized this approach to general curves [1] but in this case the computational cost rises to 9M and 2S) whereas our formulae work on any curve (as long as the characteristic of the underlying field is greater than 3).

It seems unlikely for both P_1 and P_2 to have the same z-coordinate. Fortunately the quantities $X_1 A = X_1(X_2 - X_1)^2$ and $Y_1(C - B) = Y_1(X_2 - X_1)^3$ computed during the addition can be seen as the x and y-coordinates of the point $(X_1(X_2 - X_1)^2, Y_1(X_2 - X_1)^3, Z(X_2 - X_1)) \sim (X_1, Y_1, Z)$. Thus it is possible to add P_1 and $P_1 + P_2$ with our new formulae.

Remark 1. *The same observation can be made from the doubling formulae, indeed the quantities $A = X_1(2Y_1)^2$ and $8Y_1^4 = Y_1(2Y_1)^3$ are the x and y-coordinates of the point $(X_1(2Y_1)^2, Y_1(2Y_1)^3, 2Y_1 Z_1) \sim (X_1, Y_1, Z_1)$ allowing us to compute $P + [2]P$ without additional computation.*

So we now have at our disposal an operator `NewADD` working the following way: let P_1 and P_2 be two points sharing the same z-coordinate then $\texttt{NewADD}(P_1, P_2)$ returns two points, $P_1 + P_2$ and P_1, sharing the same z-coordinate.

Example 1. *One can compute $[25]P$ in the following way:*

- *NewADD([2]P, P)=([3]P, [2]P)*
- *NewADD([2]P, [3]P)=([5]P, [2]P)*
- *NewADD([2]P, [5]P)=([7]P, [2]P)*

- $NewADD([7]P, [2]P) = ([9]P, [7]P)$
- $NewADD([9]P, [7]P) = ([16]P, [9]P)$
- $NewADD([16]P, [9]P) = ([25]P, [16]P)$

Remark 2. *The same kind of formulae can be developed in characteristic two. However we do not deal with this case in the remainder of the paper. Indeed Lopez and Dahab showed [10] that all curve can be turned into Montgomery's. Moreover many other methods, as fast doublings, point halving etc, lead to very efficient exponentiation algorithms so that our approach is no longer relevant.*

4   Point Scalar Multiplication

From the previous section we have seen that our formulae are quite efficient in terms of computational cost (more than a doubling) but cannot be used with classical double-and-add algorithms and require specific exponentiation schemes, as the one shown on example 1.

4.1   Euclidean Addition Chains

In this section we first show that the NewADD operator suits very well to Euclidean addition chains. We will then explain why finding such chains that are small is difficult.

Definition 2. *An addition chain computing an integer k is given by a sequence $v = (v_1, \ldots, v_s)$ where $v_1 = 1$, $v_s = k$ and $\forall\, 1 \leq i \leq s$, $v_i = v_{i_1} + v_{i_2}$ for some i_1 and i_2 lower than i.*

Definition 3. *An Euclidean addition chain (EAC) computing an integer k is an addition chain which satisfies $v_1 = 1, v_2 = 2, v_3 = v_2 + v_1$ and $\forall\, 3 \leq i \leq s - 1$, if $v_i = v_{i-1} + v_j$ for some $j < i-1$, then $v_{i+1} = v_i + v_{i-1}$ (case 1) or $v_{i+1} = v_i + v_j$ (case 2).*

Case 1 will be called big step (we add the biggest of the two possible numbers to v_i) and case 2 small step (we add the smallest one).

As an example, $(1, 2, 3, 4, 7, 11, 15, 19, 34)$ is an Euclidean addition chain computing 34. For instance, in step 4 we have computed 4=3+1, thus in step 5 we must add 3 or 1 to 4, in other words from step 4 we can only compute 5=4+1 or 7=4+3. In this example we have chosen to compute 7=4+3 so, at step 6, we can compute 10=7+3 or 11=7+4 etc. Another classical example of EAC is the Fibonacci sequence $(1, 2, 3, 5, 8, 13, 21, 34)$ (which is only made of big steps).

   Finding such chains is quite simple, it suffices to choose an integer g coprime with k and apply the subtractive form of Euclid's algorithm.

Example 2. *Let $k = 34$ and $g = 19$ and let apply them the subtractive form of Euclid's algorithm:*

$$34 - 19 = 15 \quad \text{(big step)}$$
$$19 - 15 = 4 \quad \text{(small step)}$$
$$15 - 4 = 11 \quad \text{(small step)}$$

$$11 - 4 = 7 \quad \textit{(big step)}$$
$$7 - 4 = 3 \quad \textit{(big step)}$$
$$4 - 3 = 1 \quad \textit{(small step)}$$
$$3 - 1 = 2$$
$$2 - 1 = 1$$
$$1 - 1 = 0$$

Reading the first number of each line gives the EAC $(1, 2, 3, 4, 7, 11, 15, 19, 34)$.

Finally, in order to simplify the writing of the algorithm, we will use the following notation : if $v = (1, 2, 3, v_4, \ldots, v_s)$ is an EAC then we only consider the chain from v_4 and we replace all the v_i's by 0 if it has been computed using a big step and by 1 for a small step.

For instance the sequence: $(1, \ 2, \ 3, \ 4, \ 7, \ 11, \ 15, \ 19, \ 34)$
will be written: $(1, 0, \ 0, \ 1, \ 1, \ 0)$.

Finally we note the chain $c = (c_4, \ldots, c_s)$ instead of v in order to prevent confusion between both representations.

We can now propose an algorithm performing a point scalar multiplication and using only the NewADD operator.

Algorithm 2. Euclid-Exp(c, P)

Data: P, $[2]P$ with $Z_P = Z_{[2]P}$ and an EAC $c = (c_4, \ldots, c_s)$
computing k ;
Result: $[k]P \in E$;
begin
    $(U_1, U_2) \leftarrow ([2]P, P)$
    for $i = 4 \ldots s$ **do**
        if $c_i = 0$ **then**
            | $(U_1, U_2) \leftarrow$ NewADD(U_1, U_2) ;
        else
            | $(U_1, U_2) \leftarrow$ NewADD(U_2, U_1) ;
        end
    end
    $(U_1, U_2) \leftarrow$ NewADD(U_1, U_2) ;
    return U_1
end

Example 3. *Let us see what happens with the chain* $c = (1, 0, 0, 1, 1, 0)$ *computing 34:*

$$\textit{first we compute} \ \ ([2]P, P)$$
$$c_4 = 1 \ \ \textit{so we compute} \ \ \textit{NewADD}(P, [2]P) = ([3]P, P)$$
$$c_5 = 0 \ \ \textit{so we compute} \ \ \textit{NewADD}([3]P, P) = ([4]P, [3]P)$$
$$c_6 = 0 \ \ \textit{so we compute} \ \ \textit{NewADD}([4]P, [3]P) = ([7]P, [4]P)$$

$$c_7 = 1 \quad \text{so we compute} \quad \textit{NewADD}([4]P, [7]P) = ([11]P, [4]P)$$

$$c_8 = 1 \quad \text{so we compute} \quad \textit{NewADD}([4]P, [11]P) = ([15]P, [4]P)$$

$$c_9 = 0 \quad \text{so we compute} \quad \textit{NewADD}([15]P, [4]P) = ([19]P, [15]P)$$

$$\text{and finally we compute } \textit{NewADD}([19]P, [15]P) \text{ which gives } [34]P$$

If we consider that the point P is given in affine coordinate (that is $Z = 1$) then the doubling step can be performed using 3M and 3S and so, the total computational cost of our algorithm is $(5s - 7)$M and $(2s - 1)$S.

Remark 3. *Some cryptographic protocols only require the x-coordinate of the point $[k]P$. In this case it is possible to save one multiplication by step of Algorithm 2 by noticing that Z does not appear during the computation of X_3' and Y_3', thus it is not necessary to compute Z_3' during the process. Appendix A shows how to recover the x-coordinate in the end.*

4.2 About Euclid's Addition Chains Length

At this point we know that Euclidean addition chains are easy to compute, however finding small chains is a lot more complicated.

We begin by a theorem proved by D. Knuth and A. Yao in 1975 [8].

Theorem 1. *Let $S(k)$ denote the average number of steps to compute $\gcd(k, g)$ using the subtractive Euclid's algorithm when g is uniformly distributed in the range $1 \leq g \leq k$. Then*

$$S(k) = 6\pi^{-2}(\ln k)^2 + O(\log k (\log \log k)^2)$$

This theorem shows that if, in order to find an EAC for an integer k, we choose an integer g at random, it will return a chain of length about $(\ln k)^2$, which is too long to be used with ECC. Indeed, for a 160-bit exponent, we will see in the last section that to be efficient, Algorithm 2 requires chains of length at most 320, whereas the previous theorem tells us that, theoretically, random chains have a length of 7000 on average (it is rather 2500 in practice).

A classic way to limit the length of EAC is to choose g close to $\frac{k}{\phi}$, where $\phi = \frac{1+\sqrt{5}}{2}$ is the golden section. This guarantees that the last steps of the EAC will be big steps. In practice this method allows to find EAC of an average length of 1100.

A second obvious way to find shorter chains is to try many g around $\frac{k}{\phi}$ and to keep the shortest chain. A more precise study can be found in [12].

Considering 160-bit integers, finding EAC of length 320 can be done by checking (on average) about 30 $g's$. Finding shorter chains is a lot more difficult, as an example finding chains of length 270 requires testing more than 45 000 $g's$. Such a computation can not be integrated into any exponentiation algorithm so, if some offline computations cannot be performed, one should not expect to use EAC whose length is shorter than 320.

5   Using Zeckendorf Representation

We have seen that finding a small Euclidean addition chain that compute a large integer is quite difficult. However if the integer k is a Fibonacci number then an optimal chain is quite easy to compute. Indeed the Fibonacci sequence is an optimal chain. The idea proposed in this section is to switch from binary to the Zeckendorf representation in order to replace doublings by Fibonacci numbers computations.

5.1   A Fibonacci-and-Add Algorithm

Theorem 2. *Let k be an integer and $(F_i)_{i \geq 0}$ the Fibonacci sequence, then k can be uniquely written in the form:*

$$k = \sum_{i=2}^{l} d_i F_i,$$

with $d_i \in \{0, 1\}$ and $d_i d_{i+1} = 0$

An integer k written in this form is said to be in Zeckendorf representation and will be denoted as $k = (d_{l-1}, \ldots, d_2)_{\mathcal{Z}}$. Such a representation is easy to compute as it can be obtained using a greedy algorithm. An equivalent of the double-and-add algorithm is proposed next.

Algorithm 3. Fibonacci-and-add(k, P)

Data: $P \in E(K)$, $k = (d_l, \ldots, d_2)_{\mathcal{Z}}$;
Result: $[k]P \in E$;
begin
  $(U, V) \leftarrow (P, P)$
  for $i = l - 1 \ldots 2$ **do**
    if $d_i = 1$ **then**
      | $U \leftarrow U + P$ (add step);
    end
    $(U, V) \leftarrow (U + V, U)$ (Fibonacci step) ;
  end
  return U
end

Example 4. *Computation of $[25]P$ with $25 = 21 + 3 + 1 = (1000101)_{\mathcal{Z}}$:*

- *initialization: $(U, V) \leftarrow (P, P)$*
- *$d_7 = 0 : (U, V) \leftarrow ([2]P, P)$*
- *$d_6 = 0 : (U, V) \leftarrow ([3]P, [2]P)$*
- *$d_5 = 0 : (U, V) \leftarrow ([5]P, [3]P)$*
- *$d_4 = 1 : U \leftarrow [6]P$ then $(U, V) \leftarrow ([9]P, [6]P)$*
- *$d_3 = 0 : (U, V) \leftarrow ([15]P, [9]P)$*
- *$d_2 = 1 : U \leftarrow [16]P$ then $(U, V) \leftarrow ([25]P, [16]P)$*
- *return $U = [25]P$*

The Zeckendorf representation needs 44% more digits in comparison with the binary method. For instance a 160-bit integer will require around 230 Fibonacci digits. However, the density of 1's in this representation is lower. From [2] we know that the density of 1's is about 0.2764. This means that representing a 160-bits integer requires, on average, 80 powers of 2 but only 64 Fibonacci numbers ($\simeq 230 \times 0.2764$).

More generally, for a n-bit integer, the classical double-and-add algorithm requires on average $1.5 \times n$ operations (n doublings and $\frac{n}{2}$ additions) and the Fibonacci-and-add requires $1.83 \times n$ operations ($1.44 \times n$ "Fibonacci" and $0.398 \times n$ additions). In other words the Fibonacci-and-add algorithms requires about 23% more operations.

5.2   Using NewADD

We want to adapt Algorithm 5.1 to elliptic curves using the NewADD operator. It is clear that, as long as U and V are two points sharing the same z-coordinate, the Fibonacci step just consists of one use of NewADD. This means that a sequence of 0's in the Zeckendorf representation of the k can be performed by a sequence of NewADD.

We need now to compute $U + P$ return $U + P$ and V with the same z-coordinate. Let us suppose that $U = (X_U, Y_U, Z)$, $V = (X_V, Y_V, Z)$ and $P = (x, y, 1)$. First we compute the point $P' = (xZ^2, yZ^3, Z) \sim P$ (3M+S) so that one can compute $U + P = (X_{U+P}, Y_{U+P}, Z_{U+P})$ using NewADD (5M+2S). Then on one hand we have $Z_{U+P} = (X_U - xZ^2)Z$. On the other hand $(X_U - xZ^2)^2$ and $(X_U - xZ^2)^3$ have been computed during the computation of $U + P$ (see the quantities A and $C - B$ in our formulae in section 3) so that updating the point V to $(X_V(X_U - xZ^2)^2, Y_V(X_U - xZ^2)^3, Z(X_U - xZ^2))$ requires only 2M.

As a conclusion the final computational cost of an add step is 10M+3S.

All this is summarized in the following algorithm:

Algorithm 4. Fibonacci-and-add(k, P)

Data: $P \in E(K)$, $k = (d_l, \dots, d_2)_{\mathcal{Z}}$;
Result: $[k]P \in E$;
begin
    $(U, V) \leftarrow (P, P)$
    for $i = l - 1 \dots 2$ **do**
        if $d_i = 1$ **then**
            update P;
            $(U, .) \leftarrow$ NewADD(U, P) ;
            update V;
        end
        $(U, V) \leftarrow$ NewADD (U, V) ;
    end
    return U
end

We have seen that this algorithm is expected to perform $1.44 \times n$ Fibonacci steps and $0.398 \times n$ add steps (where n is the bit length of k). Then the average complexity of this algorithm is $(11.18 \times n)\text{M} + (4.07 \times n)\text{S}$.

5.3   Improvements

As with the binary case, it is possible to modify the Zeckendorf representation to reduce the number add step. As an example one can use a signed version of the Zeckendorf representation. In this case the density of 1's decreases to 0.2, which means that for an n-bit integer, the number of 1's is reduced to $0.29 \times n$.

If some extra memory is available (and with minor modifications of Algorithm 5.1) one can use some kind of window methods. For instance, one can modify the Zeckendorf representation using the following properties:

- $F_{n+3} + F_n = 2F_{n+2} \rightarrow 1001_Z = 0200_Z$
- $F_{n+3} - F_n = 2F_{n+1} \rightarrow 100\bar{1}_Z = 0020_Z$
- $F_{n+4} + F_n = 3F_{n+2} \rightarrow 10001_Z = 00300_Z$
- $F_{n+6} - F_n = 4F_{n+3} \rightarrow 100000\bar{1}_Z = 0004000_Z$

Experiments seem to show that using these recoding rules allows to reduce the density of non zero digits to 0.135 so that the number of expected add steps in Algorithm 5.1 is reduced to $0.194 \times n$.

Remark 4. *Of course it is possible to find many more properties in the huge literature dedicated to Fibonacci numbers, however the four rules given previously are sufficient when dealing with 160-bit integers.*

6   Comparisons with Other Methods

In this section we give some practical results about the complexities of our point multiplication algorithms and compare them with other classical methods. More precisely in Table 1 we compare our formulae used with Euclidean addition chains to Montgomery's ladder and Euclidean chains on Montgomery's curves, and in Table 2 we compare our Fibonacci number based algorithm (and its improved version) to double-and-add, NAF and 4-NAF methods on general curves using mixed coordinates.

We assume that S=0.8M, that k is a 160-bit integer and refer to [4] for the complexity of the window method using mixed coordinate.

In Table 1 we can see that our new formulae allow to generalize the use of Euclidean chains without loss of efficiency. Moreover one can compute both the x and y-coordinates (with a little efficiency loss) which is not possible with Montgomery's formulae. However Montgomery's ladder still remains a lot more efficient than any methods.

Comparing similar algorithms in Table 2 shows that Fibonacci based algorithms are still slower than their binary equivalents. From 10 to 23 % slower for simple to window Fibonacci-and-add. However this has to be balanced by the fact that those algorithms naturally require a lot more operations than the

Table 1. Comparisons with algorithms on Montgomery curves

Algorithm	Curve type	recovery of y-coord.	Field Mult.
Mont. ladder	Montgomery	yes	1463
EAC-320	Montgomery	no	1792
EAC-270	Montgomery	no	1512
EAC-320	Weiestraß	yes	2112
EAC-270	Weiestraß	yes	1782
EAC-320	Weiestraß	no	1792
EAC-270	Weiestraß	no	1512

Table 2. Comparisons between binary and Fibonacci based algorithm

Algorithm	Coord.	Field Mult.
Double-and-add	Mixed	2104
NAF	Mixed	1780
4-NAF	Mixed	1600
Fibonacci-and-add	NewADD	2311
Signed Fib-and-add	NewADD	2088
Window Fib-and-add	NewADD	1960

binary ones. From 23 % more for the Fibonacci-and-add to 36% for the window version. So we can see that our formulae significantly reduce the additional computation cost of our Fibonacci based algorithms making then almost as efficient as the binary ones.

7   Summary

In this paper we have proposed new point addition formulae with a lower computational cost than the best known doubling. We have shown that these formulae are really well suited to a special type of addition chains: the Euclidean addition chains. Our formulae allow us to generalize the use of those chains to any elliptic curve without loss of efficiency, compared to Montgomery's formulae. In addition we have proposed a Fibonacci number based point scalar multiplication algorithm. In practice it requires a lot more operations than its binary counterpart, but coupled with our formulae the former becomes almost as efficient as the latter (the additional cost is reduced from 23 % to 10 %).

References

1. Brier, E., Joye, M.: Weierstraß elliptic curves and side-channel attacks. In: Naccache, D., Paillier, P. (eds.) PKC 2002. LNCS, vol. 2274, pp. 335–345. Springer, Heidelberg (2002)
2. Capocelli, R.M.: A generalization of fibonacci trees. In: Third In. Conf. on Fibonacci Numbers and their Applications (1988)

3. Cohen, H., Frey, G. (eds.): Handbook of Elliptic and Hyperelliptic Cryptography. Chapman & Hall, Sydney, Australia (2006)
4. Cohen, H., Miyaji, A., Ono, T.: Efficient elliptic curve exponentiation using mixed coordinates. In: Ohta, K., Pei, D. (eds.) ASIACRYPT 1998. LNCS, vol. 1514, Springer, Heidelberg (1998)
5. Doche, C., Imbert, L.: Extended double-base number system with applications to elliptic curve cryptography. In: Barua, R., Lange, T. (eds.) INDOCRYPT 2006. LNCS, vol. 4329, pp. 335–348. Springer, Heidelberg (2006)
6. Hankerson, D., Menezes, A., Vanstone, S.: Guide to Elliptic Curve Cryptography. Springer, Heidelberg (2004)
7. Knudsen, E.W.: Elliptic scalar multiplication using point halving. In: Lam, K.-Y., Okamoto, E., Xing, C. (eds.) ASIACRYPT 1999. LNCS, vol. 1716, pp. 135–149. Springer, Heidelberg (1999)
8. Knuth, D., Yao, A.: Analysis of the subtractive algorithm for greater common divisors. Proc. Nat. Acad. Sci. USA 72(12), 4720–4722 (1975)
9. Koblitz, N.: Elliptic curve cryptosystems. Mathematics of Computation 48, 203–209 (1987)
10. Lopez, J., Dahab, R.: Fast multiplication on elliptic curves over GF (2 m) without precomputation. In: Koç, Ç.K., Paar, C. (eds.) CHES 1999. LNCS, vol. 1717, pp. 316–327. Springer, Heidelberg (1999)
11. Miller, V.S.: Uses of elliptic curves in cryptography. In: Williams, H.C. (ed.) CRYPTO 1985. LNCS, vol. 218, pp. 417–428. Springer, Heidelberg (1986)
12. Montgomery, P.: Evaluating Recurrences of form $x_{m+n} = f(x_m, x_n, x_{m-n})$ via Lucas chains (1983), Available at `ftp.cwi.nl:/pub/pmontgom/Lucas.ps.gz`
13. Montgomery, P.: Speeding the pollard and elliptic curve methods of factorization. Mathematics of Computation 48, 243–264 (1987)
14. Solinas, J.A.: Improved algorithms for arithmetic on anomalous binary curves. Technical report, University of Waterloo (1999), http://www.cacr.math.uwaterloo.ca/techreports/1999/corr99-46.pdf

A   Recovery of x-Coordinate

As said in section 3 the x-coordinate of the sum of two points P_1 and P_2 can be recovered without computing the z coordinate. Or in other word the value

$P_1 + P_2 = (X_{P_1+P_2}, Y_{P_1+P_2}, Z_{P_1+P_2})$ can be recovered thanks to the the following property:

Property 1. *Let $P_1 = (X_1, Y_1, Z)$, $P_2 = (X_2, Y_2, Z)$ and $P_1 + P_2 = (X_3, Y_3, Z_3)$ be points of an elliptic curve E given in Jacobian coordinates, then*

$$Z^2 = \frac{a}{2b} \left[\frac{(X_1 - X_2)(X_3 + 2Y_2Y_1 - X_1X_2(X_1 + X_2))}{Y_1^2 - Y_2^2 + X_2^3 - X_1^3} - (X_1 + X_2) \right]$$

Proof: P_1 and P_2 satisfy $Y^2 = X^3 + aXZ^4 + bZ^6$ so

$$Y_1^2 - Y_2^2 = X_1^3 - X_2^3 + aX_1Z^4 - aX_2Z^4 + bZ^6 - bZ^6$$

which gives

$$Z^4 = \frac{Y_1^2 - Y_2^2 + X_2^3 - X_1^3}{a(X_1 - X_2)}$$

Moreover

$$\begin{aligned}
X_3 &= (Y_2 - Y_1)^2 - (X_1 + X_2)(X_2 - X_1)^2 \\
&= Y_2^2 - 2Y_2Y_1 + Y_1^2 - X_2^3 + X_2^2X_1 + X_1^2X_2 - X_1^3 \\
&= Y_2^2 - X_2^3 + Y_1^2 - X_1^3 - 2Y_2Y_1 + X_1X_2(X_1 + X_2) \\
&= aX_1Z^4 + bZ^6 + aX_2Z^4 + bZ^6 - 2Y_2Y_1 + X_1X_2(X_1 + X_2) \\
&= Z^4(a(X_1 + X_2) + 2bZ^2) - 2Y_2Y_1 + X_1X_2(X_1 + X_2)
\end{aligned}$$

and so

$$Z^2 = \frac{a}{2b}\left[\frac{(X_1 - X_2)(X_3 + 2Y_2Y_1 - X_1X_2(X_1 + X_2))}{Y_1^2 - Y_2^2 + X_2^3 - X_1^3} - (X_1 + X_2)\right]$$

Recovering the final x-coordinate can be done in 8M, 4S and one inversion.

Explicit Formulas for Real Hyperelliptic Curves of Genus 2 in Affine Representation

Stefan Erickson[1], Michael J. Jacobson Jr.[2], Ning Shang[3],
Shuo Shen[3], and Andreas Stein[4]

[1] Department of Mathematics and Computer Science, Colorado College,
14 E. Cache La Poudre, Colorado Spgs., CO. 80903, USA
Stefan.Erickson@ColoradoCollege.edu
[2] Department of Computer Science, University of Calgary,
2500 University Drive NW, Calgary, Alberta, Canada, T2N 1N4
jacobs@cpsc.ucalgary.ca
[3] Department of Mathematics, Purdue University, 150 N. University Street,
West Lafayette, IN 47907-2067, USA
nshang@math.purdue.edu, sshen@math.purdue.edu
[4] Department of Mathematics, University of Wyoming 1000 E. University Avenue,
Laramie, WY 82071-3036, USA
astein@uwyo.edu

Abstract. In this paper, we present for the first time efficient explicit formulas for arithmetic in the degree 0 divisor class group of a real hyperelliptic curve. Hereby, we consider real hyperelliptic curves of genus 2 given in affine coordinates for which the underlying finite field has characteristic > 3. These formulas are much faster than the optimized generic algorithms for real hyperelliptic curves and the cryptographic protocols in the real setting perform almost as well as those in the imaginary case. We provide the idea for the improvements and the correctness together with a comprehensive analysis of the number of field operations. Finally, we perform a direct comparison of cryptographic protocols using explicit formulas for real hyperelliptic curves with the corresponding protocols presented in the imaginary model.

Keywords: hyperelliptic curve, reduced divisor, infrastructure and distance, Cantor's algorithm, explicit formulas, efficient implementation, cryptographic key exchange.

1 Introduction and Motivation

In 1989, Koblitz [9] first proposed the Jacobian of an imaginary hyperelliptic curve for use in public-key cryptographic protocols. Hyperelliptic curves are in a sense generalizations of elliptic curves, which are an attractive option for public-key cryptography because their key-per-bit security is significantly better than RSA. This is due to the fact that the best-known attacks on elliptic curve based cryptosystems have exponential as opposed to subexponential complexity in the bit length of the key. Hyperelliptic curves can be used with the same key-per-bit

C. Carlet and B. Sunar (Eds.): WAIFI 2007, LNCS 4547, pp. 202–218, 2007.

strength as elliptic curves provided that the genus is very small. In particular, recent attacks [4,5], imply that only genus 2 and possibly genus 3 hyperelliptic curves offer the same key-per-bit security as elliptic curves.

The Jacobian is a finite abelian group which, like elliptic curve groups, has unique representatives of group elements and efficient arithmetic (divisor addition and reduction). Although the arithmetic appears more complicated than that of elliptic curves [10,16,21,1,6], there are some indications that it can in some cases be more efficient. Those results are based on optimized explicit formulas and very efficient implementations for genus 2 and 3 imaginary hyperelliptic curves.

Several years later, a key exchange protocol was presented for the real model of a hyperelliptic curve [18]. Its underlying key space was the set of reduced principal ideals in the ring of regular functions of the curve, together with its group-like infrastructure. Although the main operation of divisor class addition, which is composition followed by reduction, is comparable in efficiency to that of the imaginary model [20], the protocol in [18] was significantly slower and more complicated than its imaginary cousin [9], while offering no additional security; the same was true for subsequent modifications presented in [17].

Despite the apparent short-comings of the real model, recent work [7] shows that real hyperelliptic curves may admit protocols that are comparable in efficiency to those based on the imaginary model. The main idea is that, in addition to the divisor class addition operation, the real model has a second operation called a *baby step* that is significantly more efficient. By exploiting this operation and some reasonable heuristics, new public-key protocols for key exchange, digital signatures, and encryption have been devised that are significantly faster than all previous protocols in real hyperelliptic curves and might even be comparable in efficiency with analogous protocols in the imaginary setting. However, the protocols in [7] were based on a generic implementation and did not incorporate explicit formulas. In order to examine the efficiency of these new protocols completely, it is necessary to devise explicit formulas for divisor arithmetic in the real model of cryptographically-relevant low genus curves.

The contribution of this paper is to close this gap and for the first time present efficient explicit formulas for divisor class arithmetic on real hyperelliptic curves. We concentrate on genus 2 real hyperelliptic curves in affine coordinates for which the underlying finite field has characteristic $p > 3$. Formulas for arbitrary characteristic, that also handle all special cases, will be included in the full version of this paper, which will be submitted to a journal. We thus provide explicit formulas for the protocols in [7], thereby enabling a direct comparison with the corresponding protocols presented in the imaginary model.

Notice that although there exist easy transformations from the imaginary model to the real model of a hyperelliptic curve, the converse direction is only possible if the curve defined over $\mathbb{F}_q$ contains an $\mathbb{F}_q$-rational point. If q is odd and one uses an irreducible polynomial for the generation of the real hyperelliptic curve, one has to extend the field of constants to $\mathbb{F}_{q^{2g+2}}$ in order to be able to perform this transformation, which is unrealistic for efficient implementations.

Furthermore, complex multiplication methods for generating hyperelliptic curves of small genus often produce real hyperelliptic curves. With an efficient arithmetic, those curves can be readily used in cryptographic protocols. Finally, explicit formulas enable us to provide a real-world comparison of subexponential attacks to hyperelliptic curve cryptosystems in both the real and imaginary setting.

The analysis of the formulas presented here shows that they require a few more finite field multiplications than their imaginary counterparts. However, the baby step operation in its explicit form is significantly more efficient than divisor class addition in either setting, and as a result, the cryptographic protocols in the real setting perform almost as well as those in the imaginary case. In addition, even though the formulas are not as fast as those in the imaginary case, they are certainly more efficient than using generic algorithms. Thus, using our formulas will significantly speed other computations in the divisor class group or infrastructure of a real hyperelliptic curve, for example, computing the regulator or class number.

The paper is organized as follows. We first provide the necessary background on real hyperelliptic curves and introduce the notation. We will also present the essential, generic algorithms for real hyperelliptic curves and explain how to perform arithmetic in the degree 0 divisor class group via ideal arithmetic. In Section 3, we present the explicit formulas for the basic algorithms assuming a finite field of characteristic $p > 3$. We provide the idea for the improvements and the correctness together with a comprehensive analysis of the number of field operations. Some of the calculations can also be found in the Appendix. Section 4 contains numerical data comparing cryptographic protocols based on real hyperelliptic curves with those using imaginary hyperelliptic curves, where divisor class arithmetic is implemented using explicit formulas in both cases.

2   Background and Notation

Throughout this paper, let $\mathbb{F}_q$ be a finite field with $q = p^l$ elements, where p is a prime, and let $\overline{\mathbb{F}}_q = \bigcup_{n \geq 1} \mathbb{F}_{q^n}$ be its algebraic closure. For details on the arithmetic of hyperelliptic curves we refer to [11,6,2,7], and specifically for real hyperelliptic curves we refer to [15,20,3,7,8].

Definition 1. *A hyperelliptic curve C of genus g defined over $\mathbb{F}_q$ is an absolutely irreducible non-singular curve defined by an equation of the form*

$$C : y^2 + h(x)y = f(x), \tag{2.1}$$

where $f, h \in \mathbb{F}_q[x]$ are such that $y^2 + h(x)y - f(x)$ is absolutely irreducible, i.e. irreducible over $\overline{\mathbb{F}}_q$, and if $b^2 + h(a)b = f(a)$ for $(a, b) \in \overline{\mathbb{F}}_q \times \overline{\mathbb{F}}_q$, then $2b + h(a) \neq 0$ or $h'(a)b - f'(a) \neq 0$. A hyperelliptic curve C is called

1. *an* imaginary hyperelliptic curve *if the following hold: If q is odd, then f is monic, $\deg(f) = 2g + 1$, and $h = 0$. If q is even, then h and f are monic, $\deg(f) = 2g + 1$, and $\deg(h) \leq g$.*

2. a real hyperelliptic curve *if the following hold: If q is odd, then f is monic,* $\deg(f) = 2g+2$, *and* $h = 0$. *If q is even, then h is monic,* $\deg h = g+1$, *and either (a)* $\deg f \leq 2g+1$ *or (b)* $\deg f = 2g+2$ *and the leading coefficient of f is of the form* $\beta^2 + \beta$ *for some* $\beta \in \mathbb{F}_q^*$.

3. an unusual hyperelliptic curve *[3] if the following holds: if $\mathbb{F}_q$ has odd characteristic, then* $\deg(f) = 2g+2$ *and* $\operatorname{sgn}(f)$ *is a non-square in* $\mathbb{F}_q^*$, *whereas if $\mathbb{F}_q$ has characteristic 2, then* $\deg(h) = g+1$, $\deg(f) = 2g+2$ *and the leading coefficient of f is not of the form* $e^2 + e$ *for any* $e \in \mathbb{F}_q^*$.

The function field $K = \mathbb{F}_q(C)$ of a hyperelliptic curve C is a quadratic, separable extension of $\mathbb{F}_q(x)$ and the integral closure of $\mathbb{F}_q(x)$ in K is given by $\mathbb{F}_q[C] = \mathbb{F}_q[x,y]/(y^2 + h(x)y - f(x))$. Let S denote the set of points at infinity. Then the set $C(\overline{\mathbb{F}}_q) = \{(a,b) \in \overline{\mathbb{F}}_q \times \overline{\mathbb{F}}_q : b^2 + h(a)b = f(a)\} \cup S$ is called the set of ($\overline{\mathbb{F}}_q$-rational) points on C. For a point $P = (a,b) \in C(\overline{\mathbb{F}}_q)$, the hyperelliptic involution is given by $\iota(a,b) = (a, -b - h(a)) \in C(\overline{\mathbb{F}}_q)$.

Notice that in all three cases we can assume $h = 0$ if q is odd. The imaginary model[1] corresponds to the case where $S = \{\infty_1\}$. In the real model[2], there exist two points at infinity so that $S = \{\infty_1, \infty_2\}$. Let v_1 and v_2 be the normalized valuations of K at ∞_1 and ∞_2, respectively. It is possible to transform an imaginary model of a hyperelliptic curve into a real model. For the converse direction one needs an $\mathbb{F}_q$-rational point (see [15,6]). From now on, we only consider the real case.

Let C be a real hyperelliptic curve given as in Definition 1. A divisor on C is a finite formal sum $D = \sum_{P \in C} m_P P$ of points $P \in C(\overline{\mathbb{F}}_q)$, where $m_P \in \mathbb{Z}$ and $m_P = 0$ for almost all P. The degree of D is defined by $\deg D = \sum_P m_P$. A divisor D of $\mathbb{F}_q(C)$ is effective if $m_P \geq 0$ for all P, and a divisor D is defined over $\mathbb{F}_q$, if $D^\sigma = \sum_P m_P P^\sigma = D$ for all automorphisms σ of $\overline{\mathbb{F}}_q$ over $\mathbb{F}_q$. The set $Div(K)$ of divisors of C defined over $\mathbb{F}_q$ forms an additive abelian group under formal addition with the set $Div_0(K)$ of all degree zero divisors of C defined over $\mathbb{F}_q$ as a subgroup. For a function $G \in K$, we can associate a principal divisor $\operatorname{div}(G) = \sum_P v_P(G)P$, where $v_P(G)$ is the normalized valuation of G at P. The group of principal divisors $P(K) = \{\operatorname{div}(G) : G \in K\}$ of C forms a subgroup of $Div_0(K)$. The factor group $J(K) = Div_0(K)/P(K)$ is called the *divisor class group* of K. We denote by $\overline{D} \in J(K)$ the class of $D \in Div_0(K)$.

Since C is a real hyperelliptic curve we have $S = \{\infty_1, \infty_2\}$ and we know from [15] that every degree 0 divisor class can be represented by $\overline{D}$ such that $D = \sum_{i=1}^r P_i - r\infty_2$, where $P_i \in C(\overline{\mathbb{F}}_q)$, $P_i \neq \infty_2$, and $P_i \neq \iota P_j$ if $i \neq j$. The representative D of $\overline{D}$ is then called semi-reduced. In addition, there exists a representative D such that $r \leq g$. In this case, the representative D is called reduced. Notice that $P_i = \infty_1$ is allowed for some i. It follows that every degree 0 divisor class contains a unique representative $\overline{D}$ with

---

[1] In function field terms, the pole divisor ∞ of x in $\mathbb{F}_q(x)$ is totally ramified in K so that $\operatorname{Con}(\infty) = 2\infty_1$.

[2] In function field terms, the pole divisor ∞ of x in $\mathbb{F}_q(x)$ splits completely in K so that $\operatorname{Con}(\infty) = \infty_1 + \infty_2$.

$$D = \sum_{i=1}^{l(D)} Q_i - l(D)\infty_2 + v_1(D)(\infty_1 - \infty_2) \,,$$

where $Q_i \in C(\overline{\mathbb{F}}_q)$, $Q_i \neq \infty_1, \infty_2$, $Q_i \neq \iota Q_j$ if $i \neq j$, and $0 \leq l(D) + v_1(D) \leq g$. The regulator R of K in $\mathbb{F}_q[C]$ is defined to be the order of the degree 0 divisor class containing $\infty_1 - \infty_2$.

We know that $\mathbb{F}_q[C]$ is a Dedekind domain and the ideal class group $\mathrm{Cl}(K)$ of K in $\mathbb{F}_q[C]$ is the factor group of fractional $\mathbb{F}_q[C]$-ideals modulo principal fractional ideals. A non-zero integral ideal $\mathfrak{a}$ in $\mathbb{F}_q[C]$ is a fractional ideal such that $\mathfrak{a} \subseteq \mathbb{F}_q[C]$. It can be represented as $\mathfrak{a} = k[x] \, d(x) u(x) + k[x] \, d(x)(v(x) + y)$ where $, u, v \in k[x]$ and $u \mid f + hv - v^2$. Note that d and u are unique up to factors in $\mathbb{F}_q^*$ and v is unique modulo u. $\mathfrak{a}$ is primitive if we can take $d(x) = 1$ in which case we simply write $\mathfrak{a} = [u(x), v(x) + y]$. A primitive ideal $\mathfrak{a} = [u(x), v(x) + y]$ is *reduced* if $\deg u \leq g$. A basis $\{u(x), v(x) + y\}$ of a primitive ideal is called *adapted* or *standard* if $\deg(v) < \deg(u)$ and u is monic. For instance, $\mathbb{F}_q[C]$ is represented as $\mathbb{F}_q[C] = [1, y]$. The degree of a primitive ideal is $\deg(\mathfrak{a}) = \deg u$. We call a basis $\{u(x), v(x) + y\}$ of a primitive ideal *reduced* if $-v_1(v - h - y) < -v_1(u) = \deg(u) < -v_1(v + y)$ and u is monic.

For any two ideals $\mathfrak{a}$ and $\mathfrak{b}$ in the same ideal class, there exists $\alpha \in K^*$ with $\mathfrak{b} = (\alpha)\mathfrak{a}$. We then define the distance of $\mathfrak{b}$ with respect to $\mathfrak{a}$ as $\delta(\mathfrak{b}, \mathfrak{a}) = -v_1(\alpha) \pmod{R}$ where R is the regulator. Note that the distance is only well-defined and unique modulo R. In each ideal class, we expect up to R many reduced ideals. If we fix the principal ideal class, then we may assume that $\mathfrak{a} = \mathfrak{a}_1 = \mathbb{F}_q[C] = (1)$. Then, for any principal ideal $\mathfrak{b} = (\alpha)$, we have $\delta(\mathfrak{b}) = \delta(\mathfrak{b}, \mathfrak{a}_1) = -v_1(\alpha) \pmod{R}$. Notice that the distance defines an order on all reduced principal ideals, i.e. the set of reduced principal ideals is $\mathcal{R} = \{\mathfrak{a}_1, \mathfrak{a}_2, \ldots, \mathfrak{a}_m\}$ where $\delta(\mathfrak{a}_1) = 0 < \delta(\mathfrak{a}_2) < \ldots < \delta(\mathfrak{a}_m)$.

The following theorem gives a representation of degree 0 divisor classes in terms of reduced ideals and corresponds to the Mumford representation [13, page 317] in the imaginary model.

Theorem 1 (*Paulus-Rück, 1999*). *There is a canonical bijection between the divisor class group $J(K)$ and the set of pairs $\{(\mathfrak{a}, n)\}$, where $\mathfrak{a}$ is a reduced ideal of $\mathbb{F}_q[C]$ and n is a non-negative integer with $0 \leq \deg(\mathfrak{a}) + n \leq g$.*

The bijection is such that the unique reduced divisor D in a degree 0 divisor class $\overline{D}$ corresponds to such a pair $\{(\mathfrak{a}, n)\}$. It follows that arithmetic in $J(K)$ can be performed via arithmetic of reduced ideals. An algorithm for computing the group law in $J(K)$ based on this theorem has been presented in [18,15,20]. It consists of three steps, namely (a) composition of reduced ideals, (b) reduction of the primitive part of the product, and (c) baby steps, i.e. adjusting the output of the reduction so that the degree condition of the theorem is satisfied. Step (a) and (b) together are called a *giant step*. A giant step is the analogue of the group operation in the imaginary case. We use $[u_1, v_1] + [u_2, v_2]$

to denote the giant step operation. Elements in $J(K)$ can represented as triples $[u, v, n]$ where $[u(x), y + v(x)]$ is a reduced ideal and $0 \leq \deg(\mathfrak{a}) + n \leq g$. It can be easily seen that the arithmetic can be restricted to the special subset $\{(\mathfrak{a}, 0) : \mathfrak{a} \text{ reduced and principal}\} = \mathcal{R}$, which is not a group. We may restrict our arithmetic to the degree 0 divisor classes that correspond to the set $\mathcal{R}$. Those elements can be represented as $[u, v, 0]$ or simply as pairs $[u, v]$. We therefore assume that we only perform operations on elements of $J(K)$ which are given by a pair $\overline{D} = [u, v]$, where $u, v \in \mathbb{F}_q[x]$ such that

1. u is monic,
2. $\deg(u) \leq g$,
3. $u \mid f + hv - v^2$,
4. one of the following degree conditions is satisfied, namely
   (a) for the reduced basis: $-v_1(v - h - y) < -v_1(u) = \deg(u) < -v_1(v + y)$,
   or
   (b) for the adapted (standard) basis: $\deg(v) < \deg(u)$.

If only 1., 3., and 4. are satisfied, the ideal $[u(x), y + v(x)]$ is only primitive and the corresponding representative $D \in \overline{D}$ is semi-reduced. We also denote this element by $[u, v]$.

In [7], several optimized key-exchange protocols have been presented that use arithmetic in $\mathcal{R}$. In fact, under reasonable assumptions, one can avoid the additional adjusting steps and replace some giant steps by baby steps. Furthermore, in each giant step, it is easy to keep track of the distances of the corresponding reduced ideals. In fact, assuming certain heuristics, one can even avoid computing distances in almost all situations. We will therefore ignore the computation of distances. Even in those cases, where distances are needed, the running time for the computation of the distance is negligible. The protocols for real hyperelliptic curves are analogous to the ones in the imaginary setting, but they also make use of the additional baby step operation.

We now give all three relevant algorithms. For details on how to produce key exchange protocols with these algorithms, we refer to [18,7]. We will use additive notation in order to express the group operation in $J(K)$ even though ideal arithmetic is usually denoted multiplicatively. Note that, by using these algorithms, arithmetic in $J(K)$ is reduced to polynomial arithmetic in $\mathbb{F}_q[x]$.

Algorithm 1 (Composition)
Input: $\overline{D}_1 = [u_1, v_1]$, $\overline{D}_2 = [u_2, v_2]$, and $h(x), f(x)$ as in (2.1).
Output: $\overline{D} = [u, v]$ such that D is semi-reduced and $\overline{D} = \overline{D}_1 + \overline{D}_2$.

1. Compute $d, x_1, x_2, x_3 \in \mathbb{F}_q[x]$ such that

$$d = \gcd(u_1, u_2, v_1 + v_2 + h) = x_1 u_1 + x_2 u_2 + x_3 (v_1 + v_2 + h) .$$

2. Put $u = u_1 u_2 / d^2$ and $v = (x_1 u_1 v_2 + x_2 u_2 v_1 + x_3 (v_1 v_2 + f))/d \pmod{u}$.

For the group operation, we assume that the representatives of the degree 0 divisor classes D_1 and D_2 are reduced so that the ideals $[u_1(x), y + v_1(x)]$ and

$[u_2(x), y + v_2(x)]$ are reduced, i.e. $\deg(u_1), \deg(u_2) \leq g$. However, the algorithm also allows semi-reduced representatives D_1 and D_2 as an input. Notice that the output of this algorithm $\overline{D} = [u, v]$ corresponds to a semi-reduced divisor so that $[u(x), v(x) + y]$ is a primitive ideal which is not necessarily reduced.

For the second step, we need to precompute the principal part $H(y) = \lfloor y \rfloor$ of a root y of $y^2 + h(x)y - f(x) = 0$. The other root is $-y - h$. If $y = \sum_{i=-\infty}^{m} y_i x^i \in \mathbb{F}_q\langle x^{-1} \rangle$, then $H(y) = \sum_{i=0}^{m} y_i x^i$.

Algorithm 2 (Reduction)
Input: $\overline{D} = [u, v]$, where D is semi-reduced, and $h(x), f(x)$ as in (2.1).
Output: $\overline{D}' = [u', v']$ such that D' is reduced and $\overline{D}' = \overline{D}$.

1. Compute $a = (v + H(y))$ div u.
2. Let $v' = v - au$, $u' = (f + hv' - v'^2)/u$.
3. If $\deg(u') > g$, put $u = u'$, $v = v'$, and goto 1.
4. Make u' monic and adjust v' to a reduced/adapted basis if necessary.

If we allow the input of Algorithm 2 to be reduced and only perform 1,2, and 4, the output will be another reduced divisor D' representing a different degree 0 divisor class $\overline{D}'$. In this case, we call this operation a *baby step*[3] denoted by $[u', v'] = \rho[u, v]$.

3   Explicit Formulas

Let $[u, v]$ be a Mumford representation of a degree 0 divisor class. We present explicit formulas for divisor class addition (ideal multiplication), divisor class doubling (ideal squaring), and a baby step. We will assume that characteristic of the field is a prime $p > 3$. Under this assumption, we can transform the general equation defining the curve to one of the form

$$C : y^2 = f(x)$$

that is isomorphic to the original curve, where $f(x) = x^6 + f_4 x^4 + f_3 x^3 + f_2 x^2 + f_1 x + f_0$, i.e., we can assume that $h(x) = 0$, the leading coefficient of $f(x)$ is 1, and the x^5 term of $f(x)$ is 0. The transformation $y \mapsto y - h/2$, valid if the finite field characteristic is not 2, eliminates $h(x)$ and $x \mapsto x - f_5/6$, valid if the characteristic is not 2 or 3, eliminates the x^5 term in $f(x)$. This assumption also implies that $H(y) = x^3 + y_1 x + y_0$, with $y_1 = f_4/2$ and $y_0 = f_3/2$.

We also assume that the divisor $[u, v]$ is in reduced basis. Under our assumptions about C, this implies that v is of the form

$$v = x^3 + v_1 x + v_0,$$

i.e., the leading two coefficients of v always match that of $H(y)$. We will present formulas for the general description of a genus two real hyperelliptic curve in

---

[3] However, notice that in this case the reduced ideal $\mathfrak{a}$ corresponding to $\overline{D}$ and the reduced ideal $\mathfrak{a}'$ corresponding to $\overline{D}'$ are in the same ideal class.

affine presentation over an arbitrary finite field, using both reduced and adapted basis, in the full version of this paper.

We only count inversions, squarings and multiplications of finite field elements, which consist of the bulk of the computation when compared with additions and subtractions. In the tables below, we let I, S and M denote "inversion," "squaring," and "multiplication," respectively.

In the formulas described below, we assume that the coefficients of $f(x) = x^6 + f_4 x^4 + f_3 x^3 + f_2 x^2 + f_1 x + f_0$ and $H(y) = x^3 + y_1 x + y_0$ that define the hyperelliptic curve are available. Thus, these are not explicitly listed as input.

3.1 Baby Step

Let $[u, v]$ be the Mumford representation of a degree 0 divisor class. To compute $\rho[u, v] = [u', v']$, we apply the following formulas:

$$v' = H(y) - [(H(y) + v) \bmod u],$$

$$u' = \text{Monic}\left(\frac{f - (v')^2}{u}\right)$$

were, as mentioned above, $H(y)$ is the principal part of a root of a root y of $y^2 + h(x)y - f(x) = 0$. Explicit formulas are derived by simply expanding the operations and using the formula for reducing a degree three polynomial $(H(y) + v)$ modulo a monic polynomial of degree two (u) described in [10]. The resulting formulas are presented in Table 1.

Table 1. Explicit Formulas for a Baby Step

Baby Step, Reduced Basis, $\deg u = 2$		
Input	$u = x^2 + u_1 x + u_0, \quad v = x^3 + v_1 x + v_0$	
Output	$[u', v'] = \rho[u, v]$	
Step	Expression	Operations
1	$v' = H(y) - [(H(y) + v) \bmod u]$	1S, 1M
	$v'_1 = 2(u_0 - u_1^2) - v_1$	
	$v'_0 = -2u_0 \cdot u_1 - v_0$	
2	$u' = \text{Monic}((f - (v')^2)/u)$	1I, 1S, 3M
	$u_2 = f_4 - 2v'_1$	
	$I = u_2^{-1}$	
	$u'_1 = I \cdot (f_3 - 2v'_0 - u_1)$	
	$u'_0 = I \cdot (f_2 - (v'_1)^2) - u_0 - u'_1 \cdot u_1$	
Total		1I, 2S, 4M

3.2 Addition Formulas

Let the Mumford representations of two degree 0 divisor classes be $[u_1, v_1]$ and $[u_2, v_2]$. The main case of addition of degree 0 divisor classes occurs when the

Table 2. Explicit Formulas for Addition of Divisor Classes

Addition, Reduced Basis, $\deg u_1 = \deg u_2 = 2$, $\gcd(u_1, u_2) = 1$		
Input	$u_1 = x^2 + u_{11}x + u_{10}$, $v_1 = x^3 + v_{11}x + v_{10}$	
	$u_2 = x^2 + u_{21}x + u_{20}$, $v_2 = x^3 + v_{21}x + v_{20}$	
Output	$[u', v'] = [u_1, v_1] + [u_2, v_2]$	
Step	**Expression**	**Operations**
	Composition	
1	$inv = z_1 x + z_2$	4M
	$z_0 = u_{10} - u_{20}$, $z_1 = u_{11} - u_{21}$	
	$z_2 = u_{11} \cdot z_1 - z_0$, $z_3 = u_{10} \cdot z_1$	
	$r = z_1 \cdot z_3 - z_0 \cdot z_2$	
2	$s' = s_1' x + s_0'$	4M
	$w_0 = v_{10} - v_{20}$, $w_1 = v_{11} - v_{21}$	
	$s_1' = w_0 \cdot z_1 - w_1 \cdot z_0$, $s_0' = w_0 \cdot z_2 - w_1 \cdot z_3$	
	Reduction	
3	$k = k_2 x^2 + k_1 x + k_0$	
	$k_2 = f_4 - 2v_{21}$	
4	$s = \frac{1}{r}s' = s_1 x + s_0$	1I, 2S, 6M
	$r_2 = r^2$, $\widehat{w}_0 = r_2 - (s_1' + r)^2 (= r^2 m_4)$, $\widehat{w}_1 = (r \cdot \widehat{w}_0)^{-1}$,	
	$\widehat{w}_2 = \widehat{w}_0 \cdot \widehat{w}_1(= \frac{1}{r})$, $\widehat{w}_3 = r \cdot r_2 \cdot \widehat{w}_1(= \frac{1}{m_4})$	
	$s_1 = s_1' \cdot \widehat{w}_2$, $s_0 = s_0' \cdot \widehat{w}_2$	
5	$l = l_3 x^3 + l_2 x^2 + l_1 x + l_0$ (note that $l_3 = s_1$)	3M
	$\widetilde{w}_0 = s_0 \cdot u_{20}$, $\widetilde{w}_1 = s_1 \cdot u_{21}$, $l_2 = s_0 + \widetilde{w}_1$	
	$l_1 = (s_0 + s_1) \cdot (u_{21} + u_{20}) - \widetilde{w}_1 - \widetilde{w}_0$, $l_0 = \widetilde{w}_0$	
6	$m' = x^4 + m_3' x^3 + m_2' x^2 + m_1' x + m_0'$, $u' = x^2 + u_1' x + u_0'$	6M
	$m_3' = \widehat{w}_3 \cdot (-s_1 \cdot (s_0 + l_2) - 2s_0)(= \frac{m_3}{m_4})$	
	$m_2' = \widehat{w}_3 \cdot (k_2 - s_1 \cdot (l_1 + 2v_{21}) - s_0 \cdot l_2)(= \frac{m_2}{m_4})$	
	$u_1' = m_3' - u_{11}$, $u_0' = m_2' - u_{10} - u_{11} \cdot u_1'$	
7	$v' = x^3 + v_1' x + v_0'$	3M
	$\underline{w}_1 = u_1' \cdot (s_1 + 2)$, $\underline{w}_0 = u_0' \cdot (l_2 - \underline{w}_1)$	
	$v_1' = (u_0' + u_1') \cdot (s_1 + 2 - \underline{w}_1 + l_2) - v_{21} - l_1 - \underline{w}_0 - \underline{w}_1$	
	$v_0' = \underline{w}_0 - v_{20} - l_0$	
Total		1I, 2S, 26M

two degree 0 divisor classes consist of four points on the curve which are different from each other and their opposites. This situation occurs precisely when $\deg(u_1) = \deg(u_2) = 2$ and u_1, u_2 are relatively prime. In the rare cases when u_1 or u_2 has degree less than 2, or when u_1 and u_2 are not relatively prime, the costs are considerably less than the general case. Here, we present addition for the general case; the special cases will be presented in the full version of the paper.

To optimize the computations, we do not follow Cantor's algorithm literally; we proceed instead as described in [10]. Given two degree 0 divisor classes $[u_1, v_1]$ and $[u_2, v_2]$, the algorithm for divisor addition $[u', v'] = [u_1, v_1] + [u_2, v_2]$ is found by calculating the following subexpressions.

$r = $ resultant of u_1, u_2 $\qquad\qquad$ $inv \equiv r(u_2)^{-1} \pmod{u_1}$

$s' \equiv (v_1 - v_2) \cdot inv \pmod{u_1}$ $\qquad$ $s = \frac{1}{r} \cdot s'$

$k = \frac{f - v_2^2}{u_2}$ $\qquad\qquad\qquad\qquad$ $l = s \cdot u_2$

$m = k - s \cdot (l + 2v_2)$ $\qquad\qquad$ $m' = m/m_4 = m$ made monic

$u' = m'/u_1$ $\qquad\qquad\qquad\qquad$ $v' = H(y) - [(H(y) + v_2 + l) \pmod{u'}]$

The explicit formulas are presented in Table 2.

1. Step 1 and Step 2 calculate the coefficients of $s = s_1 x + s_0 = (v_1 - v_2) \cdot (u_2)^{-1}$ mod u_1. Instead of calculating s_1 and s_0, we calculate $s_1' = r \cdot s_1$ and $s_0' = r \cdot s_0$, thereby postponing the inversion until Step 4.
2. If $s_1' = 0$ in Step 2, one needs to modify Step 4 through Step 7. In this special case, the sum of the two divisors will be a degree one divisor. As this case only occurs very rarely, we do not describe the required modifications here, rather, they will appear in the full version of the paper.
3. The composition of divisors in Step 1 and Step 2 is exactly the same for both real and imaginary cases. These two steps can replace the first three steps of imaginary divisor addition found in [10] for a savings of one squaring. For reference, this improvement makes 1I, 2S, 22M the least known number of field operations needed for divisor addition in the imaginary case.

3.3 Doubling Formulas

Let $[u, v] = [x^2 + u_1 x + u_0, x^3 + v_1 x + v_0]$ be a degree two divisor in reduced basis with both points of the divisor not equal to their opposites. Again following [10], we compute the degree 0 divisor class $[u', v'] := [u, v] + [u, v]$ as follows.

$r = $ resultant of u and $\tilde{v}$

$\tilde{v} \equiv 2v \pmod{u}$ $\qquad\qquad\qquad$ $inv \equiv r(\tilde{v})^{-1} \pmod{u}$

$k = \frac{f - v^2}{u}$ $\qquad\qquad\qquad\qquad$ $s' \equiv k \cdot inv \pmod{u}$

$s = \frac{1}{r} \cdot s'$ $\qquad\qquad\qquad\qquad$ $\tilde{u} = s^2 + \frac{2vs - k}{u}$

$u' = \tilde{u}$ made monic $\qquad\qquad$ $v' = H(y) - [(H(y) + s \cdot u + v) \pmod{u'}]$

The resulting explicit formulas are presented in Table 3. The special cases when $s_1' = 0$ in Step 4 and when $\tilde{w}_0 = 0$ in Step 5 need to be handled separately. As these occur only rarely, we do not describe the required modifications here, rather, they will appear in the full version of the paper.

3.4 Summary of Results

The best known results for the imaginary case are found in [10]. As noted earlier, an improvement of one less squaring has been found which applies to the addition formula in the imaginary case (though not in the doubling case). Compared to the imaginary case, the addition formulas for the real case requires four more multiplications in the main case. The doubling formulas require six more

Table 3. Explicit Formulas for Doubling Divisor Classes

	Doubling, Reduced Basis, $\deg u = 2$	
Input	$[u, v], u = x^2 + u_1 x + u_0, v = x^3 + v_1 x + v_0$	
Output	$[u', v'] = 2[u, v] := [u, v] + [u, v]$	
Step	Expression	Operations
1	$\tilde{v} = \tilde{v}_1 x + \tilde{v}_0$	1S, 1M
	$w_1 = u_1^2, \tilde{v}_1 = 2(v_1 + w_1 - u_0), \tilde{v}_0 = 2(v_0 + u_0 \cdot u_1)$	
2	$r = res(\tilde{v}, u), inv = inv_1 x + inv_0$	4M
	$w_2 = u_0 \cdot \tilde{v}_1, w_3 = u_1 \cdot \tilde{v}_1$	
	$inv_1 = \tilde{v}_1, inv_0 = w_3 - \tilde{v}_0$	
	$r = \tilde{v}_0 \cdot inv_0 - w_2 \cdot \tilde{v}_1$	
3	$k' \equiv (f - v^2)/u \pmod{u} = k'_1 x + k'_0:$	1S, 3M
	$k'_2 = f_4 - 2v_1,$	
	$k'_1 = f_3 - 2v_0 - 2k'_2 \cdot u_1,$	
	$k'_0 = f_2 - v_1^2 - k'_1 \cdot u_1 - k'_2 \cdot (w_1 + 2u_0)$	
4	$s' = s'_1 x + s'_0$	4M
	$s'_1 = inv_1 \cdot k'_0 - \tilde{v}_0 \cdot k'_1, s'_0 = inv_0 \cdot k'_0 - w_2 \cdot k'_1$	
5	Inversion, $r^{-1}, s_0, s_1, \tilde{u}_2^{-1}$	I, 2S, 6M
	$r_2 = r^2, \widehat{w}_0 = (s'_1 + r)^2 - r_2(= r^2 \tilde{u}_2), \widehat{w}_1 = (r \cdot \widehat{w}_0)^{-1}$	
	$\widehat{w}_2 = \widehat{w}_0 \cdot \widehat{w}_1(= \frac{1}{r}), \widehat{w}_3 = r \cdot r_2 \cdot \widehat{w}_1(= \frac{1}{\tilde{u}_2})$	
	$s_1 = \widehat{w}_2 \cdot s'_1, s_0 = \widehat{w}_2 \cdot s'_0$	
6	$u' = x^2 + u'_1 x + u'_0$	5M
	$u'_1 = 2\widehat{w}_3 \cdot ((s_0 - u_1) \cdot s_1 + s_0)$	
	$u'_0 = \widehat{w}_3 \cdot ((s_0 - 2u_1) \cdot s_0 + \tilde{v}_1 \cdot s_1 - k'_2)$	
7	$v' = x^3 + v'_1 x + v'_0$	5M
	$z_0 = u'_0 - u_0, z_1 = u'_1 - u_1$	
	$\underline{w}_0 = z_0 \cdot s_0, \underline{w}_1 = z_1 \cdot s_1$	
	$v'_1 = 2u'_0 - v_1 + (s_0 + s_1) \cdot (z_0 + z_1) - \underline{w}_0 - \underline{w}_1 - u'_1 \cdot (2u'_1 + \underline{w}_1)$	
	$v'_0 = \underline{w}_0 - v_0 - u'_0 \cdot (2u'_1 + \underline{w}_1)$	
Total		1I, 4S, 28M

Table 4. Comparison of Operation Counts for Explicit Formulas

	Imaginary	Real
Baby Step	NA	1I, 2S, 4M
Addition	1I, 2S, 22M [10]	1I, 2S, 26M
Doubling	1I, 5S, 22M [10]	1I, 4S, 28M

multiplications but one less squaring than the imaginary case. It is worth noting that the baby step operation is the cheapest of all, and that there is no analogue for this operation in the imaginary case. Table 4 summarizes the comparison.

The main obstruction from getting more competitive formulas in the real case is the extra coefficient interfering with the inversion step. In the imaginary case, the leading coefficient of the new u is simply s_1^2, which allows one to simplify both addition and doubling formulas. In the real case, we found that computing

s_0 and s_1 explicitly was the most efficient way to compute addition and doubling of divisors.

4   Numerical Results

As cryptographic applications were one of our motivations for developing explicit formulas for divisor arithmetic on genus 2 real hyperelliptic curves, we have implemented key exchange protocols in the imaginary and real models in order to determine whether the real model can be competitive with the imaginary model in terms of efficiency. In the imaginary case, the main operation is scalar multiplication using a non-adjacent form (NAF) expansion of the multiplier, which we will refer to as SCALAR-MULT. In the real case, there are two variations of scalar multiplication described in [7] that comprise the key exchange protocol. Algorithm VAR-DIST2 is a variation of NAF-based scalar multiplication using only degree 0 divisor class doubling and baby steps, whereas Algorithm FIXED-DIST2 generalizes the usual NAF-based scalar multiplication algorithm. The costs of these each of these algorithms in terms of divisor class additions, doublings, and baby steps, assuming that the NAF representation of the corresponding scalar multiplier has $l + 1$ bits, is recalled from [7] in Table 5. All three

Table 5. Operation counts for scalar multiplication in $\mathcal{R}$

	Doubles	Adds	Baby Steps
Imaginary (SCALAR-MULT)	l	$l/3$	-
Real, Variable Distance (VAR-DIST2)	l	$l/3$	d
Real, Fixed Distance (FIXED-DIST2)	l	1	$l/3$

of these algorithms were implemented, using the explicit formulas from [10] for the imaginary case and the formulas in this paper for the real case.

We used the computer algebra library NTL [19] for finite field and polynomial arithmetic and the GNU C++ compiler version 3.4.3. The computations described below were performed on a Pentium IV 2.4 GHz computer running Linux. Although faster absolute times could be obtained using customized implementations of finite field arithmetic, our goal was to compare the relative performance of algorithms in the imaginary and real settings using exactly the same finite fields as opposed to producing the fastest times possible. Thus, NTL was sufficient for our purposes.

All three algorithms were implemented using curves defined over prime finite fields $\mathbb{F}_p$ where $p > 3$. We ran numerous examples of the three scalar multiplication algorithms using curves with genus 2 where the underlying finite field was chosen so that the size of $J(K)$, and hence the set $\mathcal{R}$, was roughly 2^{160}, 2^{224}, 2^{256}, 2^{384}, and 2^{512}. Note that $|J(K)| \approx p^2$ in this case, and that most likely $|\mathcal{R}| = |J(K)|$ for a randomly-chosen curve. Thus, curves offer 80, 112, 128, 192, and 256 bits of security for cryptographic protocols based on the corresponding DLP. NIST [14] currently recommends these five levels of security for key establishment in U.S. Government applications.

For the finite field, we chose a random prime p of appropriate length such that p^2 had the required bit length. For each finite field, we randomly selected 5000 curves and executed Diffie-Hellman key exchange once for each curve. Thus, we ran 10000 instances of Algorithm SCALAR-MULT (two instances for each participant using each curve) and 5000 instances each of Algorithm FIXED-DIST2 and VAR-DIST2 (one instance of each algorithm per participant using each curve). The random exponents used had $160, 224, 256, 384$, and 512 bits, respectively, ensuring that the number of bits of security provided corresponds to the five levels recommended by NIST (again, considering only generic attacks). In order to provide a fair comparison between the three algorithms, the same sequence of random exponents was used for each run of the key exchange protocol.

Table 6 contains the average CPU time in seconds for each of the three algorithms. The times required to generate domain parameters required for our real hyperelliptic curve protocols (see [7]), are not included in these timings, as domain parameter generation is a one-time computation that is performed when the public keys are generated. The time for Algorithm SCALAR-MULT is denoted by "Imag," the time for Algorithm FIXED-DIST2 by "Fixed" and that for Algorithm VAR-DIST2 by "Var." We also list the times required to execute Diffie-Hellman key exchange using both real and imaginary models. Note that in the imaginary case this amounts to two executions of Algorithm SCALAR-MULT, and in the real case one execution of VAR-DIST2 and one of FIXED-DIST2. The run-times achieved using the real model are slower than those using the imaginary model, but they are certainly close.

Table 6. Scalar multiplication and key exchange timings over $\mathbb{F}_p$ (in seconds)

Security Level (bits)	Imag	Fixed	Var	DH Imag	DH Real
80	0.0048	0.0050	0.0056	0.0097	0.0106
112	0.0083	0.0085	0.0096	0.0166	0.0180
128	0.0103	0.0106	0.0117	0.0206	0.0223
192	0.0220	0.0230	0.0256	0.0442	0.0485
256	0.0403	0.0411	0.0452	0.0806	0.0863

5   Conclusions

The formulas presented in this paper are the first explicit formulas for divisor arithmetic on a real hyperelliptic curve. Although they are a few field multiplications slower than their imaginary counterparts, they will certainly out-perform a generic implementation of Cantor's algorithm and will be useful for any computational tasks in the class group or infrastructure. Unfortunately cryptographic protocols using our formulas in the real model are also slower than those using the imaginary case, even with the improved protocols described in [7] in which many divisor additions are traded for significantly faster baby steps. Nevertheless, we hope the fact that we can achieve run times close to those in the imaginary case will increase interest in cryptographic protocols in this setting.

There is still much work to be done on this topic. As mentioned earlier, formulas for degree 0 divisor class arithmetic that work for the general form of the curve equation and any finite field, including characteristic 2, will be presented in the full version of this paper. As in [10], there are certain special cases that can arise in the formulas, for example, the polynomial s may have degree 1 instead of degree 2. As in the imaginary setting, this can be exploited to simplify the formulas; these cases will also be dealt with in the full version of the paper.

We continue to look for improvements to the formulas presented here. Reducing the number of field multiplications required for addition and doubling by only two or three would likely result in the cryptographic protocols in the real setting being slightly faster than the imaginary case. Another possible improvement that would improve the performance of the protocols in the real setting is compound operations. In particular, compounding the doubling and baby step operations will almost certainly save a few multiplication and require would likely require only one inversion (as opposed to two) as compared to performing them separately. This would improve the speed of the VAR-DIST2 scalar multiplication algorithm (doubling and baby steps) from [7].

Finally, a great deal of work has been done on explicit formulas in the imaginary setting including using projective coordinates to obtain inversion-free formulas, formulas for genus 3 and 4, explicit formulas via theta functions, and explicit formulas via NUCOMP. All of these topics are work in progress.

Acknowledgements. This paper is an outcome of a research project proposed at the RMMC Summer School in Computational Number Theory and Cryptography which was held at the University of Wyoming in 2006. We would like to thank the following sponsors for their support: the University of Wyoming, the National Science Foundation (grant DMS-0612103), the Rocky Mountain Mathematics Consortium, The Number Theory Foundation, The Institute for Mathematics and its Applications (IMA), The Fields Institute, The Centre for Information Security and Cryptography (CISaC) and iCORE of Canada.

References

1. Avanzi, R.M.: Aspects of hyperelliptic curves over large prime fields in software implementations. In: Joye, M., Quisquater, J.-J. (eds.) CHES 2004. LNCS, vol. 3156, pp. 148–162. Springer, Heidelberg (2004)
2. Cohen, H., Frey, G. (eds.): Handbook of Elliptic and Hyperelliptic Curve Cryptography. Discrete Mathematics and Its Applications, vol. 34. Chapman & Hall/CRC, Sydney, Australia (2005)
3. Enge, A.: How to distinguish hyperelliptic curves in even characteristic. In: Alster, K., Urbanowicz, J., Williams, H.C., (eds.) Public-Key Cryptography and Computational Number Theory, pp. 49–58, De Gruyter, Berlin (2001)

4. Gaudry, P.: On breaking the discrete log on hyperelliptic curves. In: Preneel, B. (ed.) EUROCRYPT 2000. LNCS, vol. 1807, pp. 19–34. Springer, Heidelberg (2000)

5. Gaudry, P., Thomé, E., Thériault, N., Diem, C.: A double large prime variation for small genus hyperelliptic index calculus. Mathematics of Computation 76, 475–492 (2007)

6. Jacobson Jr., M.J., Menezes, A.J., Stein, A.: Hyperelliptic curves and cryptography. In: High Primes and Misdemeanours: lectures in honour of the 60th birthday of Hugh Cowie Williams. Fields Institute Communications Series, vol. 41, pp. 255–282. American Mathematical Society (2004)

7. Jacobson Jr., M.J., Scheidler, R., Stein, A.: Cryptographic protocols on real and imaginary hyperelliptic curves. Accepted to Advances in Mathematics of Communications pending revisions (2007)

8. Jacobson Jr., M.J., Scheidler, R., Stein, A.: Fast Arithmetic on Hyperelliptic Curves Via Continued Fraction Expansions. To appear in Advances in Coding Theory and Cryptology. In: Shaaska, T., Huffman, W.C., Joyner, D., Ustimenko, V. (eds.) Series on Coding, Theory and Cryptology, vol. 2, World Scientific Publishing (2007)

9. Koblitz, N.: Hyperelliptic cryptosystems. Journal of Cryptology 1, 139–150 (1988)

10. Lange, T.: Formulae for arithmetic on genus 2 hyperelliptic curves. Applicable Algebra in Engineering, Communication, and Computing 15, 295–328 (2005)

11. Menezes, A.J., Wu, Y., Zuccherato, R.J.: An elementary introduction to hyperelliptic curves. Technical Report CORR 96-19, Department of Combinatorics and Optimization, University of Waterloo, Waterloo, Ontario, 1996. In: Koblitz, N. (ed.) Algebraic Aspects of Cryptography, Springer-Verlag, Berlin Heidelberg New York (1998)

12. Müller, V., Stein, A., Thiel, C.: Computing discrete logarithms in real quadratic congruence function fields of large genus. Mathematics of Computation 68, 807–822 (1999)

13. Mumford, D.: Tata Lectures on Theta I, II. Birkhäuser, Boston (1983/84)

14. National Institute of Standards and Technology (NIST). Recommendation on key establishment schemes. NIST Special Publication 800-56 (January 2003)

15. Paulus, S., Rück, H.-G.: Real and imaginary quadratic representations of hyperelliptic function fields. Mathematics of Computation 68, 1233–1241 (1999)

16. Pelzl, J., Wollinger, T., Paar, C.: Low cost security: explicit formulae for genus-4 hyperelliptic curves. In: Matsui, M., Zuccherato, R.J. (eds.) SAC 2003. LNCS, vol. 3006, pp. 1–16. Springer, Heidelberg (2003)

17. Scheidler, R.: Cryptography in quadratic function fields. Designs, Codes and Cryptography 22, 239–264 (2001)

18. Scheidler, R., Stein, A., Williams, H.C.: Key-exchange in real quadratic congruence function fields. Designs, Codes and Cryptography 7, 153–174 (1996)

19. V. Shoup. NTL: A library for doing number theory. Software (2001) See http://www.shoup.net/ntl.

20. Stein, A.: Sharp upper bounds for arithmetics in hyperelliptic function fields. Journal of the Ramanujan Mathematical Society 9-16(2), 1–86 (2001)

21. Wollinger, T., Pelzl, J., Paar, C.: Cantor versus Harley: optimization and analysis of explicit formulae for hyperelliptic curve cryptosystems. IEEE Transactions on Computers 54, 861–872 (2005)

A   Divisor Addition

To perform divisor addition, we compute the following expressions, then show that these formulas give the desired result.

$$s \equiv (v_1 - v_2) \cdot (u_2)^{-1} \pmod{u_1} \qquad l = s \cdot u_2$$

$$k = \frac{f - v_2^2}{u_2}$$

$$m = k - s \cdot (l + 2v_2) \qquad\qquad m' = m/m_4 = \ m \text{ made monic}$$

$$u' = m'/u_1 \qquad\qquad\qquad v' = H(y) - [(H(y) + v_2 + l) \pmod{u'}] \ .$$

Let (u_1, v_1) and (u_2, v_2) be two reduced divisors written in the Mumford representation. Assume u_1 and u_2 are both degree 2 and are relatively prime. The composition step of Cantor's Algorithm is given by

$$U_0 = u_1 u_2$$
$$V_0 \equiv v_2 + s u_2 = v_2 + l \pmod{U_0}$$

where $s \equiv u_2^{-1}(v_1 - v_2) \pmod{u_1}$ and $l = su_2$. The reduction step can be expressed as

$$V_1 = -V_0 + \left\lfloor \frac{V_0 + H(y)}{U_0} \right\rfloor \cdot U_0$$

$$U_1 = \frac{f - V_1^2}{U_0}$$

where $H(y)$ is the principal part of a root of the equation $y^2 = f(x)$. Since V_0 and d both have degree 3 and U_0 has degree 4, $\left\lfloor \frac{V_0 + H(y)}{U_0} \right\rfloor = 0$, and so

$$V_1 = -V_0 = -(v_2 + l)$$

Plugging this into the formula U_1 yields

$$U_1 = \frac{f - (v_2 + l)^2}{u_1 u_2}$$
$$= \frac{f - v_2^2 - l^2 - 2v_2 l}{u_1 u_2}$$
$$= \frac{1}{u_1}\left(\frac{f - v_2^2}{u_2} - \frac{l(l + 2v_2)}{u_2} \right)$$
$$= \frac{k - s(l + 2v_2)}{u_1}$$

where $k = \frac{f - v_2^2}{u_2}$.

The final output is $[u', v']$ transformed to reduced basis, i.e., $u' = U_1$ made monic, and $v' = H(y) - [(H(y) - V_1) \pmod{u'}]$. In the formulas, we first find $m = k - s(l + 2v_2)$, find the leading coefficient and compute its inverse, then compute $m' = m$ made monic.

B   Divisor Doubling

To perform divisor doubling, we compute the following expressions, then show that these formulas give the desired result.

$$k = \frac{f-v^2}{u} \qquad\qquad\qquad s \equiv k \cdot (2v)^{-1} \ (\mathrm{mod}\ u)$$

$$\tilde{u} = s^2 + \frac{(2v)\cdot s - k}{u} \qquad\qquad u' = \tilde{u}\ \text{made monic}$$

$$v' = H(y) - [(H(y) + s \cdot u + v)\ (\mathrm{mod}\ u')]$$

Let $(u, v) = (x^2 + u_1 x + u_0, x^3 + v_1 x + v_0)$ be a degree two reduced basis Mumford representation with both points of the divisor are not equal to their opposites. Then Cantor's Algorithm for doubling the divisor (u, v) must result in (U_1, V_1) such that

$$U_0 = u^2$$
$$V_0 \equiv v \ (\mathrm{mod}\ u)$$
$$(V_0 = v + su\ \text{for some}\ s)$$
$$V_1 = -V_0 + \left\lfloor \frac{V_0 + H(y)}{U_0} \right\rfloor U_0$$
$$U_1 = \frac{f - V_1^2}{U_0}$$

Here, s is chosen such that U_0 divides $V_0^2 - f$. Again, $\left\lfloor \frac{V_0 + H(y)}{U_0} \right\rfloor$ is zero since U_0 has degree 4 and $V_0 + H(y)$ has degree 3. Hence, $V_1 = -V_0 = -v - su$ and

$$U_1 = \frac{f - (-v - su)^2}{u^2}$$
$$= \frac{f - v^2 - 2vsu - s^2 u^2}{u^2}$$
$$= \frac{1}{u}\left(\frac{f - v^2}{u} - 2vs\right) - s^2$$
$$= \frac{1}{u}(k - 2vs) - s^2$$

where the division in $k = (f - v^2)/u$ is exact. To ensure that the division of $k - 2vs$ by u is exact, we choose $s \equiv -k \cdot (-2v)^{-1} \ (\mathrm{mod}\ u)$, and obtain

$$k - 2vs \equiv k + 2v \cdot k \cdot (-2v)^{-1} \equiv 0 \ (\mathrm{mod}\ u) \ .$$

Finally, U_1 will be made monic, to arrive at

$$u' = s^2 + \frac{2vs - k}{u} \ \text{made monic}$$
$$v' = H(y) - [(H(y) + v + su)\ \mathrm{mod}\ u']$$

where $[u', v']$ is in reduced basis.

The Quadratic Extension Extractor for (Hyper)Elliptic Curves in Odd Characteristic

Reza Rezaeian Farashahi[1,2] and Ruud Pellikaan[1]

[1] Dept. of Mathematics and Computer Science, TU Eindhoven,
P.O. Box 513, 5600 MB Eindhoven, The Netherlands
[2] Dept. of Mathematical Sciences, Isfahan University of Technology,
P.O. Box 85145 Isfahan, Iran
{r.rezaeian, g.r.pellikaan}@tue.nl

Abstract. We propose a simple and efficient deterministic extractor for the (hyper)elliptic curve $\mathcal{C}$, defined over $\mathbb{F}_{q^2}$, where q is some power of an odd prime. Our extractor, for a given point P on $\mathcal{C}$, outputs the first $\mathbb{F}_q$-coefficient of the abscissa of the point P. We show that if a point P is chosen uniformly at random in $\mathcal{C}$, the element extracted from the point P is indistinguishable from a uniformly random variable in $\mathbb{F}_q$.

Keywords: Elliptic curve, Hyperelliptic curve, Deterministic extractor.

1 Introduction

A deterministic extractor for a curve is a function that converts a random point on the curve to a bit-string of fixed length that is statistically close to uniformly random. Let $\mathcal{C}$ be an absolutely irreducible nonsingular affine curve that is defined over $\mathbb{F}_{q^2}$, where $q = p^k$, for some odd prime p and positive integer k, by the equation $\mathbf{y}^2 = f(\mathbf{x})$, where the degree of f is an odd number d. In this paper, we propose a simple and efficient deterministic extractor, called Ext, for $\mathcal{C}$. Let $\{\alpha_0, \alpha_1\}$ be a basis of $\mathbb{F}_{q^2}$ over $\mathbb{F}_q$. The extractor Ext, for a given point P on $\mathcal{C}$, outputs the *first* $\mathbb{F}_q$-coefficient of the abscissa of the point P. Similarly one could define an extractor that, for a given point on the curve, outputs a $\mathbb{F}_q$-linear combination of $\mathbb{F}_q$-coordinates of the abscissa of the point. Provided that the point P is chosen uniformly at random in $\mathcal{C}$, the element extracted from the point P is indistinguishable from a uniformly random variable in $\mathbb{F}_q$.

Gürel [7] proposed an extractor for an elliptic curve E defined over a quadratic extension of a prime field. Given a point P on $E(\mathbb{F}_{p^2})$, it extracts half of the bits of the abscissa of P. Provided that the point P is chosen uniformly at random, the statistical distance between the bits extracted from the point P and uniformly random bits is shown to be negligible [7]. We recall this extractor for E in Subsection 5.2 and we improve that result in Theorem 3. The definition of our extractor is similar, yet more general. Our extractor Ext is defined for $\mathcal{C}$.

The problem of converting random points of an elliptic curve into random bits has several cryptographic applications. Such applications are key derivation

C. Carlet and B. Sunar (Eds.): WAIFI 2007, LNCS 4547, pp. 219–236, 2007.

functions, design of cryptographically secure pseudorandom number generators and a class of key exchange protocols based on elliptic curves (e.g, the well-known Elliptic Curve Diffie-Hellman protocol). By the end of the Elliptic Curve Diffie-Hellman protocol, the parties agree on a common secret element of the group, which is indistinguishable from a uniformly random element under the decisional Diffie-Hellman assumption (denoted by DDH). However the binary representation of the common secret element is *distinguishable* from a uniformly random bit-string of the same length. Hence one has to convert this group element into a random-looking bit-string. This can be done using a deterministic extractor.

Kaliski [11] shows that if a point is taken uniformly at random from the union of an elliptic curve and its quadratic twist then the abscissa of this point is uniformly distributed in the finite field. Then Chevassut et al. [3] proposed the TAU technique. This technique allows to extract almost all the bits of the abscissa of a point of the union of an elliptic curve and its quadratic twist. Recently Farashahi et al. [5] proposed two extractors for ordinary elliptic curve E, defined over $\mathbb{F}_{2^N}$, where $N = 2\ell$ and ℓ is a positive integer. For a given point P on E, the first extractor outputs the first $\mathbb{F}_{2^\ell}$-coefficient of the abscissa of P while the second outputs the second $\mathbb{F}_{2^\ell}$-coefficient. They also propose two deterministic extractors for the main subgroup G of E, where E has minimal 2-torsion. If a point P is chosen uniformly at random in G, the bits extracted from the point P are indistinguishable from a uniformly random bit-string of length ℓ.

Sequences of x-coordinates of pseudorandom points on elliptic curves have been studied in [9,12,13,17]. On the other hand, the x-coordinate of a uniformly random point on an elliptic curve can be easily distinguished from uniformly random field element since only about 50% of all field elements are x-coordinates of points of the curve. Our extractors provide only part of the x-coordinate and thereby avoid the obvious problem; the proof shows that actual uniformity is achieved. Our approach is somewhat similar to the basic idea of pseudorandom generators proposed by Gong et al. [6] and Beelen and Doumen [2] in that they use a function that maps the set of points on elliptic curve to a set of smaller cardinality. Our aim is to extract as many bits as possible while keeping the output distribution statistically close to uniform.

We organize the paper as follows. In the next section we introduce some notations and recall some basic definitions. In Section 3, we define an affine variety $\mathcal{A}$ of dimension 2 in $\mathbb{A}^3_{\mathbb{F}_q}$ related to the affine curve $\mathcal{C}$. We show that there exists a bijection between $\mathcal{C}(\mathbb{F}_{q^2})$ and $\mathcal{A}(\mathbb{F}_q)$. Then in Section 4 we propose the extractor Ext for $\mathcal{C}$ as $\mathrm{Ext}(x,y) = x_0$, where $x = x_0\alpha_0 + x_1\alpha_1$. We show that the output of this extractor, for a given uniformly random point of $\mathcal{C}$, is statistically close to a uniformly random variable in $\mathbb{F}_q$. To show the latter we give bounds on the number of preimages $\mathrm{Ext}^{-1}(x_0)$, where $x_0 \in \mathbb{F}_q$. In fact, by using the bijection between $\mathcal{C}(\mathbb{F}_{q^2})$ and $\mathcal{A}(\mathbb{F}_q)$, we give the estimate for the number of $\mathbb{F}_q$-rational points on the intersection of $\mathcal{A}$ and the hyperplane $\mathbf{x}_0 = x_0$ in $\mathbb{A}^3_{\mathbb{F}_q}$. We show that for almost all values of x_0 in $\mathbb{F}_q$, this intersection is an absolutely irreducible nonsingular curve. Actually this problem is a special case of Bertini theorems. The classical Bertini theorems say that if an algebraic subvariety $\mathcal{X}$ of $\mathbb{P}^n$ has a certain property, then for a sufficiently general hyperplane $H \subseteq \mathbb{P}^n$, the intersection $H \cap \mathcal{X}$ has the same property (see [8,15]). Then we give two examples in Section 5. We conclude our result in Section 6.

2 Preliminaries

Let us introduce the notations and recall the basic definitions that are used throughout the paper.

Notation. Denote by $\mathbb{Z}_n$ the set of nonnegative integers less than n. A field is denoted by $\mathbb{F}$ and its algebraic closure by $\overline{\mathbb{F}}$. Denote by $\mathbb{F}^*$ the set of nonzero elements of $\mathbb{F}$. The finite field with q elements is denoted by $\mathbb{F}_q$, and its algebraic closure by $\overline{\mathbb{F}}_q$. Let C be a curve defined over $\mathbb{F}_q$, then the set of $\mathbb{F}_q$-rational points on C is denoted by $C(\mathbb{F}_q)$. The cardinality of a finite set S is denoted by $\#S$. We make a distinction between a variable $\mathbf{x}$ and a specific value x in $\mathbb{F}$.

2.1 Finite Field Notation

Consider the finite fields $\mathbb{F}_q$ and $\mathbb{F}_{q^2}$, where $q = p^k$, for some odd prime number p and positive integer k. Then $\mathbb{F}_{q^2}$ is a two dimensional vector space over $\mathbb{F}_q$. Let $\{\alpha_0, \alpha_1\}$ be a basis of $\mathbb{F}_{q^2}$ over $\mathbb{F}_q$. That means every element x in $\mathbb{F}_{q^2}$ can be represented in the form $x = x_0\alpha_0 + x_1\alpha_1$, where x_0 and x_1 are in $\mathbb{F}_q$. We recall that $\{\alpha_0, \alpha_1\}$ is a basis of $\mathbb{F}_{q^2}$ over $\mathbb{F}_q$ if and only if

$$\begin{vmatrix} \alpha_0 & \alpha_1 \\ \alpha_0^q & \alpha_1^q \end{vmatrix} \neq 0.$$

That is equivalent to $\alpha_0, \alpha_1 \in \mathbb{F}_{q^2}^*$ and $\alpha_0^{q-1} \neq \alpha_1^{q-1}$.

Let $\phi : \overline{\mathbb{F}}_q \longrightarrow \overline{\mathbb{F}}_q$ be the Frobenius map defined by $\phi(x) = x^q$. Let $\mathrm{Tr} : \mathbb{F}_{q^2} \longrightarrow \mathbb{F}_q$ be the *trace* function. Then $\mathrm{Tr}(x) = x + \phi(x)$, for $x \in \mathbb{F}_{q^2}$. Let $\mathrm{N} : \mathbb{F}_{q^2} \longrightarrow \mathbb{F}_q$ be the *norm* function. Then $\mathrm{N}(x) = x\phi(x)$, for $x \in \mathbb{F}_{q^2}$.

Remark 1. Let α be a primitive element of $\mathbb{F}_{q^2}$. So every $x \in \mathbb{F}_{q^2}^*$ is a power of α. Then $\mathrm{N}(\alpha)$ is a primitive element of $\mathbb{F}_q$. Let $x \in \mathbb{F}_{q^2}^*$. Then x is a square in $\mathbb{F}_{q^2}$ if and only if $x = \alpha^{2i}$, for some integer i. Similarly $\mathrm{N}(x)$ is a square in $\mathbb{F}_q$ if and only if $\mathrm{N}(x) = (\mathrm{N}(\alpha))^{2i}$, for some integer i. Furthermore $x = \alpha^{2i}$, for some integer i, if and only if $\mathrm{N}(x) = (\mathrm{N}(\alpha))^{2j}$, for some integer j. Obviously $\mathrm{N}(0) = 0$. Therefor x is a square in $\mathbb{F}_{q^2}$ if and only if $\mathrm{N}(x)$ is a square in $\mathbb{F}_q$.

2.2 Hyperelliptic Curves

Definition 1. *An absolutely irreducible nonsingular curve C of genus at least 2 is called* hyperelliptic *if there exists a morphism of degree 2 from C to the projective line.*

Theorem 1. *Let C be a hyperelliptic curve of genus g over $\mathbb{F}_q$, where q is odd. Then C has a plane model of the form*

$$\mathbf{y}^2 = f(\mathbf{x}),$$

where f is a square free polynomial and $2g + 1 \leq \deg(f) \leq 2g + 2$. The plane model is singular at infinity. If $\deg(f) = 2g+1$ then the point at infinity ramifies and C has only one point at infinity. If $\deg(f) = 2g + 2$ then C has zero or two $\mathbb{F}_q$-rational points at infinity.

Proof. See [1,4].

2.3   Deterministic Extractor

In our analysis we use the notion of a deterministic extractor, so let us recall it briefly. For general definition of extractors we refer to [16,18].

Definition 2. *Let X and Y be S-valued random variables, where S is a finite set. Then the statistical distance $\Delta(X, Y)$ of X and Y is*

$$\Delta(X, Y) = \tfrac{1}{2} \sum_{s \in S} |\Pr[X = s] - \Pr[Y = s]|.$$

Let U_S denote a random variable uniformly distributed on S. We say that a random variable X on S is δ-uniform, if $\Delta(X, U_S) \leq \delta$.

Note that if the random variable X is δ-uniform, then no algorithm can distinguish X from U_S with advantage larger than δ, that is, for all algorithms $D : S \longrightarrow \{0, 1\}$

$$|\Pr[D(X) = 1] - \Pr[D(U_S) = 1]| \leq \delta.$$

See [14].

Definition 3. *Let S, T be finite sets. Consider the function $\mathrm{Ext} : S \longrightarrow T$. We say that Ext is a deterministic (T, δ)-extractor for S if $\mathrm{Ext}(U_S)$ is δ-uniform on T. That means*

$$\Delta(\mathrm{Ext}(U_S), U_T) \leq \delta.$$

In the case that $T = \{0, 1\}^k$, we say Ext is a δ-deterministic extractor for S.

In this paper we consider deterministic $(\mathbb{F}_q, \delta)$-extractors. Observe that, converting random elements of $\mathbb{F}_q$ into random bit strings is a relatively easy problem. For instance, one can represent an element of $\mathbb{F}_q$ by a number in $\mathbb{Z}_q$ and use Algorithm Q_2 from [10], which was presented without analysis. It can actually be shown, however, that Algorithm Q_2 produces on average $n - 2$ bits given a uniformly distributed random number $U_{\mathbb{Z}_q}$, where n denotes the bit length of q.

Furthermore, if q is close to a power of 2, that is, $0 \leq (2^n - q)/2^n \leq \delta$ for a small δ, then the uniform element $U_{\mathbb{F}_q}$ is statistically close to n uniformly random bits.

The following simple lemma is a well-known result (the proof can be found, for instance, in [3]).

Lemma 1. *Under the condition that $0 \leq (2^n - q)/2^n \leq \delta$, the statistical distance between $U_{\mathbb{F}_q}$ and U_{2^n} is bounded from above by δ.*

3   Norm Variety

Consider an absolutely irreducible nonsingular affine curve $\mathcal{C}$ defined over $\mathbb{F}_{q^2}$. We define an affine variety $\mathcal{A}$ in $\mathbb{A}^3_{\mathbb{F}_q}$ from the curve $\mathcal{C}$. Then we show that the number of $\mathbb{F}_{q^2}$-rational points on the affine curve $\mathcal{C}$ equals the number of $\mathbb{F}_q$-rational points on the affine variety $\mathcal{A}$.

From now on, let $\mathcal{C}$ be an absolutely irreducible nonsingular affine curve that is defined over $\mathbb{F}_{q^2}$ by the equation

$$\mathbf{y}^2 = f(\mathbf{x}), \tag{1}$$

where $f(\mathbf{x}) \in \mathbb{F}_{q^2}[\mathbf{x}]$ is a monic square-free polynomial of odd degree d. Let

$$f(\mathbf{x}) = \mathbf{x}^d + \sum_{i=0}^{d-1} e_i \mathbf{x}^i = \prod_{i=1}^{d}(\mathbf{x} - \lambda_i), \tag{2}$$

where $e_i \in \mathbb{F}_{q^2}$ and $\lambda_i \in \overline{\mathbb{F}}_q$. Then $\lambda_i \neq \lambda_j$, for $i \neq j$, since $f(\mathbf{x})$ is square-free.

Define the variables $\mathbf{x}_0, \mathbf{x}_1$ by $\mathbf{x} = \mathbf{x}_0\alpha_0 + \mathbf{x}_1\alpha_1$. Then there exist two bivariate functions $f_0, f_1 \in \mathbb{F}_q[\mathbf{x}_0, \mathbf{x}_1]$, so that

$$f(\mathbf{x}) = f(\mathbf{x}_0\alpha_0 + \mathbf{x}_1\alpha_1) = f_0(\mathbf{x}_0, \mathbf{x}_1)\alpha_0 + f_1(\mathbf{x}_0, \mathbf{x}_1)\alpha_1. \tag{3}$$

Let $\phi : \overline{\mathbb{F}}_q \longrightarrow \overline{\mathbb{F}}_q$ be the Frobenius map defined by $\phi(x) = x^q$. Define the polynomial

$$\overline{f}(\mathbf{x}) = \mathbf{x}^d + \sum_{i=0}^{d-1} \phi(e_i)\mathbf{x}^i. \tag{4}$$

Define $\overline{\mathbf{x}} = \mathbf{x}_0\phi(\alpha_0) + \mathbf{x}_1\phi(\alpha_1)$. Then

$$\overline{f}(\overline{\mathbf{x}}) = \overline{f}(\mathbf{x}_0\phi(\alpha_0) + \mathbf{x}_1\phi(\alpha_1)) = f_0(\mathbf{x}_0, \mathbf{x}_1)\phi(\alpha_0) + f_1(\mathbf{x}_0, \mathbf{x}_1)\phi(\alpha_1). \tag{5}$$

Define

$$F(\mathbf{x}_0, \mathbf{x}_1) = f(\mathbf{x}_0\alpha_0 + \mathbf{x}_1\alpha_1)\overline{f}(\mathbf{x}_0\phi(\alpha_0) + \mathbf{x}_1\phi(\alpha_1)).$$

Then from equations (3) and (5), we have

$$F(\mathbf{x}_0, \mathbf{x}_1) = (f_0(\mathbf{x}_0, \mathbf{x}_1)\alpha_0 + f_1(\mathbf{x}_0, \mathbf{x}_1)\alpha_1)(f_0(\mathbf{x}_0, \mathbf{x}_1)\phi(\alpha_0) + f_1(\mathbf{x}_0, \mathbf{x}_1)\phi(\alpha_1)).$$

We note that f_0, f_1 are in $\mathbb{F}_q[\mathbf{x}_0, \mathbf{x}_1]$. Also $\alpha_i\phi(\alpha_i) = N(\alpha_i) \in \mathbb{F}_q$, for $i \in \{0, 1\}$. Furthermore $\alpha_0\phi(\alpha_1) + \phi(\alpha_0)\alpha_1 = \text{Tr}(\alpha_0)\text{Tr}(\alpha_1) - \text{Tr}(\alpha_0\alpha_1) \in \mathbb{F}_q$. Hence F is a polynomial in $\mathbb{F}_q[\mathbf{x}_0, \mathbf{x}_1]$.

Proposition 1. *The polynomial F is square-free.*

Proof. The affine curve $\mathcal{C}$ is defined by the equation $\mathbf{y}^2 = f(\mathbf{x}) = \prod_{i=1}^{d}(\mathbf{x} - \lambda_i)$, where $\lambda_i \in \overline{\mathbb{F}}_q$ and $\lambda_i \neq \lambda_j$, for $i \neq j$. Then

$$f(\mathbf{x}_0\alpha_0 + \mathbf{x}_1\alpha_1) = \prod_{i=1}^{d}(\mathbf{x}_0\alpha_0 + \mathbf{x}_1\alpha_1 - \lambda_i). \tag{6}$$

Hence $f(\mathbf{x}_0\alpha_0 + \mathbf{x}_1\alpha_1)$ is a square-free polynomial. Consider the polynomial $\overline{f}(\mathbf{x})$ (see equality (4)). Then $\overline{f}(\mathbf{x}) = \prod_{i=1}^{d}(\mathbf{x} - \phi(\lambda_i))$. Since $\lambda_i \neq \lambda_j$, for $i \neq j$, and ϕ is bijective, so $\phi(\lambda_i) \neq \phi(\lambda_j)$, for $i \neq j$. Hence the polynomial $\overline{f}(\mathbf{x})$ is a square free polynomial. Then

$$\overline{f}(\mathbf{x}_0\phi(\alpha_0) + \mathbf{x}_1\phi(\alpha_1)) = \prod_{i=1}^{d}(\mathbf{x}_0\phi(\alpha_0) + \mathbf{x}_1\phi(\alpha_1) - \phi(\lambda_i)). \tag{7}$$

So $\overline{f}(\mathbf{x}_0\phi(\alpha_0) + \mathbf{x}_1\phi(\alpha_1))$ is a square-free polynomial. Now assume that $f(\mathbf{x}_0\alpha_0 + \mathbf{x}_1\alpha_1)$ and $\overline{f}(\mathbf{x}_0\phi(\alpha_0) + \mathbf{x}_1\phi(\alpha_1))$ have a common factor. Then $\phi(\alpha_0) = \gamma\alpha_0$ and $\phi(\alpha_1) = \gamma\alpha_1$, for some $\gamma \in \mathbb{F}_{q^2}$, which is a contradiction, since $\alpha_0\phi(\alpha_1) \neq \phi(\alpha_0)\alpha_1$ (see Subsection 2.1). Therefore $f(\mathbf{x}_0\alpha_0 + \mathbf{x}_1\alpha_1)$ and $\overline{f}(\mathbf{x}_0\phi(\alpha_0) + \mathbf{x}_1\phi(\alpha_1))$ do not have a common factor. Thus F is a square-free polynomial.

In particular, Proposition 1 shows that the polynomial F is not a square in $\overline{\mathbb{F}}_q[\mathbf{x}_0, \mathbf{x}_1]$. Consider the polynomial $\mathbf{z}^2 - F(\mathbf{x}_0, \mathbf{x}_1)$ in $\mathbb{F}_q[\mathbf{x}_0, \mathbf{x}_1, \mathbf{z}]$. Then this polynomial is absolutely irreducible in $\overline{\mathbb{F}}_q[\mathbf{x}_0, \mathbf{x}_1, \mathbf{z}]$.

Definition 4. *Define the affine variety $\mathcal{A}$ over $\mathbb{F}_q$ by the equation*

$$\mathbf{z}^2 - F(\mathbf{x}_0, \mathbf{x}_1) = 0.$$

The affine variety $\mathcal{A}$ is absolutely irreducible, since the polynomial $\mathbf{z}^2 - F(\mathbf{x}_0, \mathbf{x}_1)$ is absolutely irreducible.

Remark 2. Let $P = (x, y) \in \mathcal{C}(\mathbb{F}_{q^2})$, where $x = x_0\alpha_0 + x_1\alpha_1$ and $x_0, x_1 \in \mathbb{F}_q$. So $y^2 = f(x)$. Then $\phi(y^2) = \phi(f(x)) = \overline{f}(\phi(x)) = \overline{f}(x_0\phi(\alpha_0) + x_1\phi(\alpha_1))$. Let $z = \mathrm{N}(y) = y\phi(y)$. Then

$$z^2 = f(x)\overline{f}(\phi(x)) = f(x_0\alpha_0 + x_1\alpha_1)\overline{f}(x_0\phi(\alpha_0) + x_1\phi(\alpha_1)) = F(x_0, x_1).$$

That means $(x_0, x_1, z) \in \mathcal{A}(\mathbb{F}_q)$.

In Theorem 2, we show that the number of $\mathbb{F}_{q^2}$-rational points on the affine curve $\mathcal{C}$ equals the number of $\mathbb{F}_q$-rational points on the affine variety $\mathcal{A}$. For the proof of Theorem 2, we need several lemmas and a proposition.

Lemma 2. *Define the projection map $\pi_\mathcal{C} : \mathcal{C}(\mathbb{F}_{q^2}) \longrightarrow \mathbb{A}^2(\mathbb{F}_q)$, by*

$$\pi_\mathcal{C}(x, y) = (x_0, x_1),$$

where $x = x_0\alpha_0 + x_1\alpha_1$. Assume that $\pi_\mathcal{C}^{-1}(x_0, x_1) \neq \emptyset$. If $F(x_0, x_1) = 0$, then $\#\pi_\mathcal{C}^{-1}(x_0, x_1) = 1$, otherwise $\#\pi_\mathcal{C}^{-1}(x_0, x_1) = 2$.

Proof. Let $P = (x, y) \in \pi_\mathcal{C}^{-1}(x_0, x_1)$, where $x = x_0\alpha_0 + x_1\alpha_1$. Remark 2 shows that $(\mathrm{N}(y))^2 = F(x_0, x_1)$. So $F(x_0, x_1) = 0$ if and only if $y = 0$. If $y = 0$, then $\pi_\mathcal{C}^{-1}(x_0, x_1) = \{(x, 0)\}$. If $y \neq 0$, then $-P = (x, -y) \in \pi_\mathcal{C}^{-1}(x_0, x_1)$ and $-P \neq P$. Since $P, -P$ are the only points on $\mathcal{C}(\mathbb{F}_{q^2})$, with the fixed first coordinate x, then $\pi_\mathcal{C}^{-1}(x_0, x_1) = \{P, -P\}$.

Lemma 3. *Define the projection map* $\pi_{\mathcal{A}} : \mathcal{A}(\mathbb{F}_q) \longrightarrow \mathbb{A}^2(\mathbb{F}_q)$, *by*

$$\pi_{\mathcal{A}}(x_0, x_1, z) = (x_0, x_1).$$

Assume $\pi_{\mathcal{A}}^{-1}(x_0, x_1) \neq \emptyset$. *If* $F(x_0, x_1) = 0$, *then* $\#\pi_{\mathcal{A}}^{-1}(x_0, x_1) = 1$, *otherwise* $\#\pi_{\mathcal{A}}^{-1}(x_0, x_1) = 2$.

Proof. Let $(x_0, x_1, z) \in \pi_{\mathcal{A}}^{-1}(x_0, x_1)$. Then $z^2 = F(x_0, x_1)$. If $F(x_0, x_1) = 0$, then $z = 0$ and $\pi_{\mathcal{A}}^{-1}(x_0, x_1) = \{(x_0, x_1, 0)\}$. If $F(x_0, x_1) \neq 0$, then (x_0, x_1, z) and $(x_0, x_1, -z)$ are the only points on $\mathcal{A}$, such that they have the first and second coordinates equal x_0 and x_1. Furthermore $z \neq -z$. Therefore in this case $\pi_{\mathcal{A}}^{-1}(x_0, x_1) = \{(x_0, x_1, z), (x_0, x_1, -z)\}$.

Proposition 2. *For all* $x_0, x_1 \in \mathbb{F}_q$, $\#\pi_{\mathcal{C}}^{-1}(x_0, x_1) = \#\pi_{\mathcal{A}}^{-1}(x_0, x_1)$.

Proof. First assume that $\pi_{\mathcal{C}}^{-1}(x_0, x_1) \neq \emptyset$. Then there exists a point (x, y) on $\mathcal{C}(\mathbb{F}_{q^2})$, such that $x = x_0\alpha_0 + x_1\alpha_1$. Let $z = N(y)$. Then Remark 2 shows that $(x_0, x_1, z) \in \mathcal{A}(\mathbb{F}_q)$. Therefore $(x_0, x_1, z) \in \pi_{\mathcal{A}}^{-1}(x_0, x_1)$ and $\pi_{\mathcal{A}}^{-1}(x_0, x_1) \neq \emptyset$.

Second assume that $\pi_{\mathcal{A}}^{-1}(x_0, x_1) \neq \emptyset$. Then there exists a point (x_0, x_1, z) on $\mathcal{A}(\mathbb{F}_q)$. Thus $z^2 = F(x_0, x_1)$. Let $x = x_0\alpha_0 + x_1\alpha_1$. Then from Remark 2, $z^2 = f(x)\phi(f(x)) = N(f(x))$. So $N(f(x))$ is a square in $\mathbb{F}_q$. Remark 1 implies $f(x)$ is a square in $\mathbb{F}_{q^2}$. Let $y^2 = f(x)$, where $y \in \mathbb{F}_{q^2}$. So $(x, y) \in \mathcal{C}(\mathbb{F}_{q^2})$. That means $(x, y) \in \pi_{\mathcal{C}}^{-1}(x_0, x_1)$ and $\pi_{\mathcal{C}}^{-1}(x_0, x_1) \neq \emptyset$.

Hence $\pi_{\mathcal{A}}^{-1}(x_0, x_1) \neq \emptyset$ if and only if $\pi_{\mathcal{C}}^{-1}(x_0, x_1) \neq \emptyset$. Then Lemmas 2 and 3 conclude the proof of this proposition.

Theorem 2. *The number of* $\mathbb{F}_{q^2}$-*rational points on the affine curve* $\mathcal{C}$ *equals the number of* $\mathbb{F}_q$-*rational points on the affine variety* $\mathcal{A}$.

$$\#\mathcal{C}(\mathbb{F}_{q^2}) = \#\mathcal{A}(\mathbb{F}_q).$$

Proof. Consider the projection maps $\pi_{\mathcal{C}}$ and $\pi_{\mathcal{A}}$ from Lemmas 2 and 3. Then

$$\#\mathcal{C}(\mathbb{F}_{q^2}) = \sum_{(x_0, x_1) \in \mathbb{A}^2(\mathbb{F}_q)} \#\pi_{\mathcal{C}}^{-1}(x_0, x_1),$$

and

$$\#\mathcal{A}(\mathbb{F}_q) = \sum_{(x_0, x_1) \in \mathbb{A}^2(\mathbb{F}_q)} \#\pi_{\mathcal{A}}^{-1}(x_0, x_1).$$

Proposition 2 shows that $\#\pi_{\mathcal{C}}^{-1}(x_0, x_1) = \#\pi_{\mathcal{A}}^{-1}(x_0, x_1)$, for all $x_0, x_1 \in \mathbb{F}_q$. So the proof of this theorem is completed.

Remark 3. In fact, one can show that the number of $\mathbb{F}_{q^2}$-rational points on the nonsingular projective model of $\mathcal{C}$ equals the number of $\mathbb{F}_q$-rational points on the projective closure of $\mathcal{A}$ in $\mathbb{P}^3_{\mathbb{F}_q}$.

4   The Quadratic Extension Extractor

In this section we introduce an extractor that works for the affine curve $\mathcal{C}$ as defined in Section 3. We recall that $\mathcal{C}$ is defined over the quadratic extension of $\mathbb{F}_q$. The extractor, for a given point on the curve, outputs the *first* $\mathbb{F}_q$-coordinate of the abscissa of the point. Then, we show that the output of this extractor, for a given uniformly random point of $\mathcal{C}$, is statistically close to a uniform random variable in $\mathbb{F}_q$.

Similarly one could define an extractor that, for a given point on the curve, outputs a $\mathbb{F}_q$-linear combination of $\mathbb{F}_q$-coordinates of the abscissa of the point. In more detail , let $a_0, a_1 \in \mathbb{F}_q$ be such that both are not zero. The extractor, for a given point $P = (x, y) \in \mathcal{C}(\mathbb{F}_{q^2})$, where $x = x_0\alpha_0 + x_1\alpha_1$, outputs $a_0x_0 + a_1x_1$. Interchange the basis α_0, α_1 to another basis $\widehat{\alpha}_0, \widehat{\alpha}_1$, by

$$\begin{pmatrix} \widehat{\alpha}_0 \\ \widehat{\alpha}_1 \end{pmatrix} = \begin{pmatrix} a_0 & b_0 \\ a_1 & b_1 \end{pmatrix}^{-1} \begin{pmatrix} \alpha_0 \\ \alpha_1 \end{pmatrix},$$

where $b_0, b_1 \in \mathbb{F}_q$, such that the transformation matrix is nonsingular. Then x can be represented in the form $x = \widehat{x}_0\widehat{\alpha}_0 + \widehat{x}_1\widehat{\alpha}_1$, where $\widehat{x}_0, \widehat{x}_1 \in \mathbb{F}_q$. Clearly $\widehat{x}_0 = a_0x_0 + a_1x_1$. This amounts to the extractor that outputs x_0. So without loss of generality we consider the first extractor.

4.1   The Extractor for $\mathcal{C}$

In this subsection we define the extractor for the affine curve $\mathcal{C}$ defined over $\mathbb{F}_{q^2}$ (see Section 3 equation (1)). Then we compute the number of pre-images of this extractor for an element x_0 in $\mathbb{F}_q$, in terms of the number of $\mathbb{F}_q$-rational points on a curve $\mathcal{A}_{x_0}$. In other words, we show some bounds for the number of $\mathbb{F}_{q^2}$-rational points of $\mathcal{C}$, whose abscissa have the fixed first $\mathbb{F}_q$-coordinate.

Definition 5. *The extractor* Ext *is defined as a function*

$$\text{Ext} : \mathcal{C}(\mathbb{F}_{q^2}) \longrightarrow \mathbb{F}_q$$
$$\text{Ext}(x, y) = x_0,$$

Theorem 3 gives some bounds for $\#\text{Ext}^{-1}(x_0)$, for all x_0 in $\mathbb{F}_q$. For the proof of this theorem, we need several lemmas and propositions. We define the affine curve $\mathcal{A}_{x_0}$ as the intersection of the affine variety $\mathcal{A}$ and the hyperplane $\mathbf{x}_0 = x_0$, for x_0 in $\mathbb{F}_q$. Then in Proposition 3 we show that $\#\mathcal{A}_{x_0}(\mathbb{F}_q) = \#\text{Ext}^{-1}(x_0)$, for all x_0 in $\mathbb{F}_q$. We show that the curve $\mathcal{A}_{x_0}$ is reducible if and only if $x_0 \in \mathcal{I}$ (Proposition 4) and the curve $\mathcal{A}_{x_0}$ is singular if and only if $x_0 \in \mathcal{S}$ (Proposition 5), where the sets $\mathcal{I}, \mathcal{S}$ are defined by Definition 9. If the curve $\mathcal{A}_{x_0}$ is absolutely irreducible and singular, we consider the curve $\mathcal{X}_{x_0}$, that is a nonsingular plane model of $\mathcal{A}_{x_0}$. By using the Hasse-Weil bound for the curve $\mathcal{X}_{x_0}$, we obtain the bound for $\#\mathcal{A}_{x_0}(\mathbb{F}_q)$, where $x_0 \notin \mathcal{I}$ (Proposition 8). Note that we have a trivial bound for $\#\mathcal{A}_{x_0}(\mathbb{F}_q)$, if $x_0 \in \mathcal{I}$. Then Proposition 3 concludes the proof of Theorem 3.

Consider the affine variety $\mathcal{A}$ over $\mathbb{F}_q$, as introduced in Definition 4. Fix the element x_0 in $\mathbb{F}_q$. Then the points of $\mathcal{A}$ that have the first coordinate equal to x_0 form a curve which we call $\mathcal{A}_{x_0}$.

Definition 6. *Let $x_0 \in \mathbb{F}_q$. The affine curve $\mathcal{A}_{x_0}$ is defined by the equation*

$$F_{x_0}(\mathbf{x_1}, \mathbf{z}) = \mathbf{z}^2 - F_{x_0}(\mathbf{x_1}) = 0,$$

where $F_{x_0}(\mathbf{x_1}) = F(x_0, \mathbf{x_1})$.

Therefore

$$\mathcal{A}_{x_0}(\mathbb{F}_q) = \{P = (x_1, z) : x_1, z \in \mathbb{F}_q,\ z^2 = F_{x_0}(x_1) = F(x_0, x_1)\}.$$

Note that $\mathbf{x_1}$ and $\mathbf{z}$ are variables and x_0 is a fixed element in $\mathbb{F}_q$.

Proposition 3. $\#\mathcal{A}_{x_0}(\mathbb{F}_q) = \#\mathtt{Ext}^{-1}(x_0)$*, for all x_0 in $\mathbb{F}_q$.*

Proof. Let $x_0 \in \mathbb{F}_q$. Consider the projection maps $\pi_{\mathcal{C}}$ and $\pi_{\mathcal{A}}$ from Lemmas 2 and 3. Then

$$\#\mathcal{A}_{x_0}(\mathbb{F}_q) = \sum_{x_1 \in \mathbb{F}_q} \#\pi_{\mathcal{A}}^{-1}(x_0, x_1),$$

and

$$\#\mathtt{Ext}^{-1}(x_0) = \sum_{x_1 \in \mathbb{F}_q} \#\pi_{\mathcal{C}}^{-1}(x_0, x_1).$$

Proposition 2 shows that $\#\pi_{\mathcal{C}}^{-1}(x_0, x_1) = \#\pi_{\mathcal{A}}^{-1}(x_0, x_1)$, for all $x_0, x_1 \in \mathbb{F}_q$. So the proof of this proposition is completed.

Remark 4. Let $x_0 \in \mathbb{F}_q$. Define

$$f_{x_0}(\mathbf{x_1}) = f(x_0\alpha_0 + \mathbf{x_1}\alpha_1),$$
$$\overline{f}_{x_0}(\mathbf{x_1}) = \overline{f}(x_0\phi(\alpha_0) + \mathbf{x_1}\phi(\alpha_1)).$$

We recall that $F_{x_0}(\mathbf{x_1}) = f_{x_0}(\mathbf{x_1})\overline{f}_{x_0}(\mathbf{x_1})$. Note that $f_{x_0}, \overline{f}_{x_0}$ are polynomials in $\mathbb{F}_{q^2}[\mathbf{x_1}]$ and F_{x_0} is a polynomial in $\mathbb{F}_q[\mathbf{x_1}]$. From equalities (6) and (7), we have

$$f_{x_0}(\mathbf{x_1}) = \prod_{i=1}^{d}(x_0\alpha_0 + \mathbf{x_1}\alpha_1 - \lambda_i),$$
$$\overline{f}_{x_0}(\mathbf{x_1}) = \prod_{i=1}^{d}(x_0\phi(\alpha_0) + \mathbf{x_1}\phi(\alpha_1) - \phi(\lambda_i)).$$

Definition 7. *Let $x_0 \in \mathbb{F}_q$. Define $\theta_i = \frac{\lambda_i - x_0\alpha_0}{\alpha_1}$, for $i \in \{1, 2, \ldots, d\}$.*

Then $\phi(\theta_i) = \frac{\phi(\lambda_i) - x_0\phi(\alpha_0)}{\phi(\alpha_1)}$. Furthemore

$$f_{x_0}(\mathbf{x_1}) = \alpha_1^d \prod_{i=1}^{d}(\mathbf{x_1} - \theta_i),$$
$$\overline{f}_{x_0}(\mathbf{x_1}) = \alpha_1^{qd} \prod_{i=1}^{d}(\mathbf{x_1} - \phi(\theta_i)).$$

Since $\lambda_i \neq \lambda_j$, for $i \neq j$, so $\theta_i \neq \theta_j$ and $\phi(\theta_i) \neq \phi(\theta_j)$, for $i \neq j$. Thus f_{x_0} and $\overline{f}_{x_0}$ are square free polynomials in $\overline{\mathbb{F}}_q[\mathbf{x}_1]$. Then

$$F_{x_0}(\mathbf{x}_1) = (\mathrm{N}(\alpha_1))^d \prod_{i=1}^{d} ((\mathbf{x}_1 - \theta_i)(\mathbf{x}_1 - \phi(\theta_i))).$$

Lemma 4. $F_{x_0}(\mathbf{x}_1)$ *has* $\theta \in \overline{\mathbb{F}}_q$ *as multiple root if and only if*

$$f_0(x_0, \theta) = f_1(x_0, \theta) = 0.$$

Proof. From Remark 4 and equalities (3), (5), we have

$$\begin{aligned}
f_{x_0}(\mathbf{x}_1) &= f_0(x_0, \mathbf{x}_1)\alpha_0 + f_1(x_0, \mathbf{x}_1)\alpha_1, \\
\overline{f}_{x_0}(\mathbf{x}_1) &= f_0(x_0, \mathbf{x}_1)\phi(\alpha_0) + f_1(x_0, \mathbf{x}_1)\phi(\alpha_1),
\end{aligned} \tag{8}$$

where $f_0(x_0, \mathbf{x}_1)$ and $f_1(x_0, \mathbf{x}_1)$ are polynomials in $\mathbb{F}_q[\mathbf{x}_1]$. The polynomials f_{x_0} and $\overline{f}_{x_0}$ are square free, so if $(\mathbf{x}_1 - \theta)^2$ is a factor of $F_{x_0}(\mathbf{x}_1)$, then $(\mathbf{x}_1 - \theta)$ is a common factor of both polynomials f_{x_0} and $\overline{f}_{x_0}$. Hence

$$\left(f_0(x_0, \theta) \ f_1(x_0, \theta) \right) \begin{pmatrix} \alpha_0 & \phi(\alpha_0) \\ \alpha_1 & \phi(\alpha_1) \end{pmatrix} = \left(0 \ 0 \right).$$

Since the matrix is nonsingular (see Subsection 2.1), so $f_0(x_0, \theta) = f_1(x_0, \theta) = 0$. Converse is obvious.

Definition 8. *For* $x_0 \in \mathbb{F}_q$, *let*

$$S_{x_0} = \{x_1 \in \overline{\mathbb{F}}_q : f_0(x_0, x_1) = f_1(x_0, x_1) = 0\}$$

and $d_{x_0} = \#S_{x_0}$, $g_{x_0}(\mathbf{x}_1) = \gcd(f_0(x_0, \mathbf{x}_1), f_1(x_0, \mathbf{x}_1))$.

Since $f_{x_0}(\mathbf{x}_1)$ is square free in $\overline{\mathbb{F}}_q[\mathbf{x}_1]$, it follows from equality (8) that g_{x_0} has no multiple root in $\overline{\mathbb{F}}_q$. That means $d_{x_0} = \deg(g_{x_0})$. Furthermore $0 \leq d_{x_0} \leq d$.

Remark 5. From the proof of Lemma 4, $f_{x_0}(x_1) = \overline{f}_{x_0}(x_1) = 0$, for $x_1 \in \overline{\mathbb{F}}_q$, if and only if $f_0(x_0, x_1) = f_1(x_0, x_1) = 0$. So $x_1 \in S_{x_0}$ if and only if $x_1 = \theta_i = \phi(\theta_j)$, for some indexes i and j (see Remark 4). In other words

$$S_{x_0} = \{\theta_1, \theta_2, \ldots, \theta_d\} \cap \{\phi(\theta_1), \phi(\theta_2), \ldots, \phi(\theta_d)\}.$$

Definition 9. *For* $i, j \in \{1, 2, \ldots, d\}$, *let*

$$s_{i,j} = \frac{\begin{vmatrix} \lambda_i & \alpha_1 \\ \phi(\lambda_j) & \phi(\alpha_1) \end{vmatrix}}{\begin{vmatrix} \alpha_0 & \alpha_1 \\ \phi(\alpha_0) & \phi(\alpha_1) \end{vmatrix}}.$$

Let $\mathcal{S} = \{s \in \mathbb{F}_q : s = s_{i,j}, \text{ for some indexes } i, j\}$ *and* $\mathcal{I} = \{s \in \mathcal{S} : d_s = d\}$.

Remark 6. Let $\theta_i = \phi(\theta_j)$, for some indexes i, j. Then

$$\frac{\lambda_i - x_0\alpha_0}{\alpha_1} = \frac{\phi(\lambda_j) - x_0\phi(\alpha_0)}{\phi(\alpha_1)}.$$

Thus

$$x_0 = \frac{\lambda_i\phi(\alpha_1) - \phi(\lambda_j)\alpha_1}{\alpha_0\phi(\alpha_1) - \phi(\alpha_0)\alpha_1} = s_{i,j}.$$

We note that $\alpha_0\phi(\alpha_1) - \phi(\alpha_0)\alpha_1 \neq 0$ (see Subsection 2.1).

The converse is also true. That means $x_0 = s_{i,j}$ if and only if $\theta_i = \phi(\theta_j)$. Furthermore

$$d_{x_0} = \#\{(i, j) : s_{i,j} = x_0\}.$$

So $x_0 \notin S$ if an only if $d_{x_0} = 0$.

Proposition 4. *The affine plane curve $\mathcal{A}_{x_0}$ is absolutely irreducible if and only if $x_0 \notin \mathcal{I}$.*

Proof. The affine curve $\mathcal{A}_{x_0}$ is defined by the equation $\mathbf{z}^2 = F_{x_0}(\mathbf{x}_1)$. The curve $\mathcal{A}_{x_0}$ is reducible if and only if F_{x_0} is a square in $\overline{\mathbb{F}}_q[\mathbf{x}_1]$. From equality (??) F_{x_0} is a square in $\overline{\mathbb{F}}_q[\mathbf{x}_1]$ if and only if $\{\theta_1, \theta_2, \ldots, \theta_d\} = \{\phi(\theta_1), \phi(\theta_2), \ldots, \phi(\theta_d)\}$. Remarks 5 and 6 explain that this is equivalent to $d_{x_0} = d$.

Remark 7. Assume the affine curve $\mathcal{A}_{x_0}$ is reducible. So from the proof of Proposition 4 we have, $\{\theta_1, \theta_2, \ldots, \theta_d\} = \{\phi(\theta_1), \phi(\theta_2), \ldots, \phi(\theta_d)\}$. Then $\sum_{i=1}^{d} \theta_i = \sum_{i=1}^{d} \phi(\theta_i)$. Therefore

$$\sum_{i=1}^{d} \frac{\lambda_i - x_0\alpha_0}{\alpha_1} = \sum_{i=1}^{d} \frac{\phi(\lambda_i) - x_0\phi(\alpha_0)}{\phi(\alpha_1)}.$$

Because $\sum_{i=1}^{d} \lambda_i = e_{d-1}$ (see equation (2)), we have

$$dx_0 = \frac{e_{d-1}\phi(\alpha_1) - \phi(e_{d-1})\alpha_1}{\alpha_0\phi(\alpha_1) - \phi(\alpha_0)\alpha_1}.$$

In other words, if $x_0 \in \mathcal{I}$, then

$$dx_0 = \frac{\begin{vmatrix} e_{d-1} & \alpha_1 \\ \phi(e_{d-1}) & \phi(\alpha_1) \end{vmatrix}}{\begin{vmatrix} \alpha_0 & \alpha_1 \\ \phi(\alpha_0) & \phi(\alpha_1) \end{vmatrix}}.$$

Note that the converse is not true. If d is not divisible by p, then $\#\mathcal{I} \leq 1$. Otherwise $\#\mathcal{I} \leq d$.

Proposition 5. *The affine curve $\mathcal{A}_{x_0}$ is singular if and only if $x_0 \in S$. The curve $\mathcal{A}_{x_0}$ has d_{x_0} singular points.*

Proof. The point $(x_1, z) \in \overline{\mathbb{F}}_q \times \overline{\mathbb{F}}_q$ is a singular point on $\mathcal{A}_{x_0}$ if and only if $z = 0$ and x_1 is a double root of $F_{x_0}(\mathbf{x_1})$. From Lemma 4, x_1 is a double root of $F_{x_0}(\mathbf{x_1})$ if and only if $x_1 \in S_{x_0}$. So $\mathcal{A}_{x_0}$ has d_{x_0} singular points. Remarks 5 and 6 explain that there exists $x_1 \in S_{x_0}$ if and only if $x_0 = s_{i,j}$, for some indexes i, j. Since $x_0 \in \mathbb{F}_q$, therefore $x_0 \in \mathcal{S}$ if and only if $\mathcal{A}_{x_0}$ is singular.

We recall that g_{x_0} is a square free polynomial of degree d_{x_0} in $\mathbb{F}_q[\mathbf{x_1}]$. From Lemma 4 and Remark 5, g_{x_0} is the square factor of F_{x_0}. Let

$$F_{x_0}(\mathbf{x_1}) = g_{x_0}^2(\mathbf{x_1}) H_{x_0}(\mathbf{x_1}),$$

where H_{x_0} is a square free polynomial of degree $2(d - d_{x_0})$ in $\mathbb{F}_q[\mathbf{x_1}]$.

Definition 10. *Let $\mathcal{X}_{x_0}$ be the affine curve given by the equation*

$$\mathbf{w}^2 - H_{x_0}(\mathbf{x_1}) = 0.$$

Proposition 6. *The affine curve $\mathcal{X}_{x_0}$ is absolutely irreducible and nonsingular if and only if $x_0 \notin \mathcal{I}$.*

Proof. The affine curve $\mathcal{X}_{x_0}$ is defined by the equation $\mathbf{w}^2 = H_{x_0}(\mathbf{x_1})$. Since H_{x_0} is a square free polynomial of degree $2(d - d_{x_0})$ in $\mathbb{F}_q[\mathbf{x_1}]$, it is absolutely irreducible and nonsingular if and only if H_{x_0} is not constant. Clearly H_{x_0} is constant if and only if $d_{x_0} = d$. That means H_{x_0} is constant if and only if $x_0 \in \mathcal{I}$.

Remark 8. If H_{x_0} is not constant, the affine curve $\mathcal{X}_{x_0}$ is a nonsingular plane model of $\mathcal{A}_{x_0}$.

Proposition 7. *For $x_0 \in \mathbb{F}_q$, $|\#\mathcal{A}_{x_0}(\mathbb{F}_q) - \#\mathcal{X}_{x_0}(\mathbb{F}_q)| \leq d_{x_0}$.*

Proof. The affine curves $\mathcal{A}_{x_0}$ and $\mathcal{X}_{x_0}$ are defined by the equations $\mathbf{z}^2 = F_{x_0}(\mathbf{x_1})$ and $\mathbf{w}^2 = H_{x_0}(\mathbf{x_1})$ respectively. We recall that $F_{x_0}(\mathbf{x_1}) = g_{x_0}^2(\mathbf{x_1}) H_{x_0}(\mathbf{x_1})$. Define the projection maps $\pi_{\mathcal{A}} : \mathcal{A}_{x_0}(\mathbb{F}_q) \longrightarrow \mathbb{F}_q$, by $\pi_{\mathcal{A}}(x_1, z) = x_1$ and $\pi_{\mathcal{X}} : \mathcal{X}_{x_0}(\mathbb{F}_q) \longrightarrow \mathbb{F}_q$, by $\pi_{\mathcal{X}}(x_1, w) = x_1$.

Let $x_1 \in \mathbb{F}_q$. First assume that $g_{x_0}(x_1) \neq 0$. Then

$$\#\pi_{\mathcal{A}}^{-1}(x_1) = \#\pi_{\mathcal{X}}^{-1}(x_1) = \begin{cases} 0, & \text{if } H_{x_0}(x_1) \text{ is a non-square in } \mathbb{F}_q, \\ 1, & \text{if } H_{x_0}(x_1) = 0, \\ 2, & \text{if } H_{x_0}(x_1) \text{ is a square in } \mathbb{F}_q^*. \end{cases}$$

Now assume that $g_{x_0}(x_1) = 0$. Then $\#\pi_{\mathcal{A}}^{-1}(x_1) = 1$ and $\#\pi_{\mathcal{X}}^{-1}(x_1)$ equals 0 or 2. Then

$$|\#\mathcal{A}_{x_0}(\mathbb{F}_q) - \#\mathcal{X}_{x_0}(\mathbb{F}_q)| = \left| \sum_{x_1 \in \mathbb{F}_q} \#\pi_{\mathcal{A}}^{-1}(x_1) - \sum_{x_1 \in \mathbb{F}_q} \#\pi_{\mathcal{X}}^{-1}(x_1) \right|$$

$$= \sum_{x_1 \in \mathbb{F}_q, \, g_{x_0}(x_1)=0} 1 \leq d_{x_0}.$$

Proposition 8. *Let $x_0 \in \mathbb{F}_q$. If $x_0 \notin \mathcal{I}$, then*

$$\left|\#\mathcal{A}_{x_0}(\mathbb{F}_q) - q\right| \leq 2(d - d_{x_0} - 1)\sqrt{q} + d_{x_0} + 1.$$

Proof. Let $x_0 \in \mathbb{F}_q \setminus \mathcal{I}$. Then the affine curve $\mathcal{X}_{x_0}$ is absolutely irreducible and nonsingular (see Proposition 6). The degree of $\mathcal{X}_{x_0}$ is $2(d - d_{x_0})$. Let $\widetilde{\mathcal{X}}_{x_0}$ be the nonsingular projective model of $\mathcal{X}_{x_0}$. So $\widetilde{\mathcal{X}}_{x_0}$ is a hyperelliptic curve of genus $d - d_{x_0} - 1$. Furthermore $\#\widetilde{\mathcal{X}}_{x_0}(\mathbb{F}_q) - \#\mathcal{X}_{x_0}(\mathbb{F}_q)$ equals zero or two. (see Theorem 1). By using the Hasse-Weil bound, we have

$$\left|\#\widetilde{\mathcal{X}}(\mathbb{F}_q) - (q + 1)\right| \leq 2(d - d_{x_0} - 1)\sqrt{q}.$$

Then $|\#\mathcal{X}(\mathbb{F}_q) - q| \leq 2(d - d_{x_0} - 1)\sqrt{q} + 1$. Proposition 7 concludes the proof of this proposition.

Theorem 3. *Let $x_0 \in \mathbb{F}_q$. If $x_0 \notin \mathcal{I}$, then*

$$\left|\#\texttt{Ext}^{-1}(x_0) - q\right| \leq 2(d - d_{x_0} - 1)\sqrt{q} + d_{x_0} + 1.$$

Otherwise,

$$\left|\#\texttt{Ext}^{-1}(x_0) - q\right| \leq q.$$

Proof. Let $x_0 \in \mathbb{F}_q$. Then Proposition 3 shows that $\#\mathcal{A}_{x_0}(\mathbb{F}_q) = \#\texttt{Ext}^{-1}(x_0)$. If $x_0 \notin \mathcal{I}$, then Proposition 8 gives the estimate for $\#\texttt{Ext}^{-1}(x_0)$. If $x_0 \in \mathcal{I}$, then the curve $\mathcal{A}_{x_0}$ is reducible (see Proposition 4). So in this case we have the trivial estimate for $\#\texttt{Ext}^{-1}(x_0)$.

4.2 Analysis of the Extractor

In this subsection we show that provided the point P is chosen uniformly at random in $\mathcal{C}(\mathbb{F}_{q^2})$, the element extracted from the point P by $\texttt{Ext}$ is indistinguishable from a uniformly random element in $\mathbb{F}_q$.

Let X be a $\mathbb{F}_q$-valued random variable that is defined as

$$X = \texttt{Ext}(P), \text{ for } P \in_R \mathcal{C}(\mathbb{F}_{q^2}).$$

Proposition 9. *The random variable X is statistically close to the uniform random variable $U_{\mathbb{F}_q}$.*

$$\Delta(X, U_{\mathbb{F}_q}) = O(\frac{1}{\sqrt{q}}).$$

Proof. Let $z \in \mathbb{F}_q$. For the uniform random variable $U_{\mathbb{F}_q}$, $\Pr[U_{\mathbb{F}_q} = z] = 1/q$. Also for the $\mathbb{F}_q$-valued random variable X,

$$\Pr[X = z] = \frac{\#\texttt{Ext}^{-1}(z)}{\#\mathcal{C}(\mathbb{F}_{q^2})}.$$

Hasse-Weil's Theorem gives the bound for $\#\mathcal{C}(\mathbb{F}_{q^2})$ and Theorem 3 gives the bound for $\#\mathtt{Ext}^{-1}(z)$. Hence

$$\Delta(X, U_{\mathbb{F}_q}) = \frac{1}{2}\sum_{z\in\mathbb{F}_q}\left|\Pr[X=z] - \Pr[U_{\mathbb{F}_q}=z]\right|$$

$$= \frac{1}{2}\sum_{z\in\mathbb{F}_q}\left|\frac{\#\mathtt{Ext}^{-1}(z)}{\#\mathcal{C}(\mathbb{F}_{q^2})} - \frac{1}{q}\right|$$

$$= \sum_{z\in\mathcal{I}}\frac{\left|q\#\mathtt{Ext}^{-1}(z) - \#\mathcal{C}(\mathbb{F}_{q^2})\right|}{2q\#\mathcal{C}(\mathbb{F}_{q^2})} + \sum_{z\in\mathbb{F}_q\backslash\mathcal{I}}\frac{\left|q\#\mathtt{Ext}^{-1}(z) - \#\mathcal{C}(\mathbb{F}_{q^2})\right|}{2q\#\mathcal{C}(\mathbb{F}_{q^2})}.$$

Let $r = \#\mathcal{I}$. Then

$$\Delta(X, U_{\mathbb{F}_q}) \leq \frac{r(q^2 + (d-1)q + 1) + (q-r)(2(d-1)q\sqrt{q} + dq + 1)}{2q(q^2 - (d-1)q + 1)}$$

$$= \frac{2(d-1)q\sqrt{q} + (d+r)q - 2(d-1)r\sqrt{q} - r + 1}{2(q^2 - (d-1)q + 1)} = \frac{d - 1 + \epsilon(q)}{\sqrt{q}},$$

where $\epsilon(q) = \frac{(d+r)q\sqrt{q}+2(d-1)(d-r-1)q-(r-1)\sqrt{q}-2(d-1)}{2(q^2-(d-1)q+1)}$. If $q \geq 2d^2$, then $\epsilon(q) < 1$.

Corollary 1. $\mathtt{Ext}$ *is a deterministic* $(\mathbb{F}_q, O(\frac{1}{\sqrt{q}}))$-*extractor for* $\mathcal{C}(\mathbb{F}_{q^2})$.

5   Examples

In this section we give some examples for the extractors $\mathtt{Ext}$. Our first example is the extractor for the subgroup of quadratic residues of $\mathbb{F}_{q^2}^*$. For the second example, we recall an extractor in [7] for an elliptic curve defined over $\mathbb{F}_{q^2}$. Also from the result of Theorem 3, we improve the result of [7].

5.1   The Extractor for a Subgroup of $\mathbb{F}_{q^2}^*$

In this subsection we propose a simple extractor for the subgroup of quadratic residues of $\mathbb{F}_{q^2}^*$. This extractor is the result of Theorem 3, where $f(\mathbf{x}) = \mathbf{x}$.

Let G be the subgroup of quadratic residues of $\mathbb{F}_{q^2}^*$. We recall that every element x in $\mathbb{F}_{q^2}$ is represented in the form $x = x_0\alpha_0 + x_1\alpha_1$, where $x_0, x_1 \in \mathbb{F}_q$. Define the extractor $\mathtt{ext}$ for G as the function

$$\mathtt{ext} : G \longrightarrow \mathbb{F}_q$$

$$\mathtt{ext}(x) = x_0.$$

The following proposition gives the estimate for $\#\mathtt{ext}^{-1}(z)$, where $z \in \mathbb{F}_q$.

Proposition 10. *For all* $z \in \mathbb{F}_q^*$,

$$\#\mathtt{ext}^{-1}(z) = \frac{q \pm 1}{2},$$

and for $z = 0$, $\#\mathtt{ext}^{-1}(0) = 0$ *or* $\#\mathtt{ext}^{-1}(0) = q - 1$.

Proof. Let the affine curve $\mathcal{C}$ be defined by the equation $\mathcal{C} : \mathbf{y}^2 = f(\mathbf{x}) = \mathbf{x}$. This curve is of the type considered in Section 4. Clearly for each element $x \in G$, there are exactly two points (x, y) and $(x, -y)$ on $\mathcal{C}$. In fact there is a bijection between G and the set of nonzero abscissa of points on $\mathcal{C}$. Then $\#\mathtt{Ext}^{-1}(z) = 2\#\mathtt{ext}^{-1}(z)$, for all $z \in \mathbb{F}_q^*$. It is easy to see that $\mathcal{I} = \{0\}$. Then Theorem 3 implies the proof of this proposition. Also the bound for $\#\mathtt{ext}^{-1}(0)$ is obvious.

Corollary 2. $\mathtt{ext}$ *is a deterministic* $(\mathbb{F}_q, \frac{1}{q})$*-extractor for* G.

Proof. For $d = 1$, the estimate for $\epsilon(q)$ can be made tighter (see proof of Proposition 9), so that $\epsilon(q) < \frac{1}{q}$.

5.2   The Extractor for Elliptic Curves

In this subsection we recall the extractor introduced by Gürel in [7], that works for an elliptic curve defined over $\mathbb{F}_{q^2}$. This extractor, for a given random point on elliptic curve, outputs the first $\mathbb{F}_q$-coordinate of the abscissa of the point. Then from the result of Theorem 3, we improve the bounds which are proposed in [7].

Let E be an elliptic curve defined over $\mathbb{F}_{q^2}$, where $q = p^k$, for prime number $p > 3$ and positive integer k. Then

$$E(\mathbb{F}_{q^2}) = \{(x, y) \in \mathbb{F}_{q^2} \times \mathbb{F}_{q^2} : y^2 = f(x) = x^3 + ax + b\} \cup \{\mathcal{O}_E\},$$

where a and b are in $\mathbb{F}_{q^2}$. Since E is nonsingular, then $f(\mathbf{x})$ is a square free polynomial in $\overline{\mathbb{F}}_q[\mathbf{x}]$.

Let $\alpha_0 = 1$ and $\alpha_1 = t$, where $t \in \mathbb{F}_{q^2}$, such that $t^2 = c$ and c is a nonsquare element in $\mathbb{F}_q$. So every element x in $\mathbb{F}_{q^2}$ can be represented in the form $x = x_0 + x_1 t$, where $x_0, x_1 \in \mathbb{F}_q$.

The extractor $\mathtt{ext}$ for E is defined as a function

$$\mathtt{ext} : E(\mathbb{F}_{q^2}) \longrightarrow \mathbb{F}_q$$
$$\mathtt{ext}(x, y) = x_0,$$
$$\mathtt{ext}(\mathcal{O}_E) = 0.$$

The following theorem gives the tight bounds for $\#\mathtt{ext}^{-1}(z)$, for all z in $\mathbb{F}_q$.

Proposition 11. *For all* $z \in \mathbb{F}_q^*$,

$$\left|\#\mathtt{ext}^{-1}(z) - q\right| \leq 4\sqrt{q} + 1.$$

For $z = 0$*, if* $a_1 \neq 0$ *or* $b_0 \neq 0$*, then*

$$\left|\#\mathtt{ext}^{-1}(0) - (q + 1)\right| \leq 4\sqrt{q} + 1,$$

otherwise,

$$\left|\#\mathtt{ext}^{-1}(0) - (q + 1)\right| \leq q.$$

Proof. The proof of this theorem follows from Theorems 3, in the case that $f(\mathbf{x}) = \mathbf{x}^3 + a\mathbf{x} + b$. Define the variables $\mathbf{x}_0$ and $\mathbf{x}_1$ by $\mathbf{x} = \mathbf{x}_0 + \mathbf{x}_1 t$. Then

$$f(\mathbf{x}_0 + \mathbf{x}_1 t) = f_0(\mathbf{x}_0, \mathbf{x}_1) + f_1(\mathbf{x}_0, \mathbf{x}_1)t,$$

where

$$f_0(\mathbf{x}_0, \mathbf{x}_1) = \mathbf{x}_0^3 + 3c\mathbf{x}_0\mathbf{x}_1^2 + a_0\mathbf{x}_0 + ca_1\mathbf{x}_1 + b_0$$
$$f_1(\mathbf{x}_0, \mathbf{x}_1) = c\mathbf{x}_1^3 + 3\mathbf{x}_0^2\mathbf{x}_1 + a_1\mathbf{x}_0 + a_0\mathbf{x}_1 + b_1.$$

Then we fix $\mathbf{x}_0$ by z. It is easy to see that $\mathcal{I} = \{0\}$ if and only if $f_0(z, \mathbf{x}_1) = 0$. Clearly $f_0(z, \mathbf{x}_1) = 0$, if and only if $z = a_1 = b_0 = 0$, since $p \neq 3$. Recall that p is the characteristic of $\mathbb{F}_q$. Also note that $\#\texttt{ext}^{-1}(0) = \#\texttt{Ext}^{-1}(0) + 1$, since $\texttt{ext}(\mathcal{O}_E) = 0$.

Corollary 3. $\texttt{ext}$ *is a deterministic* $(\mathbb{F}_q, \frac{3}{\sqrt{q}})$*-extractor for* $E(\mathbb{F}_{q^2})$*, if* $q \geq 18$.

Proof. The proof of this corollary is similar to the proof of Proposition 9, in the case that $d = 3$ and $r \leq 1$.

6   Conclusion

We introduce a deterministic extractor $\texttt{Ext}$, for the (hyper)elliptic curve $\mathcal{C}$, defined over $\mathbb{F}_{q^2}$, where q is some power of an odd prime. Our extractor, for a given point P on $\mathcal{C}$, outputs the first $\mathbb{F}_q$-coefficient of the abscissa of the point P. The main part of the analysis of this extractor is to compute $\#\texttt{Ext}^{-1}(z)$, where $z \in \mathbb{F}_q$. That is equivalent to counting the number of $\mathbb{F}_q$-rational points on the fibers $\mathcal{A}_z$ on the affine variety $\mathcal{A}$. Theorem 3 gives the estimates for $\#\texttt{Ext}^{-1}(z)$. Our experiments with MAGMA for $\#\texttt{Ext}^{-1}(z)$, show that the bounds in Theorem 3 are tight. Then we show that if a point P is chosen uniformly at random in $\mathcal{C}$, the element extracted from the point P is statistically close to a uniformly random variable in $\mathbb{F}_q$.

Future Work. Consider the finite field $\mathbb{F}_{q^n}$, where q is a power of a prime p and n is a positive integer. Then $\mathbb{F}_{q^n}$ is a n dimensional vector space over $\mathbb{F}_q$. Let $\{\alpha_1, \alpha_2, \dots, \alpha_n\}$ be a basis of $\mathbb{F}_{q^n}$ over $\mathbb{F}_q$. That means every element x in $\mathbb{F}_{q^n}$ can be represented in the form $x = x_1\alpha_1 + x_2\alpha_2 + \dots + x_n\alpha_n$, where $x_i \in \mathbb{F}_q$.

Let C be an absolutely irreducible nonsingular affine curve that is defined over $\mathbb{F}_{q^n}$ by the equation

$$y^m = f(x),$$

where $f(\mathbf{x}) \in \mathbb{F}_{q^n}[\mathbf{x}]$ is a monic square-free polynomial of degree d and m is a positive integer dividing $q - 1$.

We define the extractors $\texttt{ext}_\ell$ for C, where ℓ is a positive integer less than n. The extractor $\texttt{ext}_\ell$, for a given point P on the curve, outputs the ℓ *first* $\mathbb{F}_q$-coordinate of the abscissa of the point P.

Definition 11. *Let ℓ be a positive integer less than n. The extractor ext_ℓ is defined as a function*

$$\mathsf{ext}_\ell : C(\mathbb{F}_{q^n}) \longrightarrow \mathbb{A}^\ell(\mathbb{F}_q)$$
$$\mathsf{ext}_\ell(x, y) = (x_1, \ldots, x_\ell),$$

where $x \in \mathbb{F}_{q^n}$ is represented as $x = x_1\alpha_1 + x_2\alpha_2 + \cdots + x_n\alpha_n$, for $x_i \in \mathbb{F}_q$.

Let X_ℓ be a $\mathbb{F}_q^\ell$-valued random variable that is defined as

$$X_\ell = \mathsf{ext}_\ell(P), \text{ for } P \in_R C(\mathbb{F}_{q^n}).$$

Conjecture 1. The random variable X_ℓ is $\frac{c}{\sqrt{q^{n-\ell}}}$-uniform on $\mathbb{F}_q^\ell$, where c is a constant depending on m, n and d. That is

$$\Delta(X_\ell, U_{\mathbb{F}_q^\ell}) \leq \frac{c}{\sqrt{q^{n-\ell}}}.$$

We leave the proof of this conjecture for the future work.

Acknowledgment. The authors would like to thank T. Lange for her helpful comments.

References

1. Artin, E.: Algebraic Numbers and Algebraic Functions. Gordon and Breach, New York (1967)
2. Beelen, P., Doumen, J.M.: Pseudorandom sequences from elliptic curves. In: Finite Fields with Applications to Coding Theory, Cryptography and Related Areas, pp. 37–52. Springer-Verlag, Berlin Heidelberg (2002)
3. Chevassut, O., Fouque, P., Gaudry, P., Pointcheval, D.: The Twist-Augmented Technique for Key Exchange. In: Yung, M., Dodis, Y., Kiayias, A., Malkin, T.G. (eds.) PKC 2006. LNCS, vol. 3958, pp. 410–426. Springer, Heidelberg (2006)
4. Cohen, H., Frey, G.: Handbook of Elliptic and Hyperelliptic Curve Cryptography. Chapman & Hall/CRC, New York (2006)
5. Farashahi, R.R., Pellikaan, R., Sidorenko, A.: Extractors for Binary Elliptic Curves, Extended Abstract to appear at WCC (2007)
6. Gong, G., Berson, T.A., Stinson, D.R.: Elliptic Curve Pseudorandom Sequence Generators. In: Heys, H.M., Adams, C.M. (eds.) SAC 1999. LNCS, vol. 1758, pp. 34–48. Springer, Heidelberg (2000)
7. Gürel, N.: Extracting bits from coordinates of a point of an elliptic curve, Cryptology ePrint Archive, Report 2005/324, (2005), http://eprint.iacr.org/
8. Hartshorne, R.: Algebraic Geometry, Grad. Texts Math, vol. 52. Springer, Berlin Heidelberg (1977)
9. Hess, F., Shparlinski, I.E.: On the Linear Complexity and Multidimensional Distribution of Congruential Generators over Elliptic Curves. Designs, Codes and Cryptography 35(1), 111–117 (2005)

10. Juels, A., Jakobsson, M., Shriver, E., Hillyer, B.K.: How to turn loaded dice into fair coins. IEEE Transactions on Information Theory 46(3), 911–921 (2000)
11. Kaliski, B.S.: A Pseudo-Random Bit Generator Based on Elliptic Logarithms. In: Odlyzko, A.M. (ed.) CRYPTO 1986. LNCS, vol. 263, pp. 84–103. Springer, Heidelberg (1987)
12. Lange, T., Shparlinski, I.E.: Certain Exponential Sums and Random Walks on Elliptic Curves. Canad. J. Math. 57(2), 338–350 (2005)
13. —: Distribution of Some Sequences of Points on Elliptic Curves, J. Math. Crypt. 1, 1–11 (2007)
14. Luby, M.: Pseudorandomness and Cryptographic Applications. Princeton University Press, Princeton, USA (1994)
15. Poonen, B.: Bertini Theorems over Finite Fields. Annals of Mathematics 160(3), 1099–1127 (2004)
16. Shaltiel, R.: Recent Developments in Explicit Constructions of Extractors. Bulletin of the EATCS 77, 67–95 (2002)
17. Shparlinski, I.E.: On the Naor-Reingold Pseudo-Random Function from Elliptic Curves. Applicable Algebra in Engineering, Communication and Computing—AAECC 11(1), 27–34 (2000)
18. Trevisan, L., Vadhan, S.: Extracting Randomness from Samplable Distributions, IEEE Symposium on Foundations of Computer Science, pp. 32–42 (2000)

On Kabatianskii-Krouk-Smeets Signatures

Pierre-Louis Cayrel[1], Ayoub Otmani[2], and Damien Vergnaud[3]

[1] DMI/XLIM - Université de Limoges, 123 avenue Albert Thomas,
87060 Limoges, France
`pierre-louis.cayrel@xlim.fr`
[2] GREYC - Ensicaen, Boulevard Maréchal Juin, 14050 Caen Cedex, France
`Ayoub.Otmani@info.unicaen.fr`
[3] b-it COSEC - Bonn/Aachen International Center for Information Technology -
Computer Security Group, Dahlmannstr. 2, D-53113 Bonn, Germany
`vergnaud@bit.uni-bonn.de`

Abstract. Kabastianskii, Krouk and Smeets proposed in 1997 a digital signature scheme based on random error-correcting codes. In this paper we investigate the security and the efficiency of their proposal. We show that a passive attacker who may intercept just a few signatures can recover the private key. We give precisely the number of signatures required to achieve this goal. This enables us to prove that all the schemes given in the original paper can be broken with at most 20 signatures. We improve the efficiency of these schemes by firstly providing parameters that enable to sign about 40 messages, and secondly, by describing a way to extend these *few-times* signatures into classical *multi-time* signatures. We finally study their key sizes and a mean to reduce them by means of more compact matrices.

Keywords: Code-based cryptography, digital signature, random error-correcting codes, Niederreiter cryptosystem.

1 Introduction

Kabastianskii, Krouk and Smeets proposed in 1997 a digital signature scheme based on random error-correcting codes. In this paper we investigate the security and the efficiency of their proposal. We show that a passive attacker who may intercept just a few signatures can recover the private key. We give precisely the number of signatures required to achieve this goal. This enables us to prove that all the schemes given in the original paper can be broken with at most 20 signatures. We improve the efficiency of these schemes by firstly providing parameters that enable to sign about 40 messages, and secondly, by describing a way to extend these *few-times* signatures into classical *multi-time* signatures. We finally study their key sizes and a mean to reduce them by means of more compact matrices.

Related Work. In 1978, McEliece [12] proposed the first public key cryptosystem based on coding theory. His idea is to first select a particular code for which

C. Carlet and B. Sunar (Eds.): WAIFI 2007, LNCS 4547, pp. 237–251, 2007.
© Springer-Verlag Berlin Heidelberg 2007

an efficient decoding algorithm is known, and then disguise it as a general-looking linear code. A description of the original code can serve as the private key, while a description of the transformed code serves as the public key. Several proposals were made to modify McEliece's original scheme, but unfortunately most of them turn out to be insecure or inefficient. However, the original primitive, which uses Goppa codes, has remained unbroken (for appropriate system parameters).

In 1986, Niederreiter [14] proposed another (knapsack-based) scheme which relies on a linear code. The McEliece scheme uses a generator matrix while the Niederreiter scheme uses a parity-check matrix, but they were proved [10] to be equivalent in terms of security for the same parameters[1].

Compared with other public-key cryptosystems which involve modular exponentiation, these schemes have the advantage[2] of high-speed encryption and decryption. However, they suffer from the fact that the public key is very large. In 2005, Gaborit [5] proposed a method to severely decrease its size (making it almost linear – see below).

Digital signature schemes are the most important cryptographic primitive for providing authentication in an electronic world. They allow a signer with a secret key to sign messages such that anyone with access to the corresponding public key is able to verify authenticity of the message. Parallel to the efforts to build an efficient public key encryption scheme from error correcting codes, several attempts were proposed to design signature schemes based on error-correcting codes. Unfortunately, most of the proposed protocols have been proved insecure (see the survey [4] and the references therein for details).

It is well known that any trapdoor permutation permits to design digital signatures by using the unique capacity of the owner of the public key to invert this permutation. The so-called *Full Domain Hash* (FDH) approach can only be used to sign messages whose hash values lies in the range set of the trapdoor permutation. Therefore, a signature scheme based on trapdoor codes must achieve complete decoding. In 2001, Courtois, Finiasz and Sendrier [3] have presented a practical signature scheme derived from a technique allowing complete decoding of Goppa codes (for some parameter choices).

At Crypto'93, Stern [18] proposed an identification scheme based on the syndrome decoding problem. In this scheme, all users share a parity-check matrix for a binary linear code and each user secretly chooses a vector v of *small* Hamming weight (slightly below the expected minimum distance of the code). The public key of identification is the corresponding syndrome. By an interactive zero-knowledge protocol, any user can identify himself to another by proving he knows v without revealing it. A dual version of the Stern identification scheme that uses a generator matrix of the code was proposed by Véron [19]. Both protocols can give rise to digital signature schemes (though inefficient), by applying the well-known Fiat-Shamir heuristic.

---

[1] Niederreiter's original scheme relies on generalized Reed-Solomon codes and the two primitives are equivalent if we substitute these codes by Goppa codes.

[2] For the same parameters, the Niederreiter cryptosystem reveals some advantages, for example, the size of the public key and the number of operations to encrypt.

Finally, Kabastianskii, Krouk and Smeets [6] proposed, 10 years ago, a digital signature scheme based on random linear codes. They exploited the fact that for every linear code the set of its correctable syndrome contains a linear subspace of relatively large dimension. Kabatianskii *et al.* concluded their paper by asking for an analysis of the efficiency and the security of their scheme. The investigation of this issue is the main purpose of the present paper.

Organisation of the Paper. The rest of this paper is organized as follows. Section 2 begins with background material on coding theory. In section 3 and 4, we review the paradigm of *signing without decoding*, together with the schemes proposed by Kabatianskii *et al.* (KKS-1 to KKS-4). In section 5, we present an analysis of their security under a known message attack. In particular, we show that a passive attacker who may only intercept signatures can recover the private key. For the concrete parameters, just about 20 signatures are sufficient to reveal a large part of the key. Section 6 discusses a way to extend these *few-times* signatures into classical *multi-time* signatures. In section 7, we study the key sizes of the KKS signature schemes and propose a way to reduce them. Section 8 concludes the paper with efficiency considerations.

2   Notations and Definitions

Let n be a non-negative integer and q be a prime power that usually is 2. The support $\mathsf{supp}(x)$ of a vector $x \in GF(q)^n$ is the set of coordinates i such that $x_i \neq 0$. The *(Hamming) weight* $\mathsf{wt}(x)$ of $x \in GF(q)^n$ is the cardinality of $\mathsf{supp}(x)$. A $\mathscr{C}[n, n-r, d]$ code over $GF(q)$ is a linear subspace of $GF(q)^n$ of dimension $n-r$ and minimum distance d. The elements of $\mathscr{C}$ are *codewords*. A linear code can be defined either by a parity check matrix or a generator matrix. A *parity check matrix* H for $\mathscr{C}$ is an $r \times n$ matrix such that the vectors $w \in GF(q)^n$ which are solutions to the equation $Hw^T = 0$ are exactly the codewords of $\mathscr{C}$. When the first r columns of H form the $r \times r$ identity matrix, we denote H by $(I_r|M)$ where the columns of M are the last $n-r$ columns of H. A generator matrix G is an $(n-r) \times n$ matrix formed by a basis of $\mathscr{C}$. G is *systematic* if its first $n-r$ columns form I_{n-r}. For a non-negative integer t, we denote by $\mathscr{M}_{n,t}$ the set of vectors of $GF(q)^n$ of weight t and by $\mathscr{M}_{n,\leq t}$ the set $\cup_{i=0}^{t} \mathscr{M}_{n,i}$.

A *syndrome decoding algorithm* $\mathsf{dec}()$ for $\mathscr{C}$ (defined by a $r \times n$ parity check matrix H) is a process that is able to find for a given vector $s \in GF(q)^r$ a vector $e = \mathsf{dec}(s) \in GF(q)^n$ such that:

$$H \cdot e^T = s. \tag{1}$$

The vector e is seen as an *error* and the element $s \in GF(q)^r$ is called its *syndrome*. Note that a decoding algorithm does not necessarily succeed in finding an error for *any* syndrome. The algorithm $\mathsf{dec}()$ achieves a *complete decoding* if it can resolve Equation (1) for *any* $s \in GF(q)^r$, and a *t-bounded syndrome decoding algorithm* for $\mathscr{C}$ is a decoding algorithm that is able to recover any error vector of $\mathscr{M}_{n,\leq t}$. More precisely, it is an application $\mathsf{dec} : H\mathscr{M}_{n,\leq t} \to \mathscr{M}_{n,\leq t}$ such that

$\mathsf{dec}(0) = 0$ and for any $s \in H\mathcal{M}_{n,\leq t}$ where $H\mathcal{M}_{r,\leq t}$ is the set $\{Hz^T : z \in \mathcal{M}_{n,\leq t}\}$, $\mathsf{dec}(s)$ is solution to Equation (1). The set of *correctable syndromes for dec* is therefore $H\mathcal{M}_{n,\leq t}$. Note that $\mathsf{dec}()$ is well-defined when $2t + 1 < d$ because Equation (1) admits a unique solution. Finally, it has been proved in [1] that the problem of *syndrome decoding i.e.* solving Equation (1) is NP-Hard.

As mentionned in the introduction Niederreiter outlined a public-key cryptosystem based upon the difficulty of the syndrome decoding problem [14] for an arbitrary linear code. It is a modified version of the McEliece cryptosystem [12]. Each user chooses a code $\mathscr{C}[n, n - r, d]$ over $GF(q)$ for which a polynomial (in n) decoding algorithm is known. The plain text space is the set $\mathcal{M}_{n,t}$ with $2t + 1 < d$. The private key is a parity check matrix H of $\mathscr{C}$, a t-bounded syndrome decoding algorithm $\mathsf{dec}()$ for $\mathscr{C}$, an $r \times r$ invertible matrix S and a $n \times n$ permutation matrix P. The public key is the matrix $H' = SHP$. The encryption process consists of computing $c = H'm^T$ for $m \in \mathcal{M}_{n,t}$. To decrypt a cipher text $c' \in GF(q)^r$, the owner of the private key computes $\mathsf{dec}(S^{-1} \cdot c') \cdot P$.

The security of code-based cryptosystems relies upon two kinds of attacks. One type of attacks which are called *structural attacks* aims at totally breaking the cryptosystem by recovering the secret matrices. The other class of attacks try to conceive decoding algorithms for arbitrary linear codes in order to decrypt a given cipher text. Such an attack is called a *decoding attack*. The most efficient algorithms used to decode arbitrary linear codes are based on the information set decoding. A first analysis was done by McEliece in [12], then by Lee and in Brickell in [8] and also by Stern in [17] and Leon in [9] and lastly by Canteaut and Chabaud in [2] which is the best algorithm known up to now with roughly $O\left(n^3\binom{n}{t}/\binom{r}{t}\right) = O\left(n^3 2^{nh_2(\frac{t}{n})-rh_2(\frac{t}{r})+o(n)}\right)$ operations where $h_2(x) = -x\log_2(x) - (1-x)\log_2(1-x)$. Nowadays it is commonly accepted that a system is secure if the best attack known requires more than 2^{80} operations.

3   How to Sign Without Decoding?

Let k be an integer and assume that $GF(q)^k$ is the set of messages to be signed. In order to sign a message m by means of a Niederreiter cryptosystem, one has to define an $k \times n$ parity check matrix H representing a code $\mathscr{C}[n, n-k, d]$. However m has to be a correctable syndrome or in other words, there must exist $z \in \mathcal{M}_{n,t}$ such that $Hz^T = m$. This is not possible for *any* message because the decoding algorithm can only decode the set of correctable syndromes which is different from $GF(q)^r$. Thus one needs to find an application $\chi : GF(q)^k \to H\mathcal{M}_{n,t}$, and then decode $\chi(m)$ to produce $z \in \mathcal{M}_{n,t}$ that satisfies $\chi(m) = Hz^T$.

Kabatianskii *et al.* [6] presented a technique to produce code-based signatures using *any arbitrary linear code*. This means that it is not necessary to design a decoding algorithm to sign messages. The idea is to directly define a *secret* application $f : GF(q)^k \to \mathcal{M}_{n,t}$ in order to automatically generate a signature $f(m)$ for $m \in GF(q)^k$. The signer then sends $(m, f(m))$. He also publishes an application $\chi : GF(q)^k \to H\mathcal{M}_{n,t}$ to be used in the verification step which is defined for any $m \in GF(q)^k$ by $\chi(m) \stackrel{\text{def}}{=} Hf(m)^T$. A receiver checks the validity of a signed message (m, z) by verifying that:

$$\text{wt}(z) = t \quad \text{and} \quad \chi(m) = Hz^T. \tag{2}$$

The most important part of a "signing-without-decoding" scheme is to design the application f. From a practical point of view, the description of f (and χ) has to be better than an enumeration of its images for which it would need $q^k \log_2(\binom{n}{t}(q-1)^t)$ bits to store f (and also χ). Thus, a random application f would be a good choice in terms of security but a bad one for a concrete use. From a security point of view, it is necessary that the public matrix H and the public application χ *do not provide any information* about the secret application f. If this property is guaranteed then the security of the scheme is equivalent to that of the Niederreiter cryptosystem upon which it is built. Indeed to recover f from H (and χ), an opponent has to solve $\chi(m) = Hz^T$ for a given $m \in GF(q)^k$ which is actually an instance of the syndrome decoding problem.

Moreover, it should be noted that the only property needed for $\mathscr{C}$ is that it should be difficult to solve Equation (2). In other words, t should be large enough or at a pinch $\mathscr{C}$ can be a *random* linear code provided its minimum distance is large enough. The following proposition given in [6] estimates the minimum distance of a (random) linear code generated by a randomly drawn parity check matrix.

Proposition 1. *The probability $Pr\{d(\mathscr{C}) \geq d\}$ that a random $r \times n$ parity check matrix $[I_r|M]$ over $GF(q)$ defines a code $\mathscr{C}$ with a minimum distance $d(\mathscr{C})$ greater or equal to d satisfies the following inequality:*

$$Pr\{d(\mathscr{C}) \geq d\} \geq 1 - q^{-r+nh_q(\frac{d-1}{n})},$$

where $h_q(x) = x \log_q(q-1) - x \log_q(x) - (1-x) \log_q(1-x)$.

4   Kabatianskii-Krouk-Smeets Signatures

Kabatianskii *et al.* [6] proposed a signature scheme based on arbitrary linear error-correcting codes. Actually, they proposed to use a *linear* application f. Three versions are given which are presented in the sequel but all have one point in common: for any $m \in GF(q)^k$, the signature $f(m)$ is a codeword of a linear code $\mathscr{U}$. Each version of KKS proposes different linear codes in order to improve the scheme. We now give a full description of their scheme.

Firstly, we suppose that $\mathscr{C}$ is defined by a random parity check matrix H. We also assume that we have a very good estimate d of its minimum distance through Proposition 1 for instance. Next, we consider a linear code $\mathscr{U}$ of length $n' \leq n$ and dimension k defined by a generator matrix $G = [g_{i,j}]$. We suppose that there exist two integers t_1 and t_2 such that $t_1 \leq \text{wt}(u) \leq t_2$ for any non-zero codeword $u \in \mathscr{U}$.

Let J be a subset of $\{1, \ldots, n\}$ of cardinality n', $H(J)$ be the sub matrix of H consisting of the columns h_i where $i \in J$ and define an $r \times n'$ matrix

$F \overset{\text{def}}{=} H(J)G^T$. The application $f : GF(q)^k \rightarrow \mathcal{M}_{n,t}$ is then defined by $f(m) = mG^*$ for any $m \in GF(q)^k$ where $G^* = [g_{i,j}^*]$ is the $k \times n$ matrix with $g_{i,j}^* = g_{i,j}$ if $j \in J$ and $g_{i,j}^* = 0$ otherwise. The public application χ is then $\chi(m) = Fm^T$ because $HG^{*T} = H(J)G^T$. The main difference with Niederreiter signatures resides in the verification step where the receiver checks that:

$$t_1 \leq \text{wt}(z) \leq t_2 \quad \text{and} \quad F \cdot m^T = H \cdot z^T. \tag{3}$$

- Setup. The signer chooses a random matrix $H = [I_r|D]$ that represents the parity check matrix of a code $\mathscr{C}[n, n - r, \geq d]$. He also chooses a generator matrix G that defines a code $\mathscr{U}[n', k, t_1]$ such that $\text{wt}(u) \leq t_2$ for any $u \in \mathscr{U}$. He chooses a random set $J \subset \{1, \ldots, n\}$ and he forms $F = H(J)G^T$.
- Parameters.
  - Private key. The set $J \subset \{1, \ldots, n\}$ and the $k \times n'$ matrix G
  - Public key. The $r \times k$ matrix F and the $r \times n$ matrix H
- Signature. Given $m \in GF(q)^k$, the signer sends $(m, m \cdot G^*)$
- Verification. Given (m, z), the receiver verifies that:

$$t_1 \leq \text{wt}(z) \leq t_2 \quad \text{and} \quad F \cdot m^T = H \cdot z^T.$$

Fig. 1. KKS signature scheme

Note that it is not so important to have $d > 2t_2$ because it would mean otherwise that a message m may have several signatures z which are all solutions to Equation (3). Recall also that the crucial fact about $\mathscr{C}$ is that Equation (3) should be difficult to solve when the number of errors (*i.e.* the weight of z) belongs to the interval $[t_1, t_2]$. Figure 1 sums up the different steps of a KKS signature scheme.

Definition 1 (KKS-1). Let $\mathscr{U}[n', k, t]$ be an equidistant code ($t_1 = t_2 = t$) over $GF(q)$ such that $n' \leq n$ defined by a generator matrix $G = [g_{i,j}]$. It is known [11] that for such a code, $n' = \frac{q^k - 1}{q - 1}$ and $t = q^{k-1}$.

Unfortunately, KKS-1 is not practicable because it requires a code length too large. For instance in the binary case ($q = 2$) and in the FDH paradigm, k must be at least 160. It implies that $n \geq n' = 2^{160} - 1$. It is necessary to replace the equidistant code by another one for which $t_1 \neq t_2$. Two solutions are proposed in [6]: either one chooses the dual code of a binary BCH code or a random linear code thanks to Proposition 2 and Proposition 3.

Proposition 2 (Carlitz-Uchiyama Bound). Let $\mathscr{U}$ be the dual of a binary BCH code of length $n' = 2^m - 1$ and designed distance $\delta = 2s + 1$. Then for any $u \in \mathscr{U}$:

$$\left| \text{wt}(u) - \frac{n' + 1}{2} \right| \leq (s - 1)\sqrt{n' + 1}.$$

Table 1. KKS Parameters

Scheme	Version	k	n'	t_1	t_2	r	n	$nh_2(\frac{t_1}{n}) - rh_2(\frac{t_1}{r})$
KKS-2		60	1023	352	672	2808	3000	36
KKS-3	#1	60	280	50	230	990	1250	17
	#2	160	1000	90	110	1100	2000	80
KKS-4		48	180	96	96	765	1100	53

Definition 2 (KKS-2). *The signer chooses randomly: a binary $r \times (n-r)$ matrix D, a non singular $k \times k$ matrix A, an n'-subset $J \subset \{1, \ldots, n\}$. He forms a binary $r \times n$ parity check matrix $H = [I_r | D]$. He chooses a generator matrix G of a dual binary BCH code with length $n' = 2^m - 1$ and designed distance $2s + 1$. He forms $F = H(J)(AG)^T$ (he masks matrix G). The public key consists of matrices H and F, and the secret key is the set J and the matrix A.*

The following numeric values are given in [6]: $m = 10$, $s = 6$, $k = ms = 60$, $n' = 2^{10} - 1 = 1023$, $t_1 = 352$, $t_2 = 672$, $r = 2808$ and $n = 3000$. The minimum distance of $\mathscr{C}$ is at least 1024 with probability $\geq 1 - 10^{-9}$.

As for the number of bits to store, we see that the private key consists of $nh_2(\frac{n'}{n})$ bits for describing J and k^2 bits for the matrix A. For the public key, we need to store $r(n-r)$ bits for the matrix H and rk bits for the matrix F. KKS-2 can be even more improved by taking a random linear code for $\mathscr{U}$. Thanks to Proposition 3, it is possible to know the probability that a random linear code have its nonzero codeword weights inside a given interval.

Proposition 3. *Let $\mathscr{U}$ be a code defined by a random $k \times n'$ systematic generator matrix. Let δ be a real such that $0 < \delta < 1$. Then the probability that a random binary linear that its nonzero codewords have their weight inside $[\omega_1; \omega_2]$ is at least:*

$$Pr\{\omega_1 \leq wt(\mathscr{U}) \leq \omega_2\} \geq 1 - 2^{-2(n'-k)+n'h_2(\frac{\omega_1-1}{n'})+n'h_2(\frac{n'-\omega_2+1}{n'})}.$$

Definition 3 (KKS-3). *The signer follows the same steps as KKS-2 but chooses a random $k \times n'$ systematic matrix $G = [I_k | B]$. The public key consists of matrices H and F, and the secret key is the set J and the matrix B.*

Now the size of the private key of KKS-3 consists again of $nh_2(\frac{n'}{n})$ bits for J and $k(n' - k)$ bits for the matrix B. The size of the public key is not changed. We give the following numeric values: $k = 160$, $n' = 900$. The code $\mathscr{U}$ generated by G has all its weights between $t_1 = 90$ and $t_2 = 110$ with probability $\geq 1 - 2^{-749}$. The signer selects a random $1100 \times 2000)$ parity check matrix H for $\mathscr{C}$. Then $d(\mathscr{C}) > 220$ with probability $\geq 1 - 2^{-100}$. Table 1 gives the parameters given by the authors for KKS-3 called here version #1 updated with our proposition (version #2) that encounters the current security level. Unlike what was done in [6] where the security of the system is evaluated through $t = \frac{t_1+t_2}{2}$, we give a value for t_1 such that the decoding attack [2] can not cope with it.

Finally, the authors proposed a modification that helps someone to construct a KKS scheme from codes that contain codewords of low weight. The idea is to take for the code $\mathscr{U}$ the direct product of P codes $\mathscr{U}_i[n^*, k^*, t_1^*]$ over $GF(q)$ whose codewords have weight $\leq t_2^*$. Of course, $\mathscr{U}$ has also codewords of low weight. So, one has to find a way to eliminate those codewords. Assume that each code $\mathscr{U}_i$ is defined by a generator matrix G_i. The finite field $GF(q^{k^*})$ is considered as a k^*-dimensional vector space over $GF(q)$ defined by a fixed basis. We denote by M_β the matrix representing the linear map $x \mapsto x\beta$ where $\beta \in GF(q^{k^*})$. Let Q be a non-negative integer. For any $u = (u_1, \ldots, u_Q) \in GF(q^{k^*})^Q$ we define for any $x \in GF(q^{k^*})$

$$u(x) = u_1 + u_2 x + \cdots + u_Q x^{Q-1}.$$

Let $\beta_1, \ldots, \beta_P$ be non-zero elements of $GF(q^{k^*})$, $A_1, \ldots, A_P$ be non-singular $k^* \times k^*$ matrices and $J_1, \ldots, J_P$ disjoint subsets of $\{1, \ldots, n\}$ of cardinality n^* with $Pn^* < n$. The application $f : GF(q^{k^*})^Q \mapsto GF(q)^n$ sends any u to z which equals 0 on positions $\{1, \ldots, n\} \setminus \cup_{i=1}^P J_i$ and equals $u(\beta_i) A_i G_i$ on the positions of J_i. The signature equation is again $Hz^T = Fu^T$ where the public matrix is $F = (F_1, \ldots, F_Q)$ and where F_j is the $r \times k^*$ matrix that equals

$$F_j \stackrel{\text{def}}{=} \sum_{i=1}^P H(J_i) \left(M_{\beta_i^{j-1}} A_i G_i \right)^T.$$

The set of messages is now $GF(q^{k^*})^{Q-1}$ and to sign $(u_2, \ldots, u_Q)$, the sender chooses $u_1 \in GF(q^{k^*})$ such that $u_1 \notin \{\sum_{i=2}^Q u_i \beta_j^{i-1} : j = 1, \ldots, P\}$ (this is always possible when $P \leq q^{k^*}$) and so that $Pt_1^* \leq \text{wt}(f(u)) \leq Pt_2^*$. We call this scheme KKS-4.

Definition 4 (KKS-4). *The signer chooses P codes $\mathscr{U}_i[n^*, k^*, t_1]$ over $GF(q)$ whose codewords have weight $\leq t_2$, nonzero elements $\beta_1, \ldots, \beta_P$ in $GF(q^{k^*})$, non singular $k^* \times k^*$ matrices $A_1, \ldots, A_P$ and disjoint subsets $J_1, \ldots, J_P$ of $\{1, \ldots, n\}$. These quantities form the secret key. He forms matrix F as described above which constitues with matrix H the public key.*

Note that in this modified scheme $n' = Pn^*$, $k = Pk^*$, $t_1 = Pt_1^*$ and $t_2 = Pt_2^*$. The authors gave these values: $Q = 14$, $P = 12$, $k^* = 4$, $n^* = 15$. The codes $\mathscr{U}_1, \ldots, \mathscr{U}_p$ are all equal to a binary equidistant code $\mathscr{U}[15, 4, 8]$. $\mathscr{C}$ is a random code of length $n = 1100$ and dimension 335. The minimum distance $d(\mathscr{C}) \geq 193$ with probability at least $1 - 10^{-9}$. Table 1 recapitulates these values.

5   Recovering the Private Key Under a Known Message Attack

The security of KKS signatures rests on the quality of f. We have seen that if f provides no information then *a priori* KKS scheme is as secure as a Niederreiter cryptosystem. In reality, with each use of f, we do obtain a lot of information. Indeed, each signature z reveals $|\text{supp}(z)|$ positions of the secret set J. Therefore an opponent can exploit this fact to recover J. Note that once the opponent knows J, he can find secret matrix G by just solving the linear system $F = H(J)G^T$ where G represents the unknown because with high probability $H(J)$ is a full rank matrix since most of the time $r > n'$.

However in the case of KKS-4, the opponent has also to find $A_1, \ldots, A_P$ and the elements $\beta_1, \ldots, \beta_P$ that are roots of the polynomials $U(X) = \sum_{i=1}^{Q} u_i X^i$ defined by each message $(u_1, \ldots, u_Q)$. We do not treat this issue in this paper and we prefer to focus on the first scheme.

We assume that the attacker has $\ell \geq 1$ signatures (m_i, z_i) at his disposal. Each signature z_i can be seen as a result of an independent random choice, and ℓ signatures give $\left| \cup_{i=1}^{\ell} \mathsf{supp}(z_i) \right|$ elements of J. We define the random variable $U_\ell \overset{\text{def}}{=} \left| \cup_{i=1}^{\ell} \mathsf{supp}(z_i) \right|$. Thus ℓ signatures reveal on average $\mathsf{E}[U_\ell]$ positions of J where $\mathsf{E}[X]$ is the expectation of a random variable X. For any position $j \in \{1, \ldots, n'\}$, let χ_j be the Bernoulli random variable defined by $\chi_j = 1$ if $j \in \cup_{i=1}^{\ell} \mathsf{supp}(z_i)$ and by $\chi_j = 0$ otherwise. By definition $U_\ell = \sum_{j=1}^{n'} \chi_j$ and consequently $\mathsf{E}[U_\ell] = \sum_{j=1}^{n'} \mathsf{E}[\chi_j]$. Moreover, $\mathsf{Pr}\{\chi_j = 0\} = \prod_{i=1}^{\ell} \mathsf{Pr}\{j \notin \mathsf{supp}(z_i)\} = \mathsf{Pr}\{j \notin \mathsf{supp}(z_1)\}^\ell$ since the signatures z_i are considered as independent random variables:

$$
\mathsf{Pr}\{j \notin \mathsf{supp}(z_i)\} = \sum_{t=t_1}^{t_2} \mathsf{Pr}\{j \notin \mathsf{supp}(z_i), \mathsf{wt}(z_i) = t\}
$$

$$
= \sum_{t=t_1}^{t_2} \mathsf{Pr}\{j \notin \mathsf{supp}(z_i) | \mathsf{wt}(z_i) = t\} \mathsf{Pr}\{\mathsf{wt}(z_i) = t\}
$$

$$
= q^{-k} \sum_{t=t_1}^{t_2} N_t \left(1 - \frac{t}{n'}\right)
$$

where N_w is the number of codewords of $\mathcal{U}$ of weight w. The last equation is obtained thanks to the fact that $\mathsf{Pr}\{\mathsf{wt}(z_i) = t\} = \frac{N_t}{q^k}$. So we have:

$$
\mathsf{Pr}\{j \notin \mathsf{supp}(z_i)\} = 1 - \frac{q^{-k}}{n'} \sum_{t=t_1}^{t_2} t N_t.
$$

This implies that $\mathsf{Pr}\{\chi_j = 0\} = \left(1 - \frac{q^{-k}}{n'} \sum_{t=t_1}^{t_2} t N_t\right)^\ell$. Thus if we set:

$$
p_\ell \overset{\text{def}}{=} 1 - \left(1 - \frac{q^{-k}}{n'} \sum_{t=t_1}^{t_2} t N_t\right)^\ell
$$

then we have proved the following proposition.

Proposition 4. *The number U_ℓ of elements of J revealed by ℓ signatures is a random variable that follows the binomial distribution with parameters n' and p_ℓ:*

$$\Pr\{U_\ell = j\} = \binom{n'}{j} p_\ell^j (1 - p_\ell)^{n'-j}$$

$$E[U_\ell] = n'p_\ell.$$

It is necessary to know the weight distribution of the code $\mathcal{U}$ if one wishes to use Proposition 4. This property is in general difficult to calculate for an arbitrary linear code but for an equidistant code the task is easier. Corollary 1 shows that with a probability $\geq \varepsilon$ the set J can be completely determined.

Corollary 1. *Let $\mathcal{U}$ be an equidistant code $(t_1 = t_2 = t)$ and let $\varepsilon > 0$ be a positive real. Assume that $\ell \geq \dfrac{\ln(1 - \sqrt[n']{\varepsilon})}{\ln(1 - \frac{t}{n'})}$ then $\Pr\{U_\ell = n'\} \geq \varepsilon$.*

Proof. The probability p_ℓ is equal to $1 - (1 - \frac{t}{n'})^\ell$ when $\mathcal{U}$ is an equidistant code. Therefore, if $\ell \ln(1 - \dfrac{t}{n'}) \leq \ln(1 - \sqrt[n']{\varepsilon})$ then $p_\ell \geq \sqrt[n']{\varepsilon}$.

For instance, the set J is totally determined with probability ≥ 0.5 with 81 signatures for KKS-4. However, Corollary 1 is optimistic from a security point of view because we shall see that is not necessary to have so many signatures to break the scheme. An opponent can easily execute the following attack. Since $E[U_\ell] = n'p_\ell$ positions of J are known on average with ℓ signatures, the opponent has to search the $(n' - n'p_\ell)$ missing elements of J among the $(n - n'p_\ell)$ positions left. At each step, he solves k systems of r linear equations with n' unknowns and stops as soon as the system admits a solution. The cost of this attack is therefore $O(kn'^\omega \binom{n-n'p_\ell}{n'-n'p_\ell})$ where n'^ω represents the cost to solve a linear system with n' unknowns (naively $\omega = 3$). In order to apply Proposition 4 to any linear code, we need to give inequalities for p_ℓ. This can be done by remarking that:

$$1 - \left(1 - \frac{t_1}{n'}\right)^\ell \leq p_\ell \leq 1 - \left(1 - \frac{t_2}{n'}\right)^\ell.$$

Let us define $a \stackrel{\text{def}}{=} n'(1 - \frac{t_1}{n'})^\ell$ and $b \stackrel{\text{def}}{=} n - n' + n'(1 - \frac{t_1}{n'})^\ell$. We have then $\binom{n-n'p_\ell}{n'-n'p_\ell} \leq \binom{b}{a}$. We put in Table 2 the number of operations of the attack for different ℓ. These numeric results are obtained by means of Inequality (4) and by putting $\omega = 3$:

$$\binom{b}{a} \leq \frac{1}{\sqrt{2\pi a(1 - \frac{a}{b})}} 2^{bh_2(\frac{a}{b})}. \tag{4}$$

We see for instance that we need only $\ell = 13$ signatures to break KKS-2 with an amount of $O(2^{78})$ operations, and $\ell = 20$ signatures to break version #1 of the KKS-3 system with an amount of $O(2^{77})$ operations.

These numerical results are confirmed by Proposition 5 that gives a very good approximation of the maximum number of signatures allowed without compromising the security of a KKS scheme.

Table 2. Number of operations to recover J for different values of ℓ for the schemes proposed in [6]

	$\ell = 15$	$\ell = 14$	$\ell = 13$	$\ell = 12$	$\ell = 11$	$\ell = 10$
KKS-2	2^{56}	2^{65}	2^{78}	2^{97}	2^{122}	2^{160}

	$\ell = 23$	$\ell = 22$	$\ell = 21$	$\ell = 20$	$\ell = 19$	$\ell = 18$
KKS-3 (version #1)	2^{58}	2^{64}	2^{70}	2^{77}	2^{86}	2^{96}

	$\ell = 6$	$\ell = 5$	$\ell = 4$	$\ell = 3$	$\ell = 2$
KKS-4	2^{46}	2^{63}	2^{96}	2^{155}	2^{261}

Proposition 5. *Assume that n sufficiently large and let n' be such that $2n' \leq n$ and such that the security parameter λ defined by $\frac{80 - \omega \log_2 n' - \log_2 k}{n - n'}$ satisfies $0 < \lambda < 1$. Let γ be the smallest real > 0 such that $h_2(\gamma) = \lambda$. Let us define ℓ_γ by:*

$$\ell_\gamma \stackrel{\text{def}}{=} \frac{\ln \frac{\gamma}{1-\gamma} + \ln(\frac{n}{n'} - 1)}{\ln(1 - \frac{t_2}{n'})}.$$

The private key of the KKS system can be recovered with ℓ signatures if $\ell \geq \ell_\gamma$.

Proof. Let $\ell \leq \ell_\gamma$ and let $\delta_\ell \stackrel{\text{def}}{=} \frac{n' - n'p_\ell}{n - n'p_\ell}$. Note that $\delta_\ell \leq \frac{n'}{n} \leq \frac{1}{2}$. It is well-known that $\binom{n - n'p_\ell}{n' - n'p_\ell} = 2^{(n - n'p_\ell)h_2(\delta_\ell) + o(n)}$. One can check that if $\ell \leq \ell_\gamma$ then $\delta_\ell \geq \gamma$ and therefore $h_2(\delta_\ell) \geq h_2(\gamma)$. Since $p_\ell \leq 1 - (1 - \frac{t_2}{n'})^\ell$, we can write that:

$$(n - n'p_\ell)\, h_2(\delta_\ell) \geq \left(n - n' + n'\left(1 - \frac{t_2}{n'}\right)^{\ell_\gamma} \right) h_2(\gamma)$$

$$\geq \frac{1 + \gamma}{1 - \gamma}\left(80 - \omega \log_2 n' - \log_2 k\right).$$

So we have $kn'^\omega 2^{(n - n'p_\ell)h_2(\delta_\ell)} \geq 2^{80}$ because $\frac{1+\gamma}{1-\gamma} \geq 1$ ($\gamma \geq 0$) and this terminates the proof.

Proposition 5 gives $\ell_\gamma = 46$ allowed signatures obtained with $\gamma = 0.00421 \cdots$ for our parameters of KKS-3 version #2. Actually, Inequality (4) shows that we can sign at most 40 times.

6 Extension to Multi-time Signatures

From One-Time to Multi-time Signatures. Merkle trees were invented in 1979 by Merkle [13]. The original purpose was to make it possible to efficiently handle many Lamport [7] one-time signatures. A Merkle tree is a way to *commit* to n messages with a single hash value in such a way that revealing any particular message requires revelation of only $\log n$ hash values.

The underlying idea is to place the n messages at the leaves of a complete binary tree (assuming n is a power of 2 for the sake of simplicity) and then to compute the value at each non-leaf node in the tree as the hash of the values of its two children. The value at the root of the tree is the *commitment* to the n messages. To reveal a value, the user publishes it as well as the values at siblings of each ancestor of it (the so-called *authenticating path*). One can easily verify that the value was correctly revealed by simply computing hashes up the tree and checking that the ultimate hash value matches the root.

Merkle trees have been proposed to extend one-time signatures to multi-time signatures. The idea is to generate n one-time public keys, and place them in a Merkle tree. The root of the Merkle tree becomes the public key of the signature scheme. For more details, we refer the reader to Merkle's original paper [13].

From Few-Time to Multi-time Signatures. Following Merkle's idea, it is possible to extend a *few-time* signature scheme into a *multi-time* signature scheme with the same security. If the underlying signature scheme is secure against a ℓ-chosen/known message attacks, the idea is to place the n messages at the leaves of a complete ℓ-ary tree.

Let $\Sigma_\ell = (\mathsf{Setup}_\ell, \mathsf{Sign}_\ell, \mathsf{Verify}_\ell)$ be a signature scheme secure against a ℓ-chosen/known message attacks and let n be an integer (for the ease of explanation, we assume that $n = \ell^p$ is a power of ℓ). The scheme $\Sigma_{\mathsf{multi}} = (\mathsf{Setup}_{\mathsf{multi}}, \mathsf{Sign}_{\mathsf{multi}}, \mathsf{Verify}_{\mathsf{multi}})$ is defined as follows:

- $\mathsf{Setup}_{\mathsf{multi}}$: on input an integer λ (the security parameter), $\mathsf{Setup}_{\mathsf{multi}}$ calls $(\ell^{p-1} - 1)/(\ell - 1)$ times $\mathsf{Setup}_\ell(\lambda)$ in order to obtain the key pairs:

$$(\mathsf{pk}_{i,j}, \mathsf{sk}_{i,j}) \text{ for } j \in \{0, \dots, p-1\} \text{ and } i \in \{1, \dots, \ell^j\}.$$

The public key is the root $\mathsf{pk}_{1,0}$ and the private key consists of a concatenation of the key pairs $(\mathsf{pk}_{i,j}, \mathsf{sk}_{i,j})$. The user must keep a counter which contains the number of previously created signatures. In the beginning, the counter is set to zero.

- $\mathsf{Sign}_{\mathsf{multi}}$: Let i be the counter. On input a message m and the secret key, $\mathsf{Sign}_{\mathsf{multi}}$ computes $\sigma_0 = \mathsf{Sign}_\ell(m, \mathsf{sk}_{r_0,p-1})$ where $r_0 = \lfloor i/\ell \rfloor$ and then recursively $\sigma_{t+1} = \mathsf{Sign}_k(\mathsf{pk}_{r_t,p-1-t}, \mathsf{sk}_{r_{t+1},p-2-t})$, where $r_{t+1} = \lfloor r_t/\ell \rfloor$ for $t \in \{0, \dots, p-2\}$. The resulting signature on m is:

$$\sigma = \left(\sigma_0, \mathsf{pk}_{r_0,p-1}, \sigma_1, \mathsf{pk}_{r_1,p-2}, \dots, \sigma_{p-2}, \mathsf{pk}_{r_{p-2},1}\right).$$

- $\mathsf{Verify}_{\mathsf{multi}}$: on input a message m, a signature

$$\sigma = \left(\sigma_0, \mathsf{pk}_{r_0,p-1}, \sigma_1, \mathsf{pk}_{r_1,p-2}, \dots, \sigma_{p-2}, \mathsf{pk}_{r_{p-2},1}\right),$$

and a public key $\mathsf{pk}_{0,1}$, $\mathsf{Verify}_{\mathsf{multi}}$ accepts the signature σ if and only if:

$$\mathsf{Verify}_\ell(\mathsf{pk}_{r_t,p-1-t}, \mathsf{sk}_{r_{t+1},p-2-t}, \sigma_{t+1}) = 1 \text{ for } t \in \{1, \dots, p-2\}$$

and $\mathsf{Sign}_\ell(m, \mathsf{sk}_{r_0,p-1}, \sigma_0) = 1$.

The security of Σ_{multi} against an n-chosen/known message attack is trivially equivalent to the one of Σ_ℓ against a ℓ-chosen/known message attack. In the design of Σ_{multi}, tradeoffs can be made between size of the signatures and size of the public key.

This construction permits to transform the KKS signature schemes into classical *multi-time* schemes, but the resulting signatures are unfortunately very long, and in order to make the scheme more practical, it is necessary to reduce the size of the public parameters.

7   Reduction of Parameters

In this section we study the key sizes of the KKS schemes and the way to reduce them. We restrict ourselves to the binary case. Firstly, we recall the size of the different parameters for each KKS scheme. The private key consists of $nh_2(\frac{n'}{n}) + k^2$ bits in the case of KKS-2 and $nh_2(\frac{n'}{n}) + k(n'-k)$ bits in the case of KKS-3. As to the public key, both schemes need to store $r(n-r) + rk$ bits. Table 3 which gives numeric values obtained shows as such these solutions can not be used practically. However, we can improve the storage of the public key. Indeed, H can be shared by all the users. Thus each user needs only to provide his own F.

Table 3. Key sizes in bits

Scheme		Public key			Private key
		Common (H)	Personal (F)	Total public key	
KKS-2		539136	168480	707616	6378
KKS-3	version #1	257400	59400	316800	14160
	version #2	990000	176000	1166000	120385

Gaborit presented in [5] a method that reduces the key sizes of error-correcting code cryptosystems. The idea relies upon the use of almost quasi-cyclic matrices. Such matrices are completely determined if the first row is known.

Definition 5 (Almost quasi-cyclic matrix). *An $r \times n$ matrix M with $r \geq 2$ and $n \geq 2$ is a almost quasi-cyclic matrix if each row vector is rotated one element to the right relative to the preceding row vector:*

$$M = \begin{pmatrix} c_1 & c_2 & \cdots & & \cdots & c_n \\ c_n & c_1 & c_2 & & & c_{n-1} \\ c_{n-1} & c_n & c_1 & c_2 & & c_{n-2} \\ \vdots & \ddots & \ddots & \ddots & \ddots & \vdots \\ c_{n-r+2} & \cdots & c_{n-1} & c_n & c_1 & c_2 & \cdots & c_{n-r+1} \end{pmatrix}.$$

Our new scheme relies on the use of random almost quasi-cyclic codes rather than pure random linear codes. We modify KKS schemes by replacing each random matrix by a random systematic almost quasi-cyclic matrix. In other words, the parity check matrix $H = (I_r | D)$ is chosen such that D is almost quasi-cyclic. The common public key size is now $(n - r)$ bits and the personal public key still has rk bits. For KKS-2 the private key does not change but for KKS-3 the random systematic matrix G can also be a systematic almost quasi-cyclic matrix. In that case the private key has $nh_2(\frac{n'}{n}) + (n' - k)$. When applied to our proposed version of KKS-3 (number 2), this methods gives 176000 bits for the personal public key, 900 bits for common public key and only 2726 bits for the private key. The signature length is about $\log_2 \binom{n}{t_2} = \log_2 \binom{2000}{110} = nh_2(\frac{11}{200}) = 615$ bits.

8   Efficiency Issues and Conclusion

In [15], Perrig proposed a one-time signature scheme called "BiBa" (for *Bins and Balls*) whose main advantages are fast verification and short signatures. In 2002, Reyzin and Reyzin [16] presented a simpler one-time signature scheme, called HORS, that maintains BiBa's advantages and removes its main disadvantage, namely the costly generation of signatures. As the schemes studied in this paper, the HORS scheme can be used to sign a few number of messages, instead of just once (and the security decreases as this number increases). Therefore it is worth comparing its efficiency with the one of the KKS schemes.

In the table 4, we compare (for the same heuristic security) the performances of KKS-3 version #2 with our parameters and the HORS scheme implemented with the same one-way function and allowing to sign the same number of messages (namely, 40).

Table 4. Efficiency comparison of KKS and HORS for 80-bits of heuristic security

Scheme	HORS $(k, t) = (16, 23657)$	HORS $(k, t) = (20, 14766)$	HORS $(k, t) = (32, 8364)$	KKS-3
Public key size	23833000	147836000	854000	**176900**
Private key size	3785120	2362560	1338240	**2726**
Signature size	2560	3200	5120	**615**

KKS compares very favorably in performance with respect to HORS since its key sizes are much smaller and it can be used over a lower bandwidth channel. However, the generation of HORS signatures is faster since it requires only the evaluation of a hash-function. Furthermore, the security of HORS reduces to an ad-hoc (though well-defined) security assumption on the underlying hash function, whereas the scheme KKS has been proposed without any formal security analysis.

In this paper, we have quantified the variation of the security of KKS schemes against a passive attacker who may intercept a few signatures, but an interesting open issue which remains is to study their resistance to forgery in a reductionnist approach.

Acknowledgements. We thank P. Gaborit for helpful discussions.

References

1. Berlekamp, E.R., McEliece, R.J., van Tilborg, H.C.A.: On the intractability of certain coding problems. IEEE Transactions on Information Theory 24(3), 384–386 (1978)
2. Canteaut, A., Chabaud, F.: A new algorithm for finding minimum-weight words in a linear code: Application to McEliece's cryptosystem and to narrow-sense BCH codes of length 511. IEEE Transactions on Information Theory 44(1), 367–378 (1998)
3. Courtois, N.T., Finiasz, M., Sendrier, N.: How to achieve a McEliece-based digital signature scheme. In: Boyd, C. (ed.) ASIACRYPT 2001. LNCS, vol. 2248, pp. 157–174. Springer, Heidelberg (2001)
4. Engelbert, D., Overbeck, R., Schmidt, A.: A summary of McEliece-type cryptosystems and their security, Cryptology ePrint Archive, Report 2006/162 (2006), http://eprint.iacr.org/
5. Gaborit, P.: Shorter keys for code based cryptography. In: WCC 2005. LNCS, vol. 3969, pp. 81–91. Springer, Heidelberg (2006)
6. Kabatianskii, G., Krouk, E., Smeets, B.J.M.: A digital signature scheme based on random error-correcting codes. In: Darnell, M. (ed.) Cryptography and Coding. LNCS, vol. 1355, pp. 161–167. Springer, Heidelberg (1997)
7. Lamport, L.: Constructing digital signatures from a one way function, Tech. Report CSL-98, SRI International (October 1979)
8. Lee, P.J., Brickell, E.F.: An observation on the security of McEliece's public-key cryptosystem. In: Günther, C.G. (ed.) EUROCRYPT 1988. LNCS, vol. 330, pp. 275–280. Springer, Heidelberg (1988)
9. Leon, J.S.: A probabilistic algorithm for computing minimum weights of large error-correcting codes. IEEE Transactions on Information Theory 34(5), 1354–1359 (1988)
10. Li, Y.X., Deng, R.H., Wang, X.-M.: On the equivalence of McEliece's and Niederreiter's public-key cryptosystems. IEEE Transactions on Information Theory 40(1), 271–273 (1994)
11. MacWilliams, F.J., Sloane, N.J.A.: The theory of error-correcting codes, 5th edn. North–Holland, Amsterdam (1986)
12. McEliece, R.J.: A public-key system based on algebraic coding theory, pp. 114–116, Jet Propulsion Lab, DSN Progress Report 44 (1978)
13. Merkle, R.C.: A certified digital signature. In: Brassard, G. (ed.) CRYPTO 1989. LNCS, vol. 435, pp. 218–238. Springer, Heidelberg (1989)
14. Niederreiter, H.: Knapsack-type cryptosystems and algebraic coding theory. Problems Control Inform. Theory 15(2), 159–166 (1986)
15. Perrig, A.: The BiBa one-time signature and broadcast authentication protocol. In: Proceedings of the 8th ACM Conference on Computer and Communications Security, pp. 28–37. ACM Press, New York (2001)
16. Reyzin, L., Reyzin, N.: Better than BiBa: Short One-Time Signatures with Fast Signing and Verifying. In: Batten, L.M., Seberry, J. (eds.) ACISP 2002. LNCS, vol. 2384, pp. 144–153. Springer, Heidelberg (2002)
17. Stern, J.: A method for finding codewords of small weight. In: Wolfmann, J., Cohen, G. (eds.) Coding Theory and Applications. LNCS, vol. 388, pp. 106–113. Springer, Heidelberg (1989)
18. Stern, J.: A new identification scheme based on syndrome decoding. In: Stinson, D.R. (ed.) CRYPTO 1993. LNCS, vol. 773, pp. 13–21. Springer, Heidelberg (1994)
19. Véron, P.: Problème SD, opérateur trace, schémas d'identification et codes de goppa, Ph.D. thesis, Université Toulon et du Var, Toulon, France (1995)

Self-certified Signatures Based on Discrete Logarithms

Zuhua Shao

Department of Computer and Electronic Engineering
Zhejiang University of Science and Technology
No. 318, LiuHe Road, Hangzhou, Zhejiang
310023, P.R. of China
zhshao_98@yahoo.com

Abstract. In the trivial PKI, a digital signature provides the authenticity of a signed message with respect to a public key, while the authenticity of the public key with respect to a signer lies on a certificate provided by a certificate authority. To verify a signature, verifiers have to first verify the corresponding certificate. To avoid this burden, in this paper, we propose a self-certified signature scheme based on discrete logarithms to provide an implicit as well as mandatory verification of public keys. We show that this new scheme can achieve strong unforgeability in the random oracle model.

Keywords: Discrete logarithm; Self-certified public key signature; strong unforgeability.

1 Introduction

A digital signature [1] is analogous to an ordinary hand-written signature, and establishes both of signer authenticity and data integrity assurance. In a signature scheme, each signer randomly chooses a private key and publishes the corresponding public key. The signer uses his private key to compute signatures for some messages, while any other can verify the signatures with the corresponding public key. However, public key cryptosystems suffer from the well-known authentication problem [2]. If an imposter supplies a valid but incorrect public key, a user could unknowingly encipher confidential data that would be decipherable to the imposter or be tricked into accepting wrong signatures with incorrect public keys. Hence, before using a public key, users are required to first verify the binding of the public key and the identity of its holder.

The most familiar approach to certifying a public key is to use an explicit certificate such as X.509 [3]. Whenever a user wants to use a public key, he has to first verify the corresponding certificate, which incurs an additional computation burden to him. Another approach, introduced by Shamir [4], is called identity-based public key. The public key is nothing but the identity of the signer. The corresponding private key is generated by a Private Key Generator PKG. All users have to trust PKG entirely, since PKG knows their private keys. A sophisticated approach, first introduced by Girault [5], is called self-certified public key. Each user chooses his private key, computes the corresponding public key and sends it to a certificate authority. Then the certificate

C. Carlet and B. Sunar (Eds.): WAIFI 2007, LNCS 4547, pp. 252–263, 2007.
© Springer-Verlag Berlin Heidelberg 2007

authority computes the certificate (witness) for the user's public key by using a RSA cryptosystem [6], which satisfies a computationally unforgeable relationship with the public key and the identity of the holder. Verifiers can generate the public key from the identity and the certificate. The scheme offers an implicit certification, in that the authenticity of a public key is verified through the subsequent uses of the correct private key.

Compared with the identity-based approach, the self-certified public key approach can get rid of key escrow problems and secure channel problems. Compared with the traditional PKI approaches, this approach can provide an implicit validation of public keys when the system works. Hence, ordinary users without cryptography knowledge would benefit from this automatic, as well as mandatory validation of public keys.

Gentry introduced the concept of certificate-based encryption [7]. Later, Al-Riyami and Paterson introduced the concept of certificateless public key cryptography (CL-PKC) [8], including certificateless public key encryption (CL-PKE), signature (CL-PKS) and key exchange scheme from pairings. Recently, Shao proposed a self-certified scheme (SCS) from pairings [9] under the CDH assumption, which is more efficient than the CL-PKS of Al-Riyami and Paterson. Furthermore, Shao provided a formal security proof based on a new security definition in the random oracle model.

Although with different terms, the three schemes addressed the same security concerns, namely, to offer an automatic authentication for the binding public key and its holder and to get rid of the key escrow problem to be inherent in the identity-based public key cryptography (ID-PKC). Hence, they use the similar way, that is, a certificate, or, more generally, a signature of the public key and the identity of the holder, acts not only as a certificate but also as the private key.

In the three schemes, this certificate makes use of the short signatures from pairings, due to Boneh et al. [10], which is deterministic. There is single one signature corresponding to a given message. Hence, they adopt a slightly weaker security model, called single-occurrence adaptive chosen-message attack (SO-CMA), where the adversary is allowed to make at most one signature query for each message.

Since Diffie and Hellman invented the concept of the public key cryptography, discrete logarithm-based schemes have been one of the most popular signature schemes. The computation for the modular exponentiations in the discrete logarithm-based schemes is more efficient than that for the admissible pairings. Meanwhile, the underlying assumption, Discrete Logarithm problem DL, is believed to be harder than that of the short signature from pairings, Computational Diffie-Hellman CDH problem.

Petersen and Horster [11] first extended Girault's works to discrete logarithm-based cryptosystems. A problem of their self-certified public key is that it only provides implicit authentication, *i.e.*, the validity of a self-certified public key is verified only after a successful application. They illustrated the relevance of all concepts by discussing several useful applications, including proxy signatures [12]. Shao [13] proposed a self-certified signature scheme based on discrete logarithms, providing both implicit authentication and explicit authentication. Lee and Kim [14] proposed a self-certified signature scheme and a multi-certificate signature scheme. However these works did not provide a formal security proof, which results in some troublesome: there has been no consensus on the precise meaning of the security requirements for self-certified public keys based on discrete logarithms. Some schemes attempted formal security definitions and security proofs, and some schemes turned out, in fact, to be insecure [15, 16]. The main difficulty comes from the security requirement that partial private keys chosen by the adversary are unknown to simulators. To our best knowledge, no formal security definitions of strong unforgeability and security proofs for self-certified signature schemes based on discrete logarithms have been proposed to date.

In this paper, we propose a Self-Certificate Signature scheme based on Discrete Logarithms (SCSDL). We will introduce a security model and provide a formal security proof for it in the random oracle model. The underlying signature of the proposed scheme is the Schnorr signature [17]. There may be several signatures corresponding to a given message. Hence, we adopt a stronger security model, called strong unforgeability [18], where the adversary needs to forge a new signature of a message and is allowed to ask for signatures of the same message many times, and each new answer would give it some useful information. This more liberal rule makes the adversary successful when it outputs one new signature on a previously signed message.

2 Self-certified Signature Scheme

In this section, we first present a formal definition for the Self-Certified Signature schemes (SCS), which is a modification of that in [9]. The two main entities involved in a SCS scheme are a certificate authority CA and a signer S. Then we propose a concrete Self-Certified Signature scheme based on Discrete Logarithms (SCSDL).

2.1 The Definition of a Self-certified Signature Scheme

Definition 1. A Self-Certified Signature scheme (SCS) is specified as the following four probabilistic algorithms Gen, Extract, Sign, and Verify:

(1) The key generation algorithm **Gen** that when given a security parameter 1^k as input and returns two key pairs (x_{CA}, Y_{CA}) and (x_S, Y_S) of the certificate authority CA and the signer S respectively. **Gen** also outputs system parameters, including two cryptographic hash functions H and F.

(2) The certificate extracting algorithm **Extract** that when given the pair (x_{CA}, Y_{CA}) of the certificate authority CA and a certificate message CI_S in any form, which includes a serial number, the signer's identity and public key Y_S, the issuer's identity and public key Y_{CA}, a period of validity, extensions, etc., produces a signature $Cert_S$, which is the secret certificate of the certificate message CI_S.

(3) The signing algorithm **Sign** that when given the pair (x_S, Y_S) of the signer, the secret certificate $Cert_S$ and a message M as input, produces a signature σ.

(4) The verification algorithm **Verify** that on input (CI_S, M, σ), returns either *invalid* or *valid*, with the property that if $\{(x_{CA}, Y_{CA}), (x_S, Y_S)\} \leftarrow$ **Gen**(1^k), $Cert_S \leftarrow$ **Extract** (x_{CA}, Y_{CA}, CI_S) and $\sigma \leftarrow$ **Sign**$(M, x_S, Y_S, Cert_S)$, then **Verify**$(CI_S, M, \sigma) = valid$. This algorithm need not be probabilistic.

2.2 The Self-certified Signature Scheme Based on DL (SCSDL)

We use the Schnorr signature [17] as the underlying signature, which has been proven to be secure in the random oracle model. [19]

We now describe the SCSDL signature scheme in full detail.

(1) The key generation algorithm **Gen**:
Let p and q be large primes with $q|(p-1)$. Also let $G_{g,p} = \{g^0, g^1, ..., g^{q-1}\}$ be a subgroup of the multiplicative group Z_p^*, where g is a generator with the prime order q. Let H and F be (ideal) hash functions, where

$$H: \{0, 1\}^* \times Z_p^* \to Z_q^* \text{ and } F: \{0, 1\}^* \times Z_p^* \times Z_q^* \to Z_q^*$$

The certificate authority CA picks a random x_{CA} in Z_q^* as the private key and computes the corresponding public key $Y_{CA} = g^{x_{CA}} \bmod p$. The signer S generates her key pair (x_S, Y_S) similarly.

(2) The certificate extracting algorithm **Extract**:
The signer S sends her public key Y_S and her identity information to the certificate authority CA. After authenticating them, CA composes the certificate message CI_S for the signer S. The certificate authority CA computes a secret certificate $Cert_S$ that is the Schnorr signature of the certificate authority CA on the certificate message CI_S. To generate $CertS$, CA chooses a random number $k_S \in Z_q^*$ and computes

$$Cert_S = (r_S, d_S) = (g^{k_S} \bmod p, x_{CA}H(CI_S, r_s) + k_S \bmod q).$$

CA chooses a random number $r \in Z_q^*$, computes $u = g^r \bmod p$ and $w = d_S \oplus H(r_S, Y_S^r \bmod p)$. CA sends (u, r_S, w, CI_S) to the signer S.

S first recovers $Cert_S = (r_S, d_S)$ by $d_S = w \oplus H(r_S, u^{x_S} \bmod p)$. Finally, S verifies (r_S, d_S) and the certificate message CI_S by checking

$$g^{d_S} = Y_{CA}^{H(CI_S, r_S)} r_S \bmod p.$$

(3) The signing algorithm **sign**:
The signer S chooses a random number $k \in Z_q^*$ and computes

$$r = g^k \bmod p, \ h_S = H(CI_S, r_S), \ e = F(M, r, h_S), \ s = k - e(x_S h_S^2 + d_S) \bmod q.$$

The self-certified signature of the message M is $\sigma = (CI_S, s, e, r_S)$.

(4) The verification algorithm **Verify**:
The verifier first checks the validity of CI_S, and then computes $h_S = H(CI_S, r_S)$, $r = g^s (Y_S^{h_S^2} Y_{CA}^{h_S} r_S)^e \bmod p$ and $e' = F(M, r, h_S)$. If $e = e'$, outputs valid, otherwise outputs invalid.

Completeness: Because $d_S = x_{CA}H(CI_S, r_s) + k_S \bmod q$ and $s = k - e(x_S h_S^2 + d_S) \bmod q$ $= k - e(x_S h_S^2 + x_{CA}h_S + k_S) \bmod q$, then $k = s + e(x_S h_S^2 + x_{CA}h_S + k_S) \bmod q$ implies $r = g^s (Y_S^{h_S^2} Y_{CA}^{h_S} r_S)^e \bmod p$. Hence, the signature $\sigma = (s, e, CI_S, r_S)$ produced by the signing algorithm **Sign** is always *valid*.

Remark 1: By using ElGamal encryption [20] to distribute the public key certificate *Cert$_S$*, *CA* can also check that the public key Y_S chosen by the signer S is well generated [3].

Remark 2: The values $(x_S h_S^2 + d_S)$ and $(Y_S^{h_S^2} Y_{CA}^{h_S} r_S)$ can be precomputed once for all, since they are independent of messages to be signed.

Remark 3: Our results can also be carried over to other groups, such as those built on elliptic curves.

3 Security Model and Security Proof

In this section, we first describe the formal security model for the SCSDL scheme and introduce two types of adversaries. Then we provide a security proof for the SCSDL scheme in the random oracle model.

3.1 Security Model for SCSDL Scheme

Existential unforgeability against adaptive chosen message attacks (EUF-CMA) [21] is the well-accepted security model for signature schemes, where the adversary is allowed to ask the challenger to sign any message of its choice adaptively, i.e. he can adapt its queries according to previous answers. Finally, the adversary could not provide a new message-signature pair with a non-negligible advantage.

The underlying signature of the proposed scheme, the Schnorr signature, is not deterministic. The signer may generate several signatures corresponding to a given message. We adopt a stronger security model, strong unforgeability, where the adversary is allowed to ask for signatures of the same message many times, and he would obtain some useful information from each new answer. The adversary is required to forge a new signature on a previously signed message. This model gives the adversary more powers and more chances for success.

Furthermore, the security definition for Self-Certified public keys must be strengthened more. Besides ordinary adversaries of signatures, there are two types of adversaries with more powers than ordinary adversaries. A Type 1 adversary is an uncertified signer, who wants to impersonate a victim by using public keys of its choice, along with the identity of the victim. The adversary is allowed to ask for the secret certificate associated with any CI_i of its choice, including the certificate information CI_S being challenged (which is not allowed in both CL-PKS and SCS from pairings [8, 9]). We refer to such queries as certificate extraction queries. A Type 2 adversary is a malicious *CA*, who wants to impersonate a victim with a given public key. Like in PKI, however, the Type 2 adversary cannot access the corresponding private key chosen by the victim, otherwise there would be not any security at all since such malicious *CA* could know all private keys. Additionally, the self-certified signatures are with respect to two public keys, only one is chosen by the challenger, the other chosen by the adversaries is unknown to the challenger. Hence, it is a chosen key model that gives the adversaries more powers than those in ordinary signature schemes.

Type 1 attack

We say that a Self-Certified Signature scheme based on Discrete Logarithms SCSDL is strongly unforgeable against the Type 1 attack if no polynomial bounded Type 1 adversary A has a non-negligible advantage against the challenger in the following game:

Gen: The challenger takes a security parameter 1^k and runs the key-generation algorithm. It gives the Type 1 adversary the resulting system parameters $\{p, q, g\}$ and a random public key Y_{CA} of the certificate authority CA.

Queries: The Type 1 adversary A issues queries $q_1, \ldots, q_m$ adaptively where query q_i is one of:

- Certificate extraction query $<CI_i>$, where CI_i includes the public key Y_i chosen by the Type 1 adversary A besides the public key Y_{CA}.
- Self-certified sign query $<CI_i, M_i>$.

Output: Finally, the Type 1 adversary outputs a new signature σ for a message M with respect to a certificate information CI_S composed by the adversary, which includes singer's public keys Y_S chosen by the adversary besides the challenged public key Y_{CA}.

The Type 1 adversary A wins the game if the output signature (CI_S, s, e, r_S) is nontrivial, i.e. it is not an answer of a self-certified sign query for the message M and the certificate information CI_S, and r_S is not an answer of a certificate extraction query $<CI_S>$.

The probability is over the random bits used by the challenger and the Type 1 adversary.

Remark 4: If the Type 1 adversary could forge a new secret certificate for the certificate information CI_S, he would easily forge a self-certified signature for any message by using the private key of its choice.

Type 2 attack

We say that a Self-Certified Signature scheme based on Discrete Logarithms SCSDL is strongly unforgeable against the Type 2 attack if no polynomial bounded Type 2 adversary A has a non-negligible advantage against the challenger in the following game:

Gen: The challenger takes a security parameter 1^k and runs the key-generation algorithm. It gives the Type 2 adversary the resulting system parameters $\{p, q, g\}$ and a random public key Y_S of a signer.

Queries: The Type 2 adversary A issues queries $q_1, \ldots, q_m$ adaptively, where query q_i is one of:

- Ordinary sign query $<M_i>$ under the public key Y_S.
- Self-certified sign query $<CI_i, m_i>$, where CI_i includes the public key Y_{CAi} chosen by the Type 2 adversary A besides the public key Y_S.

Output: Finally, the Type 2 adversary outputs a new self-certified signature σ with respect to a certificate information CI_S composed by the adversary, which includes the public key Y_{CA} chosen by the Type 2 adversary A besides the challenged public key Y_S, or a new ordinary signature for a message M with respect to the public key Y_S.

The Type 2 adversary A wins the game if the output signature (CI_S, s, e, r_S) is nontrivial, i.e. it is not an answer of a self-certified sign query for the message M under the certificate information CI_S or the output ordinary signature is not an answer of an ordinary sign query for the message M under the challenged public key Y_S.

The probability is over the random bits used by the challenger and the Type 2 adversary.

Definition 2 (Discrete Logarithm DL assumption). A probabilistic algorithm B is said to (t, ε)-break DL in a group $G_{g,p}$, if on input $(g, p, q, y = g^a \bmod p)$ and after running in time at most t, B computes the discrete logarithm problem $a = log_{g,p}y$ with probability at least ε, where the probability is over the uniform random choices of g from the group $G_{g,p}$, of (a) from Z_q^*, and the coin tosses of B. The (t, ε)-DL assumption on the group $G_{g,p}$ is that no algorithm can (t, ε)-break DL in the group $G_{g,p}$.

3.2 Security Proof of the SCSDL Scheme

We have the following theorem about the security of the SCSDL scheme.

Theorem. Let the hash functions H, F be random oracles. Then the Self-Certified Signature scheme based on Discrete Logarithms SCSDL is strongly unforgeable under the Discrete Logarithm DL assumption.

Lemma 1: Suppose that there is a type 1 adversary A, that has advantage ε against the SCSDL scheme and A runs in time at most t. Suppose that A makes at most q_H, q_F queries to the hash functions H and F respectively, at most q_E queries to the certificate extraction oracle, and at most q_{SS} queries to the self-certified sign oracle. Then there is an algorithm B that (t', ε')-breaks DL in the group $G_{g,p}$, where:

$$\varepsilon \le q_H q_F (2^{19}\varepsilon')^{1/10} + 1/q + q_{SS}(q_F + q_{SS})/p + q_E(q_H + q_E)/p \tag{1}$$

$$t \approx t'/6 - (4q_{SS} + 2q_E)C_{exp}(G_{g,p}) \tag{2}$$

Here $C_{exp}(G_{g,p})$ denotes the computation of a long exponentiation in the group $G_{g,p}$.

Lemma 2: Suppose that there is a type 2 adversary A, that has advantage ε against the SCSDL scheme and A runs in time at most t. Suppose that A makes at most q_H, q_F queries to the hash functions H and F respectively, at most q_{OS} queries to the ordinary sign oracle, and at most q_{SS} queries to the self-certified sign oracle. Then there is an algorithm B that (t', ε')-breaks DL in the group $G_{g,p}$, where:

$$\varepsilon \le q_H q_F (2^{19}\varepsilon')^{1/10} + 1/q + q_{SS}(q_F + q_{SS})/p + q_{OS}(q_H + q_{OS})/p \tag{3}$$

$$t \approx t'/6 - (4q_{SS} + 2q_{OS})C_{exp}(G_{g,p}) \tag{4}$$

Proof: We only provide a proof for Lemma 1 since that for Lemma 2 is similar.

We show how to construct a DL algorithm B that uses A as a computer program to gain an advantage ε' for a DL problem with running time t'. The challenger takes a security parameter 1^k and runs the key-generation algorithm to obtain the group $G_{g, p}$ and Y_{CA}. Its goal to output $x_{CA} = log_{g,p}Y_{CA} \in Z_q^*$.

Algorithm B simulates the challenger and interacts with the Type 1 adversary A in the following attack games:

Algorithm B gives the adversary A the resulting parameters and Y_{CA} as the public key of the certificate authority. At any time, the adversary A can query hash oracles H or F. To response to these queries, B maintains two lists of tuples for the hash oracles H and F, respectively. We refer to these lists as H-list and F-list. The contents of the two lists are "dynamic" during the attack games. Namely, when the games start, they are initially empty, but at the end of the games, they record all pairs of queries/answers.

Answering H-oracle Queries. For a new query $<CI_i, r_i>$, algorithm B picks a random h_i in Z_q^*, and responds with $h_i = H(CI_i, r_i)$ and adds the tuple $<<CI_i, r_i>, h_i>$ to the H-list.

Answering F-oracle Queries. For a new query $<M_i, r_i, h_i>$, B checks if h_i is in the H-list and generates a random $e_i \in Z_q^*$ and responds with $e_i = F(M_i, r_i, h_i)$ and adds the tuple $<<M_i, r_i, h_i>, e_i>$ to the F-list.

Obviously, in two ways, h_i and e_i are uniform in Z_q^*, and they are independent of A's current view as required.

Answering certificate extraction queries. When A queries a new certificate extraction oracle with some certification information $<CI_i>$,

1. B checks if CI_i is a valid certificate information.
2. B generates two random integers h_i and $d_i \in Z_q^*$ and computes $r_i = g^{d_i} Y_{CAi}^{-h_i} \bmod p$, where Y_{CAi} is the public key of a certificate authority in CI_i.
3. If there exists a tuple $<<CI_i, r_i>, h_i'>$ in the H-list with $h_i \neq h_i'$, B aborts and restarts simulation (the probability of this unfortunate coincidence is at most $(q_H + q_E)/p$).
4. B answers with (r_i, d_i, CI_i), and adds the tuple $<<CI_i, r_i>, h_i>$ to the H-list.

Answering self-certified sign queries. When the adversary A queries a new self-certified signature $<CI_i, M_i>$,

1. B checks if CI_i is a valid certificate information.
2. B picks a random $r_i \in Z_p^*$ and runs the above algorithm for responding to H-queries to obtain $h_i = H(CI_i, r_i)$.
3. B chooses at random s_i and $e_i \in Z_q^*$, and computes $r = g^{s_i} (Y_i^{h_i^2} Y_{CAi}^{h_i} r_i)^{e_i} \bmod p$, where Y_{CAi} is the public key of a certificate authority and Y_i is the public key of a signer in CI_i.
4. If there exists a tuple $<<M_i, r, h_i>, e_i'>$ in the F-list with $e_i \neq e_i'$, B reports failure and terminates. (The probability of this unfortunate coincidence is at most $(q_F + q_{SS})/p$.)
5. Otherwise, B responds with (s_i, e_i, CI_i, r_i) to the adversary A and adds $<<M_i, r, h_i>, e_i>$ to the F-list.

Obviously, the outputs of the simulated oracles are indistinguishable from those in the real attacks.

Finally, the adversary A returns a new self-certified signature (CI_S, s, e, r_S) of a message M under the challenged public keys $<Y_{CA}, Y_S>$ such that

$$F(M, \ g^s (Y_S^{h_S^2} Y_{CA}^{h_S} r_S)^e \bmod p, \ h_S) = e, \text{ where } h_S = H(CI_S, r_S)$$

If the adversary A has not queried $F(M, r, h)$ or $H(CI_S, r_S)$, the probability

$$\Pr[F(M,\ g^s(Y_S^{h_S^2}Y_{CA}^{h_S}r_S)^e \bmod p, h_S) = e, \text{ where } h_S = H(CI_S, r_S)] \leq 1/q$$

since both the responses $F(M, r, H(CI_S, r_S))$ and $H(CI_S, r_S)$ are picked randomly.

Hence, with the probability

$$(1- 1/q)(\varepsilon - q_{SS}(q_F + q_{SS})/p - q_E(q_H + q_E)/p)$$
$$\geq (\varepsilon - 1/q - q_{SS}(q_F + q_{SS})/p - q_E(q_H + q_E)/p)$$

the Type 1 adversary A returns a new self-certified signature (CI_S, s, e, r_S) such that

$$F(M,\ g^s(Y_S^{h_S^2}Y_{CA}^{h_S}r_S)^e \bmod p, h_S) = e, \text{ where } h_S = H(CI_S, r_S)$$

and the responses $F(M, r, H(CI_S, r_S))$ and $H(CI_S, r_S)$ are in the F-list and the H-list.

The verification equation is equivalent to the equation

$$g^s(Y_S^{H(CI_S,r_S)^2}Y_{CA}^{H(CI_S,r_S)}r_S)^{F(M,r,H(CI_S,r_S))} = r \bmod p,$$

where the certificate information includes the public key Y_S of a signer and the public key Y_{CA} of a certificate authority. Y_S is chosen by the adversary A and Y_{CA} is chosen by the challenger.

We try to use oracle replay techniques of Pointcheval and Stern [19] to solve this DL problem, finding $x_{CA} = \log_{g,p} Y_{CA}$.

B uses six copies of the Type 1 adversary A. In the attack games, the Type 1 adversary A would choose its public key Y_S that is included in the certificate information CI_s. We first guess a fixed index $1 \leq k \leq q_H$ and hope that (CI_k, r_k) happens to be one for which A asks for $H(CI_k, r_k)$ query, i.e. $CI_S = CI_k$. Then we guess a fixed index $1 \leq j \leq q_F$ and hope that $<M_j, r_j, h_k>$ happens to be one for which A forges a self-certified signature of the message M_j. A must first asks for $H(CI_k, r_k)$ before for $F(M_j, r_j, h_k)$.

Suppose that we make two good guesses by chance, denoted by the event GoodGuess. The probability of the event GoodGuess is

$$\Pr[\text{GoodGuess}] = 1/(q_H q_F).$$

Hence, with the probability

$$\varepsilon' \geq (\varepsilon - 1/q - q_{SS}(q_F + q_{SS})/p - q_E(q_H + q_E)/p)/(q_F q_H)$$

the adversary A generates a new self-certified signature.

B gives the same system parameters, the public key Y_{CA} and same sequence of random bits to the six copies of the adversary A, and responds with the same random answers to their queries for oracles until they at the same time ask the H-oracle query for $<CI_k, r_k>$. This is the first forking point. At that point, B gives three independent random answers h_1, h_2 and h_3 to the hash queries $H(CI_k, r_k)$, the first two, gives h_1, the second two, gives h_2, and the last two, gives h_3.

Then B gives the first two copies of the adversary A same sequence of random bits, and the same random answers to their oracle queries until they both ask for $F(M_{j1}, r_{j1}, h_1)$. This is the second forking point. At that point, B gives two independent random answers e_{11} and e_{12} to the hash queries $F(M_{j1}, r_{j1}, h_1)$ in the two runs. Similarly, B gives two independent random answers e_{21} and e_{22} to the hash queries $F(M_{j2}, r_{j2}, h_2)$ (the third forking point) in the second two runs, e_{31} and e_{32} to the hash queries $F(M_{j3}, r_{j3}, h_3)$ (the 4th forking point) in the last two runs. Thus, we would obtain six self-certified signatures, satisfying the following equations:

$$g^{s_{11}}(Y_S^{h_1^2} Y_{CA}^{h_1} r_S)^{e_{11}} = r_{j1} \bmod p$$

$$g^{s_{12}}(Y_S^{h_1^2} Y_{CA}^{h_1} r_S)^{e_{12}} = r_{j1} \bmod p$$

$$g^{s_{21}}(Y_S^{h_2^2} Y_{CA}^{h_2} r_S)^{e_{21}} = r_{j2} \bmod p$$

$$g^{s_{22}}(Y_S^{h_2^2} Y_{CA}^{h_2} r_S)^{e_{22}} = r_{j2} \bmod p$$

$$g^{s_{31}}(Y_S^{h_3^2} Y_{CA}^{h_3} r_S)^{e_{31}} = r_{j3} \bmod p$$

$$g^{s_{32}}(Y_S^{h_3^2} Y_{CA}^{h_3} r_S)^{e_{32}} = r_3 \bmod p$$

From these equations, we have

$$g^{(s_{11}-s_{12})/(e_{12}-e_{11})} = Y_S^{h_1^2} Y_{CA}^{h_1} r_S \bmod p$$

$$g^{(s_{21}-s_{22})/(e_{22}-e_{21})} = Y_S^{h_2^2} Y_{CA}^{h_2} r_S \bmod p$$

$$g^{(s_{31}-s_{32})/(e_{32}-e_{31})} = Y_S^{h_3^2} Y_{CA}^{h_3} r_S \bmod p$$

Then we can derive both $\log_{g,p} Y_S$ and $\log_{g,p} Y_{CA}$, since h_1, h_2 and h_3 are different from each other.

We continue to use the "splitting lemma" [19] to compute the probability that A works as hoped. Let X be the set of possible sequences of random bits and random function values that take the adversary up to the first forking point where A asks for $H(CI_k, r_k)$; let Y be the set of possible sequences of random bits and random function values from the first forking point to the second forking point, where A asks for $F(M_{j1}, r_{j1}, h_1)$; let Z be the set of possible sequences of random bits and random function values from the second forking point. By assumption, for any $x \in X$, $y \in Y$, $z \in Z$, the probability that A, supplied the sequences of random bits and random values $(x\|y\|z)$, generates a self-certified signature is ε''.

Suppose that the sequences of random bits and random function values supplied up to the first forking point in the simulations is a. By "splitting lemma", $\Pr\{a \in$ "good" subset $\Omega\} \geq \varepsilon''/2$, and whenever $a \in \Omega$, $y \in Y$, $z \in Z$, the probability that A, supplied the sequences of random bits and random values $(a\|y\|z)$, produces a forgery is at least $\varepsilon''/2$.

Suppose that the sequences of random bits and random function values, supplied from the first forking point up to the second forking point in the simulations, is b. Thus, $\Pr\{b \in$ "good" subset $\Omega'\} \geq \varepsilon''/4$, and whenever $a \in \Omega$, $b \in \Omega'$, $z \in Z$, the probability that A, supplied the sequences of random bits and random values ($a\|b\|z$), produces a forgery is at least $\varepsilon''/4$.

By the same reason, we can compute the same probability for the other two cases.

Hence the probability that B solves the discrete logarithm through the six simulations is

$$\varepsilon' \geq (\varepsilon'')^{10}/2^{19} \geq ((\varepsilon - 1/q - q_{SS}(q_F + q_{SS})/p - q_E(q_H + q_E)/p)/(q_F q_H))^{10}/2^{19}$$

The time required to run one simulation is $t + (4q_{SS} + 2q_E)C_{exp}(G_{g,p})$. The time required to solve the discrete logarithm $\log_{g,p} Y_{CA}$ is

$$t' \leq 6(t + (4q_{SS} + 2q_E)C_{exp}(G_{g,p})). \qquad \text{Q.E.D.}$$

Therefore, from Lemma 1 and Lemma 2, we obtain the Theorem.

4 Conclusions

Compared with the trivial PKI-based signature schemes based on discrete logarithms, the proposed self-certified signature scheme avoids a multi-exponentiation for verification. If the group $G_{g, p}$ is built on GF(p), the signature size of the proposed scheme is longer than that of the trivial PKI-based signature schemes, while if the group is built on elliptic curves, the comparison result is in opposition.

Compared with the pairings-based self-certified signature schemes, such as the CL-PKS of Al-Riyami and Paterson and the SCS of Shao, the proposed scheme not only enjoys greater efficiency and easy implementation but also does not rely on the relatively new and untested hardness assumption related to bilinear maps. Furthermore the proposed scheme can achieve strong unforgeability.

In the proposed scheme, the binding of the public key and the identity of its holder can be verified implicitly at the same time as the schemes work. Afterwards the signer can use the verified private key directly to sign messages. Anyone can also use the verified public key directly to encrypt messages. This is the main advantage over the concept of self-certified keys of Petersen and Horster.

Acknowledgements. This material is based upon work funded by Zhejiang Provincial Natural Science Foundation of China under Grant No.Y104201.

References

1. Diffie, W., Hellman, M.E.: New directions in cryptography. IEEE Trans. IT-22, 644–654 (1976)
2. Kohnfelder, L.M.: A method for certificate, MIT Lab. For Computer Science, Cambridge, Mass. (1978)
3. IEEE P1363 Standard Specifications for Public Key Cryptography (2000)
4. Shamir, A.: Identity-based cryptosystem based on the discrete logarithm problem. In: Blakely, G.R., Chaum, D. (eds.) CRYPTO 1984. LNCS, vol. 196, pp. 47–53. Springer, Heidelberg (1985)
5. Girault, M.: Self-certified public keys. In: Davies, D.W. (ed.) EUROCRYPT 1991. LNCS, vol. 547, pp. 491–497. Springer, Heidelberg (1991)

6. Rivest, R.L., Shamir, A., Adelman, L.: A method for obtaining digital signatures and public-key cryptosystem. Commun. ACM 21(2), 120–126 (1978)
7. Gentry, C.: Certificated-based encryption and the certificate revocation problem. In: Biham, E. (ed.) Advances in Cryptology – EUROCRPYT 2003. LNCS, vol. 2656, pp. 272–293. Springer–Verlag, Heidelberg (2003)
8. Al-Riyami, S.S., Paterson, K.G.: Certificateless public key cryptography. In: Laih, C.-S. (ed.) ASIACRYPT 2003. LNCS, vol. 2894, pp. 452–473. Springer, Heidelberg (2003)
9. Shao, Z.: Self-certified signature scheme from pairings. Journal of System and Software 80(3), 388–395 (2007)
10. Boneh, D., Lynn, B., Shacham, H.: Short signatures from the Wail pairings. In: Boyd, C. (ed.) ASIACRYPT 2001. LNCS, vol. 2248, pp. 514–532. Springer, Heidelberg (2001)
11. Petersen, H., Hoster, P.: Self-certified keys-Concept and Applications. In: Petersen, H., Hoster, P. (eds.) Proc. Communication and Multimedia Security'97, pp. 102–116. Chapman & Hall, Sydney, Australia (1997)
12. Mambo, M., Usuda, K., Okamoto, E.: Proxy signatures: Delegation of the power to sign messages. IEICE Trans. Fundam. E79-A(9), 1338–1354 (1996)
13. Shao, Z.: Cryptographic systems using self-certified public key based on discrete logarithms. IEE Proc.-Comput. Digit. Tech. 148(6), 233–237 (2001)
14. Lee, B., Kim, K.: Self-Certified Signatures. In: Menezes, A.J., Sarkar, P. (eds.) INDOCRYPT 2002. LNCS, vol. 2551, pp. 199–214. Springer, Heidelberg (2002)
15. Wu, T.-S., Hsu, C.-L.: Threshold signature scheme using self-certified public keys. Journal of Systems and Software 67(2), 89–97 (2003)
16. Bao, H., Cao, Z., Wang, S.: Remarks on Wu-Hsu's threshold signature scheme using self-certified public keys. Journal of Systems and Software 78(1), 56–59 (2005)
17. Schnorr, C.P.: Efficient signature generation by smart cards. Journal of Cryptology 3(3), 161–174 (1991)
18. An, J., Dodis, Y., Rabin, T.: On the security of joint signature and encryption. In: Knudsen, L.R. (ed.) EUROCRYPT 2002. LNCS, vol. 2332, pp. 83–107. Springer, Heidelberg (2002)
19. Pointcheval, D., Stern, J.: Security arguments for digital signatures and blind signatures. Journal of Cryptology 13(3), 196–361 (2000)
20. ElGamal, T.: A public-key cryptosystem and a signature scheme based on discrete logarithms. IEEE Trans. Inform. Theory IT-31, 469–472 (1985)
21. Goldwasser, S., Micali, S., Rivest, R.: A digital signature scheme secure against adaptive chosen-message attacks. SIAM Journal on Computing 17(2), 281–308 (1988)

Attacking the Filter Generator over $GF(2^m)$

Sondre Rønjom and Tor Helleseth

The Selmer Center,
Department of Informatics, University of Bergen
PB 7800, N-5020 Bergen, Norway
{sondrer,torh}@ii.uib.no

Abstract. We consider the filter generator over $GF(2^m)$ consisting of a linear feedback shift register of length k that generates a maximal length linear sequence of period $2^{mk} - 1$ over $GF(2^m)$ and a Boolean function of degree d that combines bits from one element in the shift register (considered as an element in $GF(2^m)$) and creates a binary output bit z_t at any time t. We show how to extend a recent attack by the authors on the binary filter generator to the filter generator over $GF(2^m)$. The attack recovers the initial state of the filter generator from L keystream bits with complexity $O(L)$, after a pre-computation with complexity $O(L(log_2 L)^3)$, where L is the linear complexity upper bounded by $D = \sum_{i=1}^{d} \binom{n}{i}$ with $n = mk$, which is also the number of monomials of degree $\leq d$ over $GF(2)$. In addition we explain why a function of only one element of the shift register reduces the linear complexity of the keystream significantly, compared to using the function freely on bits from several words in the initial state. We also discuss implications for the WG stream cipher [4].

Keywords: Filter generators, m-sequences, Boolean functions, solving nonlinear equations.

1 Introduction

The filter generator over $GF(2)$ uses a linear feedback shift register of length n that generates a maximal linear sequence (an m-sequence) of period $2^n - 1$ in combination with a nonlinear Boolean function f of degree d that combines output from the shift register $(s_t, s_{t+1}, \ldots, s_{t+n-1})$ at any time t and produces an output bit z_t. The filter generator is an important building block in stream ciphers and some of the eSTREAM candidates make use of the filter generator as a major component.

In a recent paper [5] Rønjom and Helleseth present a new attack that reconstructs the initial state $(s_0, s_1, \ldots, s_{n-1})$ of the binary filter generator using L keystream bits with complexity $O(L)$, where L is upper bounded by $D = \sum_{i=1}^{d} \binom{n}{i}$, after a pre-computation of complexity $O(L(log_2 L)^3)$. For an introduction to results on algebraic attacks the reader is referred to [1] and [3], and [5] for a comparison of the attack described in this paper with fast algebraic attacks.

C. Carlet and B. Sunar (Eds.): WAIFI 2007, LNCS 4547, pp. 264–275, 2007.
© Springer-Verlag Berlin Heidelberg 2007

The main idea in [5] is to use the underlying structure of m-sequences to improve the existing solution techniques for the nonlinear system in the n unknowns $s_0, s_1, \ldots, s_{n-1}$ obtained from the relations

$$z_t = f_t(s_0, s_1, \ldots, s_{n-1}), \quad t = 0, 1, \ldots, L-1.$$

Then we have that $f_t(s_0, s_1, \ldots, s_{n-1}) = f(s_t, s_{t+1}, \ldots, s_{t+n-1}) \in R/J$, where $R = GF(2)[s_0, \ldots, s_{n-1}]$ is reduced over $J = \{s_0^2 + s_0, \ldots, s_{n-1}^2 + s_{n-1}\}$.

For any $U = \{u_0, u_1, \ldots, u_{r-1}\} \subset \{0, 1, \ldots, n-1\}$ let $s_U = s_{u_0} s_{u_1} \cdots s_{u_{r-1}}$. Let $K_{U,t}$ be the coefficient for the monomial s_U in the corresponding equation at time t. Then we can represent the system of equations in terms of the coefficient sequences $K_{U,t}$ as

$$z_t = \sum_U s_U K_{U,t}. \tag{1}$$

The method in [5] shows that all coefficient sequences $K_{U,t}$ where $|U| \geq 2$ corresponding to all nonlinear terms obey the same linear recursion with characteristic polynomial $p(x) = \sum_{j=0}^{L-n} p_j x^j$ with zeros β^J where the Hamming weight of the binary representation of J, denoted $wt(J)$ obeys $2 \leq J \leq d = deg(f)$. This polynomial can be constructed in the pre-computation phase and has complexity $O(L(log_2 L)^3)$ ([3]). Here β is a zero of the primitive polynomial that generates the shift register in the filter generator. The pre-computation also computes the n polynomials f_t^* for $t = 0, 1, \ldots, n-1$ defined by $f_t^*(s_0, s_1, \ldots, s_{n-1}) = \sum_{j=0}^{L-n} p_j f_{t+j}(s_0, s_1, \ldots, s_{n-1})$. Note that these are linear polynomials since all nonlinear coefficient sequences obey the recursion. Moreover, only f_0^* needs to be computed, since $f_1^*, \ldots, f_{n-1}^*$ are shifts of the equation f_0^* and the resulting $n \times n$ coefficient matrix is thus a Vandermonde type matrix.

To solve the nonlinear equation system to find the initial state $s_0, s_1, \ldots, s_{n-1}$ for a given keystream z_t of L bits, one computes the n bits $z_t^* = \sum_{j=0}^{L-n} p_j z_{t+j}$ for $t = 0, 1, \ldots, n-1$.

The initial state (secret key) $(s_0, s_1, \ldots, s_{n-1})$ can now be determined from the linear system of n equations in the n unknowns $s_0, s_1, \ldots, s_{n-1}$ given by

$$z_t^* = f_t^*(s_0, s_1, \ldots, s_{n-1}) \quad \text{for } t = 0, 1, \ldots, n-1.$$

In the case when $f_0^* \neq 0$ the coefficient matrix of the system will be non-singular. The case $f_0^* = 0$ has very small probability 2^{-n} and the attack needs some modifications and may not work so efficiently.

The method above works for a binary filter generator using a linear shift register over $GF(2)$. Some filter generators such as the WG cipher in the eSTREAM project uses a shift register over $GF(2^m)$. In this paper a filter generator over an extension field is considered. This filter generator uses a shift register of length k, where all the elements belong to $GF(2^m)$, in combination with a Boolean function f of degree d. At any time t the Boolean function is a function of the

m-bits in a single element (that is considered as an element in $GF(2)^m$) in the shift register. The WG cipher has $k = 11$, $m = 29$, $n = mk = 319$ and $d = 11$.

The focal point of this paper is to extend the attack by Rønjom and Helleseth to the filter generator over $GF(2^m)$. In order to prove this, the analog of the coefficient sequences need to be calculated. Furthermore, they are shown to possess the properties needed to extend the attack in [5]. A particular observation is that when the Boolean function is defined on a single element in the linear feedback shift register, the linear complexity of the keystream z_t will typically reduce by a factor of $e^{-d^2(k-1)/2n}$ compared with the case when the Boolean function acts on all bits in the initial state of linear feedback shift register. The attack recovers the initial state of the WG cipher in complexity $\approx 2^{45.0415}$ using the same number of keybits, after a pre-computation of complexity 2^{62}. One should, however, observe that the designers of the WG ciphers have restricted the number of keybits on a given key to 2^{45}. It should also be noted that this type of attack is applicable to any linear register, not only LFSRs, and also when the filter function is a nonlinear function of bits from several LFSRs.

2   Preliminaries

Let the sequence $\{S_t\}$ over $GF(2^m)$ obey a recursion of degree k given by

$$\sum_{j=0}^{k} g_j S_{t+j} = 0, \quad g_j \in GF(2^m)$$

where $g_0 \neq 0$ and $g_k = 1$. For cryptographic applications such as in a filter generator one normally considers the characteristic polynomial $g(x) = \sum_{j=0}^{k} g_j x^j$ of the linear recursion to be a primitive polynomial over $GF(2^m)$. The sequence $\{S_t\}$ over $GF(2^m)$ is completely determined by the initial state $(S_0, S_1, \ldots, S_{k-1})$ and the characteristic polynomial $g(x)$. We denote all 2^{mk} sequences generated by $g(x)$, corresponding to all initial states, by $\Omega(g(x))$. The nonzero sequences generated by $g(x)$ are maximal length sequences of period $2^{mk} - 1 = 2^n - 1$, where $n = mk$. The zeros of $g(x)$ are $\beta^{2^{mi}}$ for $i = 0, 1, \ldots, k-1$, where β is a primitive element in $GF(2^n)$. For further information on linear feedback shift registers the reader is referred to [2].

By repeated use of the recursion we can write S_t as a linear combination of the n elements in the initial state. Thus, we have

$$S_t = \sum_{i=0}^{k-1} S_i L_{it} \tag{2}$$

for some k sequences $\{L_{it}\}$ over $GF(2^m)$ for $i = 0, 1, \ldots, k-1$. Note that each of these k sequences obey the same recursion as $\{S_t\}$ and are thus maximal length linear sequences. This follows since it holds for all integers t that

$$0 = \sum_{j=0}^{k} g_j S_{t+j}$$

$$= \sum_{j=0}^{k} g_j \sum_{i=0}^{k-1} S_i L_{i,t+j}$$

$$= \sum_{i=0}^{k-1} S_i \sum_{j=0}^{k} g_j L_{i,t+j}.$$

This relation holds for any initial state $(S_0, S_1, \ldots, S_{n-1})$. For example, letting $S_i = 1$ and $S_l = 0$ for the remaining elements in the initial state, it follows that each $\{L_{it}\}$ obeys the recursion. Note that all the sequences $\{L_{it}\}$ are nonzero and therefore m-sequences for $i = 0, 1, \ldots, k-1$ since $L_{ii} = 1$ for $i = 0, 1, \ldots, k-1$.

Let $Tr_m^n(x)$ denote the trace mapping from $GF(2^n)$ to a subfield $GF(2^m)$, where $n = mk$, defined by

$$Tr_m^n(x) = \sum_{i=0}^{k-1} x^{2^{mi}}.$$

It is well known that we can write any sequence over $GF(2^m)$ generated by $g(x)$ in terms of the trace mapping. In particular the k sequences L_{it} for $i = 0, 1, \ldots, k-1$, can be represented in the form

$$L_{it} = Tr_m^n(A_i \beta^t) \tag{3}$$

where β is a zero (and a primitive element of $GF(2^n)$) of $g(x)$ and $A_i \in GF(2^n)$.

During the shifting of the register the elements $S_0, S_1, \ldots, S_t, \ldots$ in $GF(2^m)$ are generated. Each element in $GF(2^m)$ is identified with an m-bit binary vector using a suitable basis. Let $\{\mu_0, \mu_1, \ldots, \mu_{m-1}\}$ and $\{\alpha_0, \alpha_1, \ldots, \alpha_{m-1}\}$ denote two dual bases for $GF(2^m)$ over $GF(2)$. Thus $Tr_1^m(\mu_i \alpha_j) = \delta_{ij}$ where $\delta_{ij} = 1$ if $i = j$ and 0 otherwise. We represent the element S_t in $GF(2^m)$ as an m-bit binary vector given by $S_t = (s_{mt}, s_{mt+1}, \ldots, s_{mt+m-1})$ using the basis $\{\alpha_j\}$. Then

$$S_t = \sum_{j=0}^{m-1} s_{mt+j} \alpha_j.$$

Then using the dual basis, we can find the components of S_t by

$$s_{mt+j} = Tr_1^m(S_t \mu_j). \tag{4}$$

Since the element S_t and thus the bits s_{mt+j}, $j = 0, 1, \ldots m - 1$, are the input bits to the Boolean function at time t, it is useful to find an expression for any bit s_r in terms of the initial bits $s_0, s_1, \ldots, s_{n-1}$ in the register. In particular, the lemma below provides a relation $s_{mt+j} = \sum_{u=0}^{n-1} s_u l_{u,t}^{(j)}$ where $l_{u,t}^{(j)} = Tr_1^n(A_i \alpha_l \mu_j \beta^t)$ and $u = mi + l$, where $0 \le l < m$ and $0 \le i < k$.

Lemma 1. *Let $r = mt + j$, where $0 \le j < m$, then*

$$s_r = \sum_{u=0}^{n-1} Tr_1^n(B_u \mu_j \beta^t) s_u$$

where $u = mi + l$, $0 \le i < k$, $0 \le l < m$, and $B_u = A_i \alpha_l$.

Proof. By definition, and using the representation of $L_{it}\mu_j$ in the basis $\{\mu_l\}$, we obtain from (2), (3) and (4) that

$$s_{mt+j} = Tr_1^m(S_t \mu_j)$$

$$= Tr_1^m(\sum_{i=0}^{k-1} S_i L_{it} \mu_j)$$

$$= \sum_{i=0}^{k-1} Tr_1^m(S_i \sum_{l=0}^{m-1} Tr_1^m(L_{it}\mu_j \alpha_l)\mu_l)$$

$$= \sum_{i=0}^{k-1} \sum_{l=0}^{m-1} Tr_1^m(L_{it}\mu_j \alpha_l) Tr_1^m(S_i \mu_l)$$

$$= \sum_{i=0}^{k-1} \sum_{l=0}^{m-1} Tr_1^m(L_{it}\mu_j \alpha_l) s_{mi+l}.$$

Using the trace representation of L_{it} given in (3), and letting $r = mt + j$, we obtain

$$s_r = \sum_{i=0}^{k-1} \sum_{l=0}^{m-1} Tr_1^n(A_i \alpha_l \mu_j \beta^t) s_{mi+l}$$

$$= \sum_{i=0}^{k-1} \sum_{l=0}^{m-1} Tr_1^n(A_i \alpha_l \mu_j \beta^t) s_u$$

$$= \sum_{u=0}^{n-1} Tr_1^n(B_u \mu_j \beta^t) s_u$$

where $u = mi + l$, $0 \le i < k$, $0 \le l < m$, and $B_u = A_i \alpha_l$. $\qquad\square$

3 Finding the Coefficient Sequences

The filter function is a Boolean function $f(x_0, x_1, \ldots, x_{m-1})$ in m variables of degree d. For a subset $A = \{a_0, a_1, \ldots, a_{r-1}\}$ of $\{0, 1, \ldots, m-1\}$ we use the notation $x_A = x_{a_0} x_{a_1} \cdots x_{a_{r-1}}$. The Boolean function can then be written as

$$f(x_0, x_1, \ldots, x_{m-1}) = \sum_A c_A x_A, \quad c_A \in \{0, 1\}$$

where the summation is taken over all subsets A of $\{0, 1, \ldots, m-1\}$.

The keystream bit z_t, at time t, is computed by only selecting bits from the element $S_t = (s_{mt}, s_{mt+1}, \ldots, s_{mt+m-1})$ in the m-sequence over $GF(2^m)$ such that

$$z_t = f(s_{mt}, s_{mt+1}, \ldots, s_{mt+m-1}).$$

By expressing s_{mt}, s_{mt+1}, and s_{mt+m-1} as a linear combination of $s_0, s_1, \ldots, s_{n-1}$, we arrive at a system of equations of degree d relating the n unknowns $s_0, s_1, \ldots, s_{n-1}$ in the initial state to the keystream bits z_t. This leads to a set of nonlinear equations for $t = 0, 1, \ldots$

$$z_t = f_t(s_0, s_1, \ldots, s_{n-1}).$$

For any $U = \{u_0, u_1, \ldots, u_{r-1}\} \subset \{0, 1, \ldots, n-1\}$ let $s_U = s_{u_0} s_{u_1} \cdots s_{u_{r-1}}$. Let $K_{U,t}$ be the coefficient of s_U in the corresponding equation at time t, resulting from the keystream and the Boolean function via $z_t = f(s_{mt}, s_{mt+1}, \ldots, s_{mt+m-1})$. Then we can represent the system of equations as

$$z_t = \sum_U s_U K_{U,t}. \tag{5}$$

In this section we investigate the properties of the coefficient sequences $K_{U,t}$ of s_U. The crucial part of the attack depends on the properties of these sequences. The main observation is that these sequences have a nice structure. This is shown in [5] for the binary filter generator and will be proved also to be the case for the filter generator over an extension field.

For simplicity, we first consider the contribution to the keystream from a Boolean function consisting of a single monomial of degree r, say $f^* = x_{a_0} x_{a_1} \cdots x_{a_{r-1}}$, where $0 \leq a_0 < a_1 < \cdots < a_{r-1} < m$. Let $A = \{a_0, a_1, \ldots, a_{r-1}\}$, $u_j = m i_j + l_j$, $B_{u_j} = A_{i_j} \alpha_{l_j}$, and $l_{u_j,t}^{(a_j)} = Tr_1^n(B_{u_j} \mu_{a_j} \beta^t)$, for $j = 0, 1, \ldots, r-1$, then

$$\begin{aligned}
z_t &= f^*(s_{mt+a_0}, s_{mt+a_1}, \ldots, s_{mt+a_{r-1}}) \\
&= s_{mt+a_0} s_{mt+a_1} \cdots s_{mt+a_{r-1}} \\
&= \prod_{j=0}^{r-1} \left(\sum_{i_j=0}^{k-1} \sum_{l_j=0}^{m-1} Tr_1^n(A_{i_j} \alpha_{l_j} \mu_{a_j} \beta^t) s_{m i_j + l_j} \right) \\
&= \sum_{u_0, u_1, \ldots, u_{r-1}} s_{u_0} s_{u_1} \cdots s_{u_{r-1}} l_{u_0,t}^{(a_0)} l_{u_1,t}^{(a_1)} \cdots l_{u_{r-1},t}^{(a_{r-1})} \\
&= \sum_U s_U \sum_{U = \{u_0, u_1, \ldots, u_{r-1}\}} l_{u_0,t}^{(a_0)} l_{u_1,t}^{(a_1)} \cdots l_{u_{r-1},t}^{(a_{r-1})} \\
&= \sum_U s_U K_{U,A,t}
\end{aligned}$$

where

$$K_{U,A,t} = \sum_{U = \{u_0, u_1, \ldots, u_{r-1}\}} l_{u_0,t}^{(a_0)} l_{u_1,t}^{(a_1)} \cdots l_{u_{r-1},t}^{(a_{r-1})}. \tag{6}$$

The summation runs over all combinations of $u_0, u_1, \ldots, u_{r-1}$ where the u_j's are in $\{0, 1, \ldots, n-1\}$ and such that $U = \{u_0, u_1, \ldots, u_{r-1}\}$.

Therefore, for the general case, any Boolean function f of degree d in m variables can be written as a sum of monomials as $f = \sum_A c_A x_A$. Note in particular that each subset A of $\{0, 1, \ldots, m-1\}$ such that $|A| \geq |U|$ contributes to the coefficient sequence $K_{U,t}$. We therefore obtain

$$
\begin{aligned}
z_t &= f(s_{mt}, s_{mt+1}, \ldots, s_{mt+m-1}) \\
&= \sum_A c_A \sum_U s_U K_{U,A,t} \\
&= \sum_U s_U \sum_{A, |A| \geq |U|} c_A K_{U,A,t} \\
&= \sum_U s_U K_{U,t}
\end{aligned}
$$

where

$$
K_{U,t} = \sum_{A, |A| \geq |U|} c_A K_{U,A,t}. \tag{7}
$$

Let $U \subset \{0, 1, \ldots, n-1\}$ and $A = \{a_0, a_1, \ldots, a_{r-1}\}$. Let the sum below range over all r-tuples $(u_0, u_1, \ldots, u_{r-1})$ where the u_j's are in $\{0, 1, \ldots, n-1\}$ and such that $U = \{u_0, u_1, \ldots, u_{r-1}\}$. Let for simplicity $B_{ij} = B_i^{2^j}$, $\beta_j = \beta^{2^j}$ and $\mu_{aj} = \mu_a^{2^j}$. Then by definitions, we obtain

$$
\begin{aligned}
K_{U,A,t} &= \sum_{U = \{u_0, u_1, \ldots, u_{r-1}\}} l_{u_0,t}^{(a_0)} l_{u_1,t}^{(a_1)} \cdots l_{u_{r-1},t}^{(a_{r-1})} \\
&= \sum_U Tr_1^n(B_{u_0} \mu_{a_0} \beta^t) Tr_1^n(B_{u_1} \mu_{a_1} \beta^t) \cdots Tr_1^n(B_{u_{r-1}} \mu_{a_{r-1}} \beta^t) \\
&= \sum_U \left(\sum_{j_0=0}^{n-1} B_{u_0 j_0} \mu_{a_0 j_0} \beta_{j_0}^t \right) \left(\sum_{j_1=0}^{n-1} B_{u_1 j_1} \mu_{a_1 j_1} \beta_{j_1}^t \right) \cdots \left(\sum_{j_{r-1}=0}^{n-1} B_{u_{r-1} j_{r-1}} \mu_{a_{r-1} j_{r-1}} \beta_{j_{r-1}}^t \right) \\
&= \sum_U \sum_{j_0, j_1, \ldots j_{r-1}} B_{u_0 j_0} B_{u_1 j_1} \cdots B_{u_{r-1} j_{r-1}} \mu_{a_0 j_0} \mu_{a_1 j_1} \cdots \mu_{a_{r-1} j_{r-1}} [\beta_{j_0} \beta_{j_1} \cdots \beta_{j_{r-1}}]^t \\
&= \sum_{j_0, j_1, \ldots j_{r-1}} \left(\sum_U B_{u_0 j_0} B_{u_1 j_1} \cdots B_{u_{r-1} j_{r-1}} \right) \mu_{a_0 j_0} \mu_{a_1 j_1} \cdots \mu_{a_{r-1} j_{a-1}} [\beta_{j_0} \beta_{j_1} \cdots \beta_{j_{r-1}}]^t \\
&= \sum_{j_0, j_1, \ldots j_{r-1}} T_{U,J} \mu_{a_0 j_0} \mu_{a_1 j_1} \cdots \mu_{a_{r-1} j_{r-1}} [\beta_{j_0} \beta_{j_1} \cdots \beta_{j_{r-1}}]^t \\
&= \sum_{j_0, j_1, \ldots j_{r-1}} T_{U,J} \mu_{a_0 j_0} \mu_{a_1 j_1} \cdots \mu_{a_{r-1} j_{r-1}} \beta^{(2^{j_0} + 2^{j_1} + \cdots + 2^{j_{r-1}})t} \tag{8}
\end{aligned}
$$

where

$$
T_{U,J} = \sum_{U = \{u_0, u_1, \ldots, u_{r-1}\}} B_{u_0 j_0} B_{u_1 j_1} \cdots B_{u_{r-1} j_{r-1}}.
$$

In the following we will describe a useful lemma needed to find the minimum polynomial of the coefficient sequences $K_{U,t}$. Let j_i be integers such that

$0 \leq j_i < n$ for $i = 0, 1, \ldots, r-1$. To any r-tuple $(j_0, j_1, \ldots, j_{r-1})$ we associate the integer defined by $J = 2^{j_0} + 2^{j_1} + \cdots + 2^{j_{r-1}}$ (mod $2^n - 1$). Furthermore, the weight of J is denoted $wt(J)$ and denotes the weight of the binary representation of J.

Lemma 2. *Let $U \subseteq \{0, 1, \ldots, n-1\}$ and furthermore let $(j_0, j_1, \ldots, j_{r-1})$ and $J = 2^{j_0} + 2^{j_1} + \cdots + 2^{j_{r-1}}$ (mod $2^n - 1$). Let $wt(J) < |U|$ and let the sum below range over all r-tuples $(u_0, u_1, \ldots, u_{r-1})$ such that $U = \{u_0, u_1, \ldots, u_{r-1}\}$. Then*

$$T_{U,J} \overset{def}{=} \sum_{U=\{u_0,u_1,\ldots,u_{r-1}\}} B_{u_0 j_0} B_{u_1 j_1} \cdots B_{u_{r-1} j_{r-1}} = 0.$$

Proof. This proof is similar to the proof of Lemma 1 in [5]. We give the proof for completeness. Let $wt(J) < |U|$, where $|U|$ denotes the number of distinct elements in I considered as a subset of $\{0, 1, \ldots, n-1\}$. Since, $wt(J) < |U|$ it follows that two of the j_i's must be the same. Without loss of generality we can assume $j_0 = j_1 = j$. An important observation is that this implies that all terms cancel pairwise except the ones where $u_0 = u_1$. The reason for this is that otherwise, if $u_0 \neq u_1$, all the terms in $T_{U,J}$ will cancel pairwise since

$$B_{u_0 j} B_{u_1 j} \cdots B_{u_{r-1} j_{r-1}} = B_{u_1 j} B_{u_0 j} \cdots B_{u_{r-1} j_{r-1}}.$$

Furthermore, since if $j_0 = j_1$ we can assume that $u_0 = u_1$, it follows from the observation $B_{u,j}^2 = B_{u,j+1}$, that

$$T_{U,J} = \sum_{U=\{u_0,u_1,\ldots,u_{r-1}\}} B_{u_1 j_1+1} B_{u_2 j_2} \cdots B_{u_{r-1} j_{r-1}}.$$

Since $wt(J) < |U|$ it follows that the second indices can not all be distinct. Thus we can repeat the argument until all the remaining u_l's are distinct. The fact that $wt(J) < |U|$ implies that two of the remaining second indices must still be the same. Thus since the corresponding first indices now are different it follows that the terms in $T_{U,J}$ cancel pairwise and we obtain $T_{U,J} = 0$.   □

It follows from the expression in (8) that $K_{U,A,t}$ can be written as

$$K_{U,A,t} = \sum_{J=0}^{2^n-2} b_J \beta^{Jt}$$

and Lemma 2 combined with (8) implies that $|U| \leq wt(J) \leq |A| \leq d = deg(f)$ and some $b_J \in GF(2^n)$. Therefore $K_{U,A,t}$ is generated by the binary polynomial $p_w(x)$ with zeros β^J where $w \leq wt(J) \leq d = deg(f)$. Since $K_{U,t}$ is a linear combination of terms of the form $K_{U,A,t}$, where $|A| \geq |U|$ it holds that

$$K_{U,t} = \sum_{A,|A|\geq|U|} c_A K_{U,A,t}$$

is also generated by $p_w(x)$.

For the coefficient sequences we have therefore proved the following lemma. As a consequence of this discussion it also follows in particular that z_t is generated by $p_1(x)$.

Lemma 3. *Let $K_{U,t}$ be the coefficient sequence corresponding to s_U for a Boolean function f of degree d. Then $K_{U,t}$ is generated by the polynomial $p_w(x)$ with zeros β^J where $w = |U| \le wt(J) \le d$.*

Note that $p_1(x)| p_2(x)| \cdots | p_d(x)$. The generator polynomial $p_w(x)$ therefore generates all coefficient sequences $K_{U,t}$ for all sets of U such that $w \ge |U|$. It follows that the polynomial $p(x) = p_2(x)$ with zeros β^J where $2 \le wt(J) \le d$ generates all the coefficient sequences of degree 2 and larger. Therefore, using the recursion with characteristic polynomial $p_w(x)$ on the relation $z_t = f_t(s_0, s_1, \ldots, s_{n-1})$ leads to a linear equation in the unknowns $s_0, s_1, \ldots, s_{n-1}$ since all nonlinear terms disappear due to the recursion.

Let Z_f be the set of zeros that may occur as zeros of the minimum polynomial of the possible keystreams that may be generated by a filter generator. For a Boolean function of degree d then Z_f is contained in the set β^J where $1 \le wt(J) \le d$. Observe that the zeros of the minimum polynomial of any coefficient sequence is a subset Z_f. For example if the initial state is $(s_0, s_1, \ldots, s_{n-1}) = (1, 0, \ldots, 0)$ then $z_t = K_{\{1\},t}$. Similarly any coefficient sequence corresponding to a linear term has their zeros in Z_f. To show that this is the case for any coefficient sequence $K_{U,t}$, we proceed by induction with respect to $|U|$. In general if the initial state has $s_i = 1$ exactly when $i \in U$, then

$$z_t = \sum_{V \subset U} K_{V,t}$$

contains the term $K_{U,t}$ and terms $K_{V,t}$ corresponding to proper subsets of V of U. Since the zeros of the minimum polynomial of z_t are in Z_f and the same is true, by induction, for the zeros of the minimum polynomial of all coefficient sequences $K_{V,t}$ for all proper subsets V of U, it follows that the zeros of the minimum polynomial of $K_{U,t}$ are also in Z_f.

Thus we can perform essentially the same attack as in [5] also for the word oriented filter generator by deriving a set of linear equations in the n unknowns $s_0, s_1, \ldots, s_{n-1}$ and the bits in the keystream z_t. The complexity of the attack will be as in [5] a pre-computation of complexity $O(L(log_2 L)^3)$ and the actual attack is of complexity $O(L)$, where L is the linear complexity upper bounded by $D = \sum_{i=1}^{d} \binom{n}{i}$. For the WG cipher the linear complexity L is given in [4] to be $\approx 2^{45.0415}$. The zeros of the generator polynomial of all possible keystream z_t are known and the generator polynomial corresponding to the coefficient sequences of degree at least two can be generated similar to the techniques in [3] with complexity $O(L(log_2 L)^3)$. The initial state in the WG cipher can be found in complexity $\approx 2^{45.0415}$ using the same number of keybits, after a pre-computation done only once with complexity 2^{62}. One should, however, observe that the authors restrict the number of keybits on a given key to 2^{45}. Note that a standard algebraic attack is not possible directly and the existence of a fast algebraic attack is highly unlikely according to the authors of WG.

One additionally interesting observation in studying the word oriented filter generator that we observe from our approach is that the linear complexity is sometimes significantly smaller than it would have been if the Boolean function had acted on all the bits in the initial state instead of limiting itself to the bits within one single word.

For a typical binary filter generator involving a Boolean function of degree d, the linear complexity L will typically be of the order $\sum_{i=1}^{d} \binom{n}{i}$. This corresponds to the maximum number of possible nonzero terms of the form $b_J \beta^{Jt}$ where $1 \leq wt(J) \leq d$. In general there are no known way of showing that certain coefficients disappear for all Boolean functions. Furthermore, examples shows that frequently all coefficients are nonzero.

However, in the special case when one uses the filter generator over $GF(2^m)$ there are large families of integers J such that $b_J = 0$, no matter how the Boolean function is selected, as long as it acts on one element in the linear feedback shift register only. This observation may lead to a significant reduction of the linear complexity of z_t.

Lemma 4. *Let z_t be the keystream obtained from the filter generator of length k with elements over $GF(2^m)$. Then the linear complexity of z_t is at most*

$$\sum_{w=1}^{d} \frac{n(n-k)(n-2k)\ldots(n-(w-1)k)}{w!}$$

where $n = mk$.

Proof. The linear complexity of a keystream sequence z_t is the number of nonzero coefficients in the expression for z_t in terms of the zeros β^J given by

$$z_t = \sum_{J, 1 \leq wt(J) \leq d} b_J \beta^{Jt}.$$

The keystream z_t can be written as a sum of coefficient sequences given by

$$z_t = \sum_U s_U K_{U,t}.$$

We will show that for large families of integers J it holds that the coefficient b_J of β^{Jt} is zero for all coefficient sequences $K_{U,t}$. Let

$$J = 2^{j'_0} + 2^{j'_1} + \cdots + 2^{j'_{w-1}}$$

where the j'_i's are distinct and $0 \leq j'_i < n$. The main observation is that it turns out that $b_J = 0$ for any J whenever two of the j'_i's are the same modulo m. This holds for any Boolean function that acts on a single element in the linear feedback shift register.

To give a brief sketch of why this holds we again consider the description $K_{U,t}$ as a sum of terms of the form $K_{U,A,t}$, where $|A| \geq |U|$. It follows that it is sufficient to show that the coefficient of β^{Jt} disappears in the description of $K_{U,A,t}$ for these values of J. The main idea is to extend the arguments in Lemma 2 in combination with the key fact that the $\mu_{a_ij_i}$'s are in the subfield $GF(2^m)$. It follows from (8) that $K_{U,A,t}$ equals

$$K_{U,A,t} = \sum_{j_0, j_1, \ldots, j_{r-1}} T_{U,J} \mu_{a_0 j_0} \mu_{a_1 j_1} \cdots \mu_{a_{r-1} j_{r-1}} \beta^{(2^{j_0} + 2^{j_1} + \cdots + 2^{j_{r-1}})t}$$

where

$$T_{U,J} = \sum_{U=\{u_0, u_1, \ldots, u_{r-1}\}} B_{u_0 j_0} B_{u_1 j_1} \cdots B_{u_{r-1} j_{r-1}}.$$

Each $J = 2^{j_0} + 2^{j_1} + \cdots + 2^{j_{r-1}}$ contributing to this sum do not necessarily have distinct values of the j_i's. We assume that the unique binary representation of J is $J = 2^{j'_0} + 2^{j'_1} + \cdots + 2^{j'_{w-1}}$ where the j'_is are distinct. It follows from Lemma 2 that $T_{U,J} = 0$ whenever $wt(J) < |U|$. In the case when $wt(J) \geq |U|$, the arguments in Lemma 2 imply that $T_{U,J}$ can be represented in the form

$$T_{U,J} = \sum_{U=\{u'_0, u'_1, \ldots, u'_{w-1}\}} B_{u'_0 j'_1} B_{u'_1 j'_1} \cdots B_{u'_{w-1} j'_{w-1}}.$$

Furthermore, subsets of the j_i's are combined to form the j'_i's. Suppose $j'_1 = j'_0 + tm$. Then any of the j_i's involved in forming j'_0 are increased by tm while any of the j_i's involved in j'_1 are decreased by tm. This leads to another number J^* modulo $2^n - 1$. Since the $\mu_{a_i j_i}$'s are in $GF(2^m)$, the modifications of j_i by a multiple of m do not change the value of $\mu_{a_i j_i}$, and thus the contribution from J corresponding to $j_0, j_1, \ldots, j_{r-1}$ and J^* cancel. It follows that all terms cancel pairwise and thus the coefficient $b_J = 0$.

The number of choices of J of weight w with a nonzero coefficient b_J of β^{Jt} is therefore at most the number of choices of the $j'_1, j'_2, \ldots, j'_w$ that are distinct modulo m, which is given by $\frac{n(n-k)(n-2k)\cdots(n-(w-1)k)}{w!}$, since $n = mk$. $\square$

Observe that the fraction of J's of weight $wt(J) = w$ with $b_J \neq 0$ for this case compared with the case when the Boolean function acts on bits in the initial state (when the linear complexity contribution from terms J of weight w is typically $\binom{n}{w}$) is therefore

$$Frac(w) = \frac{n(n-k)(n-2k)\cdots(n-(w-1)k)}{n(n-1)(n-2)\cdots(n-(w-1))}$$

Hence, for large values of $n = mk$ compared to k we obtain a rough estimate of $Frac(w) = e^{-w(w-1)(k-1)/(2n)}$. Since the main contribution to the linear complexity will normally come from the highest degree d a typically reduction in complexity will be by a factor of magnitude about $e^{-d^2(k-1)/(2n)}$ under these assumptions.

4 Conclusions

The filter generator over an extension field uses a shift register of length k that generates an m-sequence $S_0, S_1, \ldots$ of period $2^{mk} - 1$ over $GF(2^m)$ in combination with a Boolean function f of degree d that acts on a single element S_t at any time t. An attack on the filter generator over $GF(2^m)$ has been described in this paper as an extension of the attack in [5]. The general attack reconstructs the initial state in complexity $O(L)$ after a pre-computation of complexity $O(L(log_2 L)^3)$, where $L \le D = \sum_{i=1}^{d} \binom{n}{i}$.

An additional observation is that this construction reduces the linear complexity L of the keystream, sometimes to a small fraction of the maximal linear complexity that would be possible by using the Boolean function on all bits instead of just using the bits confined to one single element of $GF(2^m)$.

Acknowledgements

This work was supported by the Norwegian Research Council.

References

1. Canteaut, A.: Open problems related to algebraic attacks on stream ciphers. In: Ytrehus, Ø., (ed.) WCC 2005. LNCS, vol. 3969, pp. 120–134. Springer, Heidelberg (2006)
2. Golomb, S.W., Gong, G.: Gong Signal Design for Good Correlation: For Wireless Communication, Cryptography and Radar. Cambridge University Press, Cambridge (2005)
3. Hawkes, P., Rose, G.: Rewriting variables: The complexity of fast algebraic attacks on stream ciphers. In: Franklin, M. (ed.) CRYPTO 2004. LNCS, vol. 3152, pp. 390–406. Springer, Heidelberg (2004)
4. Nawaz, Y., Gong, G.: The WG stream cipher, eSTREAM project, http://www.cosic.esat.kuleuven.ac.be/ecrypt/stream/ciphers/wg/wg.pdf .
5. Rønjom, S., Helleseth, T.: A New Attack on the Filter Generator, accepted by IEEE Transactions on Information Theory

Cyclic Additive and Quantum Stabilizer Codes

Jürgen Bierbrauer

Department of Mathematical Sciences
Michigan Technological University
Houghton, Michigan 49931, USA

Abstract. We develop the theory of additive cyclic codes and of cyclic quantum stabilizer codes.

Keywords: Cyclic codes, additive codes, quantum codes, Galois group, cyclotomic cosets, Kloosterman codes.

1   Introduction

Definition 1. *A q-linear q^m-ary code $[n, k]_{q^m}$ is a km-dimensional $\mathbb{F}_q$-subspace of $C \subseteq E^n$, where $E = \mathbb{F}_q^m$. In particular C has q^{km} codewords.*

This is a generalization of linear codes in the sense that the alphabet is not considered as field $\mathbb{F}_{q^m}$ but only as a vector space over the subfield $\mathbb{F}_q$. These codes are also collectively known as **additive codes** although it may be wiser to reserve this term for the more general case when the alphabet is considered as an abelian group.

In [8, 2, 3] we generalized the classical theory of linear cyclic codes and obtained a description of a large class of additive cyclic codes. This mechanism was then used to obtain constructions of quantum stabilizer codes. Let us call the family of cyclic additive codes as described in [8, 2, 3] **twisted codes.** The theory of twisted codes does not describe the additive cyclic codes in their generality. A small example of a good cyclic code, which is not twisted, is the $[7, 3.5, 4]_4$-code obtained by a computer construction in Blokhuis-Brouwer [6]. In [4] we give a design-theoretic construction of this code.

One of the main objectives of the present paper is to give a general theory of the additive cyclic codes of length coprime to the characteristic. It turns out that the theory of twisted codes is the main ingredient of the general theory. We obtain a parametric description of all those codes and enumerate them. A first construction result is a cyclic $[15, 4.5, 9]_4$-code. The second main objective is the theory of cyclic quantum stabilizer codes. This is the special case of the general theory when the alphabet is $E = \mathbb{F}_q^2$ and the code is self-orthogonal with respect to the symplectic form. We give a parametric description and an enumeration of those quantum stabilizer codes and of the self-dual such quantum codes. The case $q = 2$ of quaternary quantum stabilizer codes is of special interest. In fact the original paper [7] works only with quaternary codes. A justification of the extension to general self-orthogonal symplectic codes is in [1, 11].

C. Carlet and B. Sunar (Eds.): WAIFI 2007, LNCS 4547, pp. 276–283, 2007.
© Springer-Verlag Berlin Heidelberg 2007

2 Notation and Twisted Codes

Let $n \mid q^r - 1$, where r is the order of q mod n. Consider the field $F = \mathbb{F}_{q^r}$ and the trace $tr : F \longrightarrow \mathbb{F}_q$. The Galois group $G = Gal(F|\mathbb{F}_q) = \{g_0, g_1, \ldots, g_{r-1}\}$ where g_i maps $x \in F$ to x^{q^i} acts on $Z_n = \mathbb{Z}/n\mathbb{Z}$ in the natural way. The orbits of G on Z_n are the cyclotomic cosets. For a fixed cyclotomic coset Z we denote $s = |Z|$. Let $z \in Z$. The stabilizer H of z in G has order r/s and is generated by g_s. The fixed field of H is $L = \mathbb{F}_{q^s}$. Let $W \subset F$ be the subgroup of order n, generated by α.

A twisted code $\mathcal{C}$ of length n is described by two ingredients,

- a subset $A \subseteq Z_n$ and
- a tuple $\Gamma = (\gamma_1, \gamma_2, \ldots, \gamma_m) \in F^m$ such that the γ_j are linearly indendent over $\mathbb{F}_q$ (and consequently $m \leq r$).

The F-vector space $\mathcal{P}(A)$ consists of the polynomials with coefficients in F all of whose monomials have degrees in A. The codewords of the F-linear length n code $(\mathcal{B}(A), \Gamma)$ are the evaluations $ev(v, \Gamma)$ where $v \in \mathcal{P}(A)$, whose entry in coordinate $u \in W$ is the m-tuple $(\gamma_1 v, \ldots, \gamma_m v)$. Let $\langle , \rangle$ be a nondegenerate bilinear form on the vector space $E = \mathbb{F}_q^m$, extended to E^n in the natural way. The twisted codes are $tr(\mathcal{B}(A), \Gamma)$ and its dual with respect to $\langle , \rangle$. In a traditional coding context it is natural to choose $\langle , \rangle$ as the Euclidean form (the dot product). In the context of quantum codes one chooses $m = 2$ and a symplectic form, giving E the structure of a hyperbolic plane.

Let $A = \{z\}$ and Z the cyclotomic coset containing z. With the notation introduced above ($|Z| = s, H, L$) denote by Γ^H the orbit of Γ under H and by $\langle \Gamma^H \rangle_F$ the F-vector space generated by Γ^H. Also, let $\langle \gamma_1, \ldots, \gamma_m \rangle_L$ be the L-vector space generated by the γ_j.

Proposition 1. *The dimension of* $tr((\mathcal{B}(z), \Gamma)$ *equals*

$$s \cdot dim_F \langle \Gamma^H \rangle_F = s \cdot dim_L(\langle \gamma_1, \ldots, \gamma_m \rangle_L)$$

Proof. The first expression is a special case of Theorem 21 of [2]. In order to prove the second expression consider the $\mathbb{F}_q$-linear mapping $a \mapsto tr(ev(aX^z, \Gamma))$ which by definition is surjective : $F \longrightarrow tr((\mathcal{B}(z), \Gamma)$. It has a in the kernel if and only if $a \in (\langle \gamma_1, \ldots, \gamma_m \rangle_L)^\perp$, where the dual is with respect to the trace form. $\square$

We consider the ambient space $V_F = F^{mn}$ as the space of n-tuples of m-tuples: $V_F = \{(v_u) \mid u \in W\}$ where $v_u \in F^m$. For each cyclotomic coset Z let $V_F(Z) = ev(\mathcal{P}(Z)^m)$. Here the typical element $(p_1(X), \ldots, p_m(X)) \in \mathcal{P}(Z)^m$ has evaluation (v_u), where $v_u = (p_1(u), \ldots, p_m(u))$. We have $dim(V_F(Z)) = m|Z|$ and

$$V_F = \oplus_Z V_F(Z).$$

Also the $\mathbb{F}_q$-dimension of $tr(V_F(Z))$ is $m|Z|$ and

$$\mathbb{F}_q^{mn} = E^n = \oplus_Z tr(V_F(Z)).$$

It follows from the classical theory of linear cyclic codes that the dual of $tr(V_F(Z))$ with respect to $\langle , \rangle$ is $\oplus_{Z' \neq -Z} tr(V_F(Z'))$ (see Theorem 13.6 of [3]).

3   Irreducible Additive Cyclic Codes

Definition 2. *An additive cyclic code is* **irreducible** *if it does not contain proper nonzero additive cyclic subcodes.*

By Maschke's theorem the cyclic additive codes of length coprime to the characteristic are completely reducible. It suffices therefore, in a sense, to describe the irreducible cyclic codes. The irreducible cyclic codes contained in the twisted codes are easy to describe:

Proposition 2. *The irreducible cyclic codes contained in the twisted code $tr((\mathcal{B}(z), \Gamma))$ are $\{tr(ev(r_0 a X^z, \Gamma)) \mid a \in L\}$ where $r_0 \in F \setminus \langle \gamma_1, \ldots, \gamma_m \rangle_L^{\perp}$ is fixed and the dual is with respect to the trace form. The dimension of this irreducible cyclic code is $s = |Z|$.*

Proof. Because of irreducibility we can restrict to case $A = \{z\}$. Start from $0 \neq v = tr(ev(r_0 X^z))$. The fact that $v \neq 0$ is equivalent with $r_0 \notin (\langle \gamma_1, \ldots, \gamma_m \rangle_L)^{\perp}$. Let $\mathcal{C}$ be the cyclic closure of v. The cyclic shift produces $tr(ev(r_0 \alpha^z X^z))$ where $W = \langle \alpha \rangle$. This shows that $tr(ev(r_0 L X^z)) \subseteq \mathcal{C}$. As this code is additive and cyclic we have equality. The mapping from $a \in L$ to the generic codeword is $\mathbb{F}_q$-linear, and $a \neq 0$ is in the kernel of the mapping if and only if $r_0 \in \langle \gamma_1, \ldots, \gamma_m \rangle_L^{\perp}$, contradiction. It follows that each of these irreducible codes has dimension s. $\square$

Comparison with Proposition 1 shows that the twisted code itself is irreducible if and only if $dim_L(\langle \gamma_1, \ldots, \gamma_m \rangle_L) = 1$. This means that all the γ_j are L-multiples of one another.

Next we want to study how twisted codes $tr((\mathcal{B}(z), \Gamma)$ and $tr((\mathcal{B}(z), \Gamma')$ intersect. Let $v = tr(ev(a X^z, \Gamma))$. Then v is in the intersection if there exists a' such that $v = tr(ev(a' X^z, \Gamma')$. This is equivalent with $a\Gamma - a'\Gamma'$ having all its components in $L^{\perp}$.

Definition 3. *Let $c_s(\Gamma, \Gamma')$ be the L-dimension of the space $A \subseteq F$ of all a such that there exists a' satisfying $a\Gamma - a'\Gamma' \in (L^{\perp})^m$. Here the dual is with respect to the trace form.*

Clearly A is an L-vector space and $A \supseteq (\langle \gamma_1, \ldots, \gamma_m \rangle_L)^{\perp}$.

Proposition 3. *The dimension of $tr((\mathcal{B}(z), \Gamma) \cap tr((\mathcal{B}(z), \Gamma')$ is*

$$s \cdot (c_s(\Gamma, \Gamma') - dim_L(\langle \gamma_1, \ldots, \gamma_m \rangle_L^{\perp})).$$

Proof. The elements of the intersection are parametrized by the elements a of the $s \cdot c_s(\Gamma, \Gamma')$-dimensional space defined above, and a is in the kernel if it is orthogonal to the L-vector space generated by the γ_j. $\square$

Comparison with Proposition 1 shows that $tr((\mathcal{B}(z), \Gamma) \subseteq tr((\mathcal{B}(z), \Gamma')$ if and only if $c_s(\Gamma, \Gamma') = r/s$ and $tr((\mathcal{B}(z), \Gamma) \subseteq tr((\mathcal{B}(z), \Gamma')$ if and only if $c_s(\Gamma, \Gamma') = c_s(\Gamma', \Gamma) = r/s$.

A central result is the following:

Theorem 1. *Let $\mathcal{C}$ be an irreducible cyclic additive q^m-ary q-linear code of length n where $gcd(n,q) = 1$ such that the projection to each of the m coordinates is nonzero. Then $\mathcal{C}$ is contained in a twisted code.*

Proof. Let π_j be the projection from $\mathbb{F}_q^m$ onto coordinate j, where $j = 1, \ldots, m$. By assumption $\pi_j(\mathcal{C})$ is not the 0-code. As it is a q-ary linear cyclic code it can be described by a nonempty union Z_j of cyclotomic cosets. Because of irreducibility each Z_j is a cyclotomic coset. Choose $z_j \in Z_j$ and consider the stabilizers H_j and fixed fields L_j. Let $\mathcal{B}$ be the set of all m-tuples $(v_1, \ldots, v_m)$ of polynomials whose trace-evaluations are in $\mathcal{C}$. By definition and basic properties of the trace we can choose $v_j = b_j X^{z_j}$. Without restriction there is some $(aX^{z_1}, bX^{z_2}, \ldots) \in \mathcal{B}$ where $a \notin L_1^\perp, b \notin L_2^\perp$. Let α be a generator of W. The cyclicity shows that some $(a\alpha^{z_1} X^{z_1}, b\alpha^{z_2} X^{z_2}, \ldots) \in \mathcal{B}$, by induction $(a\alpha^{iz_1} X^{z_1}, b\alpha^{iz_2} X^{z_2}, \ldots) \in \mathcal{B}$ for all $i = 0, 1, \ldots$. Assume $ord(\alpha^{z_1}) \neq ord(\alpha^{z_2})$. Without restriction there is some i such that $\alpha^{iz_1} = 1, \alpha^{iz_2} \neq 1$. After subtraction this yields $(0, b(\alpha^{iz_2} - 1)(X^{z_2}, \ldots) \in \mathcal{B}$. This contradicts the irreducibility. It follows $ord(\alpha^{z_1}) = ord(\alpha^{z_2})$ and therefore $|Z_1| = |Z_2| = s$. Let $\beta = \alpha^{z_1}$. Then $\alpha^{z_2} = \beta^j$ for a j coprime with the order of β. The smallest field containing β is $L = GF(q^s)$. The irreducibility shows that $\sum c_i \beta^i = 0$ with coefficients $c_i \in \mathbb{F}_q$ if and only if $\sum c_i \beta^{ij} = 0$. This shows that raising to the j-th power is a field automorphism of L. It follows that z_2 and z_1 are in the same cyclotomic coset. $\square$

4 The Additive Cyclic Codes

By Theorem 1 each irreducible additive cyclic code is contained in a twisted code provided all its m coordinate projections are nonzero. By Proposition 2 we know the irreducible cyclic codes contained in twisted codes. If some of the coordinate projections of our irreducible cyclic code vanish, then we should forget them. The code is then contained in a twisted code for a smaller value of m. It follows that an irreducible cyclic additive code $\mathcal{C}$ is determined by the following ingredients:

- the support consisting of the coordinates $j \in \{1, 2, \ldots, m\}$ such that $\pi_j(\mathcal{C}) \neq 0$,
- an m-tuple Γ with 0-entries outside of the support and nonzero entries in the support,
- a cyclotomic coset and,
- in the case when the corresponding twisted code is not irreducible, a coset (see Proposition 2).

The first two ingredients can be combined into one: a nonzero tuple $\Gamma \in F^m$. We specialize to case $m = 2$ now.

Theorem 2. *The irreducible cyclic additive codes in case $m = 2$ are the twisted codes $tr((\mathcal{B}(z), \Gamma)$, where z varies over representatives z of cyclotomic cosets and Γ varies over representatives of points in $PG(1, L)$, where $L = GF(q^s)$ and $s = |Z|$.*

Proof. We know that all those codes are indeed irreducible, and they are precisely the twisted codes (with $m = 2$ or $m = 1$) which are irreducible cyclic codes. Consider now a code $tr((\mathcal{B}(z), \Gamma')$ where $\Gamma' = (1, \gamma')$ and $\gamma \notin L$. Let $\Gamma = (1, \gamma)$ where $\gamma \in L$. We claim that $tr((\mathcal{B}(z), \Gamma) \subset tr((\mathcal{B}(z), \Gamma')$. In order to show this it suffices by Proposition 3 to show that for each $a \in F$ we can find a' such that $a(1, \gamma) - a'(1, \gamma') \in (L^\perp)^2$, in other words $tr_L(\gamma')$ and $tr_L(a'\gamma')$ are prescribed, where tr_L is the trace : $F \longrightarrow L$. As $\gamma' \notin L$ this is possible. The same proof works when $\Gamma = (0, 1)$. We see that the nonzero elements of the $2s$-dimensional code $tr((\mathcal{B}(z), \Gamma')$ are partitioned into the nonzero elements of those irreducible twisted subcodes.   □

Proposition 4. *The number of nonzero irreducible cyclic additive codes in case $m = 2$ is $\sum_Z (q^{|Z|} + 1)$.*

We also obtain a complete picture of the additive cyclic codes, at least in case $m = 2$. Recall that the ambient space $E^n = \mathbb{F}_q^{2n}$ is written as the direct sum of the $tr(V_F(Z))$, where Z varies over the cyclotomic cosets and $dim(tr(V_F(Z))) = 2|Z| = 2s$. Each $\Gamma \in F \times F$ generates a point $P \in PG(1, F)$. Choose a representative z for each cyclotomic coset Z. The irreducible cyclic codes are precisely the $tr(\mathcal{B}(z), \Gamma)$ where z varies over the representatives of cyclotomic cosets and Γ varies over representatives of points in $PG(1, q^s)$ (and $|Z| = s$). Each Z therefore yields $q^s + 1$ irreducible cyclic codes, each of $\mathbb{F}_q$-dimension s. Their nonzero elements partition the $2s$-dimensional space $tr(V_F(Z))$, and $tr((\mathcal{B}(z), \Gamma) = tr(V_F(Z))$ whenever Γ generates a point of $PG(1, F)$ which is not in $PG(1, L)$.

Theorem 3. *The number of cyclic additive codes in case $m = 2$ is $\prod_Z (q^{|Z|} + 3)$.*

Proof. In order to describe a cyclic code we have to decide the contribution of each cyclotomic coset Z. If $|Z| = s$ there are $q^s + 3$ possibilities: the summand is either one of the $q^s + 1$ irreducible codes (dimension s each) or the 0-code or is $tr(V_F(Z))$.   □

Lengths 7 and 15

As illustrative examples consider the quaternary length 7 and 15 cases. In the first case ($q = 2, m = 2, n = 7$) we have cyclotomic cosets of lengths $1, 3, 3$ (representatives $0, 1, -1$) and therefore $3 + 9 + 9 = 21$ irreducible cyclic, $5 \times 11 \times 11 = 605$ cyclic codes alltogether. A $[7, 3.5, 4]_4$-code is obtained as the sum of the irreducible codes $tr((\mathcal{B}(1), 1, 0)$, $tr((\mathcal{B}(-1), 0, 1)$, $tr((\mathcal{B}(0), 1, 1)$. It is self-dual with respect to the dot product but not self-dual with respect to the symplectic form. In fact, there is no additive code with these parameters which is self-dual in the symplectic sense, as this would generate a quantum code $[[7, 0, 4]]_4$ which cannot exist (see [10]).

In length 15 we use the primitive equation $\epsilon^4 = \epsilon + 1$ for $F = GF(16) = \mathbb{F}_2(\epsilon)$. Representatives of the cyclotomic cosets are 0 (length 1), 5 (length 2) and $1, 3, 14$ (length 4 each). There are $3 + 5 + 3 \times 17 = 59$ irreducible cyclic codes and a total of $5 \times 7 \times 19^3$ cyclic codes. Define

$$\mathcal{C} = tr((\mathcal{B}(1), \epsilon, 1) \oplus tr((\mathcal{B}(3), 1, \epsilon^2) \oplus tr((\mathcal{B}(0), 1, 1).$$

Then $\mathcal{C}$ is a $[15, 4.5, 9]_4$-code. These are new parameters. In fact a linear $[15, 5, 9]_4$ cannot exist. Here is a concrete representation of its subcode $\mathcal{D} = tr((\mathcal{B}(1), \epsilon, 1) \oplus tr((\mathcal{B}(3), 1, \epsilon^2)$. The codewords are parametrized by pairs $a, b \in F = GF(16)$, the coordinates by $0 \neq x \in F$ and the corresponding entry is the pair

$$c_{a,b}(x) = (tr(\epsilon ax + bx^3), tr(ax + \epsilon^2 bx^3)).$$

Cases $b = 0$ and $a = 0$ describe the two irreducible subcodes. $\mathcal{C}$ is the direct sum of $\mathcal{D}$ and 11^{15}.

5   Cyclic Quantum Stabilizer Codes

This corresponds to the case when $m = 2$ and the bilinear form $\langle , \rangle$ on $E = \mathbb{F}_q^2$ is symplectic. Such a cyclic q^2-ary code $\mathcal{C}$ of length n coprime to the characteristic is a **quantum stabilizer code** if it contains its dual. If $\mathcal{C} \supseteq \mathcal{C}^\perp$ has dimension k and all codewords in $\mathcal{C} \setminus \mathcal{C}^\perp$ have weight $\geq d$, then the quantum parameters are written as $[[n, 2k - n, d]]_{q^2}$. The code is **pure** if all nonzero codewords of $\mathcal{C}$ have weight $\geq d$.

Theorem 4. *A cyclic additive q^2-ary quantum code is equivalent with a cyclic additive code in case $m = 2$ which is contained in its dual with respect to the symplectic form $\langle , \rangle$.*

Our first objective is to give a parametrized description of the self-orthogonal codes $\mathcal{B}$ as in Theorem 4. Write $\mathcal{B} = \sum_Z S_Z$. Recall from the introduction and [3] that

$$tr(V_F(Z))^\perp = \sum_{Z' \neq Z} tr(V_F(Z')).$$

It follows that $\mathcal{B}$ is self-orthogonal if and only if S_Z and S_{-Z} are orthogonal for each Z. Consider at first the generic case $Z \neq -Z$. If one of the S_Z or S_{-Z} is 0, then there is no restriction on the other. If $S_Z = tr(V_F(Z))$, then $S_{-Z} = 0$.
    We use Lemma 17.26 of [3]:

Lemma 1. $\sum_{u \in W} tr(\alpha u^{z_0}) tr(\beta u^{-z_0}) = n \times tr(\alpha Tr(\beta))$, *where Tr is the trace:*
$F \longrightarrow \mathbb{F}_{q^s}$.

Proposition 5. *Let $\Gamma = (\gamma_1, \gamma_2), \Gamma' = (\gamma_1', \gamma_2') \in L \times L$. The irreducible cyclic codes $tr((\mathcal{B}(z), \Gamma)$ and $tr((\mathcal{B}(-z), \Gamma')$ are orthogonal if and only if Γ and Γ' generate the same point in $PG(1, L)$.*

Proof. Writing out the symplectic product of typical vectors we obtain the condition

$$\sum_{u \in W} tr(\gamma_1 au^z) tr(\gamma_2' bu^{-z}) - tr(\gamma_2 au^z) tr(\gamma_1' bu^{-z}) = 0$$

for all $a, b \in F$. Because of Lemma 1 and the L-linearity of Tr an equivalent condition is $\gamma_1 \gamma_2' = \gamma_2 \gamma_1'$. $\square$

This shows what the self-orthogonality condition is: if $S_Z = tr((\mathcal{B}(z), \Gamma)$ for $\Gamma \in PG(1, q^s)$, then either $S_{-Z} = 0$ or $S_{-Z} = tr((\mathcal{B}(-z), \Gamma)$.

Consider case $Z = -Z, s > 1$. Then either $S_Z = 0$ or S_Z is a self-orthogonal irreducible code $tr((\mathcal{B}(z), \Gamma)$. By Proposition 5 this must coincide with $tr((\mathcal{B}(-z), \Gamma)$. We have $s = 2i$ and $zq^i = -z$. It follows $tr((\mathcal{B}(-z), \Gamma) = tr((\mathcal{B}(z), \Gamma^{q^i})$ and the condition is that Γ and Γ^{q^i} generate the same projective point, which means that $\Gamma \in PG(1, q^i)$. There are $q^i + 1$ choices for Γ. As $Z = \{0\}$ contributes $q + 2$ self-orthogonal and $q + 1$ self-dual codes we arrive at the following enumeration result:

Theorem 5. *The number of additive cyclic q^2-ary quantum stabilizer codes is*

$$\prod_{Z=-Z,s=1} (q+2) \prod_{Z=-Z,s>1} (q^{s/2} + 2) \prod_{Z\neq-Z} (3q^s + 6).$$

The number of self-dual such codes is

$$\prod_{Z=-Z,s=1} (q+1) \prod_{Z=-Z,s>1} (q^{s/2} + 1) \prod_{Z\neq-Z} (q^s + 3).$$

Here $s = |Z|$ and the last product is over all pairs $\{Z, -Z\}$ of cyclotomic cosets such that $Z \neq -Z$.

In case $n = 7$ we obtain $4 \times 30 = 120$ quantum codes alltogether and $3 \times 11 = 33$ self-dual ones, for $n = 15$ the number of quantum codes is $4 \times 4 \times 6 \times 54$ and there are $3 \times 3 \times 5 \times 19$ self-dual codes.

6   Codes of Kloosterman Type

The self-dual $[7, 3.5, 4]_4$-code of Section 4 is generated by the all-1-word and a code whose words are indexed by a pair a, b of elements of $\mathbb{F}_8$, with entry $(tr(ax), tr(b/x))$ in coordinate $x \in \mathbb{F}_8^*$. This is very much remeniscent of the Kloosterman or dual Mélas codes.

Definition 4. *Let $q = 2^f$. The Kloosterman code is a $2f$-dimensional length $q - 1$ binary code whose codewords are $c(a, b)$, where $a, b \in \mathbb{F}_q$, with entry $tr(ax + b/x)$ in coordinate $0 \neq x \in \mathbb{F}_q$.*

Definition 5. *For $0 \neq v \in \mathbb{F}_q$ let p_v be the number of $0 \neq x \in \mathbb{F}_q$ such that $tr(x) = tr(v/x) = 1$. Here tr is the absolute trace.*

The determination of the minimum distance of the Kloosterman codes is equivalent with the determination of the largest number p_v. A close relation with elliptic curves shows that this number is the largest integer less than $(q + 1 + 2\sqrt{q})/4$. Schoof-van der Vlugt [12] have indeed determined the weight distribution of the Kloosterman codes. In [9] we made use of this information to determine the weight distribution of certain extremal caps in $AG(4, q)$.

We may consider $tr((\mathcal{B}(1), 1, 0) \oplus tr((\mathcal{B}(-1), 0, 1)$ and its direct sum with $tr((\mathcal{B}(0), 1, 1)$ in case $q = 2, m = 2, n = 2^r - 1$ as additive versions of Kloosterman codes. Unfortunately we cannot expect good code parameters in general as the irreducible subcodes themselves have minimum distance 2^{r-1}. Consider instead the codes

$$tr((\mathcal{B}(1), 1, \gamma_1) \oplus tr((\mathcal{B}(-1), 1, \gamma_2)$$

where the γ_i are different and not in $\mathbb{F}_2$. In this case the irreducible summands have minimum weights $2^r - 2^{r-2}$. The general codeword $v(a, b)$ (where $a, b \in F = GF(2^r)$) has entry

$$v(a, b)_x = (tr((ax + b/x), tr(a\gamma_1 x + b\gamma_2/x))$$

in coordinate $x \neq 0$. We can assume $ab \neq 0$. Let $v = ab$ and replace x by vx. In order to determine the minimum distance one has to count the $x \neq 0$ such that

$$tr(x + v/x) = tr(\gamma_1 x + \gamma_2 v/x) = 0.$$

References

[1] Ashikhmin, A., Knill, E.: Nonbinary quantum stabilizer codes. IEEE Transactions on Information Theory 47, 3065–3072 (2001)
[2] Bierbrauer, J.: The theory of cyclic codes and a generalization to additive codes. Designs, Codes and Cryptography 25, 189–206 (2002)
[3] Bierbrauer, J.: Introduction to Coding Theory, Chapman and Hall/CRC Press, Boca Raton, FL (2004)
[4] Bierbrauer, J., Faina, G., Marcugini, S., Pambianco, F.: Additive quaternary codes of small length, Proceedings ACCT, Zvenigorod (Russia) (September 15-18, 2006)
[5] Bierbrauer, J., Edel, Y.: Quantum twisted codes. Journal of Combinatorial Designs 8, 174–188 (2000)
[6] Blokhuis, A., Brouwer, A.E.: Small additive quaternary codes, European Journal of Combinatorics 25, 161–167 (2004)
[7] Calderbank, A.R., Rains, E.M., Shor, P.W., Sloane, N.J.A.: Quantum error correction via codes over GF(4). IEEE Transactions on Information Theory 44, 1369–1387 (1998)
[8] Edel, Y., Bierbrauer, J.: Twisted BCH-codes. Journal of Combinatorial Designs 5, 377–389 (1997)
[9] Edel, Y., Bierbrauer, J.: Caps of order $3q^2$ in affine 4-space in characteristic 2,. Finite Fields and Their Applications 10, 168–182 (2004)
[10] Grassl, M.: Tables on, http://iaks-www.ira.uka.de/home/grassl/
[11] Rains, E.M.: Nonbinary quantum codes. IEEE Transactions on Information Theory 45, 1827–1832 (1999)
[12] Schoof, R., van der Vlugt, M.: Hecke operators and the weight distribution of certain codes. Journal of Combinatorial Theory A 57, 163–186 (1991)

Determining the Number of One-Weight Cyclic Codes When Length and Dimension Are Given

Gerardo Vega

Dirección General de Servicios de Cómputo Académico, Universidad Nacional
Autónoma de México, 04510 México D.F., Mexico
gerardov@servidor.unam.mx

Abstract. We use techniques from linear recurring sequences, exponential sums and Gaussian sums, in order to present a set of characterizations for the one-weight irreducible cyclic codes over finite fields. Without using such techniques, a subset of these characterizations was already presented in [2]. By means of this new set of characterizations, we give an explicit expression for the number of one-weight cyclic codes, when the length and dimension are given.

Keywords: One-weight cyclic codes, linear recurring sequences, exponential sums and Gaussian sums.

1   Introduction

Let C be an $[n, k]$ linear code over $\mathbb{F}_q$ whose dual weight is at least 2. It is well known (see for example [3]) that if, additionally, C is a one-weight code, then its length n must be given by

$$n = \lambda \frac{q^k - 1}{q - 1} , \tag{1}$$

for some positive integer λ. Since the minimal distance of the dual of any nonzero irreducible cyclic code over $\mathbb{F}_q$ is greater than 1, it follows that the length n of all one-weight irreducible cyclic codes of dimension k, is given by (1). For this particular case, the zeros of $x^n - 1$, which form a cyclic group, lie in the extension field $\mathbb{F}_{q^k}$ and therefore n divides $q^k - 1$. That is, for some integer s, we have $\lambda(\frac{q^k-1}{q-1})s = (q-1)(\frac{q^k-1}{q-1})$, which implies that λ divides $q - 1$. Thus, for all one-weight irreducible cyclic codes, we can always assume $\lambda | (q - 1)$. In this work we are going to consider irreducible cyclic codes and present a set of characterizations for those that are one-weight codes. Additionally, we use these characterizations in order to give an explicit expression for the number of one-weight cyclic codes, when the length and dimension are given.

In order to achieve our goal, we will use several results related to linear recurring sequences, exponential sums and Gaussian sums. All those results can be found in [1]. To avoid (an impractical) repetition, we do not include such

C. Carlet and B. Sunar (Eds.): WAIFI 2007, LNCS 4547, pp. 284–293, 2007.

results in this work. Instead of that, we will give an explicit reference for all these results.

This work is organized as follows: in Section 2 we recall the connection between linear cyclic codes and linear recurring sequences. In Section 3 we present some general results, that we will use in order to prove our characterizations. Section 4 is devoted to give a first approximation to the characterizations, whereas in Section 5, the characterizations are presented. Finally, in Section 6 we determine the number of one-weight cyclic codes, when the length and dimension are given.

2   Linear Recurring Sequences and Cyclic Codes

Let n and q be two positive integers, where q is a prime power, with $\gcd(n, q) = 1$. Let $h(x)$ and $g(x)$ be monic polynomials over $\mathbb{F}_q$, such that $h(x)g(x) = x^n - 1$ and $\deg(h(x)) = k > 0$. Without loss of generality, we suppose that:

$$h(x) = x^k - h_{k-1}x^{k-1} - h_{k-2}x^{k-2} - \cdots - h_0 .$$

Since the coefficients of the polynomial $g(x)$, can be obtained through the synthetic division of polynomials $x^n - 1$ and $h(x)$, then, if

$$g(x) = g_0 x^{n-1} + g_1 x^{n-2} + \cdots + g_{k-1} x^{n-k} + g_k x^{n-k-1} + \cdots + g_{n-1} , \qquad (2)$$

we have $g_i = 0$ for all $0 \leq i < k - 1$, $g_{k-1} = 1$ and

$$g_{m+k} = h_{k-1}g_{m+k-1} + h_{k-2}g_{m+k-2} + \cdots + h_0 g_m ,$$

with $0 \leq m < n - k$. That is, the n coefficients of $g(x)$ in (2) are the first n terms of the k-order impulse response sequence (see [1, Ch. 8, p. 402]), given by

$$g_{m+k} = h_{k-1}g_{m+k-1} + h_{k-2}g_{m+k-2} + \cdots + h_0 g_m \ \text{ for } m = 0, 1, 2, \cdots . \qquad (3)$$

In agreement with Theorem 8.27 in [1, p. 408], the previous sequence is periodic (in the sense of Definition 8.5 in [1, p. 398]), where such period, r, is equal to the *order* of $h(x)$, that is, $r = \mathrm{ord}(h(x))$.

We will use the same notation introduced in [1, Ch. 8, Secc. 5, p. 423]. Thus, $S(h(x))$ will denote the set of all homogeneous linear recurring sequences in $\mathbb{F}_q$ with characteristic polynomial $h(x)$. In fact, $S(h(x))$ is a vector space over $\mathbb{F}_q$ of dimension $k = \deg(h(x))$, where the role of the zero vector is played by the zero sequence, all of whose terms are 0. We will denote by σ the element of $S(h(x))$, which corresponds to the k-order impulse response sequence given by (3). That is, $\sigma = g_0, g_1, g_2, \cdots$. For any integer $b \geq 0$, we denote by $\sigma^{(b)}$ the shifted sequence $g_b, g_{b+1}, g_{b+2}, \cdots$. Here we are following the notation in [1, p. 426]. Clearly, $\sigma^{(b)} \in S(h(x))$, for all $b \geq 0$.

Now, if s is an integer, $0 \leq s < k$, then we have

$$x^s g(x) = g_s x^{n-1} + g_{s+1} x^{n-2} + \cdots + g_{n-1} x^s + g_0 x^{s-1} + g_1 x^{s-2} + \cdots + g_{s-1} ,$$

in the ring $\mathbb{F}_q[x]/(x^n - 1)$. But the period r, of σ, divides n, so

$$x^s g(x) = g_s x^{n-1} + g_{s+1} x^{n-2} + \cdots + g_{n-1} x^s + g_n x^{s-1} + g_{n+1} x^{s-2} + \cdots + g_{n+s-1} .$$

That is, the n coefficients of $x^s g(x)$ are the first n terms of the shifted sequence $\sigma^{(s)}$. Using this fact, and since $S(h(x))$ is a vector space over $\mathbb{F}_q$, we have that if $f(x) = \sum_{i=0}^{k-1} f_i x^i$ is a polynomial over $\mathbb{F}_q$, then the n coefficients of $f(x)g(x)$ (taking this product over $\mathbb{F}_q[x]/(x^n - 1)$) are the first n terms of the sequence $\sum_{i=0}^{k-1} f_i \sigma^{(i)}$, where the previous summations and scalar products are taken over the vector space $S(h(x))$. This allows us to establish a one-to-one relationship between the sequences in $S(h(x))$ and the codewords in $< g(x) >$ (this relationship is not new, see for example [1, Ch. 9, Secc. 2, p. 485]). Therefore, the calculation of the weight distribution of $< g(x) >$ will be equivalent to count the number of zero terms that appears in the first full period, for each sequence in $S(h(x))$. For this reason we will introduce the following notation:

Let $\tau = t_0, t_1, t_2, \cdots$ be a sequence in $S(h(x))$, also let $b \in \mathbb{F}_q$ and let N be a positive integer. Then, we will denote by $Z(\tau, b, N)$ the number of i, $0 \le i < N$, with $t_i = b$ (this notation is similar to that introduced in [1, p. 453]). Now, note that if $r|N$, then

$$Z(a\sigma^{(s)}, 0, N) = Z(\sigma, 0, N), \text{ for all } a \in \mathbb{F}_q^* \text{ and for any integer } s \ge 0. \qquad (4)$$

It is important to remark that if $\tau = t_0, t_1, t_2, \cdots$ is a sequence in $S(h(x))$, then such sequence is completely determined by its first k terms. For this reason, the vector $(t_0, t_1, t_2, \cdots, t_{k-1})$ is called the *Initial State Vector* of the linear recurring sequence τ. For sequence σ, such vector is $(0, 0, \cdots, 0, 1)$. Clearly, two sequences in $S(h(x))$ will be the same if, and only if, their initial state vectors are the same. Since there exist q^k different choices for initial state vectors, then $S(h(x))$ contains exactly q^k different sequences. This is not surprising, since we already know that there exists a one-to-one relationship between the sequences in $S(h(x))$ and the codewords in $< g(x) >$.

3   Some General Results

We will begin this section by setting some notation and giving a definition.

Notation: For a finite field $\mathbb{F}$ of characteristic p, we will denote by "Tr", the absolute trace of $\mathbb{F}$ over $\mathbb{F}_p$.

The following definition could be considered as an extension of the order, $\text{ord}(f)$, of a polynomial $f(x) \in \mathbb{F}_q[x]$.

Definition 1. *Let $h \in \mathbb{F}_q[x]$ be a polynomial of positive degree with $h(0) \ne 0$. The last positive integer ρ for which x^ρ is congruent modulo $h(x)$, to some element of $\mathbb{F}_q$, is called the quasi-order of h and it will be denoted by $\text{qord}(h(x))$.*

Clearly $\text{qord}(h(x)) \le \text{ord}(h(x))$. However, for the special case where $h(x)$ is an irreducible polynomial, even more can be said.

Theorem 2. *Let $h \in \mathbb{F}_q[x]$ be a polynomial of positive degree with $h(0) \neq 0$. Let $r = \text{ord}(h)$ and $\rho = \text{qord}(h)$. If $h(x)$ is an irreducible polynomial, then $r = \gcd(r, q - 1)\rho$.*

Proof. Let F be the splitting field of $h(x)$ over $\mathbb{F}_q$. By Theorem 3.17 in [1, p. 89] there must exist an integer d such that $d|q - 1$ and $r = d\rho$. Then we have $d|\gcd(r, q - 1)$. Suppose $d < \gcd(r, q - 1)$, then $r = \gcd(r, q - 1)\rho/t$, for some integer $t > 1$, hence $\rho > \rho/t = r/\gcd(r, q - 1)$. Let α be a fixed root of $h(x)$ in F. Clearly, α has order equal to r. Now, observe that $\alpha^{r(q-1)/\gcd(r,q-1)} = 1$, thus we have that $\alpha^{r/\gcd(r,q-1)} \in \mathbb{F}_q$, say $\alpha^{r/\gcd(r,q-1)} = a$. This means that α is root of $f(x) = x^{\rho/t} - a$ and hence $h(x)|f(x)$. But this is a contradiction since $\rho/t < \rho$. $\qquad\square$

Observe that the converse of the previous theorem is not true. For example, if $q = 3$ and $h(x) = (x + 1)^2$, then $r = 6$, $\rho = 3$ and $\gcd(r, q - 1) = 2$.

Theorem 3. *Let γ be a primitive element of $\mathbb{F}_{q^k}$, where $q = p^m$ for some positive integer m and some prime p. Let t be a fixed integer and let $S = \langle \gamma^t \rangle$. For each $a \in \mathbb{F}_p$, let $S_a = \{x \in S : \text{Tr}(x) = a\}$. If $\gcd(t, q^k - 1)$ divides $\frac{q^k-1}{p-1}$, then $|S_a| = |S_b|$ for all $a, b \in \mathbb{F}_p^*$.*

Proof. Let $a, b \in \mathbb{F}_p^*$ and let $v = \frac{q^k-1}{p-1}$. Since $\gcd(t/\gcd(t, q^k - 1), p - 1) = 1$ and $\mathbb{F}_p^*$ is a $(p - 1)$-order subgroup of $\mathbb{F}_{q^k}^*$, then we have $\mathbb{F}_p^* = \{\gamma^{sv} : 0 \leq s < (p - 1)\} = \{(\gamma^t)^{vs/\gcd(t,q^k-1)} : 0 \leq s < (p - 1)\} \subseteq S$. Let $x \in S_a$ and let $c \in \mathbb{F}_p^*$ be the uniquely determined field element such that $b = ca$. Thus, if $y = cx$, then $y \in S_b$. Finally, the correspondence $x \mapsto y = cx$ is clearly bijective. $\qquad\square$

Corollary 4. *Using the same notation as in the previous theorem, let θ be a fixed field element in $\mathbb{F}_{q^k}$. For each $a \in \mathbb{F}_p$ let $S_a^\theta = \{x \in S : \text{Tr}(\theta x) = a\}$. If $\gcd(t, q^k - 1)$ divides $\frac{q^k-1}{p-1}$, then $|S_a^\theta| = |S_b^\theta|$ for all $a, b \in \mathbb{F}_p^*$.*

For some kind of exponential sums we can now show, by means of the previous corollary, that their values are always nonzero integers.

Theorem 5. *Using the same notation as in the previous theorem, let θ be a nonzero fixed field element in $\mathbb{F}_{q^k}$ and let χ be a nontrivial additive character of $\mathbb{F}_{q^k}$. Suppose that $\gcd(t, q^k - 1)$ divides $\frac{q^k-1}{p-1}$, then, the value of the exponential sum:*

$$\sum_{c \in \mathbb{F}_{q^k}} \chi(\theta c^t) ,$$

is an integer different from zero if $\gcd(t, q^k - 1) \neq 1$, and zero otherwise.

Proof. Suppose $\gcd(t, q^k - 1) \neq 1$. Let χ_1 be the canonical additive character of $\mathbb{F}_{q^k}$, then there exists an element $d \in \mathbb{F}_{q^k}$ such that $\chi(c) = \chi_1(dc)$ for all $c \in \mathbb{F}_{q^k}$. Let γ and S be as in Theorem 3. For each $a \in \mathbb{F}_p$ let $S_a^{d\theta} = \{x \in S : \text{Tr}(d\theta x) = a\}$ and $I_a = \{c \in \mathbb{F}_{q^k}^* : \text{Tr}(d\theta c^t) = a\}$. Since S is a

$((q^k - 1)/\gcd(t, q^k - 1))$-order subgroup of $\mathbb{F}^*_{q^k}$, then $|I_a| = \gcd(t, q^k - 1)|S^{d\theta}_a|$ for all $a \in \mathbb{F}_p$. Now let $I'_0 = \{c \in \mathbb{F}_{q^k} : \text{Tr}(d\theta c^t) = 0\}$. Clearly $|I'_0| = |I_0| + 1$, and therefore, since $\gcd(t, q^k - 1) \neq 1$, it follows that $\gcd(t, q^k - 1)$ does not divide $|I'_0|$. By applying Corollary 4, we deduce that $|I_a| = |I_b|$, for all $a, b \in \mathbb{F}^*_p$. Thus,

$$\sum_{c \in \mathbb{F}_{q^k}} \chi(\theta c^t) = |I'_0| + |I_1| \sum_{a=1}^{p-1} e^{2\pi i a/p}$$

$$= |I'_0| - |I_1| \neq 0 .$$

On the other hand, if $\gcd(t, q^k - 1) = 1$, then $\mathbb{F}_{q^k} = \{\theta c^t : c \in \mathbb{F}_{q^k}\}$ and therefore, in this case, the result follows trivially. □

There is a close connection between the exponential sums and Gaussian sums:

Theorem 6. [1, Ch. 5, p. 217] Let χ be a nontrivial additive character of $\mathbb{F}_{q^k}$, $m \in \mathbb{N}$, and ψ a multiplicative character of $\mathbb{F}_{q^k}$ of order $t = \gcd(m, q^k - 1)$. Then,

$$\sum_{c \in \mathbb{F}_{q^k}} \chi(\theta c^m) = \sum_{j=1}^{t-1} \bar{\psi}^j(\theta) G(\psi^j, \chi) ,$$

for all $\theta \in \mathbb{F}^*_{q^k}$.

Since the order of any multiplicative character of $\mathbb{F}_{q^k}$ divides $q^k - 1$, then we have the following:

Corollary 7. Let χ be a nontrivial additive character of $\mathbb{F}_{q^k}$, and ψ a multiplicative character of $\mathbb{F}_{q^k}$ of order t which divides $\frac{q^k - 1}{p - 1}$. Then,

$$\sum_{j=1}^{t-1} \bar{\psi}^j(\theta) G(\psi^j, \chi) = \begin{cases} \text{an integer different from zero,} & \text{if } t \neq 1 \\ 0 & \text{otherwise} \end{cases} .$$

4   A First Approach to the Characterizations

From now on, we will assume k to be a positive integer and the length of a one-weight k-dimensional linear cyclic code over the finite field $\mathbb{F}_q$, be $n = \lambda(q^k - 1)/(q - 1)$, for some integer $\lambda > 0$, which divides $q - 1$. So, $x^\lambda - 1 | x^{q-1} - 1$, and therefore $\gcd(x^\lambda - 1, x^q - x) = x^\lambda - 1$, thus there must exist $a_1, a_2, \cdots, a_\lambda \in \mathbb{F}^*_q$, such that $x^\lambda - 1 = \prod_{i=1}^{\lambda}(x - a_i)$. That is, the splitting field of $x^\lambda - 1$ is $\mathbb{F}_q$. But $n = \lambda(q^k - 1)/(q - 1)$, therefore, we have $x^n - 1 = \prod_{i=1}^{\lambda}(x^{n/\lambda} - a_i)$. Now, if $h(x)$ is an irreducible polynomial such that $h(x)|x^n - 1$, then there must exist a $1 \leq j \leq \lambda$, such that $h(x)|x^{n/\lambda} - a_j$. Then, clearly, $\text{qord}(h(x)) \leq (q^k - 1/q - 1)$. In fact $\text{qord}(h(x))$ divides $(q^k - 1/q - 1)$:

Lemma 8. Let $h(x)$ be an irreducible polynomial over $\mathbb{F}_q$ and let $\rho = \text{qord}(h(x))$. Suppose that there exists a positive integer η and a field element $b \in \mathbb{F}^*_q$ such that $h(x)|x^\eta - b$, then $\rho|\eta$.

Proof. Clearly, $\eta \geq \rho$, then suppose $\eta = s\rho + t$ for some positive integers s and t, with $0 \leq t < \rho$. Due to the definition of ρ there must exist a uniquely determined field element $a_0 \in \mathbb{F}_q^*$ such that $a_0 \equiv x^\rho \bmod h(x)$. Thus we have $b \equiv x^\eta \bmod h(x) = x^{s\rho+t} \bmod h(x) \equiv a_0^s x^t \bmod h(x)$. That is, $x^t \equiv ba_0^{-s} \bmod h(x)$. Because of the definition of ρ this is only possible if $t = 0$.   □

For the special case when $\mathrm{qord}(h(x))$ and $(q^k - 1/q - 1)$ are equal, we have the following:

Theorem 9. *Let q, k, n and λ be as before. Let $h(x)$ be a k-degree monic irreducible polynomial over $\mathbb{F}_q$, which divides $x^n - 1$. Set $g(x) = (x^n - 1)/h(x)$. If we denote by $w_t(g(x))$ the Hamming weight of the polynomial $g(x)$, then $\mathrm{qord}(h(x)) = (q^k - 1/q - 1)$ if, and only if, $w_t(g(x)) = \lambda q^{k-1}$.*

Proof. First of all, we fix several notations; by using σ, we will denote the k-order impulse response sequence given by (3). That is, $\sigma = g_0, g_1, g_2, \cdots$. Additionally, we will denote by r and ρ, the order and the quasi-order, respectively, of the polynomial $h(x)$. Set $K = \mathbb{F}_q$, and let F be the splitting field of $h(x)$ over K. Let α be a fixed root of $h(x)$ in F; then $\alpha \neq 0$ because $h(0) \neq 0$. By Theorem 8.24 of [1, p. 406], there exists $\theta \in F$ such that

$$g_m = \mathrm{Tr}_{F/K}(\theta \alpha^m) \quad \text{for } m = 0, 1, 2, \cdots . \tag{5}$$

We clearly have $\theta \neq 0$. Let χ' be the canonical additive character of K. The character relation (5.9) of [1, p. 192] yields

$$\frac{1}{q} \sum_{c \in K} \chi'(cg_m) = \begin{cases} 1 \text{ if } g_m = 0 \\ 0 \text{ if } g_m \neq 0 \end{cases}$$

and so, together with (5),

$$Z(\sigma, 0, n) = \frac{1}{q} \sum_{m=0}^{n-1} \sum_{c \in K} \chi'(\mathrm{Tr}_{F/K}(c\,\theta\alpha^m)) .$$

If χ denotes the canonical additive character of F, then χ' and χ are related by $\chi'(\mathrm{Tr}_{F/K}(\beta)) = \chi(\beta)$ for all $\beta \in F$. Therefore,

$$Z(\sigma, 0, n) = \frac{1}{q} \sum_{c \in K} \sum_{m=0}^{n-1} \chi(c\,\theta\alpha^m) = \frac{n}{q} + \frac{1}{q} \sum_{c \in K^*} \sum_{m=0}^{n-1} \chi(c\,\theta\alpha^m) . \tag{6}$$

Now, by (5.17) of [1, p. 195],

$$\chi(\beta) = \frac{1}{q^k - 1} \sum_{\psi} G(\bar{\psi}, \chi)\psi(\beta) \quad \text{for } \beta \in F^* ,$$

where the sum is extended over all multiplicative characters ψ of F. For $c \in K^*$ it follows that

$$\sum_{m=0}^{n-1} \chi(c\,\theta\alpha^m) = \frac{1}{q^k-1} \sum_{m=0}^{n-1} \sum_{\psi} G(\bar{\psi}, \chi)\psi(c\,\theta\alpha^m)$$

$$= \frac{1}{q^k-1} \sum_{\psi} \psi(c\,\theta) G(\bar{\psi}, \chi) \sum_{m=0}^{n-1} \psi(\alpha)^m .$$

Substituting this in (6), we get

$$Z(\sigma, 0, n) = \frac{n}{q} + \frac{1}{q(q^k-1)} \sum_{c \in K^*} \sum_{\psi} \psi(c\,\theta) G(\bar{\psi}, \chi) \sum_{m=0}^{n-1} \psi(\alpha)^m$$

$$= \frac{n}{q} + \frac{1}{q(q^k-1)} \sum_{\psi} \psi(\theta) G(\bar{\psi}, \chi) \sum_{m=0}^{n-1} \psi(\alpha)^m \sum_{c \in K^*} \psi(c) .$$

Now, if the restriction ψ' of ψ to K^* is nontrivial, then $\sum_{c \in K^*} \psi(c) = 0$. Consequently, it suffices to extend the sum over the set B of characters ψ for which ψ' is trivial, so that

$$Z(\sigma, 0, n) = \frac{n}{q} + \frac{q-1}{q(q^k-1)} \sum_{\psi \in B} \psi(\theta) G(\bar{\psi}, \chi) \sum_{m=0}^{n-1} \psi(\alpha)^m .$$

Now, let γ be a fixed primitive element of F. Thus, $K^* = \{\gamma^{s(q^k-1/q-1)} : 0 \le s < (q-1)\}$, and hence, $B = \{\psi_{s(q-1)} : 0 \le s < (q^k - 1/q - 1)\}$[1]. Thus, since $\psi_{s(q-1)}(\alpha) = \psi_1(\alpha^{s(q-1)})$, for all $0 \le s < (q^k - 1/q - 1)$, we obtain:

$$Z(\sigma, 0, n) = \frac{n}{q} + \frac{q-1}{q(q^k-1)} \sum_{s=0}^{(q^k-1/q-1)-1} \psi_{s(q-1)}(\theta) G(\bar{\psi}_{s(q-1)}, \chi) \sum_{m=0}^{n-1} \psi_1(\alpha^{s(q-1)})^m .$$

The inner sum in the last expression is a finite geometric series that vanishes if $\psi_1(\alpha^{s(q-1)}) \ne 1$, because of $\psi_1(\alpha^{s(q-1)})^n = \psi_1(\alpha^{sn(q-1)}) = \psi_1(1) = 1$. On the other hand, $\psi_1(\alpha^{s(q-1)}) = 1$ if, and only if, $\alpha^{s(q-1)} = 1$. By means of Theorem 2 we know that $r = \gcd(r, q-1)\rho$. Now, the field element $\alpha^{(q-1)}$ has order $r/\gcd(r, q-1) = \rho$ and, by virtue of Lemma 8, there exists a positive integer t, such that $t = (q^k - 1/q - 1)/\rho$. Thus, $\psi_1(\alpha^{s(q-1)}) = 1$ if, and only if, $s = j\rho = (j/t)(q^k - 1/q - 1)$, for $j = 0, 1, ..., t-1$, and therefore

$$Z(\sigma, 0, n) = \frac{n}{q} + \frac{n(q-1)}{q(q^k-1)} \psi_0(\theta) G(\bar{\psi}_0, \chi) +$$

$$\frac{n(q-1)}{q(q^k-1)} \sum_{j=1}^{t-1} \psi^j_{(q^k-1)/t}(\theta) G(\bar{\psi}^j_{(q^k-1)/t}, \chi) .$$

---

[1] Here we are considering $\psi_j(\gamma^\ell) = e^{2\pi i j \ell/(q^k-1)}$.

Clearly, the multiplicative character $\psi_{(q^k-1)/t}$ has order t, thus, by Corollary 7, and since $\psi_0(\theta) = 1$ and $G(\bar{\psi}_0, \chi) = -1$, we have

$$Z(\sigma, 0, n) = \frac{(q^{k-1}-1)n}{q^k - 1} = \frac{(q^{k-1}-1)\lambda}{q - 1} \quad \Leftrightarrow \quad \rho = (q^k - 1/q - 1),$$

and hence, $w_t(g(x)) = n - Z(\sigma, 0, n) = \lambda q^{k-1} \quad \Leftrightarrow \quad \rho = (q^k - 1/q - 1).$ $\qquad\square$

5   The Characterizations

We begin this section recalling the following result that was proved in [3].

Proposition 10. *Let C be an $[n, k]$ linear code over $\mathbb{F}_q$. If C is a one-weight code with weight w and if the weight of the dual of C is at least 2, then there exists $\lambda \in \mathbb{N}$ such that $n = \lambda \frac{q^k-1}{q-1}$ and $w = \lambda q^{k-1}$.*

Finally, we are now able to present our characterizations.

Theorem 11. *Let q and k be as before. Let γ be a primitive element of $\mathbb{F}_{q^k}$. For a positive integer a, let $h_a(x) \in \mathbb{F}_{q^k}[x]$ be the minimal polynomial of γ^a. For each positive integer ℓ, such that $\ell | \gcd(a, q - 1)$, set $\lambda_\ell = \frac{(q-1)\ell}{\gcd(a,q-1)}$ and $n_\ell = \lambda_\ell(q^k - 1)/(q - 1)$. Then, the following statements are equivalent:*

A) $\gcd(a, q^k - 1/q - 1) = 1$.
B) $h_a(x)|x^{n_\ell} - 1$, $\deg(h_a(x)) = k$ and $\operatorname{qord}(h_a(x)) = (q^k - 1/q - 1)$.
C) $h_a(x)|x^{n_\ell} - 1$, $\deg(h_a(x)) = k$ and the Hamming weight of the polynomial $x^{n_\ell} - 1/h_a(x)$ is $\lambda_\ell q^{k-1}$.
D) $h_a(x)|x^{n_\ell} - 1$ and $\deg(h_a(x)) = k$. Additionally, if $g(x) = x^{n_\ell} - 1/h_a(x)$ and if we denote by σ the k-order impulse response sequence given by (3) (taking in that equation: $h(x) = h_a(x)$), then for any nonzero codeword $c(x)$ in $< g(x) >$ there exist a uniquely determined integer s, $0 \le s < q^k - 1/q - 1$, and a uniquely determined field element $d \in \mathbb{F}_q^*$, such that the n_ℓ coefficients of $c(x)$ are the first n_ℓ terms of the sequence $\tau = d\sigma^{(s)}$.
E) $h_a(x)$ is the parity-check polynomial for a one-weight cyclic code of dimension k.

Proof. Suppose statement A) holds. Then $\gcd(a, q^k-1) = \gcd(a, q-1)$ and hence $\operatorname{ord}(h_a(x))|n_\ell$, thus $h_a(x)|x^{n_\ell} - 1$. Let s be the smallest positive integer such that $aq^s \equiv a \pmod{n_\ell}$. Then $a(q^s - 1) \equiv 0 \pmod{n_\ell}$ and $(q^k - 1/q - 1)|a(q^s - 1)$, which implies that $(q^k-1)|(q^s-1)(q-1)$. The last condition is impossible if $s < k$, thus $\deg(h(x)) = k$. Now, by using Theorem 2 we have $\operatorname{qord}(h_a(x)) = q^k-1/q-1$ if and only if

$$\operatorname{ord}(h_a(x)) = (q^k - 1)/\gcd(a, q^k - 1) = \gcd(r, q - 1)(\frac{q^k - 1}{q - 1}) \quad \Leftrightarrow$$

$$\gcd(a, q^k - 1)\gcd(\frac{q^k - 1}{\gcd(a, q^k - 1)}, q - 1) = q - 1 \quad \Leftrightarrow \quad \gcd(a, q^k - 1/q - 1) = 1,$$

proving, in this way, equivalence of statements A) and B). Due to this, the equivalence of statements A), B) and C), comes from Theorem 9.

By taking $N = n_\ell$ in (4), we conclude that D) implies E). Since the weight of the dual code in E) is at least 2, then Proposition 10 and Theorem 9 proves that $\mathrm{qord}(h_a(x)) = (q^k - 1/q - 1)$. But $\mathrm{qord}(h_a(x)) = (q^k - 1/q - 1)$ if and only if $\gcd(a, q^k - 1/q - 1) = 1$, therefore E) implies A). Thus, what we now have to do is just prove that statement B) implies statement D). In order to give such a proof, we first take the sequence σ and determine the following:

$$\rho' = \min\{m \in \mathbb{N} : \sigma^{(m)} = d\sigma, \text{ for some } d \in \mathbb{F}_q^*\} .$$

Such ρ' must exist since $\sigma^{(\mathrm{ord}(h_a(x)))} = \sigma$. Thus, let d_0 be the uniquely determined field element in $\mathbb{F}_q^*$, such that $\sigma^{(\rho')} = d_0\sigma$. But, by the previous equality, it follows that $g_{m+\rho'} = d_0 g_m$, $m = 0, 1, 2, \cdots$. That is, the sequence σ is also a ρ'-order linear recurring sequence over $\mathbb{F}_q$ with characteristic polynomial $p(x) = x^{\rho'} - d_0$. Now, applying Theorems 8.42 and 8.50 in [1, pp. 418 and 422], and since $h_a(x)$ is irreducible, we have that $h_a(x)|p(x)$. Thus, by Definition 1, we conclude that $\rho' \geq \mathrm{qord}(h_a(x)) = (q^k - 1/q - 1)$. Due to this inequality, we have that if $d_1, d_2 \in \mathbb{F}_q^*$ and $0 \leq s_1, s_2 < (q^k - 1/q - 1)$, then $d_1\sigma^{(s_1)} = d_2\sigma^{(s_2)}$ if, and only if, $d_1 = d_2$ and $s_1 = s_2$. That is because if we suppose -without loss of generality- that $s_1 \leq s_2$, then $d_1\sigma^{(s_1)} = d_2\sigma^{(s_2)}$ if, and only if, $\sigma = d_2 d_1^{-1}\sigma^{(s_2-s_1)}$, and this equality is only possible if $s_2 - s_1 = 0$ and $d_2 d_1^{-1} = 1$. Considering this, we define

$$\mathcal{S} = \{\tau \in S(h_a(x)) : \tau = d\sigma^{(s)}, \text{ for some } d \in \mathbb{F}_q \text{ and } 0 \leq s < (q^k - 1/q - 1)\} .$$

Since $|\mathbb{F}_q^*| = q - 1$, then $|\mathcal{S}| = q^k$. But $\mathcal{S} \subseteq S(h_a(x))$ and $|S(h_a(x))| = q^k$, so what we have proved is that if $\tau \in S(h_a(x))$, is a nonzero sequence, then there must exist a uniquely determined integer s, $0 \leq s < (q^k - 1/q - 1)$, and a uniquely determined field element $d \in \mathbb{F}_q^*$, such that $\tau = d\sigma^{(s)}$.

Thus, if $c(x)$ is a nonzero codeword in $< x^{n_\ell} - 1/h_a(x) >$, then there must exist a uniquely determined nonzero sequence in $S(h_a(x))$, say τ, such that the n_ℓ coefficients of $c(x)$ are the first n_ℓ terms of such sequence.   □

6   Number of One-Weight Cyclic Codes When the Length and Dimension Are Given

In order to give an explicit expression for the number of one-weight cyclic codes, when the length and dimension are given, we need to keep in mind the following:

Remark: If ϕ denotes the Euler ϕ-function (see, for example, [1, p. 7]) and if ξ and m are two positive integers, then the number of integers between 1 and ξm, relatively prime to m, is $\xi\phi(m)$.

Theorem 12. *Let q and k be as before. Let $n = \lambda(q^k - 1)/(q - 1)$, for some integer $\lambda > 0$, which divides $q - 1$. The number of one-weight cyclic codes of length n and dimension k is equal to*

$$\delta_\lambda(\gcd(n, q - 1))\frac{\lambda\phi(q^k - 1/q - 1)}{k} \ ,$$

where δ is Kronecker's delta ($\delta_x(y)$ is one if $x = y$ and zero otherwise).

Proof. Let b be an integer such that $1 \le b < n$ and set $a = b(q-1)/\lambda$. Let γ and $h_a(x)$ be as in Theorem 11. Clearly $\mathrm{ord}(h_a(x))|n$ and if statement A), in Theorem 11, holds, then $\mathrm{ord}(h_a(x)) = (q - 1)/\gcd(a, q - 1)((q^k - 1)/(q - 1))$. Therefore, under these circumstances, we conclude that there must exist an integer ℓ such that $\ell | \gcd(a, q - 1)$ and $\lambda = (q - 1)\ell/\gcd(a, q - 1)$.

Taking $\xi = \lambda$ and $m = q^k - 1/q - 1$, then the result follows from above remark, the equivalence between statements A) and E), in Theorem 11, and the fact that $\gamma^{(q-1)/\lambda}$ is a primitive nth root of unity. □

For example, if $q = 31$, $k = 2$ and if n is any of the integers: $32, 64, 96, 160, 192,$ $320, 480$ or 960, then the number of one-weight cyclic codes, for each n, is, respectively, $0, 16, 0, 0, 48, 80, 0, 240$.

References

1. Lidl, R., Niederreitter, H.: Finite Fields. Cambridge Univ. Press, Cambridge (1983)
2. Vega, G., Wolfmann, J.: New Classes of 2-weight Cyclic Codes. Designs, Codes and Cryptography 42(3), 327–334 (2007)
3. Wolfmann, J.: Are 2-Weight Projective Cyclic Codes Irreducible? IEEE Trans. Inform. Theory. 51, 733–737 (2005)

Error Correcting Codes from Quasi-Hadamard Matrices

V. Álvarez, J.A. Armario, M.D. Frau, E. Martin, and A. Osuna[*]

Dpto. Matemática Aplicada I, Universidad de Sevilla, Avda. Reina Mercedes s/n
41012 Sevilla, Spain
{valvarez,armario,mdfrau,emartin,aosuna}@us.es

Abstract. Levenshtein described in [5] a method for constructing error correcting codes which meet the Plotkin bounds, provided suitable Hadamard matrices exist. Uncertainty about the existence of Hadamard matrices on all orders multiple of 4 is a source of difficulties for the practical application of this method. Here we extend the method to the case of quasi-Hadamard matrices. Since efficient algorithms for constructing quasi-Hadamard matrices are potentially available from the literature (e.g. [7]), good error correcting codes may be constructed in practise. We illustrate the method with some examples.

Keywords: Error correcting code, Hadamard matrix, Hadamard code.

1 Introduction

One of the main goals in Coding Theory is the design of optimal error correcting codes. For given length n and minimum distance d, the term optimal means a code which consists of a set of code words as large as possible. For (not necessarily linear) binary codes (n, M, d), Plotkin found out in [10] the following bounds for the number M of codewords:

$$M \leq 2\lfloor \frac{d}{2d - n} \rfloor \quad \text{if } d \text{ is even and } d \leq n < 2d, \tag{1}$$

$$M \leq 2n \quad \text{if } d \text{ is even and } n = 2d, \tag{2}$$

$$M \leq 2\lfloor \frac{d+1}{2d + 1 - n} \rfloor \quad \text{if } d \text{ is odd and } d \leq n < 2d + 1, \tag{3}$$

$$M \leq 2n + 2 \quad \text{if } d \text{ is odd and } n = 2d + 1. \tag{4}$$

Levenshtein proved in [5] that the Plotkin bounds are tight, in the sense that there exist binary codes which meet these bounds, provided that enough Hadamard matrices exist. Unfortunately, the Hadamard Conjecture about the existence of Hadamard matrices in all orders multiple of 4 remains still open. Moreover, there are infinite orders for which no Hadamard matrices have been

---

[*] All authors are partially supported by the PAICYT research project FQM–296 from Junta de Andalucía (Spain).

found. This means that, though theoretically correct, Levenshtein's method could not be useful in practise.

In the sequel a matrix for which the inner product of rows two by two is mostly zero is called a quasi-Hadamard matrix. We will use Levenshtein's method in this paper to show that "good" error-correcting codes may be analogously constructed from quasi-Hadamard matrices. Here the term "good" refers to a code formed from a significantly large number of code words, for given length and minimum distance. We must emphasize that quasi-Hadamard matrices may be straightforwardly obtained in all orders multiple of 4, so that the associated error-correcting codes may be constructed in practise.

We organize the paper as follows.

In Section 2 we introduce the notion of quasi-Hadamard matrices, and some processes to construct them, which are available in the literature. Section 3 is devoted to explain how to construct good error-correcting codes from suitable quasi-Hadamard matrices. Some examples are discussed in Section 4.

2   Quasi-Hadamard Matrices

A Hadamard matrix H of order n is an $n \times n$ matrix of $+1$'s and -1's entries such that $HH^T = nI$. That is, the inner product of any two distinct rows of H is zero.

We now generalize this notion.

We define a quasi-Hadamard matrix of order n as an $n \times n$ matrix M of $+1$'s and -1's entries such that the inner product of rows two by two is mostly zero.

Sometimes it is necessary to precise the largest number q of rows in M which are orthogonal one to each other. The larger q is, the closer M is from being a Hadamard matrix. In these circumstances, M is termed a quasi-Hadamard matrix of depth q.

In some sense, a quasi-Hadamard matrix could be thought as a Hadamard matrix in which some rows have been substituted, so that the Hadamard character is generally lost in turn.

Constructing Hadamard matrices is hard. How about constructing quasi-Hadamard matrices?

We now attend to another characterization of Hadamard matrices, in terms of cliques of graphs (that is, a collection of n vertices and $\frac{n(n-1)}{2}$ edges of a graph G which form a complete subgraph K_n of G).

Consider the graph G_{4t} whose vertices are all the tuples of length $4t$ formed from $2t$ ones and $2t$ minus ones, with the restriction that precisely t ones have to appear within the first $2t$ positions (by analogy, precisely t ones appear within the last $2t$ positions). There is an edge between two vertices if and only if the inner product of the correspondent tuples is zero. A Hadamard matrix of order $4t$ exists if and only if G_{4t} contains a clique of size $4t-2$. Furthermore, the vertices of such a clique and the normalized rows $(\overbrace{1, \ldots 1}^{4t})$ and $(\overbrace{1, \ldots, 1}^{2t}, \overbrace{-1, \ldots, -1}^{2t})$ form a Hadamard matrix. This is a particular type of Hadamard Graph, as defined in [8,9].

Unfortunately, the problem of finding out the maximum clique in a graph has been proven to be NP-hard [4]. Moreover, even its approximations within a constant factor are NP-hard [2,3]. So one should expect that finding out Hadamard matrices from G, or even quasi-Hadamard matrices for large depths close to $4t$, are to be hard problems. In fact, they are.

Hopefully, heuristic methods for the maximum clique problem can be found in the literature, which output pretty large cliques [7]. These methods can be used in turn to construct quasi-Hadamard matrices of large depth as well.

3   Quasi-Hadamard Codes

We firstly recall Levenshtein's method [5] for constructing optimal error correcting codes from suitable Hadamard matrices.

Starting from a normalized (i.e. the first row and column formed all of 1's) Hadamard matrix H of order $4t$, some codes (which are termed Hadamard codes) may be constructed (see [6], for instance). More concretely, consider the matrix A_{4t} related to H_{4t}, which consists in replacing the $+1$'s by 0's and the -1's by 1's. Since the rows of H_{4t} are orthogonal, any two rows of A_{4t} agree in $2t$ places and differ in $2t$ places, and so have Hamming distance $2t$ apart. In these circumstances, one may construct:

1. An $(4t - 1, 4t, 2t)$ code, $\mathcal{A}_{4t}$, consisting of the rows of A_{4t} with the first column deleted. This is optimal for the Plotkin bound (1).
2. An $(4t - 1, 8t, 2t - 1)$ code, $\mathcal{B}_{4t}$, consisting of $\mathcal{A}_{4t}$ together with the complements of all its codewords. This is optimal for the Plotkin bound (4).
3. An $(4t, 8t, 2t)$ code, $\mathcal{C}_{4t}$, consisting of the rows of A_{4t} and their complements. This is optimal for the Plotkin bound (2).
4. An $(4t - 2, 2t, 2t)$ code, $\mathcal{D}_{4t}$, formed from the codewords in $\mathcal{A}_{4t}$ which begin with 0, with the initial zero deleted. This is optimal for the Plotkin bound (1).

Furthermore, as explained in [6], for any $d \le n < 2d$, an optimal code attending to the Plotkin bound (1) may be obtained from a suitable combination of codes of the above type.

More concretely, given d even so that $2d > n \ge d$, define $k = \lfloor \frac{d}{2d - n} \rfloor$ and

$$a = d(2k + 1) - n(k + 1), \qquad b = kn - d(2k - 1).$$

Then a and b are nonnegative integers satisfying that $n = (2k - 1)a + (2k + 1)b$ and $d = ka + (k + 1)b$. Moreover, if n is even then so are a and b. Analogously, if n is odd and k even, then b is even. Finally, if both of n and k are odd, then a is even.

Depending on the parity of n and k, define the code $\mathcal{C}$ to be:

- If n is even, $\mathcal{C} = \dfrac{a}{2}\mathcal{D}_{4k} \oplus \dfrac{b}{2}\mathcal{D}_{4k+4}$.
- If n is odd and k even, $\mathcal{C} = a\mathcal{A}_{2k} \oplus \dfrac{b}{2}\mathcal{D}_{4k+4}$.
- If n and k are odd, $\mathcal{C} = \dfrac{a}{2}\mathcal{D}_{4k} \oplus b\mathcal{A}_{2k+2}$.

Here $\oplus$ denotes the following "summation" of codes. Suppose that $\mathcal{C}_1$ and $\mathcal{C}_2$ are (n_1, M_1, d_1) and (n_2, M_2, d_2) codes, respectively. Assume, for instance, that $M_2 \geq M_1$. For nonnegative integers a, b, the code $a\mathcal{C}_1 \oplus b\mathcal{C}_2$ consists in pasting a copies of $\mathcal{C}_1$, side by side, followed by b copies of the code obtained from $\mathcal{C}_2$ by omitting the last $M_2 - M_1$ codewords. By construction, $a\mathcal{C}_1 + b\mathcal{C}_2$ is shown to be an $(an_1 + bn_2, M_1, d)$ code, for $d \geq ad_1 + bd_2$.

This way, the code $\mathcal{C}$ defined above meets the Plotkin bound (1), since it has length n, minimum distance d, and contains $2k = 2\lfloor \dfrac{d}{2d-n} \rfloor$ codewords.

We now extend Levenshtein's method for constructing optimal error correcting codes from Hadamard matrices to the case of quasi-Hadamard matrices. The codes so obtained are termed quasi-Hadamard codes.

Consider a normalized quasi-Hadamard matrix M_{4t} of order $4t$ and depth q. We define the matrix A'_{4t} related to M_{4t} in the following way: select a q-set of rows of M_{4t} which are orthogonal one to each other (notice that there is no larger set with this property, since q is the depth of M_{4t}), and replace the $+1$'s by 0's and the -1's by 1's.

Theorem 1. *In the circumstances above, the following quasi-Hadamard codes may be constructed:*

1. *An $(4t-1, q, 2t)$ code, $\mathcal{A}'_{4t}$, consisting of the rows of A'_{4t} with the first column deleted.*
2. *An $(4t - 1, 2q, 2t - 1)$ code, $\mathcal{B}'_{4t}$, consisting of $\mathcal{A}'_{4t}$ together with the complements of all its codewords.*
3. *An $(4t, 2q, 2t)$ code, $\mathcal{C}'_{4t}$, consisting of the rows of A'_{4t} and their complements.*
4. *An $(4t - 2, h, 2t)$ code, $\mathcal{D}'_{4t}$, formed from the h codewords in $\mathcal{A}'_{4t}$ which begin with 0, with the initial zero deleted (we only know that $h \leq q$).*

Proof. It is a straightforward extension of the case of usual Hadamard codes coming from Hadamard matrices, since:

- A'_{4t} consists of q rows.
- Any two rows of A'_{4t} agree in $2t$ places and differ in $2t$ places (since they are pairwise orthogonal), and so have Hamming distance $2t$ apart.

The result follows.   □

Remark 1. Obviously, the closer q is from $4t$, the better codes $\mathcal{A}'_{4t}$, $\mathcal{B}'_{4t}$, $\mathcal{C}'_{4t}$ and $\mathcal{D}'_{4t}$ are. In the sense that the number of codewords is very close to the optimal value indicated in the Plotkin bound.

Theorem 2. *For d even so that $2d > n \geq d$, define $k = \lfloor \dfrac{d}{2d-n} \rfloor$ and*

$$a = d(2k+1) - n(k+1), \qquad b = kn - d(2k-1).$$

as before. A good error correcting code $\mathcal{C}'$ of length n and minimum distance d may be obtained, from suitable quasi-Hadamard matrices. More concretely, depending on the parity of n and k, define the code $\mathcal{C}'$ to be:

- *If n is even, $\mathcal{C}' = \dfrac{a}{2}\mathcal{D}'_{4k} \oplus \dfrac{b}{2}\mathcal{D}'_{4k+4}$.*

- *If n is odd and k even, $\mathcal{C}' = a\mathcal{A}'_{2k} \oplus \dfrac{b}{2}\mathcal{D}'_{4k+4}$.*

- *If n and k are odd, $\mathcal{C}' = \dfrac{a}{2}\mathcal{D}'_{4k} \oplus b\mathcal{A}'_{2k+2}$.*

Proof. From Levenshtein's method [5] described before, it is readily checked that $\mathcal{C}'$ consists of codewords of length n. Furthermore:

- If n is even, select a normalized quasi-Hadamard matrix $^1M_{4k}$ of order $4k$ and depth q_1, and a normalized quasi-Hadamard matrix $^2M_{4k+4}$ of order $4k+4$ and depth q_2. Denote $^i\mathcal{A}'$ a q_i-set of pairwise orthogonal rows in ^iM with their first entry dropped, and where the $+1$'s and the -1's have been replaced by 0's and 1's, respectively. Denote $^i\mathcal{D}'$ the h_i-set of rows in $^i\mathcal{A}'$ which begin with 0, for $0 \leq h_i \leq q_i$. In these circumstances, $\mathcal{C}' = \dfrac{a}{2}\left(^1\mathcal{D}'_{4k}\right) \oplus \dfrac{b}{2}\left(^2\mathcal{D}'_{4k+4}\right)$ consists in a $(n, \min\{h_1, h_2\}, d)$-code.

- If n is odd and k even, select a normalized quasi-Hadamard matrix $^1M_{2k}$ of order $2k$ and depth q_1, and a normalized quasi-Hadamard matrix $^2M_{4k+4}$ of order $4k+4$ and depth q_2. Denote $^i\mathcal{A}'$ a q_i-set of pairwise orthogonal rows in ^iM with their first entry dropped, and where the $+1$'s and the -1's have been replaced by 0's and 1's, respectively. Denote $^2\mathcal{D}'$ the h_2-set of rows in $^2\mathcal{A}'$ which begin with 0, for $0 \leq h_2 \leq q_2$. In these circumstances, $\mathcal{C}' = a\left(^1\mathcal{A}'_{2k}\right) \oplus \dfrac{b}{2}\left(^2\mathcal{D}'_{4k+4}\right)$ consists in a $(n, \min\{q_1, h_2\}, d)$-code.

- If n and k are odd, select a normalized quasi-Hadamard matrix $^1M_{4k}$ of order $4k$ and depth q_1, and a normalized quasi-Hadamard matrix $^2M_{2k+2}$ of order $2k+2$ and depth q_2. Denote $^i\mathcal{A}'$ a q_i-set of pairwise orthogonal rows in ^iM with their first entry dropped, and where the $+1$'s and the -1's have been replaced by 0's and 1's, respectively. Denote $^1\mathcal{D}'$ the h_1-set of rows in $^1\mathcal{A}'$ which begin with 0, for $0 \leq h_1 \leq q_1$. In these circumstances, $\mathcal{C}' = \dfrac{a}{2}\left(^1\mathcal{D}'_{4k}\right) \oplus b\left(^2\mathcal{A}'_{2k+2}\right)$ consists in a $(n, \min\{h_1, q_2\}, d)$-code.

The "goodness" of the code $\mathcal{C}'$ depends on the choices of q_i and h_i, so that the number of codewords is not far from the Plotkin bound (1).   □

4   Examples

The examples below illustrate that suitable quasi-Hadamard matrices give raise to good error correcting codes, even optimal ones.

In the sequel we write "$-$" instead of "-1" for simplicity.

4.1   Example 1: An Optimal Quasi-Hadamard Code

Consider the Hadamard matrices

$$H_8 = \begin{pmatrix} 1 & 1 & 1 & 1 & 1 & 1 & 1 & 1 \\ 1 & - & 1 & 1 & - & 1 & - & - \\ 1 & - & - & 1 & 1 & - & 1 & - \\ 1 & - & - & - & 1 & 1 & - & 1 \\ 1 & - & 1 & - & - & - & 1 & 1 \\ 1 & 1 & - & - & - & 1 & 1 & - \\ 1 & 1 & - & 1 & - & - & - & 1 \\ 1 & 1 & 1 & - & 1 & - & - & - \end{pmatrix}, \quad H_{12} = \begin{pmatrix} 1 & 1 & 1 & 1 & 1 & 1 & 1 & 1 & 1 & 1 & 1 & 1 \\ 1 & - & 1 & - & 1 & 1 & 1 & - & - & - & 1 & - \\ 1 & - & - & 1 & - & 1 & 1 & 1 & - & - & - & 1 \\ 1 & 1 & - & - & 1 & - & 1 & 1 & 1 & - & - & - \\ 1 & - & 1 & - & - & 1 & - & 1 & 1 & 1 & - & - \\ 1 & - & - & 1 & - & - & 1 & - & 1 & 1 & 1 & - \\ 1 & - & - & - & 1 & - & - & 1 & - & 1 & 1 & 1 \\ 1 & 1 & - & - & - & 1 & - & - & 1 & - & 1 & 1 \\ 1 & 1 & 1 & - & - & - & 1 & - & - & 1 & - & 1 \\ 1 & 1 & 1 & 1 & - & - & - & 1 & - & - & 1 & - \\ 1 & - & 1 & 1 & 1 & - & - & - & 1 & - & - & 1 \\ 1 & 1 & - & 1 & 1 & 1 & - & - & - & 1 & - & - \end{pmatrix}$$

As it is shown in [6], Levenshtein's method provide a $(27, 6, 16)$ Hadamard code $\mathcal{C}$ from H_8 and H_{12}. Assuming the notation of the precedent section, this code is constructed as the summation $2\mathcal{D}_{12} \oplus \mathcal{A}_8$, so that

$$\mathcal{C} = \begin{pmatrix} 0\,0\,0\,0\,0\,0\,0\,0\,0\,0\,0 & 0\,0\,0\,0\,0\,0\,0\,0\,0\,0\,0 & 0\,0\,0\,0\,0\,0\,0 \\ 1\,1\,0\,1\,0\,0\,0\,1\,1\,1 & 1\,1\,0\,1\,0\,0\,0\,1\,1\,1 & 1\,0\,0\,1\,0\,1\,1 \\ 1\,1\,1\,0\,1\,1\,0\,1\,0\,0 & 1\,1\,1\,0\,1\,1\,0\,1\,0\,0 & 1\,1\,0\,0\,1\,0\,1 \\ 0\,1\,1\,1\,0\,1\,1\,0\,1\,0 & 0\,1\,1\,1\,0\,1\,1\,0\,1\,0 & 1\,1\,1\,0\,0\,1\,0 \\ 0\,0\,1\,1\,1\,0\,1\,1\,0\,1 & 0\,0\,1\,1\,1\,0\,1\,1\,0\,1 & 1\,0\,1\,1\,1\,0\,0 \\ 1\,0\,0\,0\,1\,1\,1\,0\,1\,1 & 1\,0\,0\,0\,1\,1\,1\,0\,1\,1 & 0\,1\,1\,1\,0\,0\,1 \end{pmatrix}$$

Taking into account Theorem 2, the same optimal code $\mathcal{C}$ may be obtained as the summation $2({}^1\mathcal{D}'_{12}) \oplus ({}^2\mathcal{A}'_8)$ from the following quasi-Hadamard matrices:

- A quasi-Hadamard matrix ${}^1M_8$ of order 8 and depth 6, which consists in randomly substituting the last two rows of H_8.
- A quasi-Hadamard matrix ${}^{12}M_{12}$ of order 12 and depth 6, which consists in randomly substituting those rows of H_{12} which begin with $(1 - \ldots)$.   □

4.2   Example 2: A Good (Non Optimal) Quasi-Hadamard Code

The section "Finding out a liar" in ([1], chap. 17) has provided inspiration for this example.

Suppose that someone thinks of a number between 1 and 10, and that you are supposed to guess which number it is. The rules of the game let you to ask 8 questions (with "yes" or "no" answers), and no more than one lie is allowed.

In order to win, it suffices to get a code capable of correcting up to 1 error, formed from at least 10 codewords (one for every number in the given range). Writing 1 for "yes" and 0 for "no", now choose the questions so that the binary tuple that the answers generates in each case coincides with the corresponding codeword. This requires that the length of the code should coincide with the number of questions. Summing up, you need a (n, M, d) code so that $n = 8$, $M \geq 10$ and d allows to correct at least 1 error.

Your elementary background on the Theory of Codes indicates that in order to correct e errors you need to use a code of minimum distance d such that $\lfloor \frac{d-1}{2} \rfloor \geq e$. Since $e = 1$, you need $d \geq 3$.

Assume that $d = 4$. Taking into account the Plotkin bound (2), it follows that $n = 2d = 8$ and the number M of codewords is always $M \leq 2n = 16$.

Attending to Levenshtein's method, the code $\mathcal{C}_8$

$$
\mathcal{C}_8 = \begin{pmatrix}
0\,0\,0\,0\,0\,0\,0\,0 \\
0\,1\,0\,0\,1\,0\,1\,1 \\
0\,1\,1\,0\,0\,1\,0\,1 \\
0\,1\,1\,1\,0\,0\,1\,0 \\
0\,1\,0\,1\,1\,1\,0\,0 \\
0\,0\,1\,1\,1\,0\,0\,1 \\
0\,0\,1\,0\,1\,1\,1\,0 \\
0\,0\,0\,1\,0\,1\,1\,1 \\
1\,1\,1\,1\,1\,1\,1\,1 \\
1\,0\,1\,1\,0\,1\,0\,0 \\
1\,0\,0\,1\,1\,0\,1\,0 \\
1\,0\,0\,0\,1\,1\,0\,1 \\
1\,0\,1\,0\,0\,0\,1\,1 \\
1\,1\,0\,0\,0\,1\,1\,0 \\
1\,1\,0\,1\,0\,0\,0\,1 \\
1\,1\,1\,0\,1\,0\,0\,0
\end{pmatrix}
$$

related to the matrix H_8 above is optimal for given length 8 and minimum distance 4. Since $\mathcal{C}_8$ consists of 16 codewords, $\mathcal{C}_8$ may be used to solve the game.

In spite of this fact, a smaller $(8, M, 4)$ code may be used as well, provided $M \geq 10$.

Consider the quasi-Hadamard matrix M_8 obtained from H_8 by randomly changing the entries located at the 1st, 7th and 8th rows,

$$
M_8 = \begin{pmatrix}
* & * & * & * & * & * & * & * \\
1 & - & 1 & 1 & - & 1 & - & - \\
1 & - & - & 1 & 1 & - & 1 & - \\
1 & - & - & - & 1 & 1 & - & 1 \\
1 & - & 1 & - & - & - & 1 & 1 \\
1 & 1 & - & - & - & - & 1 & 1 & - \\
* & * & * & * & * & * & * & * \\
* & * & * & * & * & * & * & *
\end{pmatrix}
$$

Taking into account Theorem 1, we may construct the $(8, 10, 4)$ code C'_8,

$$
C'_8 =
\begin{pmatrix}
0\,1\,0\,0\,1\,0\,1\,1 \\
0\,1\,1\,0\,0\,1\,0\,1 \\
0\,1\,1\,1\,0\,0\,1\,0 \\
0\,1\,0\,1\,1\,1\,0\,0 \\
0\,0\,1\,1\,1\,0\,0\,1 \\
1\,0\,1\,1\,0\,1\,0\,0 \\
1\,0\,0\,1\,1\,0\,1\,0 \\
1\,0\,0\,0\,1\,1\,0\,1 \\
1\,0\,1\,0\,0\,0\,1\,1 \\
1\,1\,0\,0\,0\,1\,1\,0
\end{pmatrix}
$$

related to the matrix M_8 above.

Map every integer i in the range $[1, 10]$ to the i-th codeword c_i in C'_8. Now you should ask the following questions:

1. Is the number greater than 5?
2. Is it less or equal to 4 modulo 10?
3. Is it in the set $\{2, 3, 5, 6, 9\}$?
4. Is it in the range $[3, 7]$?
5. Is it in the set $\{1, 4, 5, 7, 8\}$?
6. Is it even?
7. Is it in the set $\{1, 3, 7, 9, 10\}$?
8. Is it in the set $\{1, 2, 5, 8, 9\}$?

Assume that the vector of answers is $a = (a_1, \ldots, a_8)$. Select the unique codeword c_i in C'_8 whose summation with a modulo 2 produces a tuple with at most one non zero entry. Then the correct number is i, and the player lied precisely when he answered the question which corresponds to the column with the non zero entry. □

Remark 2. Notice that any quasi-Hadamard matrix of order 8 and depth 5 could have been used as well in order to solve the game. The only variation is the questions to ask. In fact, the questions should be formulated so that if the number to guess is i, then the answer to the j-th question is the i-th entry of the j-th codeword of the code.

Summarizing, depending on the needs of the user, suitable quasi-Hadamard matrices have to be constructed in order to perform the desired error correcting code. Notice that working with quasi-Hadamard matrices and codes instead of Hadamard ones does not mean that functionality is lost (see example 1, for instance). In fact, it often occurs that not all codewords of a given code are actually used for transmissions in practise (see example 2 above). So quasi-Hadamard matrices and quasi-Hadamard codes may suffice to perform transmissions at the entire satisfaction of users, including optimal detection and correction affairs.

Acknowledgments

The authors want to express their gratitude to the referees for their suggestions, which have led to a number of improvements of the paper.

References

1. Cameron, P.J.: Combinatorics: topics, techniques, algorithms. Cambridge University Press, Cambridge (1994)
2. Feige, U., Goldwasser, S., Safra, S., Lovász, L., Szegedy, M.: Approximating clique is almost NP-complete. Proceedings 32nd Annual Symposium on the Foundations of Computer Science, FOCS, pp. 2–12 (1991)
3. Hastad, J.: Clique is hard to approximate within $n^{1-\epsilon}$. Proceedings 37th Annual IEEE Symposium on the Foundations of Computer Science, FOCS, pp. 627–636 (1996)
4. Karp, R.M.: Reducibility among combinatorial problems. Complexity of Computer Computations, pp. 85–103 (1972)
5. Levenshtein, V.I.: Application of the Hadamard matrices to a problem in coding. Problems of Cybernetics 5, 166–184 (1964)
6. MacWilliams, F.J., Sloane, N.J.A.: The theory of error-correcting codes. North Holland, New York (1977)
7. Marchiori, E.: Genetic, Iterated and Multistart Local Search for the Maximum Clique Problem. In: Cagnoni, S., et al. (ed.) EvoIASP 2002, EvoWorkshops 2002, EvoSTIM 2002, EvoCOP 2002, and EvoPlan 2002. LNCS, vol. 2279, pp. 112–121. Springer, Heidelberg (2002)
8. Noboru, I.: Hadamard Graphs I. Graphs Combin. 1 1, 57–64 (1985)
9. Noboru, I.: Hadamard Graphs II. Graphs Combin. 1 4, 331–337 (1985)
10. Plotkin, M.: Binary codes with specified minimum distances. IEEE Trans. Information Theory 6, 445–450 (1960)

Fast Computations of Gröbner Bases and Blind Recognitions of Convolutional Codes*

Peizhong Lu and Yan Zou

Fudan University, Shanghai 200433, P.R. China
pzlu@fudan.edu.cn

Abstract. This paper provides a fast algorithm for Gröbner bases of homogenous ideals of the ring $\mathbb{F}[x,y]$ over a field $\mathbb{F}$. The computational complexity of the algorithm is $O(N^2)$, where N is the maximum degree of the input generating polynomials. The new algorithm can be used to solve a problem of blind recognition of convolutional codes. This is a new generalization of the important problem of synthesis of a linear recurring sequence.

Keywords: Gröbner basis, sequence synthesis, Berlekamp-Massey algorithm, blind recognition of convolutional code.

1 Introduction

Let $\mathbb{F}$ be a field, and $\mathbb{F}[X] = \mathbb{F}[x_1, ..., x_n]$ the polynomial ring with n unknown variables. Let I be an ideal of $\mathbb{F}[X]$. The theory of Gröbner basis suggested by B.Buchberger [4] (1965) is a powerful tool used in the theoretical researches of algebraic decoding, algebraic attacks in cryptanalysis [11]. It is a critical problem to find a fast computation of Gröbner bases for the polynomial ideals related to practical applications. However, for a general ideal I, the computation of Gröbner basis is very complex. The upper boundary [15] of computational complexity is $O(N^{2^n})$, where N is the maximum degree of the generating polynomials of I.

Many of the algorithmic developments in Gröbner basis computation have been modifications of Buchberger's original algorithm to reduce the number of unnecessary S-polynomials that are processed. We must mention two results of important practical significance. The first is Faugere's F4 algorithm [7], which exploits sparse linear algebra to allow multiple pairs to be processed simultaneously. The second is the 'FGLM' algorithm of [6], which allows one to rapidly convert a Gröbner basis for a zero dimensional ideal from one term ordering to another with computational complexity $O(D^3)$, where $D = \dim_{\mathbb{F}} \mathbb{F}[X]/I$. However, how to precisely determine the computational complexity of F4, even in the special cases of zero dimensional ideals, is still an intractable open problem.

* This work was supported by the National Natural Science Foundation of China (60673082,90204013),and Special Funds of Authors of Excellent Doctoral Dissertation in China(200084).

C. Carlet and B. Sunar (Eds.): WAIFI 2007, LNCS 4547, pp. 303–317, 2007.

This paper provides a fast algorithm (Algorithm2) for Gröbner basis of homogenous ideals of the ring $\mathbb{F}[x, y]$ over a field $\mathbb{F}$. We show that the computational complexity of our new algorithm is $O(N^2)$ where N is the maximum degree of the input generating polynomials of the ideal.

As one of the most important applications of our new algorithm, we demonstrate that Algorithm 2 can solve the problem of fast blind recognition of convolutional code, which is a new important topic in adaptive communications, communication acquisitions [19]. We also show that the fast blind recognition of convolutional code is a novel generalization of synthesis problem of linearly recurring sequence (LRS). It is well-known that the problem of synthesis of LRS is a very important object to study in algebraic coding, cryptography, and signal processing. Berlekamp [3](1968) and Massey [20](1969) found the famous BM algorithm that carries out the fast computation of synthesis problem by solving Key Equation (KE) with computational complexity $O(N^2)$. Due to the importance of BM algorithm, a great deal of attention has been given to the problem of finding new algorithms or relationships between the existence algorithms [1],[2],[5],[12],[13],[14],[16], [17].

G.L.Feng(1991) [9] generalized KE and successfully provided an efficient algorithm of decoding the algebraic geometry codes. S.Sakata [21] generalized KE in another way, which can solve the synthesis problem of linear recurring array(LRA). Both of their algorithms were generalizations of BM algorithm basically.

In this paper, we further generalize KE by Homogenous Key Module Equation (HKME). The fast computation of Gröbner basis is used for blind recognition of convolutional code which turns out to be a new generalization of the synthesis problem. Our generalization has a new direction different from that suggested by Feng and Sakata. Our new direction also has a promising future. As a special application, our new algorithm can be used to solve the HKME with computational complexity $O(N^2)$. P.Fitzpatrick [10] also solved KE by using Gröbner bases derived from FGLM algorithm with computational complexity $O(D^3)$, where $D = \dim_{\mathbb{F}} \mathbb{F}[X]/I$. In general, due to $O(\dim_{\mathbb{F}} \mathbb{F}[X]/I) \geq O(N)$, our algorithm to solve HKME is much faster than FGLM's.

2   Fast Computation of Gröbner Basis of Homogenous Ideals with Two Variables

We give a brief introduction to Gröbner basis. Details can be found in [4].

Let $\mathbb{N}$ be the set of natural number. For arbitrary $\boldsymbol{i} = (i_1, \ldots, i_n) \in \mathbb{N}^n$, we call $X^{\boldsymbol{i}} = x_1^{i_1} \cdots x_n^{i_n}$ a power product of $\mathbb{F}[X]$. Let T^n be a set of power products of $\mathbb{F}[X]$, namely

$$T^n = \left\{ x_1^{\beta_1} \cdots x_n^{\beta_n} \, | \beta_i \in \mathbb{N}, i = 1, \ldots, n \right\}.$$

For a given term ordering over $\mathbb{F}[X]$, and arbitrary $0 \neq f \in \mathbb{F}[X]$, then f can be represented as

$$f = a_1 X^{\alpha_1} + a_2 X^{\alpha_2} + \cdots + a_r X^{\alpha_r}, \tag{1}$$

where $0 \neq a_i \in R, X^{\alpha_i} \in T^n$, and $X^{\alpha_1} > X^{\alpha_2} > \cdots > X^{\alpha_r}$.

Definition 1. *In the representation of f in (1), let $\mathrm{lp}(f) = X^{\alpha_1}$ be the leading power product of f , $\mathrm{lc}(f) = a_1$ the leading coefficient of f , and $\mathrm{lt}(f) = a_1 X^{\alpha_1}$ the leading term of f.*

In the sequel, any discussions about $\mathrm{lp}, \mathrm{lc}, \mathrm{lt}$ are related to a fixed term ordering.

Definition 2. *Given f, h in $\mathbb{F}[X]$ and a polynomial set $G = \{g_1, \ldots, g_s\}$, we say that f is reduced to h by modulo G in one step, written $f \xrightarrow{G} h$, if and only if*

$$h = f - (c_1 X_1 g_1 + \cdots + c_s X_s g_s)$$

where $c_1, \ldots, c_s \in F$, $X_1, \ldots, X_s \in T^n$, $\mathrm{lp}(f) = \mathrm{lp}(X_i)\mathrm{lp}(g_i)$, and $\mathrm{lt}(f) = c_1 X_1 \mathrm{lt}(g_1) + \cdots + c_s X_s \mathrm{lt}(g_s)$.

Definition 3. *Let $f, h \in \mathbb{F}[X]$, and nonzero polynomial set $G = \{g_1, \ldots, g_s\} \subset \mathbb{F}[X]$. we call f is reduced to h by module G, denoted as $f \xrightarrow{G}_+ h$, if there exist polynomials $h_1, \ldots, h_{t-1} \in \mathbb{F}[X]$ such that $f \xrightarrow{F} h_1 \xrightarrow{G} \cdots \xrightarrow{G} h_{t-1} \xrightarrow{G} h$.*

Definition 4. *Let $0 \neq f, g \in \mathbb{F}[x_1, \cdots, x_n]$, $L = \mathrm{lcm}(\mathrm{lp}(f), \mathrm{lp}(g))$. The following polynomial $S(f, g) = \frac{L}{\mathrm{lt}(f)} f - \frac{L}{\mathrm{lt}(g)} g$ is called the S-polynomial of f and g.*

Theorem 1. *([4]) Let I be an ideal of $\mathbb{F}[X]$. Let $G = \{g_1, \ldots, g_t\}$ be a subset of I. The following statements are equivalent.*

1. *$f \in I$ if and only if $f \xrightarrow{G}_+ 0$.*
2. *$f \in I$ if and only if*

$$f = h_1 g_1 + \cdots + h_t g_t, \tag{2}$$

   where $h_1, \ldots, h_t \in \mathbb{F}[X]$ and $\mathrm{lp}(f) = \max_{1 \leq i \leq t}(\mathrm{lp}(h_i)\mathrm{lp}(g_i))$.
3. *G is a generating set of I, and $S(g_i, g_j) \xrightarrow{G}_+ 0$, for arbitrary $1 \leq i < j \leq t$.*

Definition 5. *Let I be an ideal of $\mathbb{F}[X]$, G a subset of I. Then G is called a Gröbner basis of I, if G satisfies any equivalent condition in theorem 1. We call the subset G of $\mathbb{F}[X]$ as Gröbner basis if G is a Gröbner basis of $\langle G \rangle$.*

If B is a subset of $\mathbb{F}[X]$, then we denote $\mathrm{Lt}(B)$ an ideal generated by the leading terms of the polynomials in B.

Definition 6. *Let I be an ideal of $\mathbb{F}[X]$, $G = \{g_1, g_2, \cdots, g_r\}$ a Gröbner basis of I. If for each $j = 1, \cdots, r$, $\mathrm{lt}(g_j) \notin \mathrm{Lt}(G \setminus \{g_j\})$, G is called a minimal Gröbner basis.*

In the following we study the computation of Gröbner bases of ideals of the polynomial ring $\mathbb{F}[x, y]$ with two variables.

Definition 7. f *is called a nonzero monomial over* $\mathbb{F}[x,y]$, *if there exists* $0 \neq c \in F$ *such that* $f = cx^n y^m$. *In this case, we denote* $\deg_x f = n$, $\deg_y f = m$.

Let $G = \{g_1, \cdots, g_l\}$, where each g_i is a homogenous polynomial of $\mathbb{F}[x,y]$. In the sequel, we chose lexicographic order as the fixed term order on $\mathbb{F}[x,y]$, and $y < x$.

Example 1. $G = \{g_1 = x^5 + x^4 y + x^2 y^3, g_2 = x^7 + x^2 y^5 + y^7, g_3 = x^6 y + xy^6\}$, each monomials of polynomial g_1 is ordered as : $x^5 > x^4 y > x^2 y^3$.

Lemma 1. *Let* $G = \{g_1, \cdots, g_l\}$ *be a minimal Gröbner basis. Let* $\mathrm{lp}(g_i) = x^{n_i} y^{m_i}, i = 1, 2, \cdots, l$ *be the leading term of* g_i *and* $n_1 \geq n_2 \geq \cdots \geq n_l$. *Then* $n_1 > n_2 > \cdots > n_l$, *and* $m_1 < m_2 < \cdots < m_l$.

Definition 8. *Let* $G = \{g_1, \cdots, g_l\}$ *be a subset of* $\mathbb{F}[x,y]$. *With a fixed term order, if* $\mathrm{lp}(g_1) > \mathrm{lp}(g_2) > \cdots > \mathrm{lp}(g_l)$, *and* $\mathrm{lp}(g_i) \nmid \mathrm{lp}(g_j), i \neq j$, *then the leading terms of* G *are called to be strictly ordered.*

Corollary 1. *Let* $G = \{g_1, \cdots, g_l\}$ *be a minimal Gröbner basis, and* $\mathrm{lp}(g_i) = x^{n_i} y^{m_i}$ *be the leading term of* g_i. *If* $n_1 \geq n_2 \geq \cdots \geq n_l$, *then the leading terms of* G *are strictly ordered.*

Let K be a finite subset of homogenous polynomials in $\mathbb{F}[x,y]$. The following algorithm converts K to a strictly ordered finite set G in lexicographic order, where $y <_T x$, and $\langle K \rangle = \langle G \rangle$.

Algorithm 1. *Finding a strictly ordered homogenous generator set for an ideal.*

  INPUT: a homogenous polynomial set $K = \{g_1, \cdots, g_l\}$
  OUTPUT: a strictly ordered generator set G *such that* $\langle G \rangle = \langle K \rangle$.
  INITIAL: $G = \emptyset$
  WHILE $K \neq \emptyset$
    $I = \{g \in K \mid \deg_y \mathrm{lp}(g) = \min_{f \in K} \{\deg_y \mathrm{lp}(f)\}\}$.
    Find $g \in I$ *such that* $\deg_x \mathrm{lp}(g) = \min_{f \in I} \{\deg_x \mathrm{lp}(f)\}$,
    $K' := K \setminus \{g\}$, $G := G \cup \{g\}$, $K = \emptyset$
    WHILE $K' \neq \emptyset$
      Choose $f \in K'$, $K' := K' \setminus \{f\}$
      WHILE $\deg_x \mathrm{lp}(f) \geq \deg_x \mathrm{lp}(g)$
        Reduced f *by* g *to produce a reduced canonical form*
        $f' \in \mathbb{F}[x,y]$ *such that* $f \to_G f'$
        IF $f' \neq 0$ *THEN* $f = f'$
      $K := K \cup \{f\}$
  RETURN G

Proposition 1. *Let* $K = \{f_1, \ldots, f_k\}$ *be a homogenous polynomial subset of* $\mathbb{F}[x,y]$ *and* $G = \{g_1, \ldots, g_\ell\}$ *the finite set resulting from Algorithm 1. Then the leading terms of* G *are strictly ordered. Moreover* $\ell \leq k$, *and* $\langle K \rangle = \langle G \rangle$.

Proof. Let $K_0 = K, G_0 = \emptyset$, and let K_i, G_i be the sets established in i-th step in Algorithm 1. Assume that the Algorithm 1 has generated all the subsets $K_0, \ldots, K_i, G_0, \cdots, G_i$ such that for all $0 \le j \le i$, the polynomials in G_j are strictly ordered, and

$$\langle G_j \cup K_j \rangle = \langle K \rangle, 0 \le j \le i.$$

Thus the assumption is true when $j = 0$. By Algorithm 1, $G_{i+1} = \{g_{i+1}\} \cup G_i$, where g_{i+1} is a polynomial g form K_i such that $(\deg_x \mathrm{lp}(g), \deg_y \mathrm{lp}(g))$ is minimal. Here, the ordering rule of a pair of positive integers is that $(n_1, m_1) < (n_2, m_2)$ if and only if $m_1 < m_2$, or $m_1 = m_2$ and $n_1 < n_2$. Since K_{i+1} is the set of canonical form of the elements from $K_i \setminus \{g_{i+1}\}$ reduced by g_{i+1}, thus

$$\langle K_i \rangle = \langle K_{i+1} \cup \{g_{i+1}\}\rangle.$$

Therefore we have

$$\langle K_{i+1} \cup G_{i+1} \rangle = \langle K_{i+1} \cup \{g_{i+1}\} \cup G_i \rangle = \langle K_i \cup G_i \rangle = \langle K \rangle.$$

In the following we prove that $\deg_x \mathrm{lp}(g_{i+1}) < \deg_x \mathrm{lp}(g_i)$. By the assumption, the leading terms of G_i are strictly ordered, and $G_{i+1} = \{g_{i+1}\} \cup G_i$. We know that $g_{i+1} \in K_i$. If g_{i+1} is a polynomial reduced from some f in K_{i-1} by g_i, then $\deg_x \mathrm{lp}(g_{i+1}) < \deg_x \mathrm{lp}(g_i)$. If g_{i+1} is a polynomial f in K_{i-1} such that it can not be nontrivially reduced by g_i, then $\deg_y \mathrm{lp}(g_{i+1}) < \deg_y \mathrm{lp}(g_i)$, or $\deg_y \mathrm{lp}(g_{i+1}) \ge \deg_y \mathrm{lp}(g_i)$ and $\deg_x \mathrm{lp}(g_{i+1}) < \deg_x \mathrm{lp}(g_i)$. The former contradicts with the rule in the algorithm to select g_i. Therefore $\deg_x \mathrm{lp}(g_{i+1}) < \deg_x \mathrm{lp}(g_i)$.

We further prove that $\deg_y \mathrm{lp}(g_{i+1}) > \deg_y \mathrm{lp}(g_i)$. Obviously $\deg_y \mathrm{lp}(g_{i+1}) \ge \deg_y \mathrm{lp}(g_i)$. We show that $\deg_y \mathrm{lp}(g_{i+1}) \ne \deg_y \mathrm{lp}(g_i)$.

From the construction of g_{i+1}, there are two cases need to consider.

(a) g_{i+1} is a polynomial f of K_{i-1}, and can not be nontrivial reduced by g_i. In this case, $\deg_x \mathrm{lp}(g_{i+1}) < \deg_x \mathrm{lp}(g_i)$. If $\deg_y \mathrm{lp}(g_{i+1}) = \deg_y \mathrm{lp}(g_i)$, it contradicts the minimum property of g_i.

(b) g_{i+1} is the nontrivial reduced polynomial f of K_{i-1} by g_i. Since $\deg_y \mathrm{lp}(f) \ge \deg_y \mathrm{lp}(g_i)$, and f is reduced by g_i to g_{i+1}, therefore

$$\deg_y \mathrm{lp}(g_{i+1}) > \deg_y(\mathrm{lp}(f)) \ge \deg_y \mathrm{lp}(g_i).$$

Thus $\deg_y \mathrm{lp}(g_{i+1}) > \deg_y \mathrm{lp}(g_i)$. Hence G_{i+1} is strictly ordered. Thus, by induction on i, we have proved that G is strictly ordered.

Due to

$$|K_{i+1}| \le |K_i| - 1 \le k - i - 1, \ |G_{i+1}| \le |G_i| + 1 \le i + 1, \ i = 0, 1, \cdots, k - 1,$$

thus $K_k = \emptyset$, and $\langle G_k \rangle = \langle G_k \cup K_k \rangle = \langle K \rangle$. Therefore $\ell = |G| \le |G_k| \le k$. $\quad\square$

Theorem 2. *Let $G = \{g_1, \cdots, g_\ell\}$ be a homogenous strictly ordered subset of $\mathbb{F}[x, y]$. Then G is a minimal Gröbner basis if and only if*

$$S(g_i, g_{i+1}) \xrightarrow{G}_+ 0, i = 1, \cdots, \ell - 1$$

Proof. Necessity: If G is a minimal Gröbner basis, then, by theorem 1, we have $S(g_i, g_{i+1}) \xrightarrow{G}_+ 0$, for $i = 1, \cdots, \ell - 1$.

Sufficiency: Let $S(g_i, g_{i+1}) \xrightarrow{G}_+ 0$, for $i = 1, \cdots, \ell - 1$. We prove that G is a Gröbner basis. From theorem 1, we only need to prove that for arbitrary $f \in \langle g_1, g_2, \cdots, g_\ell \rangle$, f can be written as $f = h_1 g_1 + \cdots + h_t g_\ell$, where $h_1, \ldots, h_\ell \in \mathbb{F}[x, y]$, and

$$\mathrm{lp}(f) = \max_{1 \le i \le \ell}(\mathrm{lp}(h_i)\mathrm{lp}(g_i)). \tag{3}$$

In fact, since $f \in \langle g_1, g_2, \cdots, g_\ell \rangle$, f can be written as

$$f = h_1 g_1 + \cdots + h_\ell g_\ell, \tag{4}$$

where $h_1, \ldots, h_\ell \in \mathbb{F}[x, y]$.

There is a representation of f as (4), such that $X = \max_{1 \le i \le \ell}(\mathrm{lp}(h_i)\mathrm{lp}(g_i))$ is minimal, and moreover the number of elements in the following set

$$S = \{i | 1 \le i \le \ell, \mathrm{lp}(h_i)\mathrm{lp}(g_i) = X\},$$

namely $|S|$, is minimal.

If $X \le \mathrm{lp}(f)$, then formula (3) is true.

Now let $\mathrm{lp}(f) < X$.

If $t = |S| = 1$, then in the right side of formula (4) there only exists one maximal term in the summations that can not be eliminated by the other terms. Thus $\mathrm{lp}(f) = X$, which contradicts with the hypothesis. So let $t = |S| \ge 2$. Let us suppose that the interval between the least two elements in S is minimal. Hence, if $S = \{i_1, \cdots, i_t\}$, where $1 \le i_1 < i_2 < \cdots < i_t \le \ell$, then t and $i_2 - i_1$ are both minimal.

For convenience in description, we suppose $1, j \in S$, where $1 < j \le \ell$, and j is minimal. From $X = \mathrm{lp}(h_1)\mathrm{lp}(g_1)) = \mathrm{lp}(h_j)\mathrm{lp}(g_j))$, we know that X is a common multiple of $\mathrm{lp}(g_1), \mathrm{lp}(g_j)$. Let $\deg_x g_i = n_i, \deg_y g_i = m_i$. Then $n_1 > n_2 > \cdots > n_\ell$, $m_1 < m_2 < \cdots < m_\ell$. Thus the least common multiple of $\mathrm{lp}(g_1)$ and $\mathrm{lp}(g_j)$ is $x^{n_1} y^{m_j}$. Let $\triangle = \mathrm{lc}(h_1)$ and $\delta = \frac{\mathrm{lc}(g_1)\mathrm{lc}(h_1)}{\mathrm{lc}(g_2)}$. From the definition of S-polynomial, we have $S(g_1, g_2) = y^{m_2 - m_1} g_1 - (\mathrm{lc}(g_1)/\mathrm{lc}(g_2))x^{n_1 - n_2} g_2$.

Thus

$$h_1 g_1 + h_2 g_2 - \triangle \frac{X}{x^{n_1} y^{m_2}} S(g_1, g_2) = (h_1 - \mathrm{lt}(h_1))g_1 + (h_2 + \delta \frac{X}{x^{n_2} y^{m_1}})g_2.$$

If $j = 2$, then

$$f = h_1 g_1 + h_2 g_2 + \cdots + h_\ell g_\ell$$

$$= h_1 g_1 + h_2 g_2 + \sum_{i \in S, i \ne 1, i \ne 2} h_i g_i + \sum_{i \notin S} h_i g_i$$

$$= \triangle \frac{X}{x^{n_1} y^{m_2}} S(g_1, g_2) - \triangle \frac{X}{x^{n_1} y^{m_2}} S(g_1, g_2) + h_1 g_1 + h_2 g_2$$

$$+ \sum_{i \in S, i \ne 1, i \ne 2} h_i g_i + \sum_{i \notin S} h_i g_i$$

$$= \triangle \frac{X}{x^{n_1}y^{m_2}}S(g_1,g_2) + (h_1 - \mathrm{lt}(h_1))g_1 + (h_2 + \delta\frac{X}{x^{n_2}y^{m_1}})g_2$$

$$+ \sum_{i\in S, i\neq 1, i\neq 2} h_i g_i + \sum_{i\notin S} h_i g_i.$$

Thus

$$f = (h_2 + \delta\frac{X}{x^{n_2}y^{m_2}})g_2 + \sum_{i\in S, i\neq 1, i\neq 2} h_i g_i + \sum_{i\notin S} h_i g_i + \omega, \tag{5}$$

where $\omega = \triangle\frac{X}{x^{n_1}y^{m_2}}S(g_1,g_2) + (h_1 - \mathrm{lt}(h_1))g_1$. Due to $S(g_1,g_2) \xrightarrow{G}_+ 0$, we have $\triangle\frac{X}{x^{n_1}y^{m_2}}S(g_1,g_2) \xrightarrow{G}_+ 0$, and

$$\triangle\frac{X}{x^{n_1}y^{m_2}}S(g_1,g_2) = h_1' g_1 + \cdots + h_\ell' g_\ell,$$

where $X > \mathrm{lp}(\triangle\frac{X}{x^{n_1}y^{m_2}}S(g_1,g_2)) = \max_{1\le i\le \ell}(\mathrm{lp}(h_i')\mathrm{lp}(g_i))$. Since $\mathrm{lp}(h_1 - \mathrm{lt}(h_1))g_1 < \mathrm{lp}(h_1 g_1) = X$, then ω can be written as

$$\omega = h_1'' g_1 + \cdots + h_\ell'' g_\ell,$$

where $X > \mathrm{lp}(\omega) = \max_{1\le i\le\ell}(\mathrm{lp}(h_i'')\mathrm{lp}(g_i))$. In the right side of (5) we can see in the first term that $\mathrm{lp}(\frac{X}{x^{n_2}y^{m_2}}) = \mathrm{lp}(h_2)$, and thus

$$\mathrm{lp}((h_2 + \delta\frac{X}{x^{n_2}y^{m_2}}))\mathrm{lp}(g_2) \le X.$$

Therefore, from (5) we conclude that f can be represented as a summation in which the number of maximal monomials not exceed $|S| - 1$, which contradicts the hypothesis that $|S|$ is minimal. So it's true in the condition of $j = 2$.

Now consider $j > 2$. For convenience in description, we suppose $\mathrm{lc}(g_1) = \mathrm{lc}(g_j) = \mathrm{lc}(h_j) = 1$. Since

$$f = h_1 g_1 + h_j g_j + \sum_{i\in S, i\neq 1, i\neq j} h_i g_i + \sum_{i\notin S} h_i g_i$$

$$= \frac{X}{x^{n_{j-1}}y^{m_j}}S(g_{j-1},g_j) - \frac{X}{x^{n_{j-1}}y^{m_j}}S(g_{j-1},g_j) + h_1 g_1 + h_j g_j$$

$$+ \sum_{i\in S, i\neq 1, i\neq j} h_i g_i + \sum_{i\notin S} h_i g_i$$

$$= \frac{-X}{x^{n_{j-1}}y^{m_j}}S(g_{j-1},g_j) + h_1 g_1 + \frac{X}{x^{n_{j-1}}y^{m_{j-1}}}g_{j-1} + (h_j - \mathrm{lp}(h_j))g_j$$

$$+ \sum_{i\in S, i\neq 1, i\neq j} h_i g_i + \sum_{i\notin S} h_i g_i,$$

thus

$$f = h_1 g_1 + \frac{X}{x^{n_{j-1}}y^{m_{j-1}}}g_{j-1} + \sum_{i\in S, i\neq 1, i\neq 2} h_i g_i + \sum_{i\notin S} h_i g_i$$

$$- \frac{X}{x^{n_{j-1}}y^{m_j}}S(g_{j-1},g_j) + (h_j - \mathrm{lp}(h_j))g_j. \tag{6}$$

In (6), due to $\frac{X}{x^{n_j-1}y^{m_j}}S(g_{j-1},g_j) \to_G 0$ and $\mathrm{lp}(\frac{X}{x^{n_j-1}y^{m_j}}S(g_{j-1},g_j)) < X$, $\mathrm{lp}((h_j - \mathrm{lp}(h_j))g_j) < X$, the number of maximal monomials X in f equal to $|S|$. But the position set of maximum monomials X is $\{1, j-1\} \cup S \setminus \{j\}$, which contradicts with the hypothesis that j is minimal. Here the conclusions are all proved.   □

Based on theorem 2 and proposition 1 above, we give the following algorithm to compute a minimal Gröbner basis for a homogenous ideal over $\mathbb{F}[x,y]$.

Algorithm 2. *Computing a minimal Gröbner basis for a homogenous ideal of* $\mathbb{F}[x,y]$

> *INPUT: A homogenous set* $K = \{g_1, \cdots, g_l\}$
> *OUTPUT: Gröbner basis* G *such that* $\langle G \rangle = \langle K \rangle$.
> *INITIAL: Call subroutine Algorithm 1 such that homogenous set* K
> *is changed into a strictly ordered set.* $i = 0, G_i = \emptyset$
> WHILE $K \neq \emptyset$
> > $i := i + 1$
> > $I = \{g \in K \mid \deg_y \mathrm{lp}(g) = \min_{f \in K}\{\deg_y \mathrm{lp}(f)\}\}$.
> >   *Find* $g \in I$ *that* $\deg_x \mathrm{lp}(g) = \min_{f \in I}\{\deg_x \mathrm{lp}(f)\}$,
> > $K' := K - \{g\}, G_i := G_{i-1} \cup \{g\}, g_i = g, K = \emptyset$
> > $h := S(g_i, g_{i-1}), K' := K' \cup \{h\}$,
> >   WHILE $K' \neq \emptyset$
> > >   *Find* $f \in K', K' := K' - \{f\}$
> > >   WHILE $\deg_x \mathrm{lp}(f) \geq \deg_x \mathrm{lp}(g)$
> > > >   *Reduced* f *by* g *gives the canonical form* $f' \in \mathbb{F}[x,y]$
> > > >   *such that* $f \to_{G_i} f'$, *namely* $f \equiv f' \mod G_i$
> > > >   IF $f' \neq 0$ THEN $f = f'$
> > > $K := K \cup \{f\}$
> $G = G_i$
>   RETURN G

3   Analysis of Computational Complexity

Lemma 2. *Let* $f, g \in \mathbb{F}[x]$, g *be a monic polynomial, and* $\deg f \geq \deg g$. *To find polynomials* $q, r \in \mathbb{F}[x]$ *such that* $f = qg + r$, *and* $\deg r < \deg g$, *by division algorithm, the computational complexity is* $O((\deg f - \deg g + 1)\deg g)$ *operations over* $\mathbb{F}$.

Proposition 2. *Let* $K = \{f_1, \ldots, f_k\}$ *be a homogenous polynomial subset of* $\mathbb{F}[x,y]$ *and* $G = \{g_1, \ldots, g_\ell\}$ *the finite set in Algorithm 1, and* $\deg_x g_i = n_i, i = 1, \cdots, \ell$. *Let* $N = n_0 = \max_{f \in K} \deg_x \mathrm{lp}(f)$, *then the computational complexity of Algorithm 1 is* $O(kN^2)$.

Proof. In the Algorithm 1, let S, S_i be the numbers of all operations taken place in all steps and in step i respectively . Let $G_i = G_{i-1} \cup \{g_i\}, n_i = \deg_x \mathrm{lp}(g_i), m_i = \deg_y \mathrm{lp}(g_i)$. The computational complexity to find g_{i+1} in K_i

is less than $2(|K_i| - 1) \leq 2(k - i - 1)$ times comparing operations. Let f be an arbitrary element in K_i, then $\deg_y \mathrm{lp}(f) \geq \deg_y \mathrm{lp}(g_{i+1})$. From the constructing process of K_{i+1}, we know that for $f \in K_i$, if $\deg_x \mathrm{lp}(f) < \deg_x \mathrm{lp}(g_{i+1})$, then f can be put into K_{i+1}. If $\deg_x \mathrm{lp}(f) \geq \deg_x \mathrm{lp}(g_{i+1})$, then, after dividing f by g_{i+1}, the reduced canonical form $f' = f - h \cdot g_{i+1}$ is put into K_{i+1}. By Lemma 2, the number of addition and multiplication required in the above process is

$$(\deg_x \mathrm{lp}(f) - \deg_x g_{i+1} + 1) \deg_x \mathrm{lp}(g_{i+1}) = (\deg_x \mathrm{lp}(f) - n_{i+1} + 1)n_{i+1}.$$

Then the computational complexity to construct K_{i+1} from K_i satisfies

$$S_0 \leq \sum_{f \in K - \{g_1\}} (\deg_x \mathrm{lp}(f) - n_1 + 1) \cdot n_1 + 2k \leq k(n_0 - n_1 + 1)n_1 + 2k,$$

where $n_0 = \max_{f \in K}\{\deg_x f\}$, and

$$S_{i+1} \leq \sum_{f \in K_i - \{g_{i+1}\}} (\deg_x \mathrm{lp}(f) - n_{i+1} + 1) \cdot n_{i+1} + 2(k - i - 1).$$

If $f \in K_i$, then $\deg_x \mathrm{lp}(f) \leq \deg_x \mathrm{lp}(g_i) = n_i$, and thus

$$\begin{aligned} S_{i+1} &\leq \sum_{f \in K_i - \{g_{i+1}\}} (\deg_x f - n_{i+1} + 1) \cdot n_{i+1} + 2(k - i - 1) \\ &\leq (n_i - n_{i+1} + 1) \cdot n_{i+1}(|K_i| - 1) + 2(k - i - 1) \\ &\leq ((n_i - n_{i+1} + 1) \cdot n_{i+1} + 2)(k - i - 1). \end{aligned}$$

Therefore, the total computation S in Algorithm 1 satisfies

$$\begin{aligned} S &\leq S_0 + \sum_{i=1}^{\ell}((n_i - n_{i+1} + 1) \cdot n_{i+1} + 2)(k - i - 1) \\ &\leq S_0 + k \sum_{i=1}^{\ell}((n_i - n_{i+1} + 1) \cdot n_{i+1} + 2) \\ &\leq S_0 + k \sum_{i=1}^{\ell}(n_i - n_{i+1} + 1) \cdot n_{i+1} + 2k\ell \\ &\leq S_0 + k \sum_{i=1}^{\ell}(n_i - n_{i+1} + 1) \cdot n_1 + 2k\ell \\ &\leq S_0 + k(n_1^2 + \ell n_1 + 2\ell) \\ &= O(kN^2). \end{aligned}$$

Here we suppose $n_{\ell+1} = 0$. The conditions of $\ell \leq k$, and $n_1 > \cdots > n_\ell > n_{\ell+1} = 0$ are needed. $\qquad\square$

Theorem 3. *Let $K = \{g_1, \cdots, g_k\}$ be a homogenous subset of $\mathbb{F}[x, y]$, then the finite set G from Algorithm 2 is a strictly ordered Gröbner basis. The computational complexity is $O(kN^2)$, where N the maximal degree of variable x in K.*

Proof. 1) The degree $\deg_x g$ of g selected in each step in Algorithm 2 is strictly descending. Thus the algorithm ends in finite steps. 2) Since G is strictly ordered, and satisfies the conditions of Lemma 2, G is a Gröbner basis. 3) Let S_i be the number of operations needed in step i in Algorithm 2. From the proof of Proposition 2, we know that

$$\begin{aligned} S_{i+1} = \ &\text{numbers of operations in selecting } g_{i+1} \text{ from } K_i \\ &+\text{numbers of operations of computing S-polynomial} \\ &+\text{numbers of operations of } K' \bmod g_i \\ \leq\ &2(k - 1) + n_{i+1} + \sum_{f \in K_i} (\deg_x \mathrm{lp}(f) - n_{i+1})n_{i+1}. \end{aligned}$$

Thus the total computation is

$$S \leq \sum_{i=0}^{\ell}(2(k-1) + n_{i+1} + \sum_{f \in K_i}(\deg_x \mathrm{lp}(f) - n_{i+1})n_{i+1})$$
$$\leq 2(k-1)(n_0 + 2) + n_0(n_0 + 1) + k(n_0 + 1)n_0,$$

where ℓ is steps of Algorithm 2. Since $\deg_x \mathrm{lp}(g_{i+1}) \leq \deg_x \mathrm{lp}(g_i) - 1$, thus $\ell \leq n_0 + 1$. Therefore the computational complexity of Algorithm 2 does not exceed $O(kN^2)$. $\qquad\square$

4   A New Generalization of Sequence Synthesis

In this section we show that the fast computation of Gröbner basis of homogenous ideals of $\mathbb{F}[x, y]$ can be used to solve the famous problem of sequence synthesis, and its new generalization for blind recognition of convolutional code.

4.1   Synthesis of LRS and Blind Recognition of Convolutional Code

Let $\boldsymbol{a} = (a_0, a_1, \cdots, a_N)$ be a sequence with finite length over $\mathbb{F}$. Let $f(x) = f_0 + f_1 x + \cdots + f_L x^L$ be a polynomial in $\mathbb{F}[x]$, where $\deg f(x) \leq L, f_0 = 1$. If this polynomial satisfies the following linear recurrence relation

$$a_{i+L} + f_1 a_{i+L-1} + f_2 a_{i+L-2} + \cdots + f_L a_i = 0, \ i \geq 0,$$

then $(f(x), L)$ is called linear recurrence relation of sequence $\boldsymbol{a}$. If $(f(x), L)$ is linear recurrence relation of sequence $\boldsymbol{a}$, and L is minimal, then $(f(x), L)$ is called the minimal linear recurring relation of $\boldsymbol{a}$ and L is called linear complexity of $\boldsymbol{a}$, namely $\ell(\boldsymbol{a})$.

The so called synthesis problem of LRS is to find the minimal linear recurrence relation that can generate sequence $\boldsymbol{a}$.

Let a convolutional code with coding rate $1/2$ be

$$C = \{(a(x)g_1(x), a(x)g_2(x)) \,|\, a(x) \in \mathbb{F}[[x]]\} \tag{7}$$

where $a(x) = a_0 + a_1 x + a_2 x^2 + \cdots$ is the signal sequence, $g_1(x), g_2(x)$ are the generating polynomials, and $\mathbb{F}[[x]]$ is Laurent series ring over $\mathbb{F}$. The problem of blind recognition of convolutional codes is to find the unknown $g_1(x), g_2(x)$ from a finite partial known convolutional subsequence $C(x)$ with errors, namely, $C(x) = (C_1(x), C_2(x))$

$$C_i(x) + E_i(x) = (c_{i0} + c_{i1}x + \cdots + c_{iN}x^N) + (e_{i0} + e_{i1}x + \cdots + e_{iN}x^N), \tag{8}$$

$i = 1, 2$, where the error code is $E_i(x) = e_{i0} + e_{i1}x + \cdots + e_{iN}x^N$.

We only consider the noiseless case. For computing $g_i(x)$, we first estimate the restricted length $k = \max(\deg g_1(x), \deg g_2(x))$. If the length N of the partial known convolutional subsequence is so large that $N > 3 \times (k+1)$, then we can solve the following linear equations

$$
\begin{pmatrix}
c_{1,k} & c_{1,k-1} & \cdots & c_{1,0} & c_{2,k} & c_{2,k-1} & \cdots & c_{2,0} \\
c_{1,k+1} & c_{1,k} & & c_{1,1} & c_{2,k+1} & c_{2,k} & & c_{2,1} \\
\\
\ddots & \ddots & \ddots & & \ddots & \ddots \\
\\
c_{1,N} & c_{1,N-1} & \cdots & c_{1,N-k} & c_{2,N} & c_{2,N-1} & \cdots & c_{2,N-k}
\end{pmatrix}
\begin{pmatrix}
g_{2,0} \\
g_{2,1} \\
\vdots \\
g_{2,k} \\
g_{1,0} \\
g_{1,1} \\
\vdots \\
g_{1,k}
\end{pmatrix}
= 0. \quad (9)
$$

To guarantee the solvability of equations (9), we need to set k a large integer. But if k is too large, the computational complexity becomes higher. To solve equations (9) by Gauss elimination, the computational complexity is $O(N^3)$.

When $g_2(x) = 0$, the problem of blind recognition of convolutional code becomes synthesis problem of LRS.

4.2 Key Equation and Key Module Equation

Let $A(x) = a_0 + a_1 x + \cdots + a_N x^N$ be a polynomial representation of the sequence a. Synthesis problem of LRS can be converted to how to solve the following key equation efficiently.

Key equation (KE): Find an element $(f(x), L)$ in

$$
\Phi^{(1)} = \left\{ (f(x), L) \in \mathbb{F}[x] \times \mathbb{N} \,\middle|\, \begin{array}{c} \deg f(x) \le L, \exists b(x) \in \mathbb{F}[x], \deg b(x) < L \text{ such that} \\ f(x)A(x) \equiv b(x) \bmod x^N \end{array} \right\}.
$$

such that $f(0) \ne 0$ and L is minimal.

Since the problem of blind recognition of convolutional code is a generalization of synthesis problem of LRS, we propose the following generalized key equation. Let

$$
C_i(x) = c_{i0} + c_{i1}x + c_{i2}x^2 + \cdots + c_{iN}x^N, i = 1, 2
$$

be the partial known convolutional subsequences over $\mathbb{F}$.

Key module equation (KME): Find an element pair $(h_1(x), h_2(x), L)$ in

$$
\Phi^{(2)} = \left\{ (h_1(x), h_2(x), L) \in R[x]^2 \times \mathbb{N} \,\middle|\, \begin{array}{c} \exists d(x) \in \mathbb{F}[x], \text{ such that} \\ h_1(x)C_1(x) + h_2(x)C_2(x) \equiv d(x) \bmod x^{N+1}, \\ \deg d(x) < L, \max(\deg h_1(x), \deg h_2(x)) \le L \end{array} \right\}
$$

such that L be minimal and $(h_1(0), h_2(0)) \ne (0, 0)$.

4.3 Homogenous Key Equation and Homogenous Key Module Equation

In reference[17], we originally proposed homogenous key equation for modelling the synthesis problem of LRS. The homogeneity of $A(x)$ is denoted as $A(x, y)$, namely $A(x, y) = a_0 y^N + a_1 xy^{N-1} + \cdots + a_N x^N$. Let $\mathcal{I} = \langle x^{N+1}, y^{N+1} \rangle$ be a homogenous ideal of $\mathbb{F}[x, y]$.

Homogenous key equation (HKE): Find a minimal generator of the homogenous ideal

$$\Gamma^{(1)} = \{b(x, y) \in \mathbb{F}[x, y] \,|\, b(x, y)A(x, y) \equiv 0 \bmod \mathcal{I}\} \tag{10}$$

Similar to synthesis of LRS, we propose homogenous key module equation for blind recognition of convolutional code. Firstly, we need to homogenize the convolutional code subsequence $C_i(x)$ to

$$C_i(x, y) = c_{i0} y^N + c_{i1} xy^{N-1} + c_{i2} x^2 y^{N-2} + \cdots + c_{iN} x^N, i = 1, 2$$

Denote $\mathcal{I} = \langle x^{N+1}, y^{N+1} \rangle$ as a homogenous ideal of $\mathbb{F}[x, y]$ generated from x^{N+1}, y^{N+1} .

Homogenous key module equation (HKME): Find a minimal generator of $\mathbb{F}[x, y]$-module

$$\Gamma^{(2)} = \left\{ (H_1, H_2) \in \mathbb{F}[x, y]^2 \,|\, H_1 C_1(x, y) + H_2 C_2(x, y) \equiv 0 \bmod \mathcal{I} \right\} \tag{11}$$

Blind recognition of convolutional code is a problem to compute a Gröbner basis of finite generating module $\Gamma^{(2)}$. The following proposition show the tight relationship between KME and HKME.

Theorem 4. *[18] Let $C_i(x), C_i(x, y)$ be polynomials in KME and HKME respectively. Let $h_i(x) = h_{i0} + h_{i1}x + \cdots + h_{im}x^m, i = 1, 2$ be polynomials over $\mathbb{F}[x]$ with corresponding homogenous polynomials $H_i(x, y) = h_{i0}y^m + h_{i1}xy^{m-1} + \cdots + h_{im}x^m, i = 1, 2$ respectively. Then $(h_1(x), h_2(x), m) \in \Phi^{(2)}$ in KME if and only if $(H_1(x, y), H_2(x, y)) \in \Gamma^{(2)}$ in HKME.*

Obviously, HKME has a nice algebraic structure. By theorem 4, we have the following conclusions:

1. Synthesis problem of LRS can be carry out by finding a Gröbner basis of homogenous ideal

$$I^{(1)} = \langle A(x, y), x^{N+1}, y^{N+1} \rangle$$

and the Syzygy of $I^{(1)}$, where $A(x, y)$ is defined in HKE.

2. The problem of blind recognition of convolutional code can be fast carry out by finding a Gröbner basis of homogenous ideal

$$I^{(2)} = \langle C_1(x, y), C_2(x, y), x^{N+1}, y^{N+1} \rangle$$

and the Syzygy of $I^{(2)}$, where $C_1(x, y), C_2(x, y)$ are defined in HKME.

The definition of Syzygy of ideal can be found in reference [4]. If only the Gröbner basis is found, the corresponding Syzygy can be obtained at the same time. See the example in the next subsection.

4.4 Computational Example of Blind Recognition of Convolutional Code

Example 2. The received convolutional code subsequence are denoted as $F_1 = x^{25} + x^{21}y^4 + x^{20}y^5 + x^{19}y^6 + x^{18}y^7 + x^{15}y^{10} + x^{13}y^{12} + x^{12}y^{13} + x^{11}y^{14} + x^{10}y^{15} + x^7y^{18} + x^2y^{23} + xy^{24} + y^{25}$, and $F_2 := x^{25} + x^{24}y + x^{22}y^3 + x^{19}y^6 + x^{18}y^7 + x^{17}y^8 + x^{16}y^9 + x^{15}y^{10} + x^{14}y^{11} + x^{13}y^{12} + x^{11}y^{14} + x^{10}y^{15} + x^9y^{16} + x^3y^{22} + x^2y^{23} + y^{25}$.

We choose the partial subsequences of length 18 which are written as $C_1(x,y) = x^{17} + x^{16}y + x^{15}y^2 + x^{14}y^3 + x^{13}y^4 + x^{11}y^6 + x^{10}y^7 + x^9y^8 + x^3y^{14} + x^2y^{15} + y^{17}$, $C_2(x,y) = x^{15}y^2 + x^{13}y^4 + x^{12}y^5 + x^{11}y^6 + x^{10}y^7 + x^7y^{10} + x^2y^{15} + xy^{16} + y^{17}$. We compute the Gröbner basis G of the homogenous ideal $I = \langle C_1(x,y), C_2(x,y), x^{18}, y^{18} \rangle$ by Algorithm 2, and record the coefficient matrix in the process of reductions. The coefficient matrix is just the Syzygy relation.

Initial: Let $(f_{0,-1}, f_{1,-1}, f_{2,-1}, f_{3,-1}) = (C_1(x,y), x^{18}, C_2(x,y), y^{18})$ based on the term order. Let

$(1,0,0,0)$	$f_{0,-1} = x^{17} + x^{16}y + x^{15}y^2 + x^{14}y^3 + x^{13}y^4 + x^{11}y^6 + x^{10}y^7 + x^9y^8 + x^3y^{14} + x^2y^{15} + y^{17}$
$(0,1,0,0)$	$f_{1,-1} = x^{18}$
$(0,0,1,0)$	$f_{2,-1} = x^{15}y^2 + x^{13}y^4 + x^{12}y^5 + x^{11}y^6 + x^{10}y^7 + x^7y^{10} + x^2y^{15} + xy^{16} + y^{17}$
$(0,0,0,1)$	$f_{3,-1} = y^{18}$

The vectors in the first row of above table compose the matrix $A^{(k)}$ such that

$$(f_{0,k}, f_{1,k}, f_{2,k}, f_{3,k}) = (f_{0,-1}, f_{1,-1}, f_{2,-1}, f_{3,-1})A^{(k)}.$$

Thus $A^{(k)}$ records the changes of corresponding coefficients in the process that computing Gröbner basis from input generators in algorithm 2. But for blind recognition of convolutional code, we only need to compute the generators set of the module satisfying the following equation:

$$h_0 f_{0,-1} + h_2 f_{2,-1} \equiv 0 \bmod (x^{18}, y^{18}).$$

Hence, we do not record the second row (corresponding to the coefficients x^{18}) and the forth row (corresponding to the coefficient of y^{18}) of $A^{(k)}$. We list

$(1,0)$	$f_{0,-1} = x^{17} + x^{16}y + x^{15}y^2 + x^{14}y^3 + x^{13}y^4 + x^{11}y^6 + x^{10}y^7 + x^9y^8 + x^3y^{14} + x^2y^{15} + y^{17}$
$(0,0)$	$f_{1,-1} = x^{18}$
$(0,1)$	$f_{2,-1} = x^{15}y^2 + x^{13}y^4 + x^{12}y^5 + x^{11}y^6 + x^{10}y^7 + x^7y^{10} + x^2y^{15} + xy^{16} + y^{17}$
$(0,0)$	$f_{3,-1} = y^{18}$

Step 0: $f_{0,0} = f_{0,-1} = x^{17} + x^{16}y + x^{15}y^2 + x^{14}y^3 + x^{13}y^4 + x^{11}y^6 + x^{10}y^7 + x^9y^8 + x^3y^{14} + x^2y^{15} + y^{17}$,

$(1,0)$	$f_{0,0} = f_{0,-1} = x^{17} + x^{16}y + x^{15}y^2 + x^{14}y^3$ $+ x^{13}y^4 + x^{11}y^6 + x^{10}y^7 + x^9y^8 + x^3y^{14} + x^2y^{15} + y^{17}$
$(0,0)$ $-(x+y)(1,0)$	$f_{1,0} = f_{1,-1} - (x+y)f_{0,-1} = x^{13}y^5 + x^{12}y^6 + x^9y^9 + x^4y^{14} + x^2y^{16} + xy^{17}$
$(0,1)$	$f_{2,0} = f_{2,-1} = x^{15}y^2 + x^{13}y^4 + x^{12}y^5 + x^{11}y^6 + x^{10}y^7 + x^7y^{10} + x^2y^{15} + xy^{16} + y^{17}$
$(0,0)$	$f_{3,0} = f_{3,-1} = y^{18}$

Step 1: $f_{0,1} = f_{2,0} = x^{15}y^2 + x^{13}y^4 + x^{12}y^5 + x^{11}y^6 + x^{10}y^7 + x^7y^{10} + x^2y^{15} + xy^{16} + y^{17}$,

$(0,1)$	$f_{0,1} = f_{2,0} =$ $x^{15}y^2 + x^{13}y^4 + x^{12}y^5 + x^{11}y^6 + x^{10}y^7 + x^7y^{10} + x^2y^{15} + xy^{16} + y^{17}$
$(x+y,0)$	$f_{1,1} = x^{13}y^5 + x^{12}y^6 + x^9y^9 + x^4y^{14} + x^2y^{16} + xy^{17}$
$y^2(1,0)$ $+(x^2+xy)(0,1)$	$f_{2,1} = y^2f_{0,0} - (x^2+xy)f_{2,0}$ $= x^{14}y^5 + x^{13}y^6 + x^{10}y^9 + x^8y^{11} + x^4y^{15} + x^3y^{16} + x^2y^{17}$
$(0,0)$	$f_{3,1} = f_{3,0} = y^{18}$

Step 2: $f_{0,2} = f_{1,1} = x^{13}y^5 + x^{12}y^6 + x^9y^9 + x^4y^{14} + x^2y^{16} + xy^{17}$,

$(x+y,0)$	$f_{0,2} = f_{1,1} = x^{13}y^5 + x^{12}y^6 + x^9y^9 + x^4y^{14} + x^2y^{16} + xy^{17}$
$y^3(0,1) + (x^2+xy)(x+y,0)$	$f_{1,2} = y^3f_{0,1} - (x^2+xy)f_{1,1} =$ $x^{12}y^8 + x^7y^{13} + x^6y^{14} + x^4y^{16} + x^5y^{15}$
$(y^2, x^2+xy) - x(x+y,0)$	$f_{2,2} = f_{2,1} - xf_{1,1} = x^8y^{11} + x^4y^{15} - x^5y^{14}$
$(0,0)$	$f_{3,2} = f_{3,1} = y^{18}$

Step 3: $f_{0,3} = f_{1,2} = x^{12}y^8 + x^7y^{13} + x^6y^{14} + x^4y^{16} + x^5y^{15}$

(x^3+xy^2, y^3)	$f_{0,3} = f_{1,2} = x^{12}y^8 + x^7y^{13} + x^6y^{14} + x^5y^{15} + x^4y^{16}$
$y^3(x+y,0) - (x+y)(x^3+xy^2, y^3)$	$f_{1,3} = y^3f_{0,2} - (x+y)f_{1,2} = x^9y^{12} - x^8y^{13}$
$(y^2 - x^2 - xy, x^2+xy)$	$f_{2,3} = f_{2,2} = x^8y^{11} + x^4y^{15} - x^5y^{14}$
$(0,0)$	$f_{3,3} = f_{3,2} = y^{18}$

Step 4: $f_{0,4} = f_{2,3} = x^8y^{11} + x^4y^{15} - x^5y^{14}$

$(y^2 - x^2 - xy, x^2+xy)$	$f_{0,4} = f_{2,3} = x^8y^{11} + x^4y^{15} + y^{19} - x^5y^{14}$
$(y^4 - x^4 - x^2y^2 - yx^3, -xy^3 + y^4)$ $-(xy+y^2)(y^2 - x^2 - xy, x^2+xy)$	$f_{1,4} = f_{1,3} - (xy+y^2)f_{2,3} = -x^4y^{17} + x^6y^{15}$
$y^3(x^3+xy^2, y^3) -$ $(x^4+xy^3+y^4)(y^2-x^2-xy, x^2+xy)$	$f_{2,4} = y^3f_{0,3} - (x^4+xy^3+y^4)f_{2,3} = y^{16}x^7$
$(0,0)$	$f_{3,4} = f_{3,3} = y^{18}$

Step 5: $f_{0,5} = f_{1,4} = x^6y^{15} - y^{17}x^4$

$(-x^4 - x^2y^2, y^4 - yx^3)$	$f_{0,5} = f_{1,4} = x^6y^{15} - x^4y^{17}$
$y^4(y^2 - x^2 - xy, x^2+xy) - (x^2+y^2)(-x^4 - x^2y^2, y^4 - yx^3)$	$f_{1,5} = y^4f_{0,4} - (x^2+y^2)f_{1,4} = 0$
$(xy^5 - x^4y^2 + x^6 + x^5y - y^6, y^6 - x^6 - x^5y - y^3x^3 - y^5x)$ $-xy(-x^4 - x^2y^2, y^4 - yx^3)$	$f_{2,5} = f_{2,4} - xyf_{1,4} = 0$
$(0,0)$	$f_{3,5} = f_{3,4} = y^{18}$

Step 6: $f_{0,6} = f_{3,5} = y^{18}$

$(0,0)$	$f_{0,6} = f_{3,5} = y^{18}$
$(x^5 + xy^5 + y^6, x^5y + x^3y^3 + xy^5 + y^6)$	$f_{1,6} = f_{1,5} = 0$
$(x^6 + x^4y^2 + x^3y^3 + xy^5 + y^6, x^6 + x^5y + x^4y^2 + x^3y^3 + y^6)$	$f_{2,6} = f_{2,5} = 0$
$y^3(-x^4 - x^2y^2, y^4 - yx^3)$	$f_{3,6} = y^3f_{0,5} - (x^6 - x^2y^4)f_{3,5} = 0$

Thus we have the Gröbner basis $G = \{f_{0,0}, f_{0,1}, f_{0,2}, f_{0,3}, f_{0,4}, f_{0,5}, f_{0,6}\}$, and the generating polynomials of the convolutional code are $H_1 = x^6 + x^4y^2 + x^3y^3 + xy^5 + y^6$, $H_2 = x^6 + x^5y + x^4y^2 + x^3y^3 + y^6$.

References

1. Althaler, J., Dür, A.: Finite Linear Recurring Sequences and Homogeneous Ideals. AAECC 7, 377–390 (1996)
2. Althaler, J., Dür, A.: A Generalization of the Massey-Ding Algorithm. AAECC 9, 1–14 (1998)
3. Berlekamp, E.R.: Algebraic Coding Theory. McGrw-Hill, New York (1968)
4. Buchberger, B.: Gröbner Bases: An Algorithmic Method in Polynomial Ideal Theory, in Multidimensional Systems Theory, Ed. by N.K.Bose (1984)
5. Cheng, M.H.: Generalised Berlekamp-Massey Algorithm. IEE Proc.Comm. 149(4), 207–210 (2002)
6. Faugére, J.C., Gianni, P., Lazard, D., Mora, T.: Efficient Computation of Zero-dimensional Gröbner bases by Change of Ordering. J.Symb. Comp. 16, 329–344 (1993)
7. Faugére, J.: A New Efficient Algorithm for Computing Gröbner Bases (F4). J. Pure and Applied Algebra 139, 61–83 (1999)
8. Faugére, J.: A New Efficient Algorithm for Computing Gröbner Bases Without Reduction to Zero (F5). In: Proc. of ISSAC 02, pp. 75–83. ACM Press, New York (2002)
9. Feng, K.L.: A Generation of the Berlekamp-Massey Algorithm for Mutisequence Shift-register Synthesis with Applications to Decoding Cyclic Codes. IEEE Trans. on Inform. Theory 37, 1274–1287 (1991)
10. Fitzpatrick, P.: Solving a Multivariable Congruence by Change of Term Order. J. Symb.Comp. 11, 1–15 (1997)
11. Golic, J.D.: Vectorial Boolean Functions and Induced Algebraic Equations. IEEE Trans. on Inform. Theory 52(2), 528–537 (2006)
12. Heydtmann, A.E., Jensen, J.M.: On the Equivalence of the Berlekamp-Massey and the Euclidean Algorithms for Decoding. IEEE Trans. Inform.Theory 46(7), 2614–2624 (2000)
13. Kailath, T.: Encounters with the Berlekamp-Massey Algorithm. In: Blahut, R.E., et al. (ed.) Communication and Cryptography, pp. 209–220. Kluwer Academic Publisher, Boston, MA (1994)
14. Kuijper, M., Willems, J.C.: On Constructing a Shortest Linear Recurrence Relation. IEEE Trans. Automat. Contr. 42(11), 1554–1558 (1997)
15. Lazard, D.: A Note on Upper Bounds for Ideal-Theoretic Problems. J.Symbolic Computation 13(3), 231–233 (1992)
16. Lu, P.Z., Liu, M.L.: Gröbner Basis of Characteristic Ideal of LRS over UFD, Science in China(Series A), vol. 28(6) (1998)
17. Lu, P.Z.: Synthesis of Sequences and Efficient Decoding for a Class of Algebraic Geometry Codes. Acta. Electronic Sinica 21(1), 1–10 (1993)
18. Lu, P.Z.: Structure of Groebner Bases for the Characteristic Ideal of a Finite Linear Recursive Sequence, AAECC-13, HI,USA (November 14-November 19, 1999)
19. Lu, P.Z., Shen, L., Zou, L., Luo, X.Y.: Blind Recognition of Punctured Concolutional Codes. Science in China(Series F) 48(4), 484–498 (2005)
20. Massey, J.L.: Shift-Register Synthesis and BCH Decoding. IEEE Trans. Info. Theory 15(1), 122–127 (1969)
21. Sakata, S.: Synthesis of Two-Dimensional Linear Feedback Shift-Registers and Groebner Basis. LNCS, vol. 356. Springer, Heidelberg (1989)

A Twin for Euler's ϕ Function in $\mathbb{F}_2[X]^{\star}$

R. Durán Díaz[1], J. Muñoz Masqué[2], and A. Peinado Domínguez[3]

[1] Departamento de Automática
Universidad de Alcalá de Henares
Carretera de Madrid-Barcelona, km. 33.6
28871-Alcalá de Henares, Spain
raul.duran@uah.es
[2] CSIC, Instituto de Física Aplicada
C/ Serrano 144, 28006-Madrid, Spain
jaime@iec.csic.es
[3] Departamento de Ingeniería de Comunicaciones
E.T.S. de Ingenieros de Telecomunicación
Universidad de Málaga
Campus de Teatinos, 29071-Málaga, Spain
apeinado@ic.uma.es

Abstract. In this paper, we present a function in $\mathbb{F}_2[X]$ and prove that several of its properties closely resemble those of Euler's ϕ function. Additionally, we conjecture another property for this function that can be used as a simple primality test in $\mathbb{F}_2[X]$, and we provide numerical evidence to support this conjecture. Finally, we further apply the previous results to design a simple primality test for trinomials.

Keywords: Characteristic-2 field, Euler ϕ function, polynomial factorization.

Mathematics Subject Classification 2000: Primary 13P05; Secondary 11T06, 12E05, 15A04.

1 Statement of the Main Result

Let V, W be finite-dimensional k-vector spaces. We recall (e.g., see [3, II, §1, 13]) that a semi-linear map is a pair (A, σ) consisting of a field automorphism $\sigma \colon k \to k$, and a homomorphism $A \colon (V, +) \to (W, +)$ of the underlying additive groups such that $A(\lambda v) = \sigma(\lambda)A(v)$, $\forall \lambda \in k$, $\forall v \in V$. If $k \subset F$ is a finite extension, then we consider the structure of F-vector space on $F \otimes_k V$ uniquely defined by $x \cdot (y \otimes v) = (xy) \otimes v$, for all $x, y \in F$, $v \in V$. Moreover, if $L \colon V \to W$ is a k-linear map, then $I \otimes L \colon F \otimes_k V \to F \otimes_k W$ is a F-linear map, where I stands for the identity map of F, and if $L \colon V \to V$ is an endomorphism then

---

* Supported by Ministerio de Educación y Ciencia of Spain under grant number MTM2005–00173 and Consejería de Educación y Cultura de la Junta de Castilla y León under grant number SA110A06.

both, L and $I \otimes L$ have the same invariant polynomials (e.g., see [4, VII, §5, 1, Corollaire 2]), although not the same elementary divisors. In particular they have the same characteristic polynomial.

Throughout the paper, the ground field considered is the characteristic-2 prime field $\mathbb{F}_2$. We use the standard properties of finite field extensions (e.g., see [10]). In particular we use the properties of conjugate roots of an irreducible polynomial over a finite field.

Theorem 1. *Given a polynomial $f \in \mathbb{F}_2[X]$, let R_f be the finite-dimensional $\mathbb{F}_2$-algebra $R_f = \mathbb{F}_2[X]/(f)$, and let $L_f \colon R_f \to R_f$ be the linear map*

$$L_f(x) = x^2 + \alpha x, \quad \forall x \in R_f,$$
$$\alpha = X \pmod{f}.$$

Let χ_f be the characteristic polynomial of L_f, that is, $\chi_f(X) = \det(X \cdot I + L_f)$, I being the identity map of R_f. Then,

(i) *If f is irreducible and e is a positive integer, then $\chi_{f^e} = f^{e-1}(f + 1)$.*
(ii) *If f and g are two coprime polynomials in $\mathbb{F}_2[X]$, then $\chi_{fg} = \chi_f \chi_g$.*
(iii) *If $\chi_f = f + 1$ for $f \in \mathbb{F}_2[X]$, then f is squarefree.*

2 Proof of Theorem 1

In order to prove the item (i), we proceed by recurrence on e.

Assume $e = 1$. We thus need to prove that if f is irreducible, then $\chi_f = f + 1$. As f is irreducible, R_f is a finite extension of the ground field, i.e., $R_f \cong \mathbb{F}_{2^n}$, $n = \deg f$. As is well known the Galois group of the extension $\mathbb{F}_2 \subset \mathbb{F}_{2^n}$ is cyclic and generated by the Frobenius automorphism $\varphi \colon \mathbb{F}_{2^n} \to \mathbb{F}_{2^n}$, $\varphi(x) = x^2$. Then we have $L_f = \varphi + T_\alpha$, where $T_\alpha \colon \mathbb{F}_{2^n} \to \mathbb{F}_{2^n}$ is the $\mathbb{F}_2$-linear map: $T_\alpha(x) = \alpha x$. The characteristic polynomial of T_α is f; i.e., $\det(X \cdot I + T_\alpha) = f(X)$, as the matrix $A = (a_{ij})_{i,j=1,\dots,n}$ of T_α in the basis $(1, \alpha, \alpha^2, \dots, \alpha^{n-1})$ is given by

$$A = \begin{bmatrix} 0 & 0 & \dots & 0 & a_0 \\ 1 & 0 & \dots & 0 & a_1 \\ \vdots & \vdots & \ddots & \vdots & \vdots \\ 0 & 0 & \dots & 0 & a_{n-2} \\ 0 & 0 & \dots & 1 & a_{n-1} \end{bmatrix},$$

where

$$f(X) = X^n + a_{n-1}X^{n-1} + a_{n-2}X^{n-2} + \dots + a_1 X + a_0 \in \mathbb{F}_2[X].$$

Accordingly, the eigenvalues of

$$I \otimes T_\alpha \colon \mathbb{F}_{2^n} \otimes_{\mathbb{F}_2} \mathbb{F}_{2^n} \to \mathbb{F}_{2^n} \otimes_{\mathbb{F}_2} \mathbb{F}_{2^n}$$

are $\alpha, \alpha^2, \alpha^{2^2}, \dots, \alpha^{2^{n-1}}$, and so $I \otimes T_\alpha$ is diagonalizable. Let v_0 be a non-vanishing eigenvector of the eigenvalue α, i.e.,

$$v_0 = x_0 \left(1 \otimes 1 \right) + x_1 \left(1 \otimes \alpha \right) + x_2 \left(1 \otimes \alpha^2 \right) + \ldots + x_{n-1} \left(1 \otimes \alpha^{n-1} \right),$$

where $(x_0, x_1, x_2, \ldots, x_{n-1}) \in (\mathbb{F}_{2^n})^n$ is a non-trivial solution of the linear system

$$\sum_{j=1}^{n} a_{ij} x_{j-1} = \alpha x_{i-1}, \quad i = 1, \ldots, n. \tag{1}$$

As $a_{ij} \in \mathbb{F}_2$, by applying φ successively to (1), we obtain

$$\sum_{j=1}^{n} a_{ij} \left(x_{j-1} \right)^{2^h} = \alpha^{2^h} \left(x_{i-1} \right)^{2^h}, \quad h = 0, \ldots, n - 1,$$

thus showing that the vector

$$v_h = (\varphi \otimes I)^h (v_0)$$
$$= (x_0)^{2^h} (1 \otimes 1) + (x_1)^{2^h} (1 \otimes \alpha) + (x_2)^{2^h} (1 \otimes \alpha^2)$$
$$+ \ldots + (x_{n-1})^{2^h} \left(1 \otimes \alpha^{n-1} \right)$$

satisfies

$$(I \otimes T_\alpha)(v_h) = \alpha^{2^h} v_h, \quad h = 0, \ldots, n - 1. \tag{2}$$

The vectors $(v_0, v_1, \ldots, v_{n-1})$ are a basis of the $\mathbb{F}_{2^n}$-vector space $\mathbb{F}_{2^n} \otimes_{\mathbb{F}_2} \mathbb{F}_{2^n}$ since they are eigenvectors corresponding to pairwise distinct eigenvalues, and the matrix of $I \otimes T_\alpha$ in this basis is

$$\begin{bmatrix} \alpha & 0 & \cdots & 0 \\ 0 & \alpha^2 & \cdots & 0 \\ \vdots & \vdots & \ddots & \vdots \\ 0 & 0 & \cdots & \alpha^{2^{n-1}} \end{bmatrix}. \tag{3}$$

Set $w_i = (I \otimes \varphi)(v_i)$, $i = 0, \ldots, n - 1$. By remarking that $\varphi \circ T_\alpha = T_{\alpha^2} \circ \varphi$, and using the formula (2) we obtain

$$(I \otimes T_{\alpha^2})(w_i) = (I \otimes \varphi) \left(\alpha^{2^i} v_i \right) = \alpha^{2^i} w_i, \quad i = 0, \ldots, n - 1. \tag{4}$$

Moreover, again from formula (2), we have

$$(I \otimes T_{\alpha^2})(v_i) = \alpha^{2^{i+1}} v_i, \quad i = 0, \ldots, n - 1.$$

From the formulas (2) and (4), we conclude that there exist scalars $\lambda_i \in \mathbb{F}_{2^n}$, $\lambda_i \neq 0$, such that $w_i = \lambda_i v_{n-1+i}$, $i = 0, \ldots, n-1$, where the subindices are taken mod n. Hence

$$(I \otimes \varphi)(v_i) = \lambda_i v_{n-1+i}, \quad i = 0, \ldots, n - 1,$$

and recalling that $v_i = (\varphi \otimes I)(v_{i-1})$ and $(\varphi \otimes I, \varphi)$ is a semi-linear map from $\mathbb{F}_{2^n} \otimes_{\mathbb{F}_2} \mathbb{F}_{2^n}$ onto itself, we have

$$
\begin{aligned}
\lambda_i v_{n-1+i} &= (I \otimes \varphi)(v_i) \\
&= (I \otimes \varphi)(\varphi \otimes I)(v_{i-1}) \\
&= (\varphi \otimes I)((I \otimes \varphi)(v_{i-1})) \\
&= (\varphi \otimes I)(\lambda_{i-1} v_{n-2+i}) \\
&= \lambda_{i-1}^2 v_{n-1+i}.
\end{aligned}
$$

Hence $\lambda_i = \lambda_{i-1}^2$, and setting

$$
\mu_i = (\lambda_0)^{-\left(2^1 + 2^2 + 2^3 + \ldots + 2^i\right)}, \quad i = 0, \ldots, n-1,
$$

we have $(I \otimes \varphi)(\mu_i v_i) = \mu_{n-1+i} v_{n-1+i}$, thus finally obtaining a basis

$$
(\mu_0 v_0, \mu_1 v_1, \ldots, \mu_{n-1} v_{n-1})
$$

on which the matrix of $I \otimes T_\alpha$ is the same matrix as the one given in formula (3), whereas the matrix of $I \otimes \varphi$ is the following:

$$
\begin{bmatrix}
0 & 1 & 0 & \cdots & 0 \\
0 & 0 & 1 & \cdots & 0 \\
\vdots & \vdots & \vdots & \ddots & \vdots \\
0 & 0 & 0 & \cdots & 1 \\
1 & 0 & 0 & \cdots & 0
\end{bmatrix}
$$

Therefore,

$$
\chi_f(X) = \begin{vmatrix}
X + \alpha & 1 & 0 & \cdots & & 0 \\
0 & X + \alpha^2 & 1 & \cdots & & 0 \\
\vdots & \vdots & & \ddots & \ddots & \vdots \\
0 & 0 & 0 & & \ddots & 1 \\
1 & 0 & 0 & \cdots & & X + \alpha^{2^{n-1}}
\end{vmatrix}
$$

$$
= (X + \alpha)(X + \alpha^2) \cdots \left(X + \alpha^{2^{n-1}}\right) + 1
$$

$$
= f(X) + 1.
$$

Next assume $e > 1$. Then we have an exact sequence of $\mathbb{F}_2$-vector spaces

$$
0 \longrightarrow R_f \xrightarrow{\iota_e} R_{f^e} \xrightarrow{\pi_e} R_{f^{e-1}} \longrightarrow 0,
$$

where

$$
\iota_e (F \pmod{f}) = F f^{e-1} \pmod{f^e},
$$
$$
\pi_e (F \pmod{f^e}) = F \pmod{f^{e-1}},
$$

for every $F \in \mathbb{F}_2[X]$. Furthermore, $\iota_e(R_f)$ remains invariant under L_{f^e} and the restriction of L_f to R_f is nothing but T_α. Hence we have a commutative diagram

$$0 \longrightarrow R_f \xrightarrow{\iota_e} R_{f^e} \xrightarrow{\pi_e} R_{f^{e-1}} \longrightarrow 0$$
$$\downarrow T_\alpha \qquad \downarrow L_{f^e} \qquad \downarrow L_{f^{e-1}}$$
$$0 \longrightarrow R_f \xrightarrow{\iota_e} R_{f^e} \xrightarrow{\pi_e} R_{f^{e-1}} \longrightarrow 0$$

Accordingly, $\chi_{f^e}(X) = \det(X \cdot I + T_\alpha) \cdot \chi_{f^{e-1}}(X)$, and we can conclude by virtue of the recurrence hypothesis.

Next, we prove the item (ii). If f and g are coprime polynomials in $\mathbb{F}_2[X]$, then from Chinese Remainder Theorem we obtain an isomorphism

$$R_{fg} = \mathbb{F}_2[X]/(fg) \cong R_f \times R_g$$

such that $\alpha = X \pmod{fg}$ corresponds to the pair (α_f, α_g), with $\alpha_f = X \pmod{f}$, $\alpha_g = X \pmod{g}$. Hence $L_{fg}(x, y) = (L_f(x), L_g(y))$.

Finally, we prove the item (iii). If $f = p_1^{e_1} \cdots p_r^{e_r}$ is the factorization of f into irreducible factors, then the equation $\chi_f = f + 1$ means

$$p_1^{e_1-1} \cdots p_r^{e_r-1}(p_1 + 1) \cdots (p_r + 1) = p_1^{e_1} \cdots p_r^{e_r} + 1.$$

If $e_i \geq 2$ for some index $i = 1, \ldots, r$, then the equation above leads us to a contradiction as p_i should divide the unit.

Moreover, for $r = 2$ we have $f = p_1 p_2$ and the equation $\chi_f = f + 1$ yields $p_2 = p_1$, thus leading to a contradiction.

3 Trinomials

Conjecture 1. If a reducible polynomial $f \in \mathbb{F}_2[X]$ satisfies the equation $\chi_f = f + 1$, then the number of terms of f is $2r - 1$ at least, r being its number of factors.

Corollary 1. *If a polynomial $f \in \mathbb{F}_2[X]$ satisfies the equation $\chi_f = f + 1$ and the number of its terms is ≤ 4, then f is irreducible.*

Let $f = X^n + X^m + 1$, $n > m$, be a trinomial in $\mathbb{F}_2[X]$. We can further assume $m \leq \frac{n}{2}$, as

$$X^n + X^m + 1 = X^n \left[(X^{-1})^n + (X^{-1})^{n-m} + 1 \right],$$

and hence $X^n + X^m + 1$ is reducible if and only if $X^n + X^{n-m} + 1$ is reducible.

With the same notations as in Theorem 1, the matrix of T_α in the basis $(1, \alpha, \ldots, \alpha^{n-1})$ is the following:

$$A = \begin{bmatrix} 0\,0 \ldots 0\,1 \\ 1\,0 \ldots 0\,0 \\ \vdots\ \vdots\ \ddots\ \vdots\ \vdots \\ 0\,0 \ldots 0\,1 \\ \vdots\ \vdots\ \ddots\ \vdots\ \vdots \\ 0\,0 \ldots 0\,0 \\ 0\,0 \ldots 1\,0 \end{bmatrix} \leftarrow (m+1)\text{-th row.}$$

Next, we compute the matrix S of the map $\varphi(x) = x^2$ in the same basis.

We first remark that the inverse of α^m is $(\alpha^m)^{-1} = \alpha^{-m} = \alpha^{n-m} + 1$. Let $q = q(n, m)$, $r = r(n, m)$ be the quotient and the remainder of the Euclidean division on n over m, i.e., $n = mq + r$, $0 \le r \le m - 1$.

We are led to distinguish the following cases:

1. m and n are even, say $m = 2m'$, $n = 2n'$.
   Then, we have

$$\varphi\left(\alpha^i\right) = \alpha^{2i}, \quad 0 \le i \le n' - 1, \tag{5}$$

$$\varphi\left(\alpha^{n'+i}\right) = \alpha^{m+2i} + \alpha^{2i}, \quad 0 \le i \le n' - m' - 1. \tag{6}$$

Moreover, we have

$$\varphi\left(\alpha^{n-m'+i}\right) = \left(\alpha^m + \alpha^{m(q-1)+r} + 1\right)\alpha^{2i}. \tag{7}$$

As $m \le \frac{n}{2}$, we have $q \ge 2$ necessarily, and then there are two subcases:

(a) If $n = 2m$, then the trinomial under consideration is reducible; indeed, one has $X^n + X^m + 1 = X^{2m} + X^{2m'} + 1 = (X^m + X^{m'} + 1)^2$. Hence, this subcase is not interesting.

(b) If $n > 2m$, then $m(q - 1) + r = n - m > m$. From the formula (7) we thus obtain

$$\varphi\left(\alpha^{n-m'+i}\right) = \alpha^{n-m+2i} + \alpha^{m+2i} + \alpha^{2i}, \quad 0 \le i \le m' - 1. \tag{8}$$

2. m is even and n is odd, say $m = 2m'$, $n = 2n' + 1$.
   We have

$$\varphi\left(\alpha^i\right) = \alpha^{2i}, \quad 0 \le i \le n', \tag{9}$$

$$\varphi\left(\alpha^{n'+1+i}\right) = \alpha^{m+1+2i} + \alpha^{1+2i}, \quad 0 \le i \le n' - m' - 1, \tag{10}$$

$$\varphi\left(\alpha^{n-m'+i}\right) = \alpha^{n-m+2i} + \alpha^{m+2i} + \alpha^{2i}, \quad 0 \le i \le m' - 1. \tag{11}$$

3. m is odd and n is even, say $m = 2m' + 1$, $n = 2n'$.
   Then, we have

$$\varphi\left(\alpha^i\right) = \alpha^{2i}, \quad 0 \le i \le n' - 1, \tag{12}$$

$$\varphi\left(\alpha^{n'+i}\right) = \alpha^{m+2i} + \alpha^{2i}, \quad 0 \le i \le n' - m' - 1. \tag{13}$$

Moreover, we have

$$\varphi\left(\alpha^{n-m'+i}\right) = \left(\alpha^m + \alpha^{m(q-1)+r} + 1\right)\alpha^{2i+1}. \tag{14}$$

As in the first case, there are two subcases:

(a) If $n = 2m$, then from the formula (14) we obtain

$$\varphi\left(\alpha^{n-m'+i}\right) = \alpha^{2i+1}, \quad 0 \le i \le m' - 1. \tag{15}$$

(b) If $n > 2m$, then from the formula (14) we obtain

$$\varphi\left(\alpha^{n-m'+i}\right) = \alpha^{n-m+1+2i} + \alpha^{m+1+2i} + \alpha^{2i+1}, \quad 0 \le i \le m' - 1.$$

4. m and n are odd, say $m = 2m' + 1$, $n = 2n' + 1$.
   We have

$$\varphi\left(\alpha^i\right) = \alpha^{2i}, \quad 0 \le i \le n', \tag{16}$$

$$\varphi\left(\alpha^{n'+1+i}\right) = \alpha^{m+1+2i} + \alpha^{1+2i}, \quad 0 \le i \le n' - m' - 1, \tag{17}$$

$$\varphi\left(\alpha^{n-m'+i}\right) = \alpha^{n+1-m+2i} + \alpha^{m+1+2i} + \alpha^{2i+1}, \quad 0 \le i \le m' - 1. \tag{18}$$

Theorem 2. *If Conjecture 1 holds and* $\operatorname{rk} L_f = \deg f - 1$, *then* f *is irreducible.*

Proof. We have $\operatorname{rk} L_f = \sum_{i=1}^{r} e_i \deg p_i - r$. Hence, if f is a polynomial with no repeated factors (i.e., $e_1 = \ldots = e_r = 1$) then $\operatorname{rk} L_f = n - r$, and the result follows.

Remark 1. Note that the matrix of L_f in the basis $B = (1, \alpha, \alpha^2, \ldots, \alpha^{n-1})$ does not coincide with Berlekamp's matrix (cf. [10, IV, §1]).

Remark 2. If f is irreducible, then χ_f coincides with the minimal polynomial for L_f. In fact, as a computation shows, the entry $(1,1)$ of the matrix of $(L_f)^k$, for $k = 0, \ldots, n - 1$, in the basis B is equal to α^k.

Remark 3. It is not difficult to see that the polynomials $f \in \mathbb{F}_2[X]$ satisfying the relationship $\chi_f = f$ are exactly the polynomials $f(X) = X^h(X + 1)^k$, $h, k \in \mathbb{N}$.

The cases above complete the computation of χ_f. In Appendix II we provide a program that computes this determinant.

4   Concluding Remarks

We have presented the function χ_f in $\mathbb{F}_2[X]$ and we have shown how some of its properties closely resemble those of Euler's ϕ function. We have conjectured another property of χ_f and we have provided numerical evidence to support this conjecture. This property can be used as a simple primality test in $\mathbb{F}_2[X]$, and it is specially well suited for the case of trinomials.

Determining the primality of a trinomial is not a trivial problem and has been paid a lot of attention in the literature (see, for example, [2], [5], [6], [7], [8], [11], [13], [14]). Even the "simple" case $X^n + X + 1$ is not easy to deal with: see, for example, [12, Table I] that contains all 33 values of $n \leq 30000$, for which $X^n + X + 1$ is irreducible, along with some other information.

Our result boils down the problem of trinomial primality to the computation of the determinant of a matrix, which can be explicitly written. This matrix is extremely sparse and hence, the size of the compressed format of the matrix should be small. Accordingly, the computation of the determinant should not be long even for high degrees of the trinomial.

In a forthcoming paper we shall try to compare the results above with those appeared in the literature on the topic; among other authors and approaches, we should specially mention [1], [2], [6], [7], [8], [11], [12], [13], and [14].

References

1. Blake, I., Gao, S., Lambert, R.: Constructive problems for irreducible polynomials over finite fields. In: Gulliver, T.A., Secord, N.P. (eds.) Information Theory and Applications. LNCS, vol. 793, pp. 1–23. Springer, Heidelberg (1994)
2. —: Construction and distribution problems for irreducible trinomials over finite fields, Applications of finite fields, Inst. Math. Appl. Conf. Ser. New Ser., vol. 59, pp. 19–32, Oxford University Press, New York (1996)
3. Bourbaki, N.: Éléments de Mathématique, Algèbre, Chapitres 1 à 3, Hermann, Paris (1970)
4. —: Éléments de Mathématique, Livre II, Algèbre, Chapitres 6–7, Deuxième Édition, Hermann, Paris (1964)
5. Ciet, M., Quisquater, J.-J., Sica, F.: A Short Note on Irreducible Trinomials in Binary Fields. In: Macq, B., Quisquater, J.-J. (eds.) 23rd Symposium on Information Theory in the BENELUX, Louvain-la-Neuve, Belgium, pp. 233–234 (2002)
6. Fredricksen, H., Wisniewski, R.: On trinomials $x^n + x^2 + 1$ and $x^{8l \pm 3} + x^k + 1$ irreducible over $GF(2)$. Inform. and Control 50, 58–63 (1981)
7. von zur Gathen, J.: Irreducible trinomials over finite fields. In: Proceedings of the 2001 International Symposium on Symbolic and Algebraic Computation (electronic), pp. 332–336. ACM, New York (2001)
8. —: Irreducible trinomials over finite fields. Math. Comp. 72, 1987–2000 (2003)
9. von zur Gathen, J., Panario, D.: Factoring Polynomials over Finite Fields: A Survey. J. Symbolic Computation 31, 3–17 (2001)
10. Lidl, R., Niederreiter, H.: Introduction to finite fields and their applications. Cambridge University Press, Cambridge, UK (1994)
11. Vishne, U.: Factorization of trinomials over Galois fields of characteristic 2. Finite Fields Appl. 3, 370–377 (1997)
12. Zierler, N.: On $x^n + x + 1$ over $GF(2)$. Information and Control 16, 502–505 (1970)
13. Zierler, N., Brillhart, J.: On primitive trinomials (mod 2). Information and Control 13, 541–554 (1968)
14. —: On primitive trinomials (mod 2), II. Information and Control 14, 566–569 (1969)

Appendix I

Below we present a MAPLE source code computing some of the polynomials with four prime factors $f = P_1 \cdots P_4$ satisfying the equation $\chi_f = f + 1$ in

Theorem 1-(iii). The integer appearing on the right hand of each output denotes the number of—non-vanishing—terms of f. Observe that this number is always equal or greater than $2r - 1$, where r is the number of factors in the polynomial, 4 for the present case.

```
> L:=[]:
> f := x:
> for n from 0 to 500 do
>   L:=[op(L),[f,degree(f)]]:
>   f:=Nextprime(f,x) mod 2:
> end do:
> max_deg:=op(2,L[nops(L)]):
> ji:=1: jf:=1:
>
> for i from 1 to max_deg do
>   while (jf <= nops(L) and op(2,L[jf]) = i) do
>     jf:=jf+1:
>   end do:
>
>   jf:=jf-1:
>
>   for p1 from ji to jf do
>    for p2 from p1+1 to jf do
>     for p3 from p2+1 to nops(L) do
>      P1:=op(1,L[p1]): P2:=op(1,L[p2]): P3:=op(1,L[p3]):
>      if P1*P2+P1*P3+P1+P2*P3+P2+P3+1 mod 2 <> 0 then
>       if prem(P1*P2*P3+P1*P2+P1*P3+P1+P2*P3+P2+P3 mod 2,
>        P1*P2+P1*P3+P1+P2*P3+P2+P3+1 mod 2, x) mod 2 = 0 then
>        Q1:=simplify(expand(P1*P2*P3+
>                    P1*P2+P1*P3+P1+P2*P3+P2+P3)) mod 2;
>        Q2:=simplify(expand(P1*P2+P1*P3+
>                    P1+P2*P3+P2+P3+1)) mod 2;
>        P4:=(Factor(Q1) mod 2)/(Factor(Q2) mod 2);
>        if type(P4, 'polynom') then
>         if irreduc(P4) then
>         n:=nops(simplify(expand(P1*P2*P3*P4)) mod 2);
>         if not assigned(tabla[n]) then
>          tabla[n]:=0:
>         end if:
>         tabla[n]:=tabla[n]+1:
>         print(P1,P2,P3,P4,n);
>         end if;
>        end if;
>       end if;
>      end if;
>     end do;
```

```
>    end do;
>    end do:
>    ji:=jf+1:
>    jf:=ji:
> end do:
```

$$\left.\begin{array}{l} x^5 + x^3 + 1,\, x^5 + x^4 + x^3 + x^2 + 1,\, x^6 + x^5 + x^4 + x + 1 \\ \qquad\qquad\qquad\qquad\qquad x^9 + x^8 + x^4 + x + 1 \end{array}\right\} 11$$

$$\left.\begin{array}{l} x^5 + x^3 + 1,\, x^5 + x^4 + x^3 + x^2 + 1,\, x^6 + x^5 + x^4 + x^2 + 1 \\ \qquad\qquad\qquad\qquad\qquad x^9 + x^4 + 1 \end{array}\right\} 9$$

$$\left.\begin{array}{l} x^6 + x + 1, \qquad\qquad x^6 + x^5 + x^2 + x + 1, \\ x^7 + x^6 + x^4 + x^2 + 1,\, x^{13} + x^{12} + x^8 + x^6 + x^3 + x^2 + 1 \end{array}\right\} 15$$

$$\left.\begin{array}{l} x^6 + x^4 + x^2 + x + 1,\, x^7 + x^5 + x^4 + x^3 + x^2 + x + 1 \\ x^6 + x^5 + 1, \qquad\qquad x^{13} + x^9 + x^8 + x^6 + x^5 + x^4 + x^3 + x^2 + 1 \end{array}\right\} 15$$

$$\left.\begin{array}{l} x^7 + x^3 + x^2 + x + 1 \\ x^7 + x^5 + x^4 + x^3 + x^2 + x + 1 \\ x^9 + x^8 + x^6 + x^5 + x^3 + x^2 + 1 \\ x^{15} + x^{14} + x^{12} + x^{11} + x^{10} + x^9 + x^8 + x^6 + x^4 + x^3 + 1 \end{array}\right\} 23$$

$$\left.\begin{array}{l} x^7 + x^4 + x^3 + x^2 + 1,\, x^7 + x^5 + x^3 + x + 1, \\ x^9 + x^8 + x^5 + x^4 + 1,\, x^{13} + x^{12} + x^{11} + x^{10} + x^7 + x + 1 \end{array}\right\} 17$$

$$\left.\begin{array}{l} x^7 + x^6 + 1,\, x^7 + x^6 + x^5 + x^2 + 1 \\ x^9 + x^5 + 1,\, x^{13} + x^{11} + x^7 + x^6 + x^4 + x^2 + 1 \end{array}\right\} 15$$

$$\left.\begin{array}{l} x^7 + x^6 + x^4 + x^2 + 1 \\ x^7 + x^6 + x^5 + x^4 + x^2 + x + 1 \\ x^9 + x^6 + x^5 + x^3 + x^2 + x + 1 \\ x^{15} + x^{13} + x^{12} + x^9 + x^8 + x^7 + x^6 + x^5 + x^4 + x^3 + 1 \end{array}\right\} 15$$

$$\left.\begin{array}{l} x^7 + x^6 + x^4 + x^2 + 1 \\ x^7 + x^6 + x^5 + x^4 + x^2 + x + 1 \\ x^9 + x^6 + x^5 + x^3 + x^2 + x + 1 \\ x^{15} + x^{13} + x^{12} + x^9 + x^8 + x^7 + x^6 + x^5 + x^4 + x^3 + 1 \end{array}\right\} 15$$

$$\left.\begin{array}{l} x^9 + x^6 + x^5 + x^2 + 1 \\ x^9 + x^7 + x^6 + x^4 + x^3 + x + 1 \\ x^{11} + x^8 + x^5 + x^3 + 1 \\ x^{19} + x^{18} + x^{17} + x^{16} + x^{15} + x^{14} + x^{12} + x^{11} + x^{10} + x^7 + x^6 \\ \qquad\qquad\qquad\qquad\qquad +x^5 + x^4 + x + 1 \end{array}\right\} 23$$

$$\left.\begin{array}{l} x^9 + x^6 + x^5 + x^2 + 1 \\ x^9 + x^7 + x^6 + x^4 + x^3 + x + 1 \\ x^{11} + x^8 + x^5 + x^3 + 1 \\ x^{19} + x^{18} + x^{17} + x^{16} + x^{15} + x^{14} + x^{12} + x^{11} + x^{10} + x^7 + x^6 \\ \qquad\qquad\qquad\qquad\qquad +x^5 + x^4 + x + 1 \end{array}\right\} 23$$

$$\left.\begin{array}{l} x^9 + x^7 + x^5 + x^3 + x^2 + x + 1 \\ x^9 + x^8 + x^7 + x^5 + x^4 + x^3 + 1 \\ x^{10} + x^9 + x^7 + x^5 + x^4 + x^2 + 1 \\ x^{25} + x^{22} + x^{21} + x^{20} + x^{19} + x^{18} + x^{16} + x^{15} + x^{12} + x^{11} + x^6 \\ \qquad\qquad\qquad\qquad\qquad\qquad\qquad +x^4 + x^2 + x + 1 \end{array}\right\}\ 29$$

$$\left.\begin{array}{l} x^9 + x^7 + x^6 + x^4 + 1 \\ x^9 + x^8 + x^7 + x^6 + x^2 + x + 1 \\ x^{10} + x^9 + x^7 + x^6 + x^4 + x^3 + x^2 + x + 1 \\ x^{25} + x^{24} + x^{22} + x^{21} + x^{20} + x^{19} + x^{18} + x^{17} + x^{16} + x^{15} + x^{14} + x^{13} \\ \qquad\qquad\qquad\qquad\qquad +x^9 + x^7 + x^6 + x^4 + x^3 + x^2 + 1 \end{array}\right\}\ 25$$

$$\left.\begin{array}{l} x^9 + x^8 + x^6 + x^5 + 1 \\ x^9 + x^8 + x^7 + x^5 + x^4 + x^2 + 1 \\ x^{11} + x^{10} + x^9 + x^5 + x^4 + x + 1 \\ x^{19} + x^{15} + x^{13} + x^{12} + x^8 \\ \qquad\qquad\qquad\qquad\qquad +x^5 + x^4 + x + 1 \end{array}\right\}\ 23$$

$$\left.\begin{array}{l} x^{10} + x^3 + x^2 + x + 1 \\ x^{10} + x^8 + x^6 + x^5 + x^2 + x + 1 \\ x^{12} + x^9 + x^8 + x^7 + x^5 + x^4 + x^3 + x + 1 \\ x^{25} + x^{22} + x^{20} + x^{14} + x^{13} + x^8 + x^6 + x + 1 \end{array}\right\}\ 27$$

$$\left.\begin{array}{l} x^{10} + x^6 + x^5 + x^2 + 1 \\ x^{10} + x^8 + x^3 + x^2 + 1 \\ x^{12} + x^9 + x^8 + x^7 + x^6 + x + 1 \\ x^{25} + x^{24} + x^{22} + x^{18} + x^{17} + x^{16} + x^{14} + x^{13} + x^{10} \\ \qquad\qquad\qquad\qquad +x^6 + x^5 + x^4 + x^2 + x + 1 \end{array}\right\}\ 35$$

$$\left.\begin{array}{l} x^{10} + x^7 + 1 \\ x^{10} + x^9 + x^6 + x + 1 \\ x^{11} + x^{10} + x^9 + x^8 + x^5 + x^4 + x^2 + x + 1 \\ x^{27} + x^{26} + x^{25} + x^{23} + x^{22} + x^{20} + x^{18} + x^{16} + x^{12} + x^{11} + x^8 \\ \qquad\qquad\qquad\qquad\qquad\qquad +x^6 + x^4 + x^3 + 1 \end{array}\right\}\ 23$$

$$\left.\begin{array}{l} x^{10} + x^7 + x^6 + x^5 + x^3 + x^2 + 1 \\ x^{10} + x^9 + x^8 + x^6 + x^5 + x^4 + x^3 + x + 1 \\ x^{11} + x^8 + x^5 + x^3 + 1 \\ x^{26} + x^{24} + x^{22} + x^{18} + x^{17} + x^{16} + x^{12} + x^{11} + x^{10} + x^9 + x^6 \\ \qquad\qquad\qquad\qquad\qquad\qquad +x^5 + x^3 + x + 1 \end{array}\right\}\ 33$$

$$\left.\begin{array}{l} x^{10} + x^8 + x^7 + x^4 + x^2 + x + 1 \\ x^{10} + x^9 + x^8 + x^6 + x^5 + x^4 + x^3 + x^2 + 1 \\ x^{11} + x^{10} + x^9 + x^5 + x^4 + x + 1 \\ x^{26} + x^{22} + x^{20} + x^{18} + x^{17} + x^{12} + x^{11} + x^{10} + x^5 + x^2 + 1 \end{array}\right\}\ 27$$

$$\left.\begin{array}{l} x^{10} + x^9 + x^6 + x^4 + 1 \\ x^{11} + x^5 + x^3 + x^2 + 1 \\ x^{10} + x^8 + x^7 + x^6 + x^5 + x^4 + x^3 + x + 1 \\ x^{27} + x^{24} + x^{23} + x^{21} + x^{20} + x^{18} + x^{17} + x^{12} + x^{10} + x^9 + x^7 \\ \qquad\qquad\qquad\qquad\qquad\qquad +x^6 + x^5 + x + 1 \end{array}\right\}\ 33$$

The following table shows the number N of polynomials in the list above with n terms:

Appendix II

Below we present a MAPLE source code computing χ_f for every $f = x^n + x^m + 1$, with $n \leq 12$, $m \leq \frac{n}{2}$, which shows the Corollary 1 holds true for this list. The output provides the following four fields: $[\chi_f, n, m, irreduc]$.

```
> with(PolynomialTools):
> f:=x^n+x^m+1:
>
> Listan:=[]:
> for n from 2 to 12 do
>   for m from 1 to n/2 do
>     S:=array(1..n,1..n):
>     for i from 1 to n do
>       for j from 1 to n do
>         S[i,j]:=0:
>       end do:
>     end do:
>     if type(n, even) and type(m, even) then
>       mp:= m/2;
>       np:= n/2;
>       for i from 0 to np-1 do
>         S[2*i+1,i+1]:=1:
>       end do:
>       for i from 0 to np-mp-1 do
>         S[m+2*i+1,np+i+1]:=1:
>         S[2*i+1,np+i+1]:=1:
>       end do:
>       if n = 2*m then
>         for i from 0 to mp-1 do
>           S[2*i+1,n-mp+i+1]:=1:
>         end do:
>       else
>         for i from 0 to mp-1 do
>           S[n-m+2*i+1,n-mp+i+1]:=1:
>           S[m+2*i+1,n-mp+i+1]:=1:
>           S[2*i+1,n-mp+i+1]:=1:
>         end do:
>       end if:
>     elif type(n, odd) and type(m, even) then
>       mp:=m/2:
>       np:=(n-1)/2:
```

```
>    for i from 0 to np do
>      S[2*i+1,i+1]:=1:
>    end do:
>    for i from 0 to np-mp-1 do
>      S[m+2*i+2,np+i+2]:=1:
>      S[2*i+2,np+i+2]:=1:
>    end do:
>    for i from 0 to mp-1 do
>      S[n-m+2*i+1,n-mp+i+1]:=1:
>      S[m+2*i+1,n-mp+i+1]:=1:
>      S[2*i+1,n-mp+i+1]:=1:
>    end do:
>  elif type(n, even) and type(m, odd) then
>    np:=n/2:
>    mp:=(m-1)/2:
>    for i from 0 to np-1 do
>      S[2*i+1,i+1]:=1:
>    end do:
>    for i from 0 to np-mp-1 do
>      S[m+2*i+1,np+i+1]:=1:
>      S[2*i+1,np+i+1]:=1:
>    end do:
>    if mp > 0 then
>      if n = 2*m then
>        for i from 0 to mp-1 do
>          S[2*i+2,n-mp+i+1]:=1:
>        end do:
>      else
>        for i from 0 to mp-1 do
>          S[n-m+2*i+2,n-mp+i+1]:=1:
>          S[m+2*i+2,n-mp+i+1]:=1:
>          S[2*i+2,n-mp+i+1]:=1:
>        end do:
>      end if:
>    end if:
>  else
>    np:=(n-1)/2:
>    mp:=(m-1)/2:
>    for i from 0 to np do
>      S[2*i+1,i+1]:=1:
>    end do:
>    for i from 0 to np-mp-1 do
>      S[m+2*i+2,np+i+2]:=1:
>      S[2*i+2,np+i+2]:=1:
>    end do:
```

```
>     for i from 0 to mp-1 do
>      S[n-m+2*i+2,n-mp+i+1]:=1:
>      S[m+2*i+2,n-mp+i+1]:=1:
>      S[2*i+2,n-mp+i+1]:=1:
>     end do:
>    end if:
>
>    T:=array(1..n,1..n);
>    for i from 1 to n do
>     for j from 1 to n do
>      if i-1 = j then
>       T[i,j] := 1:
>      else
>       T[i,j] := 0:
>      end if:
>     end do:
>    end do:
>    T[1,n]:=1:
>    T[m+1,n]:=1:
>    identidad:=Matrix(n,n,shape=identity):
>    C:=evalm(x*identidad+S+T):
>    Listan:=[op(Listan),[Irreduc(f) mod 2,
>    (det(C) + x^n + x^m) mod 2, n, m, det(C) mod 2]]:
>    unassign('S', 'T', 'identidad');
>   end do:
> end do:
>
> for i from 1 to nops(Listan) do
>  n:=op(3,op(i,Listan)):
>  m:=op(4,op(i,Listan)):
>  det:=op(5,op(i,Listan)):
>  irreduc:=op(1,op(i,Listan)):
>  print(det, n, m, irreduc);
> end do:
```

$\chi_f:$	$x + x^2$	$x + x^3$	$x + x^4$	$x + x^4$	$x^3 + x^5$	$x^2 + x^5$	$x + x^6$
$n:$	2	3	4	4	5	5	6
$m:$	1	1	1	2	1	2	1
$irreduc:$	true	true	true	false	false	true	true

$\chi_f:$	$x + x^2 + x^3 + x^6$	$x^3 + x^6$	$x + x^7$	$x^2 + x^4 + x^5 + x^7$	$x^3 + x^7$
$n:$	6	6	7	7	7
$m:$	2	3	1	2	3
$irreduc:$	false	true	true	false	true

$\chi_f:$	$x^3+x^5+x^6+x^8$	$x+x^2+x^4+x^8$	$x^2+x^3+x^5+x^8$
$n:$	8	8	8
$m:$	1	2	3
$irreduc:$	false	false	false

$\chi_f:$	$x+x^3+x^4+x^5+x^6+x^8$	$x+x^9$	x^5+x^9	$x^4+x^5+x^6+x^9$
$n:$	8	9	9	9
$m:$	4	1	2	3
$irreduc:$	false	true	false	false

$\chi_f:$	x^4+x^9	$x^4+x^5+x^7+x^{10}$	$x^3+x^4+x^5+x^6+x^8+x^{10}$	x^3+x^{10}
$n:$	9	10	10	10
$m:$	4	1	2	3
$irreduc:$	true	false	false	true

$\chi_f:$	$x^2+x^4+x^5+x^{10}$	$x^5+x^7+x^8+x^{10}$	$x^3+x^5+x^6+x^8+x^9+x^{11}$
$n:$	10	10	11
$m:$	4	5	1
$irreduc:$	false	false	false

$\chi_f:$	x^2+x^{11}	$x^2+x^3+x^6+x^{11}$	$x^2+x^3+x^4+x^5+x^6+x^8+x^9+x^{11}$
$n:$	11	11	11
$m:$	2	3	4
$irreduc:$	true	false	false

$\chi_f:$	$x^2+x^3+x^4+x^6+x^8+x^{11}$	$x^6+x^7+x^9+x^{12}$	$x+x^2+x^6+x^{12}$
$n:$	11	12	12
$m:$	5	1	2
$irreduc:$	false	false	false

$\chi_f:$	x^3+x^{12}	$x+x^2+x^4+x^5+x^6+x^7+x^9+x^{12}$	x^5+x^{12}	x^3+x^{12}
$n:$	12	12	12	12
$m:$	3	4	5	6
$irreduc:$	true	false	true	false

Discrete Phase-Space Structures and Mutually Unbiased Bases

A.B. Klimov[1], J.L. Romero[1], G. Björk[2], and L.L. Sánchez-Soto[3]

[1] Departamento de Física, Universidad de Guadalajara, Revolución 1500, 44420 Guadalajara, Jalisco, Mexico
[2] School of Information and Communication Technology, Royal Institute of Technology (KTH), Electrum 229, SE-164 40 Kista, Sweden
[3] Departamento de Óptica, Facultad de Física, Universidad Complutense, 28040 Madrid, Spain

Abstract. We propose a unifying phase-space approach to the construction of mutually unbiased bases for an n-qubit system. It is based on an explicit classification of the geometrical structures compatible with the notion of unbiasedness. These consist of bundles of discrete curves intersecting only at the origin and satisfying certain additional conditions. The effect of local transformations is also studied.

Keywords: Mutually unbiased bases, quantum state estimation, Galois fields, Abelian curves.

1 Introduction

The notion of mutually unbiased bases (MUBs) emerged in the seminal work of Schwinger [1] and it has turned into a cornerstone of the modern quantum information. Indeed, MUBs play a central role in a proper understanding of complementarity [2,3,4,5,6], as well as in approaching some relevant issues such as optimum state reconstruction [7,8], quantum key distribution [9,10], quantum error correction codes [11,12], and the mean king problem [13,14,15,16,17].

For a d-dimensional system (also known as a qudit) it has been found that the maximum number of MUBs cannot be greater than $d + 1$ and this limit is reached if $d = p$ is prime [18] or power of prime, $d = p^n$ [19]. It was shown in Ref. [20] that the construction of MUBs is closely related to the possibility of finding of $d + 1$ disjoint classes, each one having $d - 1$ commuting operators, so that the corresponding eigenstates form sets of MUBs. Since then, different explicit constructions of MUBs in prime power dimensions have been suggested in a number of papers [21,22,23,24,25,26,27].

The phase space of a qudit can be seen as a $d \times d$ lattice whose coordinates are elements of the finite Galois field $GF(d)$ [28]. The use of elements of $GF(d)$ as coordinates allow us to endow the phase-space grid with a similar geometric properties as the ordinary plane. There are several possibilities for mapping quantum states onto this phase space [29,30,31]. However, the most elaborate

C. Carlet and B. Sunar (Eds.): WAIFI 2007, LNCS 4547, pp. 333–345, 2007.

approach was developed by Wootters and coworkers [32,33], which has been used to define a discrete Wigner function (see Refs. [34,35,36] for picturing qubits in phase space). According to this approach, we can put in correspondence lines in the $d \times d$ phase space with states in a Hilbert space in such a way that different sets of $d + 1$ parallel lines (striations) become associated with orthogonal bases which results in a set of mutually unbiased bases.

In this paper, we start by considering the geometrical structures in phase space that are compatible with the notion of unbiasedness and classify these admissible structures into rays and curves (and the former also in regular and exceptional, depending on the degeneracy). Some properties of regular curves are shown. To each curve, we associate a set of commuting operators, and we show how different curves are related by local transformations on the corresponding operators that do not change the entanglement properties of their eigenstates.

2   Constructing Mutually Unbiased Bases

When the space dimension $d = p^n$ is a power of a prime it is natural to view the system as composed of n subsystems, each of dimension p [38]. We briefly summarize a simple construction of MUBs for this case, according to the method introduced in Ref. [27], although focusing on the particular case of n qubits. The main idea consists in labeling both the states of the subsystems and the elements of the generalized Pauli group with elements of the finite field $GF(2^n)$, instead of natural numbers. In particular, we shall denote as $|\alpha\rangle$ with $\alpha \in GF(2^n)$ an orthonormal basis in the Hilbert space of the system. Operationally, the elements of the basis can be labeled by powers of a primitive element (that is, a root of the minimal irreducible polynomial), so that the basis reads

$$\{|0\rangle, |\sigma\rangle, \ldots, |\sigma^{2^n-1} = 1\rangle\} . \tag{1}$$

These vectors are eigenvectors of the generalized position operators Z_β

$$Z_\beta = \sum_{\alpha \in GF(2^n)} \chi(\alpha\beta) \, |\alpha\rangle\langle\alpha| , \tag{2}$$

where henceforth we assume $\alpha, \beta \in GF(2^n)$. Here $\chi(\theta)$ is an additive character

$$\chi(\theta) = \exp\left[i\pi \, \mathrm{tr}(\theta)\right] , \tag{3}$$

and the trace operation, which maps elements of $GF(2^n)$ onto the prime field $GF(2) \simeq \mathbb{Z}_2$, is defined as $\mathrm{tr}(\theta) = \theta + \theta^2 + \ldots + \theta^{2^{n-1}}$.

The diagonal operators Z_β are conjugated to the generalized momentum operators X_β

$$X_\beta = \sum_{\alpha \in GF(2^n)} |\alpha + \beta\rangle\langle\alpha| , \tag{4}$$

through the finite Fourier transform

$$F X_\beta F^\dagger = Z_\beta , \tag{5}$$

with

$$F = \frac{1}{2^n} \sum_{\alpha,\beta \in GF(2^n)} \chi(\alpha\beta) \, |\alpha\rangle\langle\beta| . \tag{6}$$

The operators $\{Z_\alpha, X_\beta\}$ are the generators of the generalized Pauli group

$$Z_\alpha X_\beta = \chi(\alpha\beta) \, X_\beta Z_\alpha . \tag{7}$$

In consequence, we can form $2^n + 1$ sets of commuting operators (which will be called displacement operators) as follows,

$$\{X_\beta\}, \qquad \{Z_\alpha X_{\beta=\mu\alpha}\} , \tag{8}$$

with $\mu \in GF(2^n)$. The operators (8) can be factorized into products of powers of single-particle operators σ_z and σ_x, whose expression in the standard basis of the two-dimensional Hilbert space is

$$\sigma_z = |1\rangle\langle1| - |0\rangle\langle0|, \qquad \sigma_x = |0\rangle\langle1| + |1\rangle\langle0| . \tag{9}$$

This factorization can be carried out by mapping each element of $GF(2^n)$ onto an ordered set of natural numbers [33], $\alpha \leftrightarrow (a_1, \ldots, a_n)$, where a_j are the coefficients of the expansion of α in a field basis θ_j

$$\alpha = a_1\theta_1 + \ldots + a_n\theta_n . \tag{10}$$

A convenient basis is that in which the finite Fourier transform is factorized into a product of single-particle Fourier operators. This is the so-called self-dual basis, defined by the property $\mathrm{tr}(\theta_i\theta_j) = \delta_{ij}$ and leads to the following factorizations

$$Z_\alpha = \sigma_z^{a_1} \otimes \ldots \otimes \sigma_z^{a_n}, \qquad X_\beta = \sigma_x^{b_1} \otimes \ldots \otimes \sigma_x^{b_n} . \tag{11}$$

The simplest geometrical structure in the discrete phase space are the straight lines; i.e., the set of points $(\alpha, \beta) \in GF(2^n) \times GF(2^n)$ satisfying the relation

$$\zeta\alpha + \eta\beta = \vartheta, \tag{12}$$

where ζ, η, ϑ are fixed elements of $GF(2^n)$. Two lines

$$\zeta\alpha + \eta\beta = \vartheta, \qquad \zeta'\alpha + \eta'\beta = \vartheta', \tag{13}$$

are parallel if they have no common points, which implies that $\eta\zeta' = \zeta\eta'$. If the lines (13) are not parallel they cross each other at a single point. A *ray* is a line passing through the origin, so that its equation has the form

$$\alpha = 0, \qquad \text{or} \qquad \beta = \mu\alpha. \tag{14}$$

The equation (14) can be rewritten in the parametric form:

$$\alpha(\kappa) = \eta\kappa, \qquad \beta(\kappa) = \zeta\kappa, \tag{15}$$

where $\kappa \in \mathrm{GF}(2^n)$ is a parameter running through the field. The rays are the simplest nonsingular (that is, with no self intersection) Abelian substructures in the phase space, in the sense that

$$
\begin{aligned}
\alpha(\kappa + \kappa') &= \alpha(\kappa) + \alpha(\kappa'), \\
\beta(\kappa + \kappa') &= \beta(\kappa) + \beta(\kappa').
\end{aligned}
\tag{16}
$$

where $\alpha(\kappa)$ and $\beta(\kappa)$ are parameterized as in (15). The Abelian condition (16) can be interpreted as follows: by summing two points belonging to a ray we obtain some other point of the same ray. In particular, this leads to the possibility of introducing some specific (displacement) operators, which generate "translations" along such rays [2]. The important properties of the rays consists in that the monomials $Z_\alpha X_\beta$ labeled with the points of the phase space belonging to the same ray commute:

$$
Z_{\alpha_1} X_{\beta_1 = \mu\alpha_1} Z_{\alpha_2} X_{\beta_2 = \mu\alpha_2} = Z_{\alpha_2} X_{\beta_2 = \mu\alpha_2} Z_{\alpha_1} X_{\beta_1 = \mu\alpha_1},
\tag{17}
$$

and thus, have a common system of eigenvectors $\{|\psi_\nu^\mu\rangle\}$, with $\mu, \nu \in \mathrm{GF}(2^n)$:

$$
Z_\alpha X_{\mu\alpha} |\psi_\nu^\mu\rangle = \exp(i\xi_{\mu,\nu}) |\psi_\nu^\mu\rangle,
\tag{18}
$$

where μ is fixed and $\exp(i\xi_{\mu,\nu})$ is the corresponding eigenvalue, so that $|\psi_\nu^0\rangle = |\nu\rangle$ are eigenstates of Z_α (displacement operators labeled by the points of the ray $\beta = 0$, which we take as horizontal axis).

This means that each ray defines a set of $2^n - 1$ commuting operators (that form "lines") and the whole bundle of $2^n + 1$ rays (which is obtained by varying the "slope" index μ and adding the set X_β labeled by points of the vertical axis $\alpha = 0$) allows one to construct a complete set of MUBs operators arranged in a $(2^n - 1) \times (2^n + 1)$ table.

3   Curves in Phase Space

3.1   General Form of Abelian Curves

The rays are not the only Abelian structures that exist in the discrete phase space. It is easy to see that the parametrically defined curves (which obviously pass through the origin)

$$
\alpha(\kappa) = \sum_{m=0}^{n-1} \mu_m \kappa^{2^m}, \qquad \beta(\kappa) = \sum_{m=0}^{n-1} \nu_m \kappa^{2^m},
\tag{19}
$$

satisfy the Abelian condition (16). If we demand that the displacement operators labeled by points of the curves (19) commute with each other; i.e.,

$$
\mathrm{tr}(\alpha\beta') = \mathrm{tr}(\alpha'\beta),
\tag{20}
$$

where $\alpha' = \alpha(\kappa'), \beta' = \beta(\kappa')$, the coefficients μ_m, ν_m should satisfy the restriction

$$\sum_{m \neq k} \mathrm{tr}(\mu_m \nu_k) = 0 \,. \tag{21}$$

If the condition (20) is satisfied, we can associate to each curve (19), with given coefficients $\boldsymbol{\mu}, \boldsymbol{\nu}$, a state $|\psi_{\boldsymbol{\mu}, \boldsymbol{\nu}}\rangle$. The curves satisfying Eqs. (19) will be called Abelian curves. If such a curve contains exactly $n+1$ different points, the monomials $Z_\alpha X_\beta$, where (α, β) are coordinates of the points on the curve, will form a set of commuting operators. Thus, the problem of finding sets of all the possible MUB operators can be reduced to the problem of arranging commutative non-singular curves in the discrete phase space in bundles of $n+1$ mutually non-intersecting curves. It is worth noting that the parametrization (19) is not unique: each parametrization determines a certain order of the points on the curve.

3.2   Non-singularity Condition

Let us try to classify all the Abelian curves. First, we discuss the no self-intersection condition. To this end, let us consider a curve defined in the parametric form (19) and note that a self-intersection means that there exists $\kappa' \neq \kappa \in \mathrm{GF}(2^n)$, so that

$$\alpha(\kappa) = \alpha(\kappa'), \qquad \beta(\kappa) = \beta(\kappa'). \tag{22}$$

We can find a necessary condition of no self-intersection. It is clear that if $\alpha(\kappa) = \alpha(\kappa')$, then $\sigma^m[\alpha(\kappa + \kappa')] = 0$, for $m = 0, \ldots, n-1$, where $\sigma^m[\alpha] = \alpha^{2^m}$ are the Frobenius automorphism operators [28]. Let us introduce the following matrices

$$M(\mu) = \begin{bmatrix} \mu_0^{2^0} & \cdots & \mu_{n-1}^{2^0} \\ \mu_{n-1}^{2^1} & \cdots & \mu_{n-2}^{2^1} \\ \vdots & \ddots & \vdots \\ \mu_1^{2^{n-1}} & \cdots & \mu_0^{2^{n-1}} \end{bmatrix}, \qquad N(\nu) = \begin{bmatrix} \nu_0^{2^0} & \cdots & \nu_{n-1}^{2^0} \\ \nu_{n-1}^{2^1} & \cdots & \nu_{n-2}^{2^1} \\ \vdots & \ddots & \vdots \\ \nu_1^{2^{n-1}} & \cdots & \nu_0^{2^{n-1}} \end{bmatrix}, \tag{23}$$

where the lines in the above matrices are formed by the coefficients appearing in the expansion (19) and the corresponding expansions of $\sigma^m[\alpha(\kappa)]$ and $\sigma^m[\beta(\kappa)]$. If $\det M$ and $\det N$ do not vanish simultaneously, the corresponding curve (19) has no self-intersection. Nevertheless, the condition

$$\det M = \det N = 0 \,, \tag{24}$$

means that the ranks of the matrices M and N are lesser than the dimension of the system, but this does not guarantee that there exist $\kappa' \neq \kappa$ satisfying (22), because the solutions of $\alpha(\kappa) = \alpha(\kappa')$ and $\beta(\kappa'') = \beta(\kappa''')$ can form disjoint sets. Therefore, the condition (24) is necessary, but not sufficient to determine if a curve is singular. Another necessary, but not sufficient, condition of a curve singularity is $\det(M + N) = 0$. A curve with either $\det M \neq 0$ or $\det N \neq 0$ will be called a *regular* curve. A peculiarity of such a curve is that the parameter α (if $\det M \neq 0$) or β (if $\det N \neq 0$) takes all the values in the field. Non-singular curves satisfying (24) will be called *exceptional* curves.

It is worth noting that the conditions (24) mean that $\sigma^m[\alpha]$ $(m = 0, \ldots, n-1)$ as well as $\sigma^m[\beta]$ $(m = 0, \ldots, n-1)$ are not linearly independent, so that not one of α and β run through the whole field (in other words, the values of both α and β are degenerate). The number of linearly independent powers of α (β) equals to the rank of the matrix M (N) and the quantities $n - \mathrm{rank} M$ and $n - \mathrm{rank} N$ determine the degree of degeneration of each allowed value of α and β, respectively.

It is interesting to note that the determinants of matrices of type (23) over $\mathrm{GF}(2^n)$ take only values zero and one; i.e., $\det M \in \mathbb{Z}_2$, which can be easily seen by observing that $(\det M)^2 = \det M$.

4   Regular Curves

4.1   Explicit Forms

Given a regular curve, we can invert one of the relations (19) and substitute it into the other one to find an explicit equation of the curve:

a) if $\det M \neq 0$, then the equation of the curve can always be written as

$$\beta = \sum_{m=0}^{n-1} \phi_m \alpha^{2^m} ; \tag{25}$$

b) if $\det N \neq 0$, then the equation of the curve reads as

$$\alpha = \sum_{m=0}^{n-1} \psi_m \beta^{2^m} . \tag{26}$$

Nevertheless, if $\det M \neq 0$ but $\det N = 0$, then the coordinate β is degenerate and the curve equation cannot be expressed in the form (26). We shall refer to the corresponding curve as an α-curve. Similarly, when $\det N \neq 0$ but $\det M = 0$, the coordinate α is degenerate and the corresponding curve will be called a β-curve.

The general commutativity condition (21) can be essentially simplified for regular curves. When $\det M \neq 0$ (or $\det N \neq 0$) we obtain, by direct substitution of the explicit forms (25) or (26) into (20), the following restrictions on the coefficients ϕ_m (or ψ_m):

$$\phi_j = \phi_{n-j}^{2^j}, \qquad \psi_j = \psi_{n-j}^{2^j}, \tag{27}$$

where $j = 1, \ldots, [(n-1)/2]$. For even values of n, the additional conditions $\phi_{n/2} = \phi_{n/2}^{2^{n/2}}$ and $\psi_{n/2} = \psi_{n/2}^{2^{n/2}}$, should be fulfilled.

Note that, because the regular curves are automatically non-singular, we do not have to carry out the whole analysis involving the parametric forms and the properties of the corresponding M and N matrices, but rather to write down explicit expressions directly using (27).

4.2 Examples

1. Completely non-degenerate regular curve.
   Consider the curve given in the parametric form in $GF(2^3)$

$$\alpha = \sigma^2 \kappa + \kappa^2 + \sigma^4 \kappa^4,$$
$$\beta = \sigma^3 \kappa + \sigma^6 \kappa^2 + \sigma^6 \kappa^4, \tag{28}$$

where σ is the primitive element, which fulfills $\sigma^3 + \sigma + 1 = 0$. The associated matrices are now

$$M = \begin{bmatrix} \sigma^2 & 1 & \sigma^4 \\ \sigma & \sigma^4 & 1 \\ 1 & \sigma^2 & \sigma \end{bmatrix}, \qquad N = \begin{bmatrix} \sigma^3 & \sigma^6 & \sigma^6 \\ \sigma^5 & \sigma^6 & \sigma^5 \\ \sigma^3 & \sigma^3 & \sigma^5 \end{bmatrix}, \tag{29}$$

with $\det M = \det N = 1$, thus leading to the following explicit forms

$$\beta = \sigma^6 \alpha + \sigma^3 \alpha^2 + \sigma^5 \alpha^4, \qquad \text{or} \qquad \alpha = \sigma^6 \beta + \sigma^3 \beta^2 + \sigma^5 \beta^4, \tag{30}$$

whose coefficients satisfy the general condition (27). The set of commuting operators corresponding to this curve is:

$$Z_{\sigma^3} X_{\sigma^3},\ Z_{\sigma^6} X_{\sigma^5},\ Z_{\sigma^5} X_{\sigma^6},\ Z_{\sigma^4} X_{\sigma^2},\ Z_\sigma X_\sigma,\ Z_{\sigma^7} X_{\sigma^7},\ Z_{\sigma^2} X_{\sigma^4}. \tag{31}$$

2. β-curve.
   To the curve given in the parametric form

$$\alpha = \sigma^2 \kappa + \kappa^2 + \sigma \kappa^4,$$
$$\beta = \sigma^2 \kappa^4, \tag{32}$$

correspond the following matrices

$$M = \begin{bmatrix} \sigma^2 & 1 & \sigma \\ \sigma^2 & \sigma^4 & 1 \\ 1 & \sigma^4 & \sigma \end{bmatrix}, \qquad N = \begin{bmatrix} 0 & 0 & \sigma^2 \\ \sigma^4 & 0 & 0 \\ 0 & \sigma & 0 \end{bmatrix}, \tag{33}$$

with $\det M = 0, \det N = 1$, thus leading to the following explicit form of the β-curve

$$\alpha = \sigma^6 \beta + \sigma^5 \beta^2 + \sigma^6 \beta^4. \tag{34}$$

Note that the above equation cannot be inverted and represented in the form (25). The set of commuting operators corresponding to this curve is:

$$Z_{\sigma^5} X_{\sigma^2},\ X_{\sigma^6},\ Z_{\sigma^2} X_{\sigma^3},\ Z_{\sigma^5} X_{\sigma^7},\ Z_{\sigma^2} X_{\sigma^4},\ Z_{\sigma^3} X_\sigma,\ Z_{\sigma^3} X_{\sigma^5}. \tag{35}$$

5 Exceptional Curves

The analysis of exceptional curves is more involved, but the number of such curves is substantially lesser than the number of regular curves. As it was pointed above, the points on the curve do not take all the values in the field, and their admissible values are fixed by the structural equations

$$\sum_{m=0}^{r_M} v_m \alpha^{2^m} = 0, \qquad \sum_{m=0}^{r_N} \tau_m \beta^{2^m} = 0, \tag{36}$$

where $r_M = \operatorname{rank} M \leq n-1$ and $r_N = \operatorname{rank} N \leq n-1$, which are consequence of the linear dependence of α^{2^m} and β^{2^m} ($m = 0, \ldots, n-1$). The coordinates α and β of the points on a exceptional curve are 2^{n-r_M} and 2^{n-r_N} times degenerate, respectively. The non-singularity condition means that there are 2^n different pairs (α, β) belonging to the curve, so that the condition $r_M + r_N \geq n$ should be satisfied. When a curve equation; i.e., a relation of the type $F(\alpha) = G(\beta)$, with $F(\alpha)$ and $G(\beta)$ being polynomials of degrees 2^{r_M-1} and 2^{r_N-1}, can be found, it establishes a correspondence between the roots of (36). Nevertheless, such a relation cannot always be established, as we will see below.

There are two ways to approach the classification of exceptional curves. The first would be a direct analysis of an arbitrary curve given in parametric form, whose coefficients satisfy the commutativity relation (21) and the corresponding determinants turn to zero. We first have to determine the rank of the matrices M and N and find the structural relations (36), which determine admissible values of the "coordinates" α and β along the curve and check the non-singularity condition. Afterwards, we should find the curve equation of the type $F(\alpha) = G(\beta)$, which establishes a relation between the values of α and β determined by the structural relations. Nevertheless, the main difficulty of this approach consists in the complicated form of the commutativity condition (21), related to the fact that there is no one-to-one correspondence between a parametric form of a curve and points in the discrete phase space of such a curve.

The other approach consists in constructing the possible exceptional curves by imposing from the beginning the non-singularity and commutativity conditions. In this construction we need information about the degree of degeneration in α and β directions. As an example, consider the following Abelian curve,

$$\alpha = \sigma \kappa^2 + \kappa^4,$$
$$\beta = \sigma^4 \kappa + \sigma^5 \kappa^2. \tag{37}$$

The corresponding matrices are

$$M = \begin{bmatrix} 0 & \sigma & 1 \\ 1 & 0 & \sigma^2 \\ \sigma^4 & 1 & 0 \end{bmatrix}, \qquad N = \begin{bmatrix} \sigma^4 & \sigma^5 & 0 \\ 0 & \sigma & \sigma^3 \\ \sigma^6 & 0 & \sigma^2 \end{bmatrix}, \tag{38}$$

so that $\det M = \det N = 0$ and the structural equations turn out to be

$$\alpha + \sigma^5 \alpha^2 + \sigma \alpha^4 = 0, \qquad \beta + \sigma^4 \beta^2 + \sigma^5 \beta^4 = 0, \tag{39}$$

which define the admissible values of α and β, in particular, $\alpha = \{0, \sigma^3, \sigma^6, \sigma^4\}$ and $\beta = \{0, \sigma^7, \sigma^5, \sigma^4\}$. The coordinates over the curve are related by

$$\beta^2 + \sigma^5 \beta = \sigma^2 \alpha^2 + \sigma^6 \alpha, \tag{40}$$

while the set of commuting operators corresponding to this curve is

$$Z_{\sigma^3}X_{\sigma^7},\ Z_{\sigma^6}X_{\sigma^4},\ Z_{\sigma^6}X_{\sigma^7},\ Z_{\sigma^4}X_{\sigma^5},\ X_{\sigma^5},\ Z_{\sigma^3}X_{\sigma^4},\ Z_{\sigma^4}. \tag{41}$$

6   Local Transformations

When local transformations are applied to a set of commuting MUB operators, nontrivial transformations of the curve form are introduced, although they preserve the factorization property (11) in a given basis:

$$Z_\alpha X_\beta = \bigotimes_{j=1}^{n} (\sigma_z^{a_j}\sigma_x^{b_j}) \equiv \prod_{j=1}^{n}(a_j, b_j). \tag{42}$$

Under local transformations applied to the j-th particle ($\pi/2$ rotations around $z, x,$ and y axis, which we denote as z-, x-, and y-rotations), we have

$$\begin{aligned}
z &: (a_j, b_j) \rightarrow (a_j + b_j, b_j),\\
x &: (a_j, b_j) \rightarrow (a_j, b_j + a_j),\\
y &: (a_j, b_j) \rightarrow (a_j + a_j + b_j, b_j + a_j + b_j) = (b_j, a_j).
\end{aligned} \tag{43}$$

In terms of the field elements this is equivalent to

$$\begin{array}{llll}
z: & \alpha \rightarrow \alpha + \theta_j \operatorname{tr}(\beta\theta_j), & \beta \rightarrow \beta,\\
x: & \alpha \rightarrow \alpha, & \beta \rightarrow \beta + \theta_j \operatorname{tr}(\alpha\theta_j),\\
y: & \alpha \rightarrow \alpha + \theta_j \operatorname{tr}[(\alpha + \beta)\theta_j], & \beta \rightarrow \beta + \theta_j \operatorname{tr}[(\alpha + \beta)\theta_j].
\end{array} \tag{44}$$

The above transformations are non-linear in terms of the field elements. In particular, this means that starting with a standard set of MUB operators related to rays, we get another set of MUB operators parameterized by points of curves, but leading to the same factorization structure.

7   Curves over GF(2^2)

In the case of GF(2^2) a full analysis of curves can be carried out by studying the parametric forms

$$\alpha(\kappa) = \mu_0\kappa + \mu_1\kappa^2, \qquad \beta(\kappa) = \nu_0\kappa + \nu_1\kappa^2. \tag{45}$$

The commutativity condition impose the following restrictions on the coefficients μ_j and ν_j :

$$\mu_1\nu_0 + (\mu_1\nu_0)^2 = \mu_0\nu_1 + (\mu_0\nu_1)^2. \tag{46}$$

All the possible curves satisfying condition (46) can be divided into two types:
  a) regular curves

$$\begin{array}{lll}
\alpha-\text{curves}: & \alpha = \sigma\kappa, & \beta = \nu\kappa + \sigma^2\kappa^2,\\
\beta-\text{curves}: & \beta = \sigma\kappa, & \alpha = \nu\kappa + \sigma^2\kappa^2,
\end{array} \begin{array}{l}(47)\\(48)\end{array}$$

where ν takes all values in GF(2^2).

b) exceptional curves

$$\alpha = \mu(\kappa + \kappa^2), \qquad \beta = \mu^2(\sigma\kappa + \sigma^2\kappa^2), \tag{49}$$

where μ takes all values in GF(2^2).

The regular curves are nondegenerate, in the sense that α or β (or both) are not repeated in any set of four points defining a curve. In other words, α or β (or both) take all the values in the field GF(2^2). This allows us to write down explicit relations between α and β as follows

$$\alpha\text{–curves}: \quad \beta = \nu\sigma^2\alpha + \alpha^2, \tag{50}$$
$$\beta\text{–curves}: \quad \alpha = \nu\sigma^2\beta + \beta^2. \tag{51}$$

Every point of exceptional curves is doubly degenerate and can be obtained from equations that relate powers of α and β:

$$\alpha^2 = \mu\alpha, \qquad \beta^2 = \mu^2\beta. \tag{52}$$

It is impossible to write an explicit nontrivial equation of the form $f(\alpha, \beta) = 0$ for them. However, it is possible to obtain all the curves of the form (47) and (49) from the rays after some (nonlinear) operations (44), corresponding to local transformations of operators. The families of such transformations are the following [41]: First, 8 rays and curves can be obtained from a single ray $\alpha = 0, \beta = \sigma^2\kappa$ (vertical axis). Second, another 5 different rays and curves can be obtained from the ray $\alpha = \sigma\kappa, \beta = \sigma^2\kappa$ ($\beta = \sigma\alpha$).

8   Curves over GF(2^3)

A generic Abelian curve over GF(2^3) is defined in parametric form as

$$\alpha = \mu_0\kappa + \mu_1\kappa^2 + \mu_2\kappa^4,$$
$$\beta = \nu_0\kappa + \nu_1\kappa^2 + \nu_2\kappa^4. \tag{53}$$

The commutativity condition in this case is much more complicated than for GF(2^2), so that a full analysis of all the possible curves becomes extremely cumbersome if we start with (53). Instead, we just find all the possible regular curves. A generic regular α-curve has the following form

$$\beta = \phi_0\alpha + \phi^2\alpha^2 + \phi\alpha^4. \tag{54}$$

There are two kind of exceptional curves: a) when both α and β coordinates of each point is doubly degenerate and b) when one of the coordinate is doubly degenerate while the other one is four times degenerate. All the double degenerate exceptional curves over GF(2^3) are parametrized by two numbers and have the form of two "parallel" lines

$$\beta_j^{(1)} = \upsilon_1\upsilon_0^{-1}\alpha_j^{-1}\alpha, \qquad \beta_j^{(2)} = \upsilon_1\upsilon_0^{-1}(\alpha_j^{-1}\alpha + 1), \tag{55}$$

where $\alpha_j = \alpha_1, \alpha_2, \alpha_1 + \alpha_2$ and the admissible values of α are $0, \alpha_1, \alpha_2, \alpha_1 + \alpha_2$. So, for fixed α_1 and α_2 we have three exceptional curves

1. $- (\alpha_1, 0), (\alpha_1, \delta), (\alpha_2, \delta(\alpha_1^{-1}\alpha_2 + 1)), (\alpha_2, \delta\alpha_1^{-1}\alpha_2), (\alpha_1 + \alpha_2, \delta\alpha_1^{-1}\alpha_2),$
   $(\alpha_1 + \alpha_2, \delta(\alpha_1^{-1}\alpha_2 + 1)), (0, \delta);$

2. $- (\alpha_2, 0), (\alpha_2, \delta), (\alpha_1, \delta(\alpha_2^{-1}\alpha_1 + 1)), (\alpha_1, \delta\alpha_2^{-1}\alpha_1), (\alpha_1 + \alpha_2, \delta\alpha_2^{-1}\alpha_1),$
   $(\alpha_1 + \alpha_2, \delta(\alpha_2^{-1}\alpha_1 + 1)), (0, \delta);$

3. $- (\alpha_1 + \alpha_2, 0), (\alpha_1 + \alpha_2, \delta), (\alpha_1, \delta(\alpha_1^{-1}\alpha_2 + 1)^{-1}), (\alpha_1, \delta(\alpha_1^{-1}\alpha_2 + 1)^{-1} + \delta),$
   $(\alpha_2, \delta\left(\alpha_2^{-1}\alpha_1 + 1\right)^{-1}), (\alpha_2, \delta(\alpha_2^{-1}\alpha_1 + 1)^{-1} + \delta), (0, \delta);$

where

$$\delta = \frac{1}{\alpha_1 + \alpha_2} + \frac{\alpha_1 + \alpha_2}{\alpha_1\alpha_2}. \tag{56}$$

Note that α_1, α_2 and $\alpha_3 = \alpha_1 + \alpha_2$ can be considered as roots of the following structural equation

$$v_0\alpha + v_1\alpha^2 + \alpha^4 = 0, \tag{57}$$

where $v_{0,1}$ are symmetrical functions of the roots $\alpha_{1,2,3}$.

The other possibility to form exceptional curves is when one of the curve coordinate (let us say, α) is still double degenerate while the other one (β) is four times degenerate. In this case, the coordinate β takes only two values: 0 and δ, while the allowed values of α are $0, \alpha_1, \alpha_2, \alpha_1 + \alpha_2$. Thus, any such curve has the form

$$(0, 0), (\alpha_1, 0), (\alpha_2, 0), (\alpha_1 + \alpha_2, 0), (\alpha_1, \delta), (\alpha_2, \delta), (\alpha_1 + \alpha_2, \delta), (0, \delta). \tag{58}$$

The commutativity condition (20) leads to the following structural equation for the α coordinate:

$$\delta\alpha + \delta^2\alpha^2 + \delta^4\alpha^4 = 0, \tag{59}$$

so that $\delta^5 = \alpha_1\alpha_2 + \alpha_1^2 + \alpha_2^2$.

Acknowledgements

This work was supported by the Grant 45704 of Consejo Nacional de Ciencia y Tecnologia (CONACyT), Mexico, the Swedish Foundation for International Cooperation in Research and Higher Education (STINT), the Swedish Research Council (VR), the Swedish Foundation for Strategic Research (SSF), and the Spanish Research Directorate (DGI), Grant FIS2005-0671.

References

1. Schwinger, J.: The geometry of quantum states. Proc. Natl. Acad. Sci. USA 46, 257–265 (1960)
2. Wootters, W.K.: A Wigner-function formulation of finite-state quantum mechanics. Ann. Phys (NY) 176, 1–21 (1987)

3. Kraus, K.: Complementary observables and uncertainty relations. Phys. Rev. D 35, 3070–3075 (1987)
4. Lawrence, J., Brukner, Č., Zeilinger, A.: Mutually unbiased binary observable sets on N qubits. Phys. Rev. A 65, 32320 (2002)
5. Chaturvedi, S.: Aspects of mutually unbiased bases in odd-prime-power dimensions. Phys. Rev. A 65, 44301 (2002)
6. Wootters, W.K.: Quantum measurements and finite geometry. Found. Phys. 36, 112–126 (2006)
7. Wootters, W.K., Fields, B.D.: Optimal state-determination by mutually unbiased measurements. Ann. Phys (NY) 191, 363–381 (1989)
8. Asplund, R., Björk, G.: Reconstructing the discrete Wigner function and some properties of the measurement bases. Phys. Rev. A 64, 12106 (2001)
9. Bechmann-Pasquinucci, H., Peres, A.: Quantum cryptography with 3-State systems. Phys. Rev. Lett. 85, 3313–3316 (2000)
10. Cerf, N., Bourennane, M., Karlsson, A., Gisin, N.: Security of quantum key distribution using d-level systems. Phys. Rev. A 88, 127902 (2002)
11. Gottesman, D.: Class of quantum error-correcting codes saturating the quantum Hamming bound. Phys. Rev. A 54, 1862–1868 (1996)
12. Calderbank, A.R., Rains, E.M., Shor, P.W., Sloane, N.J.A.: Quantum error correction and orthogonal geometry. Phys. Rev. Lett. 78, 405–408 (1997)
13. Vaidman, L., Aharonov, Y., Albert, D.Z.: How to ascertain the values of σ_x, σ_y, and σ_z of a spin-1/2 particle. Phys. Rev. Lett. 58, 1385–1387 (1987)
14. Englert, B.-G., Aharonov, Y.: The mean king's problem: prime degrees of freedom. Phys. Lett. A 284, 1–5 (2001)
15. Aravind, P.K.: Solution to the king's problem in prime power dimensions. Z. Naturforsch. A. Phys. Sci. 58, 85–92 (2003)
16. Schulz, O., Steinhübl, R., Weber, M., Englert, B.-G, Kurtsiefer, C., Weinfurter, H.: Ascertaining the values of σ_x, σ_y, and σ_z of a polarization qubit. Phys. Rev. Lett. 90, 177901 (2003)
17. Kimura, G., Tanaka, H., Ozawa, M.: Solution to the mean king's problem with mutually unbiased bases for arbitrary levels. Phys. Rev. A 73, 50301 (R) (2006)
18. Ivanović, I.D.: Geometrical description of quantal state determination. J. Phys. A 14, 3241–3246 (1981)
19. Calderbank, A.R., Cameron, P.J., Kantor, W.M., Seidel, J.J.: $\mathbb{Z}_4$-Kerdock codes, orthogonal spreads, and extremal Euclidean line-sets. Proc. London Math. Soc. 75, 436–480 (1997)
20. Bandyopadhyay, S., Boykin, P.O., Roychowdhury, V., Vatan, V.: A new proof for the existence of mutually unbiased bases. Algorithmica 34, 512–528 (2002)
21. Klappenecker, A., Rötteler, M.: Constructions of mutually unbiased bases. In: Mullen, G.L., Poli, A., Stichtenoth, H. (eds.) Finite Fields and Applications. LNCS, vol. 2948, pp. 137–144. Springer, Heidelberg (2004)
22. Lawrence, J.: Mutually unbiased bases and trinary operator sets for N qutrits. Phys. Rev. A 70, 12302 (2004)
23. Parthasarathy, K.R.: On estimating the state of a finite level quantum system. Infin. Dimens. Anal. Quantum Probab. Relat. Top. 7, 607–617 (2004)
24. Pittenger, A.O., Rubin, M.H.: Wigner function and separability for finite systems. J. Phys. A 38, 6005–6036 (2005)
25. Durt, T.: About mutually unbiased bases in even and odd prime power dimensions. J. Phys. A 38, 5267–5284 (2005)
26. Planat, M., Rosu, H.: Mutually unbiased phase states, phase uncertainties, and Gauss sums. Eur. Phys. J. D 36, 133–139 (2005)

27. Klimov, A.B., Sánchez-Soto, L.L., de Guise, H.: Multicomplementary operators via finite Fourier transform. J. Phys. A 38, 2747–2760 (2005)
28. Lidl, R., Niederreiter, H.: Introduction to Finite Fields and their Applications. Cambridge University Press, Cambridge (1986)
29. Buot, F.A.: Method for calculating $\mathrm{Tr}\,\mathcal{H}^n$ in solid-state theory. Phys. Rev. B 10, 3700–3705 (1974)
30. Galetti, D., De Toledo Piza, A.F.R.: An extended Weyl-Wigner transformation for special finite spaces. Physica A 149, 267–282 (1988)
31. Cohendet, O., Combe, P., Sirugue, M., Sirugue-Collin, M.: A stochastic treatment of the dynamics of an integer spin. J. Phys. A 21, 2875–2884 (1988)
32. Wootters, W.K.: Picturing qubits in phase space. IBM J. Res. Dev. 48, 99–110 (2004)
33. Gibbons, K.S., Hoffman, M.J., Wootters, W.K.: Discrete phase space based on finite fields. Phys. Rev. A 70, 62101 (2004)
34. Paz, J.P., Roncaglia, A.J., Saraceno, M.: Qubits in phase space: Wigner-function approach to quantum-error correction and the mean-king problem. Phys. Rev. A 72, 12309 (2005)
35. Durt, T.: About Weyl and Wigner tomography in finite-dimensional Hilbert spaces. Open Syst. Inf. Dyn. 13, 403–413 (2006)
36. Klimov, A.B., Munoz, C., Romero, J.L.: Geometrical approach to the discrete Wigner function in prime power dimensions. J. Phys. A 39, 14471–14497 (2006)
37. Romero, J.L., Björk, G., Klimov, A.B., Sánchez-Soto, L.L.: On the structure of the sets of mutually unbiased bases for N qubits. Phys. Rev. A 72, 62310 (2005)
38. Vourdas, A.: Quantum systems with finite Hilbert space. Rep. Prog. Phys. 67, 267–320 (2004)
39. Englert, B.-G., Metwally, N.: Separability of entangled q-bit pairs. J. Mod. Opt. 47, 2221–2231 (2000)
40. Björk, G., Romero, J.L., Klimov, A.B., Sánchez-Soto, L.L.: Mutually unbiased bases and discrete Wigner functions. J. Opt. Soc. Am. B 24, 371–379 (2007)
41. Klimov, A.B., Romero, J.L., Björk, G., Sánchez-Soto, L.L.: J. Phys. A 40, 3987–3998 (2007)

Some Novel Results of p-Adic Component of Primitive Sequences over $Z/(p^d)$

Yuewen Tang and Dongyang Long

Department of Computer Science, Sun Yat-Sen University, Guangzhou 510275, PRC
yw811@yahoo.com.cn, issldy@mail.sysu.edu.cn

Abstract. Some novel results of the p-adic components of primitive sequences over ring Z mod p^d ($Z/(p^d)$) are given. An improving result of Dai Zongdao formula is presented. Moreover, we characterize the minimal polynomials and trace expressions for 0, 1 level components of primitive sequences over $Z/(p^d)$ for any prime number p.

1 Introduction

The sequences over ring have many properties different from the ones over field [1-9]. The study of sequences over ring has been one of hotspot topics of modern cryptography research recently [4-9].

Many significant results are given [4-9] for the sequences over ring. Huang Minqiang characterized the algebraic structures of components of primitive sequences over $Z/(2^d)$ [5]. He obtained the minimal polynomials and trace expressions for 0, 1, 2 level component of primitive sequence over $Z/(2^d)$ [5]. In this paper, we investigate the algebraic structures of components of primitive sequences over $Z/(p^d)$ for any prime number p. The methods using in the paper are different from Huang's method in [5]. An improving result of the Dai Zongdao formula [5] is presented. At the same time, we characterize the minimal polynomials and trace expressions for 0, 1 level component of primitive sequences over $Z/(p^d)$ for any prime number p. In particular, when $p = 2$ we easily reduce some main results of Huang's work in [5].

We first introduce the necessary concepts and notations. For additional details and definitions, see the references, in particular [1-5].

Let p is a prime number and d a positive integer, $f(x) \in P_n[Z/(p^d), x]$ If $\text{per}(f(x))_{p^d} = p^{d-1}(p^n - 1), n = \deg f(x)$, then $f(x)$ is said to be a primitive polynomial for module p^d [4]. If $\text{per}(\alpha)_{p^d} = p^{d-1}(p^n - 1)$, then we call α a primitive sequence for module p^d.

For each $f(x)$ and α over $Z/(p^d)$, there exists an unique p-adic decomposition

C. Carlet and B. Sunar (Eds.): WAIFI 2007, LNCS 4547, pp. 346–353, 2007.

$$\alpha = \sum_{i \geq 0} p^i \alpha_i = \alpha_0 + p\alpha_1 + p^2\alpha_2 + \ldots\ldots,$$

$$f(x) = \sum_{i \geq 0} p^i f_i(x) = f_0(x) + p f_1(x) + p^2 f_2(x) + \ldots\ldots,$$

where α_i is said to be the i-th level component sequence of α , by $g_i(x)$ denotes the minimal polynomial of α_i and by $L(\alpha_i) = \deg g_i(x)$ denotes the linear complexity of α_i .

2 A Key Theorem

Lemma 1 ([3]). Let $n = a_h p^h + a_{h-1} p^{h-1} + \ldots + a_1 p + a_0$ where $1 \leq a_h < p$,

$0 \leq a_j < p$, $j = 0,1,\ldots,h-1$, $A(n,p) = \sum_{k=0}^{h} a_k$ .Then we have

(1) $pot_p(n!) = [n - A(n,p)]/(p-1)$;

(2) $pot_p\binom{n}{r} = [A(r,p) + A(n-r,p) - A(n,p)]/(p-1)$.

Theorem 1. Let $n = a_h p^h + a_{h-1} p^{h-1} + \ldots + a_1 p + a_0$, where $1 \leq a_h < p$, $0 \leq a_j < p$, $j = 0,1,\ldots,h-1$. Then we have

$$a_k = \binom{n}{p^k} \bmod p$$

Proof. (1) If $a_k = 0$, let $a_{k+1} = a_{k+2} = \ldots = a_{k+j} = 0$, $a_{k+j+1} \neq 0$, then

$$n - p^k = a_0 + a_1 p + \ldots + a_{k-1} p^{k-1} + (p-1)p^k + \ldots$$

$$+ (p-1)p^{k+j} + (a_{k+j+1} - 1)p^{k+j+1} + a_{k+j+2} p^{k+j+2} + \ldots + a_h p^h$$

By Lemma 1, we have $pot_p\binom{n}{p^j} = j + 1 (j \geq 0)$

so that $a_k \equiv \binom{n}{p^k} \equiv 0 \bmod p$.

(2) If $a_k \neq 0$, assume that

$$\binom{n}{p^k} = c_0 + c_1 p + c_2 p^2 + \ldots\ldots \tag{2.1}$$

is a p-adic decomposition of $\begin{pmatrix} n \\ p^k \end{pmatrix}$, then we easily obtain

$$n! = p^k!(n - p^k)![c_0 + c_1 p + c_2 p^2 + \ldots\ldots] \qquad (2.2)$$

By Lemma 1 and $a_k \neq 0$, then we have $pot_p\begin{pmatrix} n \\ p^j \end{pmatrix} = 0$ so that $c_0 \neq 0$. Since

$pot_p(n!) = pot_p[p^k!(n - p^k)!]$, we can easily calculate every element module p of Equation (2.2) and then we have

$$a_0! a_1! \ldots a_k! \ldots a_h! [(p-1)!]^x \equiv a_0! a_1! \ldots a_h! [(p-1)!]^y c_0 \bmod p \qquad (2.3)$$

where

$$x = a_1 + (p+1)a_2 + (p^2 + p + 1)a_3 + \ldots + (p^{h-1} + p^{h-2} + \ldots + p + 1)a_h,$$

$$y = a_1 + (p+1)a_2 + (p^2 + p + 1)a_3 + \ldots + (p^{k-1} + p^{k-2} + \ldots + p + 1)$$

$$(a_k - 1) + \ldots + (p^h + p^{h-1} + \ldots + p + 1)a_h + (p^{k-1} + p^{k-2} + \ldots + p + 1)$$

Since $x = y$, by Equations (2.3) and (2.1), then we have

$$a_k = c_0 \equiv \begin{pmatrix} n \\ p^k \end{pmatrix} \bmod p$$

3 An Improvement of Dai Zongduo Formula

Lemma 2 (Dai Zongduo Formula [5]). Let $f(x)$ is a polynomial over $Z/(p^d)$ and α is a sequence over $Z/(p^d)$. If $f(x) = \sum_{i \geq 0} p^i f_i(x)$,

$\alpha = \sum_{j \geq 0} p^j \alpha_j$, then $f(x)\alpha = \sum_{r \geq 0} p^r \beta_{r0}$. where β_{ij} is defined by

$$f_0(x)\alpha_0 = \beta_{00} + p\beta_{01} + p^2\beta_{02} + \ldots\ldots$$

and $\beta_{00}, \beta_{01}, \beta_{02}, \ldots\ldots$, are p-adic sequences. Based on definition of $\beta_{ij}(i < r)$, we define β_{rj} as follow

$$\sum_{\substack{i+j=r}} f_i(x)\alpha_j + \sum_{\substack{i+j=r \\ i<r}} \beta_{ij} = \sum_{j \geq 0} p^j \beta_{rj}$$

Definition 1. Let $y_1, y_2,..., y_t$ be p-adic variables. Then their p-adic function is defined by f

$$M_k(y_1, y_2,..., y_t) = \sum_{\substack{i_1+i_2+...+i_t=p^k \\ 0\le i_m \le y_m < p \\ m=1,2,...,t}} \binom{y_1}{i_1}\binom{y_2}{i_2}\cdots\binom{y_t}{i_t}$$

Theorem 2. Let $f_i(x) = \sum_{k\in s^i} x^k$, then we have

$$\beta_{0k} = M_k(L^i\alpha_0 | i \in s^0),$$

$$\beta_{rk} = M_k(L^m\alpha_j : \beta_{uv} | m \in s^i, i+j=r, u+v=r, u<r), r \ge 1.$$

Proof. By Theorem 1 and Lemma 2, we have

$$\beta_{0k} = \binom{f_0\alpha_0}{p^k} \bmod p, \quad \beta_{rk} = \left(\frac{\sum_{i+j=r} f_i\alpha_j + \sum_{\substack{i+j=r \\ r<r}} \beta_{ij}}{p^k}\right) \bmod p$$

Since $(1+x)^{\sigma_1+\sigma_2+...+\sigma_t} = (1+x)^{\sigma_1}(1+x)^{\sigma_2}\cdots(1+x)^{\sigma_t}$,

$$\binom{\sigma_1+\sigma_2+...+\sigma_t}{p^k} = \sum_{\substack{i_1+i_2+...+i_t=p^k \\ 0\le i_m \le \sigma_m}} \binom{\sigma_1}{i_1}\binom{\sigma_2}{i_2}\cdots\binom{\sigma_t}{i_t}$$

and the definition of $M_k(y_1, y_2,..., y_t)$, we can easily complete the proof of Theorem 2.

Theorem 3. Let $\beta_{01} = (\beta_1, \beta_2,..., \beta_t,...)$, then we have

$$\beta_t \equiv \sum_{\substack{i_1+...+i_k=p \\ 0\le i_1,...,i_k < p}} C_{i_1...i_k}[\sum_{v=0}^{n-1} \pi^{p^v(i_1t_1+...+i_kt_k)}\pi^{p^{v+1}t} + \sum_{s\ne p^w} d_s\pi^{st}]$$

where $C_{i_1\cdots i_k} = (p-1)!/(i_1!...i_t!)$, $s = p^{v_1} + p^{v_2} +\cdots+ p^{v_r}$, d_s is a constant over GF(p^n).

Proof. By Theorem 2, we have

$$\beta_{01} = \binom{f_0 \alpha_0}{p} \mod p,$$

$$\beta_t = \binom{\sum_{u=1}^{k} \sigma_{t+t_u}}{p} \mod p \equiv \sum_{\substack{i_1 + \cdots + i_k = p \\ 0 \le i_1, \cdots, i_k < p}} \binom{\sigma_{t+t_1}}{i_1} \binom{\sigma_{t+t_2}}{i_2} \cdots \binom{\sigma_{t+t_k}}{i_k}$$

$$\equiv - \sum_{(i_1, \cdots, i_k)} C_{i_1 \cdots i_k} \prod_{u=1}^{k} \prod_{j=0}^{i_u - 1} (\sigma_{t+t_u} - j) \pmod{p}.$$

The remainder of proof is the same as Ref. [5], so we omit the details.

4 Algebraic Structures of Components α_0 and α_1

Lemma 3 ([4]). Let p be a odd prime and $f_0(x) = x^{t_1} + x^{t_2} + \cdots x^{t_k}$, where $0 \le t_1 \le t_2 \le \ldots \le t_k = n$, then we have $per(f)_{p^d} = p^{d-1}(p^n - 1)$ iff $f_0(x)$ is a primitive polynomial in GF(p), and

$$f_1(\pi) \ne \left(\sum_{\substack{i_1 + i_2 + \ldots + i_k = p \\ 0 \le i_1, \cdots, i_k < p}} C_{i_1 \cdots i_k} \pi^{i_1 t_1 + \cdots + i_k t_k} \right)^{1/p}$$

where $C_{i_1 \cdots i_k} = (p-1)!/(i_1! \ldots i_k!)$, and define

$$\theta(\pi) = \sum_{\substack{i_1 + i_2 + \ldots + i_k = p \\ 0 \le i_1, \cdots, i_k < p}} C_{i_1 \cdots i_k} \pi^{i_1 t_1 + \cdots + i_k t_k}.$$

Definition 2. Assume that π is a root of $f_0(x) = \sum_{i \in S^0} x^i$ over GF(p^n), we define $E_t(x)$ and $D_t(x)$ as follows:

$$E_t(x) = \prod_{1 \le w(\zeta) \le t} (x - \zeta) \quad \in F_p[x],$$

$$D_t(x) = \prod_{w(\zeta) = t} (x - \zeta) \quad \in F_p[x].$$

Lemma 4 ([2], [4-5]). For $i \geq 1$, we define p-adic sequences a_i by $a_i = (a_{it})_t$, $a_{it} = \begin{pmatrix} t \\ i-1 \end{pmatrix}$ mod p. then (1) (L-1) $a_{i+1} = a_i$, $m_{a_i}(x) = (x-1)^i$; (2) If b is a p-adic sequence and $m_b(x)$ has not multi-root, then $m_{a,b}(x) = (m_b(x))^i$;

(3) Suppose $p(x) = \sum_i d_i x^i \in F_p[x]$, then

$$p(x)(a_i b) = a_i p(x)b + a_{i-1} x p'(x)b + \dots$$
$$= \sum_{0 \leq k \leq i} a_{i-k} x^k p^{[\kappa]}(x)b$$

where $p^{[k]}(x) = \sum_i d_i \begin{pmatrix} i \\ k \end{pmatrix} x^{i-k}$ is k-th differential coefficient of $p(x)$.

Theorem 4. Let $f(x)$ be a primitive polynomial with degree n over $Z/(p^d)$ and α is a primitive sequence generated by $f(x)$ over $Z/(p^d)$, then we have

(1) $g_0(x) = f_0(x)$, $L(\alpha_0) = n$.

(2) α_0 has a trace expression $\alpha_0 = \{t_r(\delta_0 \pi^t)\}_{t \geq 0} = \{\sum_{u=0}^{n-1} \delta_0^{p^u} \pi^{p^u t}\}_t$, where π is a root of $f_0(x)$ over $GF(p^n)$.

Proof. By Lemma 2, we have seen that $f \cdot \alpha = 0$ iff nether two equations hold over F_p

$$f_0 \alpha_0 = 0$$
$$\sum_{i+j=r} f_i \alpha_j + \sum_{\substack{i+j=r \\ i<r}} \beta_{ij} = 0 \qquad 1 \leq r \leq d-1$$

Obviously, we can get the conclusion of Theorem 4.

Theorem 5. Let $f(x) = f_0(x) + p f_1(x) + p^2 f_2(x) + \dots \in P_n[Z/(p^d), x]$ be a primitive polynomial, $\alpha = \alpha_0 + p\alpha_1 + p^2 \alpha_2 + \dots \in G[Z/(p^d), f(x)]$ be a primitive sequence, then we have

(1) $\alpha_0 = j_1(x)E_p(x)\alpha_1$, where $j_1(x) \in F_p[x]$,

$\quad j_1(\pi) = [(\theta(\pi))^{1/p} - f_1(\pi))D_2(\pi) \cdots D_p(\pi)]^{-1}$;

(2) $g_1(x) = f_0(x)E_p(x)$, $L(\alpha_1) = 2n + C_n^2 + C_n^3 + \dots + C_n^p$;

(3) If $\alpha_0 = \{t_r(\delta_0 \pi^t)\}_t$, then there exists $\delta_s \in GF(p^n)$ which satisfy

$$\alpha_1 \equiv \{a_2 t_r(\delta_0 \Delta_1 \pi^t) + \sum_{s \neq p^w} \delta_s \pi^{s.t}\}_t \quad \text{mod } G(f_0(x)),$$

where $\quad \Delta_1 = (\pi f_0'(\pi))^{-1}[\theta(\pi)^{1/p} - f_1(\pi)]\quad$, $\quad s = p^{v_1} + p^{v_2} + \cdots + p^{v_r}\quad$,

$0 \le v_1, \cdots, v_r < n$, $2 \le r < p$, $\delta_s \in \mathrm{GF}(p^n)$, w is a positive integer.

Proof. (1) By Theorem 3, we have

$$\beta_{01} \equiv \{-\sum_{(i_1,\ldots,i_k)} C_{i_1\ldots i_k}[\sum_{v=0}^{n-1} \pi^{p^v(i_1 t_1 + \ldots + i_k t_k)} \pi^{p^{v+1}t} + \sum_{s \neq p^w} d_s \pi^{st}]\}_t$$

Let $g(x) = D_2(x)D_3(x)\cdots D_p(x)$, we get $g(\pi^s) = 0$, and

$$g(x)\beta_{01} = -\{t_r[\theta(\pi)^{1/p} g(\pi)\pi^t]\}_t,$$

By Theorem 4 and $f_0\alpha_1 + f_1\alpha_0 + \beta_{01} = 0$, we easily get $f_0\alpha_1 = -f_1\alpha_0 - \beta_{01}$, thus

$$\begin{aligned}
f_0(x)g(x)\alpha_1 &= \{t_r[\theta(\pi)^{1/p} g(\pi)\pi^t] - t_r[f_1(\pi)g(\pi)\pi^t]\}_t \\
&= \{t_r[(\theta(\pi)^{1/p} - f_1(\pi))g(\pi)\pi^t]\}_t \\
&= [\theta(x)^{1/p} - f_1(x)]g(x)\alpha_0
\end{aligned}$$

By Lemma 3, $j(x) = [\theta(x)^{1/p} - f_1(x)]g(x)$ there exists a element $j_1(x)$ which satisfy $j_1(x) = [j(x)]^{-1}$, namely, $j_1(x) = = [(\theta(x)^{1/p} - f_1(x))g(x)]^{-1}$, then

$$\alpha_0 = j_1(x)E_p(x)\alpha_1.$$

(2) Since $f_0(x)g(x)\alpha_1 = j(x)\alpha_0$, we get $f_0^2(x)g(x)\alpha_1 = 0$. namely,

$g_1(x) = f_0(x)E_p(x)$, $L(\alpha_1) = 2n + C_n^2 + C_n^3 + \cdots + C_n^p$.

(3) Let $\alpha_0 = \{t_r(\delta_0\pi^t)\}_t$, since $f_0\alpha_1 = -f_1\alpha_0 - \beta_{01}$, we have

$$\begin{aligned}
f_0\alpha_1 &= \{t_r[\delta_0(\theta(\pi)^{1/p} - f_1(\pi))\pi^t] + \sum_{s \neq p^w} \delta_s'\pi^{s.t}\}_t \\
&= \{t_r[\delta_0\pi f_0'(\pi)\Delta_1\pi^t] + \sum_{s \neq p^w} \delta_s'\pi^{s.t}\}_t ;
\end{aligned}$$

By Lemma 4, $\quad \alpha_1 \equiv \{a_2 t_r(\delta_0\Delta_1\pi^t) + \sum_{s \neq p^w} \delta_s\pi^{s.t}\}_t \quad$ mod $G(f_0(x))$ where

$$\Delta_1 = (\pi f_0'(\pi))^{-1}[\theta(\pi)^{1/p} - f_1(\pi)],$$

$$S = p^{v_1} + p^{v_2} + \cdots + p^{v_r}, 0 \le v_1, \cdots; v_r < n, 2 \le r < p,$$

$$\delta_s = f_0'(\pi)\delta_s' \in \mathrm{GF}(p^n), \text{ which is a stream of constants.}$$

5 Conclusion

Although we only give the algebraic structures for 0, 1 level components of primitive sequences over $Z/(p^d)$ in this paper, the similar results for higher level components are easily shown by the same way, the details of the proof are omitted here.

Acknowledgments

This work was partially sponsored by the National Natural Science Foundation of China (Project No. 60273062, 60573039) and the Guangdong Provincial Natural Science Foundation (Project No. 04205407, 5003350)

References

[1] Lidl, R., Niedereiter, H.: Finite Fields, Encyclopedia of Mathematics and Its Applications. vol. 20 (1983)
[2] Zhexian, W.: Algebra and Coding Theory. Science Press, Beijing (1980) (in Chinese)
[3] Zhao, K., Qi, S.: Lecture on the Number Theory. Higher Education Press, Beijing (1988) (in Chinese)
[4] Minqiang, H.: Maximal Period Polynomials over $Z/(p^d)$ Science in China(series A) 35(3), 270–275 (1992) (in Chinese)
[5] Minqiang, H.: Structures of Binary Component of Primitive Sequences over Z mod 2^d Postdoctoral Thesis of China (3), 18–23 (1990)
[6] Xiangang, L.: The Linear Recurring Sequences on Information Theory, San Diego, California (January 1990)
[7] Wenfeng, Q., Zongduo, D.: The Trace Representation of Sequences and the Space of Nonlinear Filtered Sequences over. $Z/(p^d)$ Acta. Mathematicae Applicatae Sinica 20(1), 128 (1997) (in Chinese)
[8] Wenfeng, Q., Xuanyong, Z.: Injectivness of compression Mappings on Primitive Sequence over Galois Rings. Acta Mathematica Sinica 44(3), 445–452 (2001) (in Chinese)
[9] Xuanyong, Z., Wenfeng, Q.: Uniqueness of the Distribution of Zeroes of Primitive. Level Sequences over $Z/(p^e)$ (II). Acta. Mathematicae Applicatae Sinica 27(4), 731–743 (2004) (in Chinese)

Author Index

Lecture Notes in Computer Science

For information about Vols. 1–4451

please contact your bookseller or Springer